P9-AGA-526

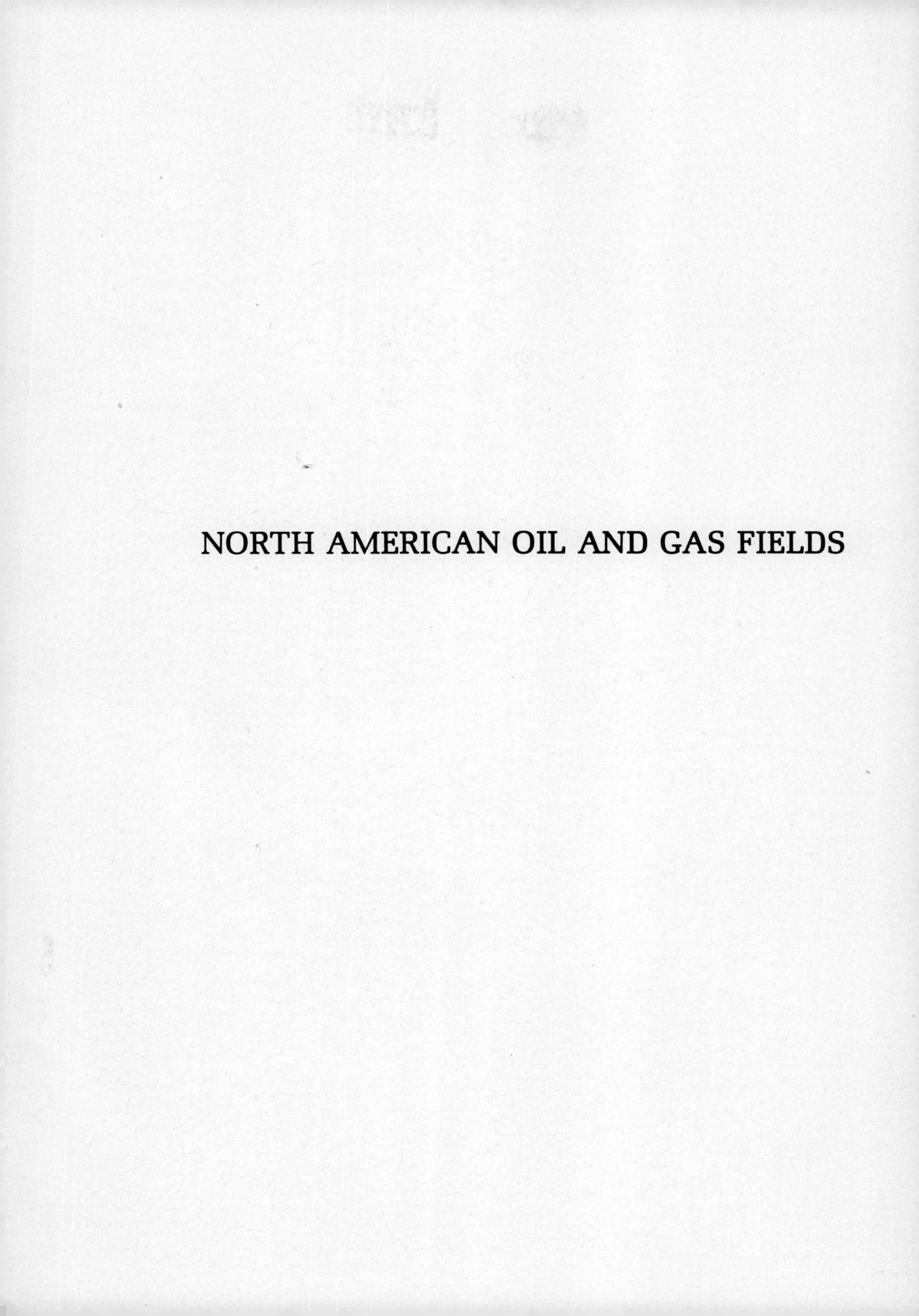

NORTH AMERICAN OIL AND GAS FIELDS

Memoir 24

NORTH AMERICAN OIL AND GAS FIELDS

Edited by JULES BRAUNSTEIN

Published by The American Association of Petroleum Geologists
Tulsa, Oklahoma, U.S.A., 1976

Published March 1976

Library of Congress Catalog Card No. 76-258

ISBN 0-89181-300-4

The AAPG staff responsible for editing and production of this book was:

Peggy Rice, Book Editor
E. M. Tidwell, Assistant Editor
assisted by
Deborah Zikmund, Assistant Editor
Nancy Greeson, Secretary-Typist
Sharon Robinson, Keyboarder

Printed by Banta West, Inc., Sparks, Nevada

Contents

Foreword

The Association has published three volumes called *Structure of Typical American Oil Fields.* Volumes I and II, with prefatory notes by Sidney Powers, were derived from the program of the 12th Annual Convention of the Association (March 1927) and published in 1929. Volume III, edited by J. V. Howell and dedicated to Alexander Watts McCoy, was published in 1948. Each volume carries the subtitle "A Symposium on the Relation of Oil Accumulation to Structure."

The Sidney Powers Memorial Volume, *Problems of Petroleum Geology*, published in 1934, was subtitled "A Sequel to Structure of Typical American Oil Fields." Subsequently, the Association has published several volumes in which oil and gas fields are described, each one devoted to some special attribute of the fields (e.g., size, trapping mechanism, predominance of gas accumulation).

When I conceived of this volume, my criteria for the selection of fields to be described were the relative importance of the fields and the lack of an adequate description of them in literature readily available to Association members.

To judge by the wide range of types of fields described in the early volumes, there are really no "typical" fields. Similarly, the relationship of accumulation to structure is so well understood that it need no longer be emphasized; hence the simple declarative title, "North American Oil and Gas Fields."

We are dealing herein with fields distributed in space from Alaska and the McKenzie Delta area of Canada on the northwest, to the Gulf of Mexico (offshore from Louisiana) and southern Florida on the southeast. I had hoped to include one or more papers on recently discovered fields in Mexico, but was unable to obtain any.

The diversity of geologic conditions which make possible the accumulation of hydrocarbons in the fields described in this volume (Table 1) suggests that any given field is "typical" only of itself. In these 17 fields, the reservoirs range in age from Devonian to Pleistocene; their lithology is sandstone, limestone, or dolomite; and the trapping mechanism itself is structural or stratigraphic or a combination of the two. In the case of several fields, the structure-maker has been the Louann Salt (Jurassic), moving at depth to arch the overlying strata or penetrating the entire section to form a shallow piercement dome. As to reserves, we have descriptions of the largest fields in Alaska (Prudhoe Bay the largest field in North America), Mississippi (Tinsley), Alabama (Citronelle), and Florida (Jay).

In order to make this book more valuable to the reader (and to justify the all-inclusive title), I have included an index to those North American oil and gas fields which have been described in publications issued by the Association. Some exclusions were necessary, and those are noted in the introduction to the bibliography. Most of the credit for compiling this portion of the book belongs to E. M. Tidwell, Assistant Editor in the AAPG book department in Tulsa, and it appears under her authorship.

Jules Braunstein
New Orleans, Louisiana
November 25, 1975

Table 1. Significant Parameters of all 17 Fields

Field	State or Province	Age of Reservoir	Lithology of Reservoir	Trap Mechanism	Discovery Date
1. Middle Ground Shoal	Alaska	Oligocene-Miocene	Sandstone	Anticline	1963
2. Prudhoe Bay	Alaska	*Permo-Triassic	Sandstone	Strat.-Struct.	1969
3. Taglu	N.W.T.	Eocene	Sandstone	Fault block	1971
4. Mitsue	Alberta	Middle Devonian	Sandstone	Stratigraphic	1964
5. Kaybob	Alberta	Middle Devonian	Limestone	Reef	1957
6. Big Piney-La Barge	Wyoming	Triassic-Paleocene	Sandstone	Strat.-Struct.	1924
7. Altamont	Utah	Eocene-Paleocene	Sandstone	Stratigraphic	1970
8. Wattenberg	Colorado	Cretaceous	Sandstone	Stratigraphic	1970
9. Big Wells	Texas	Late Cretaceous	Sandstone	Stratigraphic	1969
10. Fairway	Texas	Early Cretaceous	Limestone	Reef	1960
11. Walker Creek	Arkansas	Jurassic	Limestone	Strat.-Struct.	1968
12. East Cameron Block 270	Offshore Louisiana	Pleistocene	Sandstone	Anticline	1971
13. Grand Isle Block 16	Offshore Louisiana	Late Miocene	Sandstone	Piercement salt dome	1948
14. Tinsley	Mississippi	Late Cretaceous	Sandstone	Anticline	1939
15. Citronelle	Alabama	Early Cretaceous	Sandstone	Domal uplift	1955
16. Jay	Florida	Jurassic	Dolomite	Anticline, Strat.	1970
17. Sunoco-Felda	Florida	Early Cretaceous	Limestone	Reefoid mound, Strat.	1964

*Main reservoir; smaller accumulations in Miss.-Penn., L. Cret., U. Cret.; gas pockets at base of permafrost.

Volume (Listed in Order of Their Inclusion)

th to uction nd (m)	Oil Column	Surface Area	Cum. Prod. to 1/1/74 (million bbl)	(Bcf)	Ultimate Prod. (million bbl)	(Bcf)
00-9,700 55-2,955)	2,800	9.5 x 0.75 mi (15.2 x 1.2 km)	78.6	37.3	185	
35-9,780 25-2,980)	450	—	4.5	3.5	9,600	26,000
00-10,190 80-3,160)	1,700	12 sq mi (31 km^2)				3,000
08 00)	—	10 x 3 mi (16 x 4.8 km)	86	45.6	134	
00 85)	257	18,000 acres (72.5 km^2)	49	2.45	126	
)-9,700)-2,960)	—	—	65	1,200	75	2,500
00-17,000 40-5,180)	8,000	675 sq mi (1,748 km^2)	9.6	11.4	250	
18 05)	—	978 sq mi (2,533 km^2)	166	9.6		1,100
00-5,800 85-1,770)	600	30,500 acres (123 km^2)	17.5	20.8	44	64
36 76)	—	—	106.5	49.5	200	
,850-11,900 310-3,625)	322	—	10.5	23	30	100
70-8,710 970-2,655)	—	2,500 acres (13km^2)	1.1	258		
500-15,000 370-4,570)	—	—	198.5	220	277	
400-6,115 340-1,865)	—	7.5 x 3 mi (12 x 4.8 km)	195		200	
,870-11,415 310-3,480)	—	16,400 acres (66 km^2)	107		159	
,470 715)	420	7 x 3 mi (11 x 4.8 km)	40.6	33.5	346	350
,475 500)	34	4,500 acres (18 km^2)	6.8	.045	31	

Middle Ground Shoal Oil Field, Alaska[1]

R. F. BOSS, R. B. LENNON, and B. W. WILSON[2]

Abstract Middle Ground Shoal oil field, in upper Cook Inlet, Alaska, is located beneath water with an average depth of 100 ft (30 m). The Shell MGS State No. 1, drilled in 1963, was the first offshore oil completion in Alaska. The field produces oil from a gross interval of about 2,800 ft (850 m) in the Tertiary lower Tyonek Formation. The productive interval has been separated into seven pools; the A pool is produced separately, but production from the B, C, and D and the E, F, and G pools, respectively, is commingled. Three production platforms are in use, and the field contains 31 producing wells, 23 injection wells, 1 shut-in gas well, and 8 abandoned or suspended wells. As of January 1, 1974, the field had produced 78,662,670 bbl of oil, 37,270,730 Mcf of gas, and 9,162,874 bbl of water. Because of declining reservoir pressures, pressure maintenance by water injection was started in 1969.

The structure at Middle Ground Shoal is a narrow anticlinal feature which strikes N10°E. Little or no paleostructural growth is thought to have occurred during deposition of the oil-bearing sandstone sequence. Channel fills and braided-stream deposits provide the reservoirs. The main productive interval in the field is the G pool, which is in the Hemlock Sandstone Member of the Tyonek.

Introduction

Middle Ground Shoal oil field is in upper Cook Inlet. Cook Inlet, a very large estuary in southern Alaska (Fig. 1), has a tidal range of 30± ft (9± m). Anchorage, Alaska's largest city, is located at its northeastern end.

This geologic discussion is divided into two parts. The first part is an overview of the major structural features and stratigraphic units of the upper Cook Inlet basin and the related tectonic and depositional history (Kirschner and Lyon, 1973; Calderwood and Fackler, 1972). The second part discusses the structure, stratigraphy, and oil accumulation in the Middle Ground Shoal field.

Basin Framework

Structural Setting

The Cook Inlet basin trends NNE-SSW and is approximately 200 mi (320 km) long and 60 mi (96.5 km) wide (Fig. 1). It is bounded by the Castle Mountain and Bruin Bay faults on the north and west, the Kenai Mountains (Border Ranges) fault zone on the east, and a line from Cape Douglas to the Barren Islands on the south. The basin is about the same size as California's San Joaquin basin, and the gross structural settings of the two have many similarities.

Cook Inlet appears to be a basin of the trench–arc gap type (Dickinson, 1971). The volcanic arc is represented by the present-day and older Quaternary volcanoes and Mesozoic granitic batholiths and volcanic intrusives in the southern Alaska Range. This relationship continues in Shelikof Strait, the southwest extension of Cook Inlet basin, where the adjacent Alaska Peninsula contains Quaternary volcanoes, thick Tertiary volcanic rocks, and Mesozoic granitic intrusives and volcano-related dikes and sills (Burk, 1965). The original trench apparently occupied a position parallel with the present Aleutian Trench, but more landward. The Kenai and Kodiak Mountains, on the southeast side of the basin, consist of very steeply dipping and severely deformed isoclinal folds, largely in Upper Cretaceous turbidite deposits. Although the turbidite sequences have been referred to by a variety of names—Chugach, Valdez, Kodiak, and Shumagin—they all have essentially the same characteristics. The sandstones and shales of the turbidite sequences have been metamorphosed into metasandstone and sericite-chlorite phyllites. The Kenai Mountains fault zone and its southern extension separate the nonmetamorphosed strata in open folds of the Cook Inlet basin from the metamorphosed, isoclinally folded rocks of the Kenai Mountains and Kodiak Island. This zone is thought to mark the most landward position of the ancestral submarine trench. The Upper Cretaceous turbidites and other associated turbidite sequences were deposited on oceanic crust, were metamorphosed during subduction, and then were severely squeezed against the much thicker adjacent continental crust into steeply dipping isoclinal folds. The deposits of the Cook Inlet basin were laid down above much more competent continental crust. Because of the rigidity of the continental crust,

[1]Manuscript received, July 17, 1975. Published with permission of Shell Oil Company, Atlantic Richfield Company, and Standard Oil Company of California.

[2]Shell Oil Company, Houston, Texas 77001.

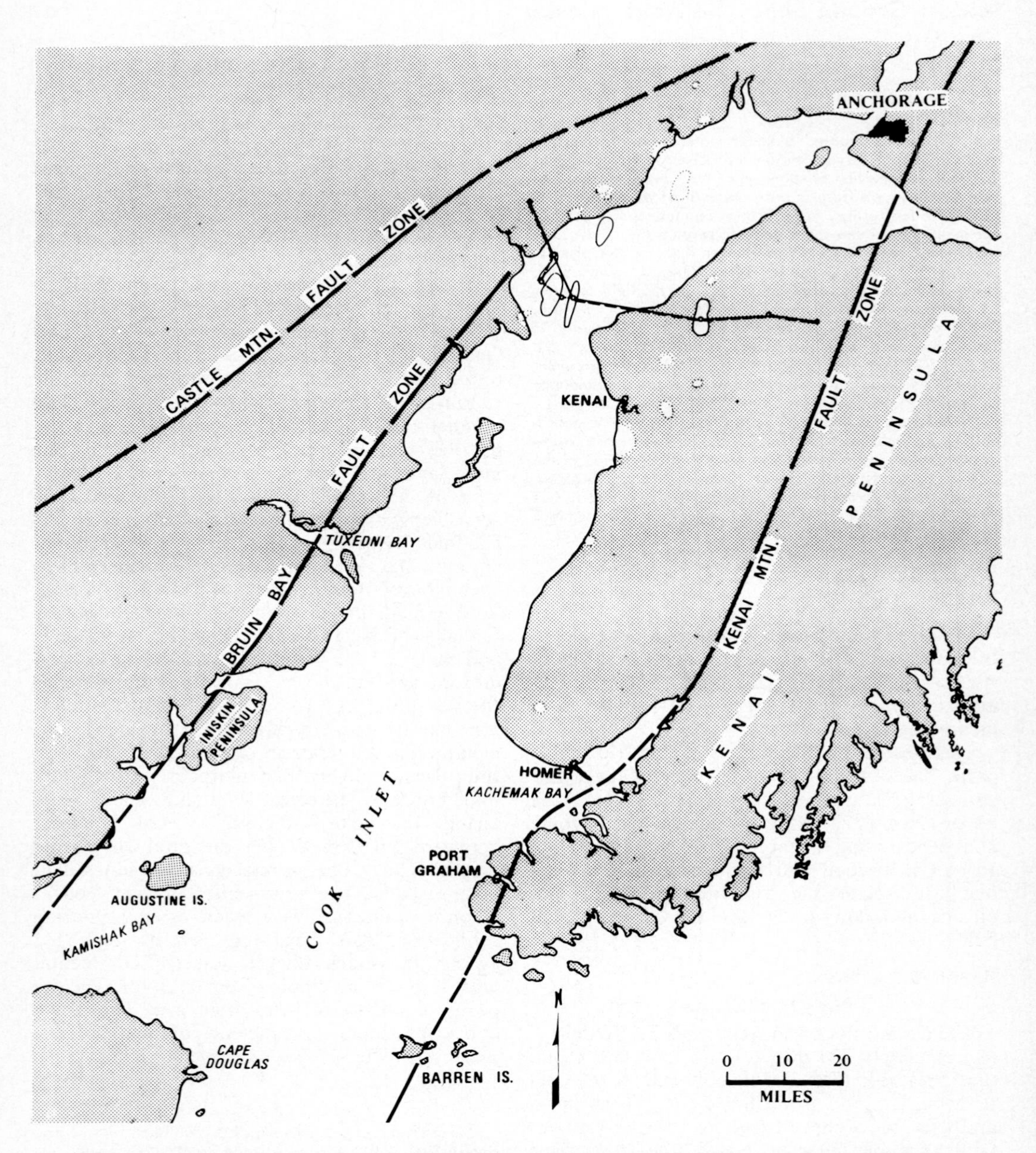

FIG. 1—Index map of upper and lower Cook Inlet showing oil and gas fields and lines of section for Figures 3 (solid) and 4 (dashed).

lateral compression was fairly minor and folding was relatively mild.

The time of inception of the ancestral Aleutian Trench and the duration and continuity of subduction are not known. The extensive Upper Cretaceous turbidite sequences, which are interpreted as trench fill (Moore, 1972, 1973), require a trench at least as old as Late Cretaceous. No older trench deposits have been reported, but indirect evidence suggests that a trench was present at an earlier date. The presence of extensive Lower Jurassic volcanic rocks and Middle Jurassic granitic intrusions, if they are products of a volcanic arc, indicates active subduction and a correlative trench in Jurassic time. The occurrence of Upper Triassic volcanic rocks could indicate that the trench-arc system was active still earlier. Isoclinally folded Tertiary turbidites, such as the Paleocene-Eocene Sitkalidak Formation on the southeast side of Kodiak Island and the Orca Group on Montague Island, may represent later trench-fill deposits and continuing continental accretion to the southeast during the Tertiary.

The Kenai Mountains, Castle Mountain, and Bruin Bay fault zones are the major boundary features of the Cook Inlet basin (Fig. 1). The first two are believed to be lateral faults, but only the Castle Mountain has measurable right-lateral displacement. Near the southern margin of the Talkeetna Moutains, 45 mi (72 km) northeast of Anchorage, the Castle Mountain fault separates the Paleocene arkosic sandstones on the north side from very lithic Chickaloon sandstones and conglomerates of the same age on the south side (Grantz and Wolfe, 1961). These strikingly different lithologies indicate displacement of considerable magnitude. The Susitna Flats region is the northern extension of the post-Eocene Cook Inlet basin across the Castle Mountain fault zone, and little lateral displacement between the two is indicated. The Copper River basin is considered to have been the northern extension of Cook Inlet during the Mesozoic. If so, a chiefly Paleocene right-lateral displacement of nearly 150 mi (240 km) is indicated. The Bruin Bay fault zone is pronounced along lower Cook Inlet and extends southwest into the Alaska Peninsula as the major boundary fault of the basin. It terminates to the northwest in upper Cook Inlet, where it is replaced by several small, high-angle faults. An age at least as old as Middle Jurassic is indicated for the Bruin Bay fault zone on the Alaska Peninsula, where Middle Jurassic granite on the upthrown northwest side supplied boulders to sediments deposited on the downthrown side during latest Middle Jurassic and Late Jurassic time (Burk, 1965). The small, high-angle faults in upper Cook Inlet show evidence of growth in the lower Kenai Group (Tyonek Formation), indicating that there was movement along the Bruin Bay fault zone into the Miocene. No significant lateral displacement is associated with this fault. The Kenai Mountains frontal fault is not exposed where it involves the Tertiary, but southeasterly thinning of the Cook Inlet Tertiary sedimentary package suggests vertical displacement throughout Tertiary time and into the Quaternary, as indicated by present physiographic relief. The amount of lateral displacement, if any, is unknown. During much of the Mesozoic, the Kenai Mountains fault may have been the suture between continental crust and the subducting oceanic crust.

The major oil and gas fields—Swanson River, Middle Ground Shoal, McArthur River, Granite Point, Kenai, North Cook Inlet, and Beluga River—are on north-south-trending anticlines, as are most of the nonproducing structural features (Figs. 2, 3). Although some structural features, such as that at Swanson River, show evidence of structural growth throughout the Tertiary, the major growth was very late—late Pliocene and/or early Pleistocene. The Plio-Pleistocene structural growth was accompanied by strong vertical uplift and truncation of the entire basin, with the possible exception of a limited portion in the basin center. The megastructural control of the north-south-trending anticlines is not understood, but a simple explanation would be that of a couple produced by right-lateral movement on the Castle Mountain and Kenai Mountains boundary fault zones (Fig. 1). If such a couple was responsible, the accompanying lateral displacement appears to have been slight.

Stratigraphy

The stratigraphic section of upper Cook Inlet basin consists of a Mesozoic package and a Tertiary package. The two are separated by a major angular unconformity (Fig. 3). The Lower Jurassic, which is in large part volcanic, is considered to be economic "basement." The Middle and Upper Jurassic sections are relatively thick, but Lower Cretaceous beds are generally absent. They probably were present at one time, but have been removed by mid-Cretaceous erosion. Upper Cretaceous strata normally overlie the Jurassic and are also thick. No commercial hydrocarbon production has been established from the Mesozoic

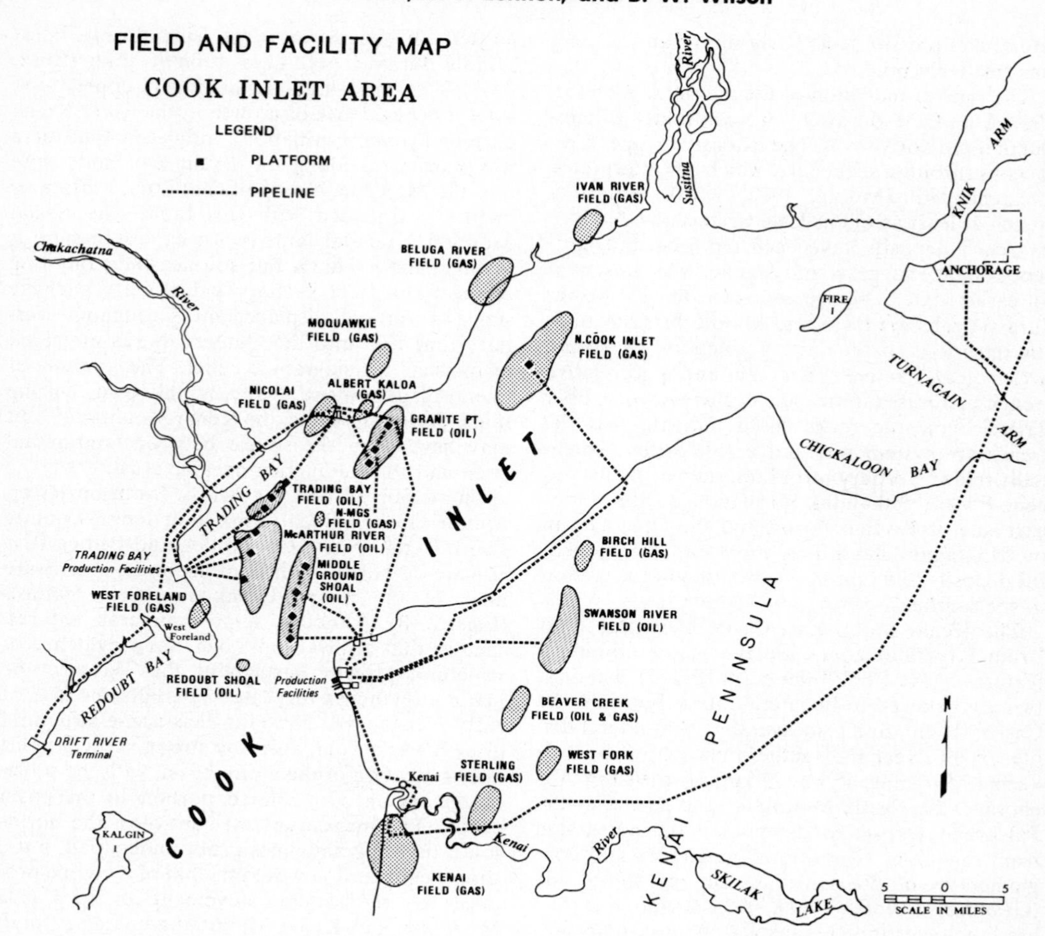

FIG. 2—Location of oil and gas fields and accompanying facilities, Cook Inlet, Alaska.

section, although potential source rocks have been reported.

Lower Jurassic

Talkeetna Formation—Lower Jurassic rocks, which rim the Cook Inlet basin, crop out in the Iniskin-Tuxedni area on the west side of the Cook Inlet and on the southern end of the Kenai Peninsula near Port Graham. At Port Graham, more than 4,500 ft (1,371 m) of massive volcanic conglomerate, tuff, and sandstone unconformably overlies the Upper Triassic beds. In the Iniskin-Tuxedni region, the Lower Jurassic interval is represented by more than 9,000 ft (2,750 m) of marine volcanic breccia, agglomerates, flows, bedded tuffs, and some fluvial tuffaceous sandstones. Detterman and Hartsock (1966) correlated these rocks with similar rocks from the Talkeetna Mountains and proposed extending the Talkeetna Formation to the Iniskin-Tuxedni region.

In the Cook Inlet basin, numerous wells have penetrated the Lower Jurassic "basement." In most of the wells, tuffs and volcanic conglomerates are the major Lower Jurassic rock types.

Middle Jurassic

Tuxedni Group—In the northern Alaska Peninsula and Iniskin-Tuxedni areas, the strata of the Middle Jurassic Tuxedni Group overlie the Lower Jurassic Talkeetna Formation with minor discordance and slight hiatus. Marine sandstone, conglomerate, siltstone, and shale are the major rock types in outcrops of the Tuxedni Group. In the Iniskin-Tuxedni area the 10,000± ft (3,000± m) of Tuxedni strata appear to have been deposited during a series of marine transgressions and regressions; they contain abrupt lateral facies changes normal to the present northeast-southwest shoreline trend.

Oil seeps have been reported from outcrops of the Tuxedni on the Iniskin Peninsula, and oil in noncommercial quantities has been found in wells drilled into these Middle Jurassic rocks. Comparative analyses of crude oil from this area and crude oil from the producing Tertiary reservoirs of the Cook Inlet basin lend support to the thesis that Middle Jurassic beds are the source rocks for the oil (Osment et al, 1967).

In the subsurface of Cook Inlet basin there has been very little penetration of the Middle Jurassic section. Only two or possibly three wells have been drilled into it. However, the results of this drilling suggest the presence of probable source rocks below the Tertiary over or near the productive area of the Cook Inlet basin.

Upper Jurassic

Chinitna Siltstone and Naknek Formation—The contact between the Middle and Upper Jurassic in the Cook Inlet basin is generally conformable. In contrast, continued uplift and erosion of marginal areas on the northwest eventually exposed Middle Jurassic granite. These rocks were the predominant source of sediments throughout the Late Jurassic. In the Iniskin-Tuxedni region, 2,000± ft (600± m) of marine slope-facies siltstones, the Chinitna Siltstone (Kirschner and Minard, 1948), overlies the Middle Jurassic Tuxedni Group. Conformably overlying the Chinitna is 5,000± ft (1,525± m) of Upper Jurassic Naknek Formation. However, just south of the north end of the Alaska Peninsula, there is a more complete section of Naknek, between 5,000 and 10,000 ft (1,525–3,050 m), overlain by Lower Cretaceous rocks. In this area the lower part of the Naknek consists of boulder conglomerates and interbedded coarse sandstones deposited in environments ranging from high-gradient alluvial fan to high-energy shallow marine. The Naknek constitutes a grossly "fining-upward" sequence, and at the top consists of shallow-marine, crossbedded, wave-agitated, fine-grained sandstone overlain by shallow-marine, sub-wave-base siltstones of Early Cretaceous age. Detterman and Jones (1974) have reported fossiliferous Jurassic rocks on Augustine Island that are correlated with the Naknek.

Few wells in the basin have been drilled to the Upper Jurassic section. However, where it has been penetrated, the rocks are marine but finer grained than those of the outcrop area. Also, the percentage of fine-grained sandstones and shales is far greater, and these lithologies are probably predominant.

Upper Cretaceous

Matanuska Formation and equivalents—In the Cook Inlet basin, the Upper Cretaceous is represented by the Matanuska Formation. It lies with angular unconformity on the Jurassic section—Upper Jurassic in the central part of the basin and Middle and Lower Jurassic toward the margins (Figs. 3, 4). This unconformity represents not only extensive erosion of Jurassic and very probably Lower Cretaceous strata, but also a considerable hiatus caused by nondeposition during early Late Cretaceous time. The Upper Cretaceous section penetrated by the drill is largely of marine shale but includes some sandstone. The depositional environment of the shales was relatively deep water, about 1,000 ft (300 m). Consequently, the sandstones are most likely turbidites.

More or less equivalent Upper Cretaceous strata are exposed in very limited outcrops on the northwest side of Cook Inlet and in extensive outcrops on the Alaska Peninsula (Burk, 1965). They consist of fluvial conglomerates, sandstones and shales with interbedded coals, shallow-water shoreline sandstones, and marine shales with a few turbidites. Original thickness was of the order of 5,000 ft (1,525 m) or more. The probable shoreline trend was northeast-southwest, parallel with the basin axis; water depths increased toward the southeast.

Tertiary

In Cook Inlet, Tertiary strata unconformably overlie Mesozoic strata—commonly Upper Cretaceous in the central basin but, progressively toward the margins, they lie on Upper, Middle, and Lower Jurassic strata (Fig. 3). Considerable deformation and presumably subaerial erosion preceded Tertiary deposition, and many thousands of feet of strata apparently were removed in some areas.

The Tertiary section consists of a thick sequence of alluvial deposits; no marine beds have been identified. Extensive uplift and erosion oc-

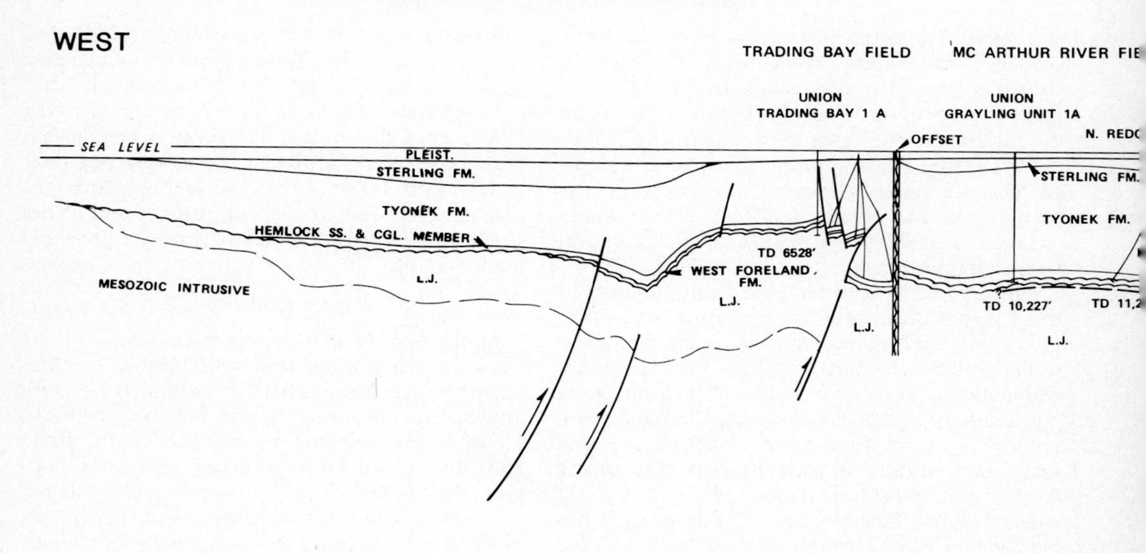

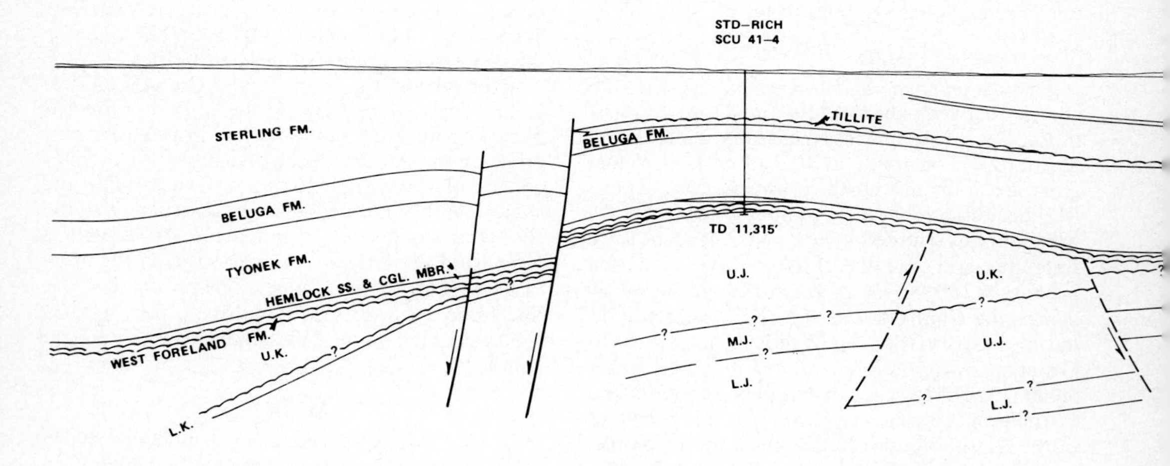

FIG. 3—Regional geological cross section across central Cook Inlet bas

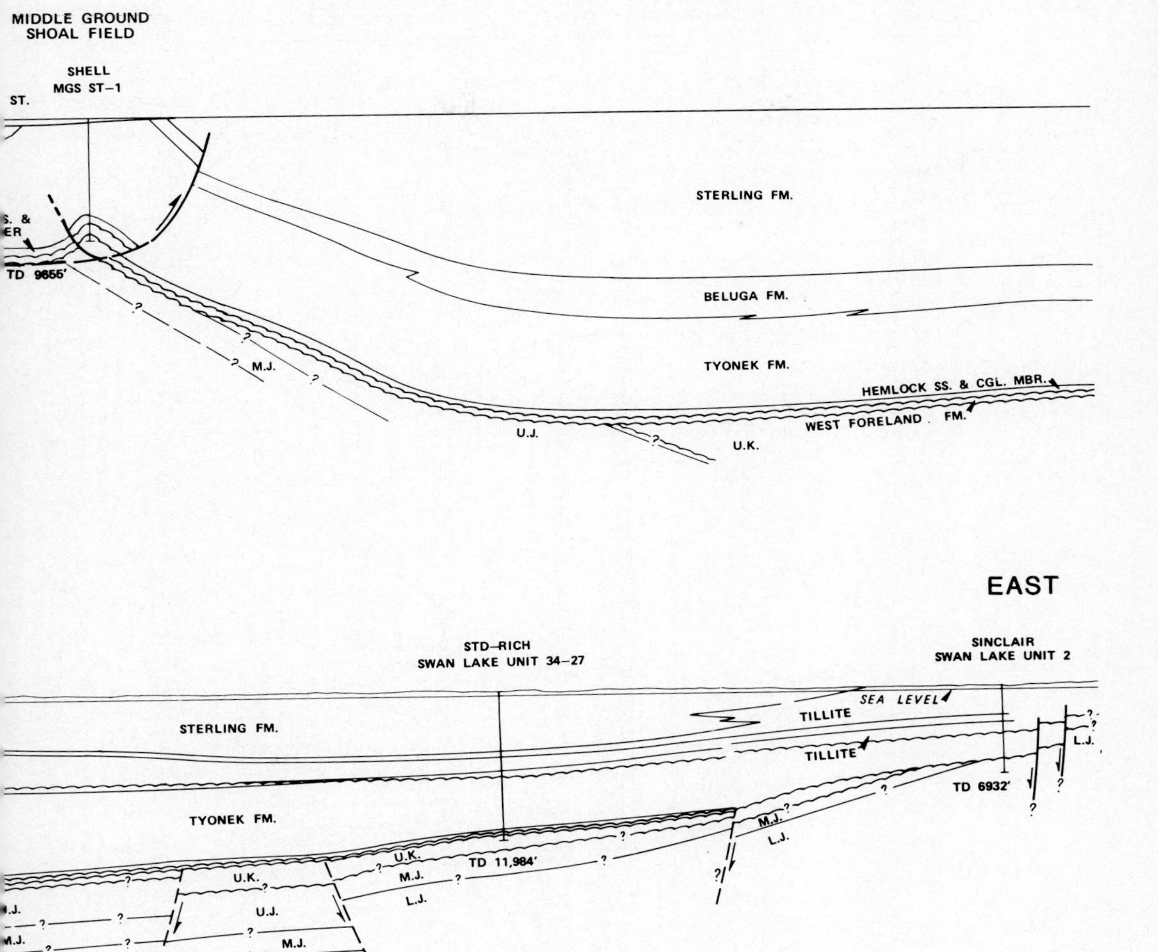

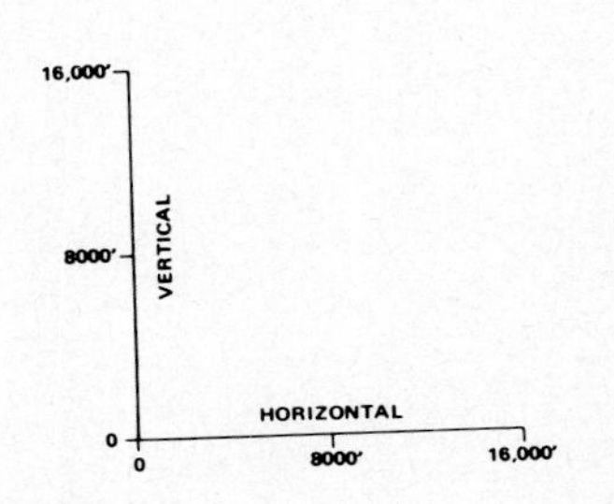

cation is shown on Figure 1. Constructed using seismic and well control.

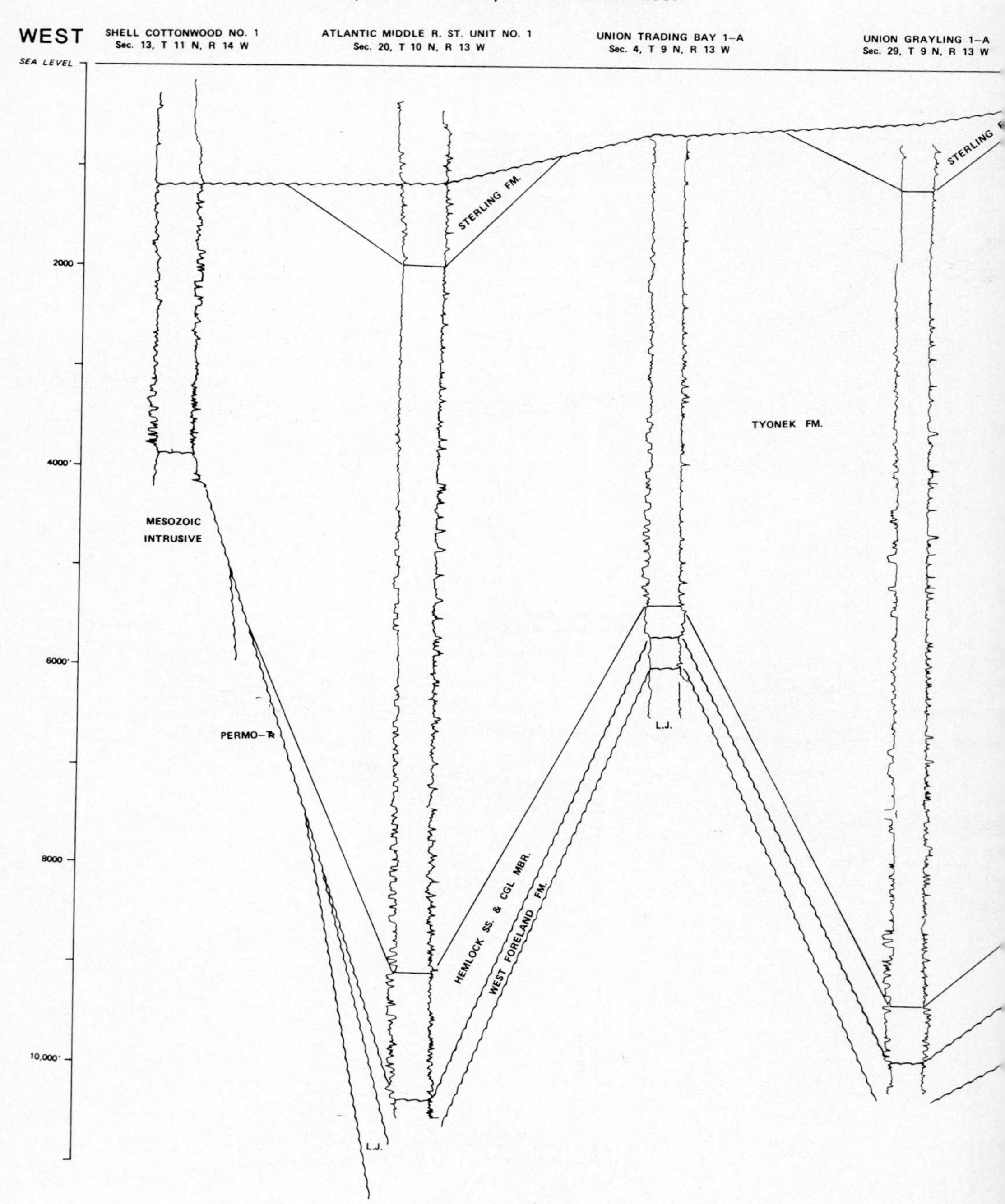

FIG. 4—Regional E-log correlation across ce

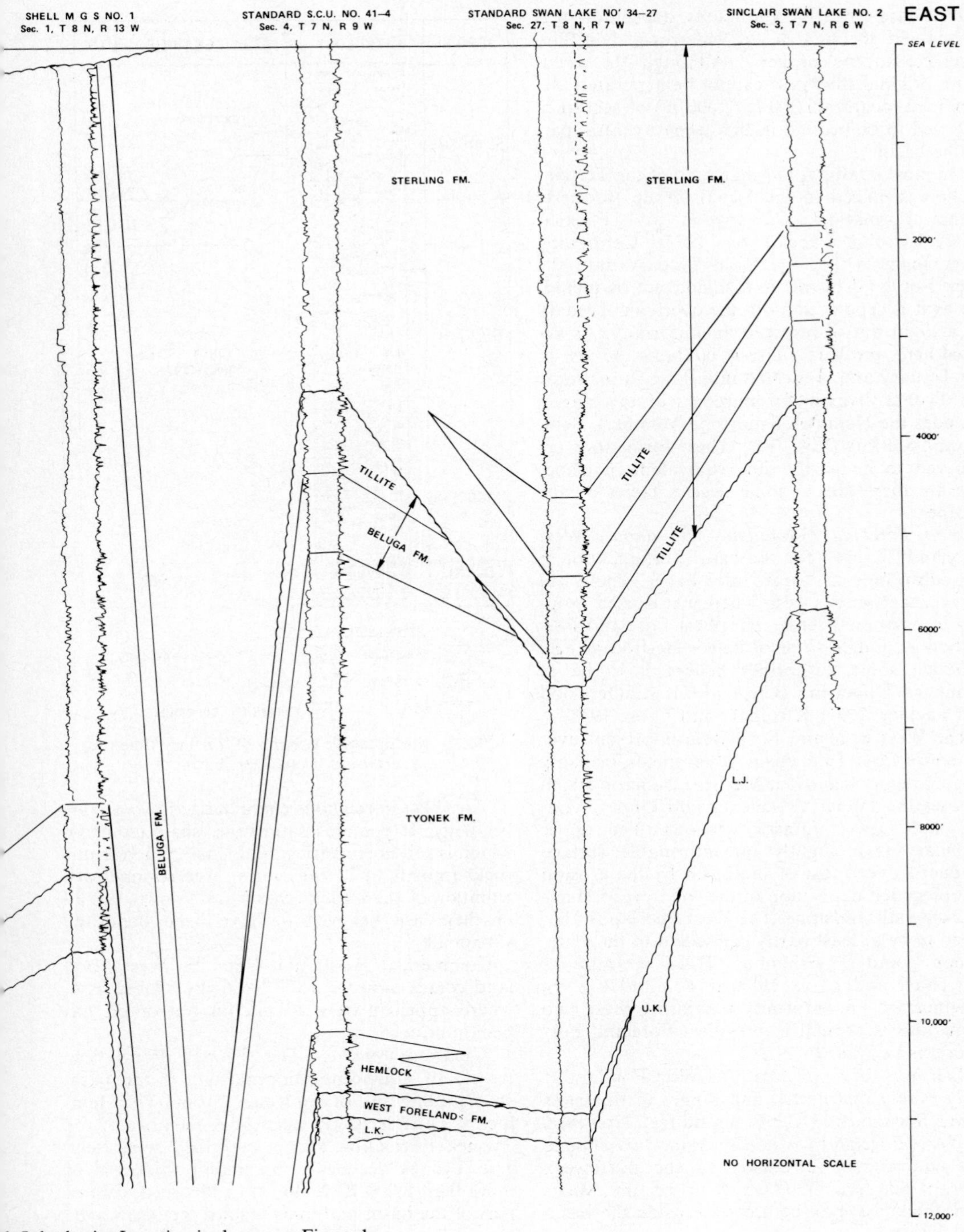

k Inlet basin. Location is shown on Figure 1.

curred when nearly continuous deposition was terminated abruptly by the widespread late Pliocene–Pleistocene orogeny. Although the maximum original thickness cannot be accurately determined, nearly 25,000 ft (7,600 m) of section is believed to be present in the deepest central part of the basin.

On most stratigraphic charts, all of the Tertiary section is placed in the Kenai Group (formerly Kenai Formation; Calderwood and Fackler, 1972; Kirschner and Lyon, 1973). Continuing work suggests that the basal Tertiary unit, the West Foreland Formation, should not be included, as it is separated from the overlying Tertiary by a slight but widespread unconformity. As defined here, the Kenai Group contains the rest of the Tertiary and is divided into three formations: the Tyonek (which, for purposes of this paper, includes the Hemlock Sandstone Member), Beluga, and Sterling (Figs. 3–5). These formations are believed to be partly time equivalent in places and are therefore, to some degree, facies of one another.

West Foreland Formation—The name "West Foreland" is used for the basal Tertiary unit in the subsurface of Cook Inlet basin. There are many uncertainties as to whether it is used properly in a regional sense. The West Foreland Formation is thought to be of Paleocene-Eocene age, although some authorities believe it to be as young as Oligocene (Crick, 1971; Calderwood and Fackler, 1972; Kirschner and Lyon, 1973).

The West Foreland Formation is present over the entire Cook Inlet basin. It lies unconformably on a markedly truncated Mesozoic subcrop which exposes the Upper Cretaceous and Upper, Middle, and Lower Jurassic sections. The upper boundary is a slightly unconformable surface produced over most of the basin by the erosion that preceded deposition of the widespread Hemlock alluvial sandstone. The West Foreland is believed to be at least partly equivalent to the Chickaloon and Wishbone Hill Formations (Kirschner and Lyon, 1972, p. 402 and Fig. 9). Whether the unconformity separating these two formations is present in the West Foreland Formation is not known.

Over most of Cook Inlet, the West Foreland is fairly evenly distributed and ranges in thickness from a few hundred to a thousand feet. However, the West Foreland (or its Chickaloon–Wishbone Hill equivalents) is thicker to the northwest, where it may reach 3,000± ft (915± m). Maximum thickness may be greater outside the basin than within it.

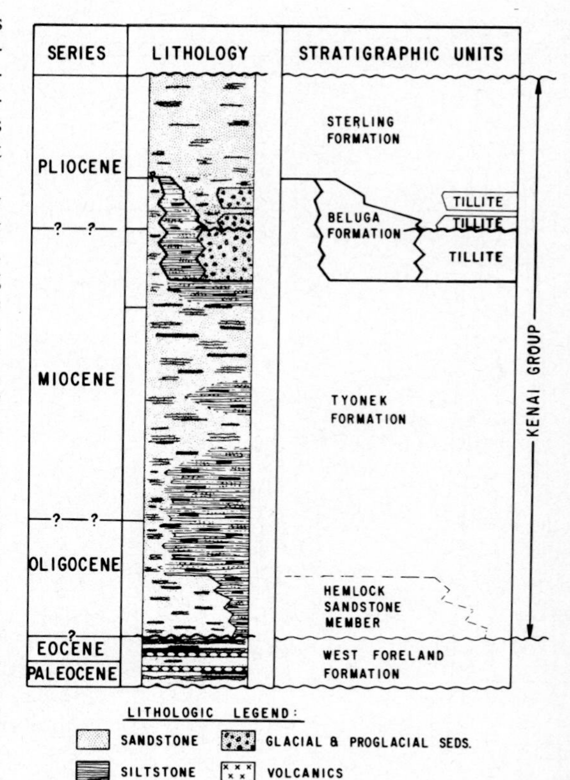

FIG. 5—Stratigraphic column of Tertiary time-rock relations, Cook Inlet basin.

The West Foreland is characterized by variable lithology. It contains sandstone, shale (some of which is red-green variegated), coal, and volcanic rocks (mostly or entirely tuffs). The regional distribution of these rock types is not known, but all of them are believed to have been deposited subaerially.

Commercial production from the West Foreland comes only from the McArthur River field, where approximately 10 million bbl of oil has been produced.

Tyonek Formation—The Tyonek Formation, largely of Oligocene-Miocene age, is the basal stratigraphic unit of the Kenai Group. The Hemlock is considered by us to be a member of the Tyonek Formation and is described separately. The Tyonek reaches a maximum thickness of more than 8,000 ft (2,440 m) in the deep, central part of the basin and thins toward the basin margins, where it subcrops below the Quaternary.

The lower contact is the unconformity at the top of the West Foreland Formation. The upper contact is more variable. In the northwest part of the basin where the Tyonek is overlain by the Sterling, the contact is gradational with only slight intertonguing. In the central and southeastern parts, it intertongues with the generally overlying Beluga Formation through an interval of 1,000 ft (300 m) or more.

The Tyonek Formation is entirely nonmarine; it consists of massive fluvial sandstones interbedded with floodplain shales and thick coal beds. The sandstones, commonly conglomeratic, are medium grained and well to moderately well sorted. The sandstones are more abundant on the northwest side of the basin (Fig. 6). They are oil bearing in all of the northwest-side oil fields and are expected to produce over 100 million bbl.

The *Hemlock Sandstone Member* occupies the lower part of the Tyonek Formation and is probably Oligocene in age. The Hemlock in this region is a sandstone that is genetically similar to the gradationally overlying Tyonek sandstones and in places difficult to distinguish from them. Therefore, for the purposes of this paper, we refer to the "Hemlock Sandstone Member" of the Tyonek Formation rather than use the formation status formerly assigned to the Hemlock (i.e., Hemlock Conglomerate; Calderwood and Fackler, 1972).

The interval is 500–600 ft (150–180 m) thick in the central part of the basin and even thicker locally. The Hemlock is a widespread blanket sandstone that covers most of the basin. It normally consists of 50–70 percent sandstone; the remainder is floodplain shale and a few thin coal seams (Fig. 7). Inasmuch as the Hemlock is the lowest unit in the Tyonek Formation, its basal contact is the slight unconformity above the West Foreland. The upper and lateral contacts are gradational into typical Tyonek strata, which generally are not more than 25 percent sandstone.

The Hemlock sandstone is believed to have been deposited by braided streams of low to moderate gradient flowing from the north and west. It is medium grained and well to moderately well sorted. Conglomeratic sandstones are abundant but the coarsest deposits, pebble conglomerates, are much less abundant. Porosity ranges from 10 to 25 percent; the lower values (15 percent or less) are most common. The Hemlock sandstone is the major reservoir in all of the Cook Inlet oil fields except Granite Point. Of the more than 1 billion bbl of producible oil in Cook Inlet, 90 percent will come from the Hemlock sandstone.

Beluga Formation—The Beluga Formation is the middle lithologic unit of the Kenai Group. It is thought to be chiefly of late Miocene age. The Beluga is 3,000–4,000 ft (900–1,200 m) thick in the central part of the basin. Marginal thinning is largely the result of facies change to the west and truncation to the northeast.

The Beluga Formation is a "waste basket" stratigraphic unit. It is of wholly continental origin and, although highly variable lithologically, it is mappable. Genetically, it consists largely of floodplain shales with minor interbedded coal seams and thin channel sandstones. On Kenai Peninsula, adjacent to the Kenai Mountains, the fluvial and related deposits are replaced by glacial-moraine tillites and outwash-plain sandstones (Figs. 3–5). The glacially derived sediments indicate that the Kenai Mountains were high in late Miocene time, as they are today. In distribution, the glacial deposits appear to be similar to those of the youngest Pleistocene (Karlstrom, 1964), and they presumably were laid down by piedmont glaciers which flowed out over the Kenai lowlands.

The Beluga is not considered to be economically important. Although many of the gas-field reservoirs are assigned to the Beluga, they may be thought of more properly as sandstone intertongues of the overlying Sterling Formation.

Sterling Formation—The Sterling Formation is the uppermost lithologic unit of the Kenai Group. It probably is of mainly Pliocene age. The Sterling consists of at least 50 percent sandstone, and commonly 75 percent or more is sandstone. The sandstones are medium grained, well to fairly well sorted, and slightly conglomeratic. Porosity values of 30 percent are characteristic. The Sterling has been markedly uplifted and truncated over most, if not all, of the basin. The maximum preserved thickness is nearly 14,000 ft (4,270 m), but the original maximum thickness must have been greater.

The upper contact of the formation is the very pronounced, widespread angular unconformity which is overlain by Quaternary deposits (Figs. 3, 4). The lower contact is generally one of intertonguing with the Beluga, except in the northeast, where the two formations are separated by a local angular unconformity.

The Sterling sandstones are the major gas reservoirs. They contain more than 75 percent of the 4 Tcf of dry gas known to be present in Cook Inlet gas fields.

Deposition of the Sterling Formation probably was related to the uplift of the Alaska Range. The

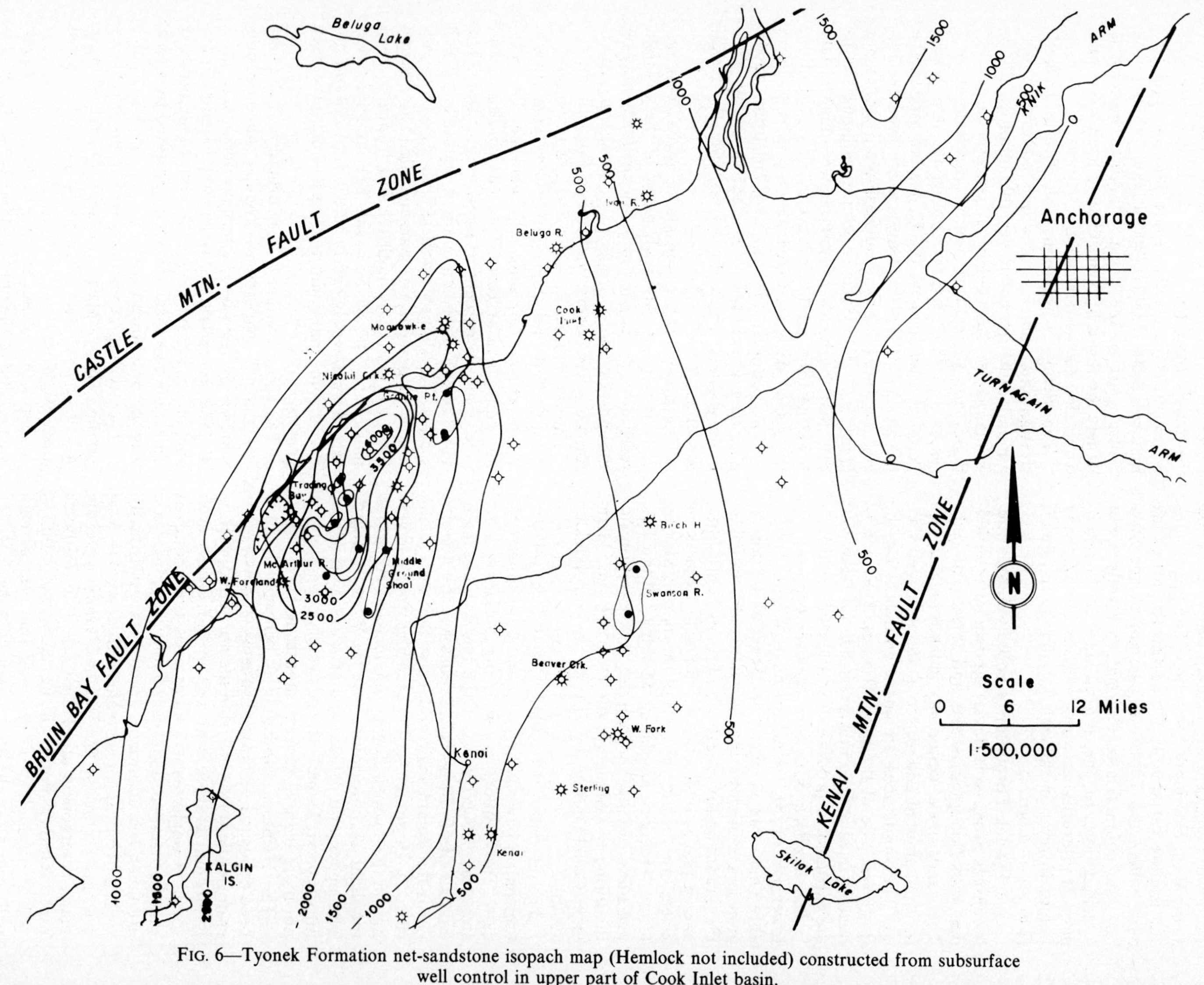

FIG. 6—Tyonek Formation net-sandstone isopach map (Hemlock not included) constructed from subsurface well control in upper part of Cook Inlet basin.

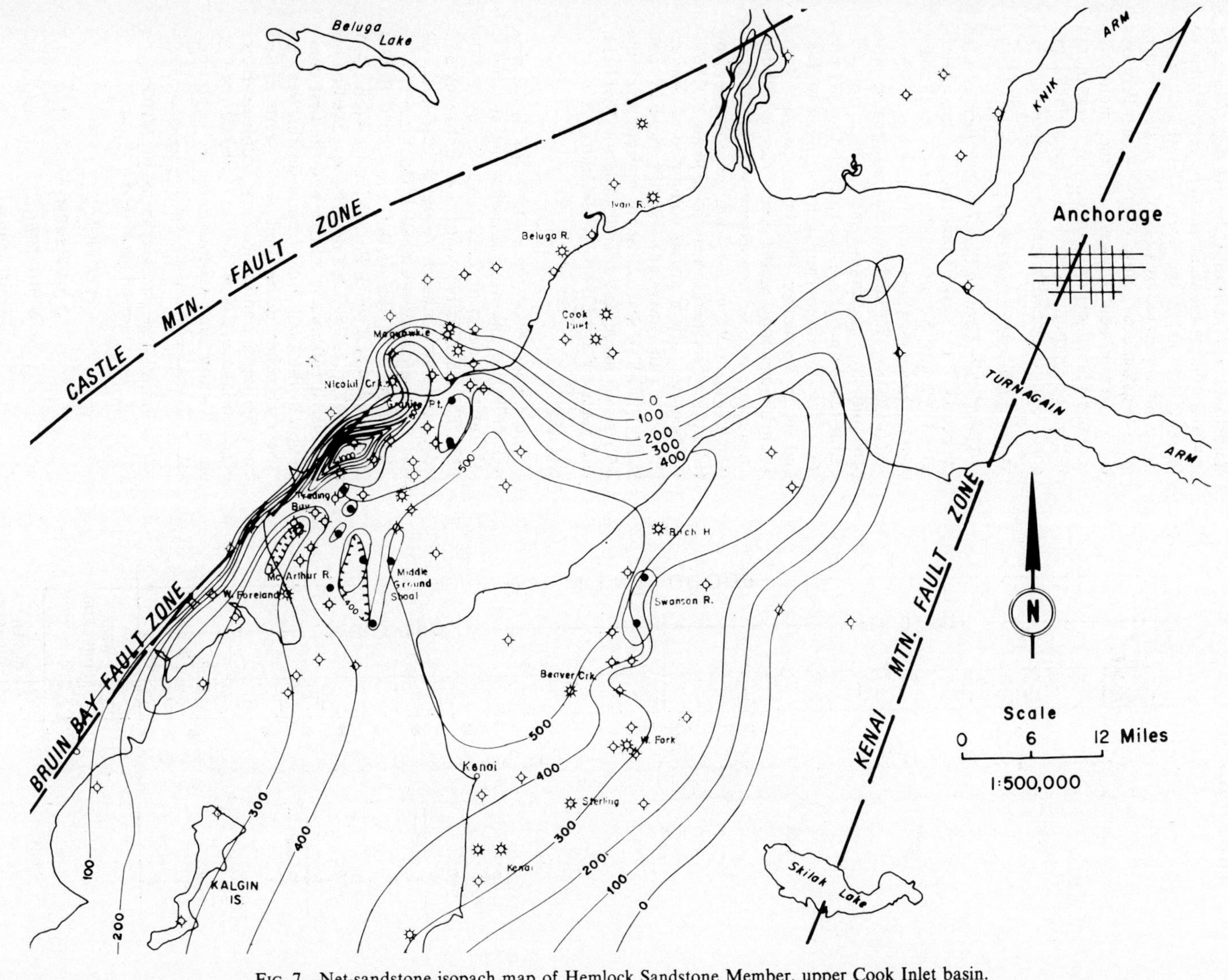

FIG. 7—Net-sandstone isopach map of Hemlock Sandstone Member, upper Cook Inlet basin.

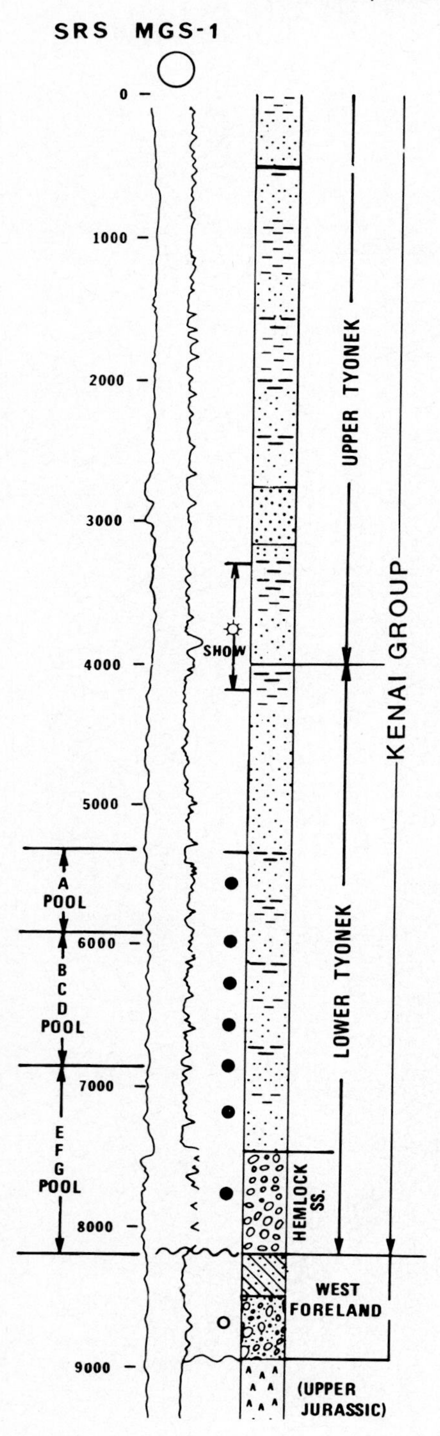

sandstones were deposited by braided streams of low to moderate gradient, perhaps very similar to present streams in the area. The fluvial sandstones are separated by floodplain shales containing a few thin coal seams. On the Kenai Peninsula, a narrow belt adjacent to the mountains contains glacial deposits similar to those in the Beluga Formation (Figs. 3–5). These glacial deposits intertongue with the fluvial deposits.

Middle Ground Shoal Field

The Middle Ground Shoal field is located near the center of the upper Cook Inlet, approximately 60 mi (97 km) by air southwest of Anchorage, Alaska, beneath water with an average depth of 100 ft (30 m). The field was named because of its proximity to a prominent shoal which rises 80–120 ft (24–37 m) above the surrounding sea bottom. Gas was discovered in the summer of 1962 and oil in September 1963 in sandstones of the Tyonek Formation of the Kenai Group. Hydrocarbons are trapped in a narrow anticlinal feature which is up to 9.5 mi (15.3 km) long and 0.75 mi (1.2 km) wide. Operators in the field are (1) Amoco, operating for a group composed of Amoco (formerly Pan American), Phillips, Sinclair, and Skelly (PaPSS), and (2) Shell, operating for the SAS group made up of Shell, Atlantic Richfield, and Standard Oil Company of California.

History

The Amoco MGS State 17595 No. 1, the first well drilled on the structure, blew out at 1,100 ft (335 m) and again at 5,200 ft (1,585 m) in July 1962. The final gas blowout cratered the well and led to its abandonment. About a year later, oil was discovered by the Shell MGS State No. 1, which flowed on test at the rate of 850 BOPD from the lower Tyonek Formation between 7,475 and 8,180 ft (2,278–2,493 m). This was the first oil discovered offshore in Alaska and the first completion in the Cook Inlet from a floating vessel.

Several other delineation wells were drilled from floating rigs and temporary platforms. However, a tidal range of 30 ft (9 m), tidal currents of up to 5 knots, and severe winter ice conditions

←Fig. 8—Type stratigraphic log for Middle Ground Shoal field.

—LEGEND—

☼ GAS SHOW

● OIL ZONES

○ OIL SHOW

Table 1. Characteristics and Statistics of Producing Intervals, Middle Ground Shoal Field*

	Pool A	*Pools B, C, D*	*Pools E, F, G*
Gross thickness	500 ft (152 m)	1,000 ft (305 m)	1,350 ft (410 m)
Average net productive interval	70 ft (21 m)	150 ft (46 m)	400 ft (122 m)
Productive area	500 acres (2 km^2)	600 acres (2.4 km^2)	4,000 acres (16 km^2)
Average porosity (%)	16	16	11
Average permeability (md)	60	60	10
Waterflood started	July 1969	July 1969	Feb. 1969
Water injected (bbl)	5,523,709	4,500,367	71,905,663
Oil produced (bbl)	1,810,492	6,361,307	78,662,670
Gas produced (Mcf)	4,083,776	4,127,027	37,270,730
Water produced (bbl)	916,122	972,222	9,162,874
Number of producing wells (including dual)	2	5	29
Number of injection wells (including dual)	2	3	22
Gravity of oil (API)	36 - 38°	36 - 38°	36 - 38°

*All production and injection figures are cumulative to January 1, 1974 (from State of Alaska Dept. of Natural Resources Statistical Report 1973).

made this type of development impractical, and the first permanent drilling and production platform, Platform B, was installed by Amoco during the summer of 1964. Shortly thereafter, Platform A was set by the Shell-operated SAS group. By the latter part of 1967, two other platforms (Amoco "D" and Shell "C") had been installed and drilling operations from them had commenced. The primary development of the field was completed in mid-1969 with the drilling of 55 wells (32 by Shell and 23 by Amoco). However, prior to completion of the primary drilling, reservoir pressures started declining rapidly, and it was decided to augment the natural reservoir energy (depletion drive) with pressure maintenance by water injection, which was started in 1969. Presently the field contains 31 producing wells, 23 injection wells, 1 shut-in gas well, and 8 abandoned or suspended wells. As of January 1, 1974, the field had produced 78,662,670 bbl of oil, 37,270,730 Mcf of gas, and 9,162,874 bbl of water.

GEOLOGY

Stratigraphy

The Middle Ground Shoal field produces oil from a gross interval of approximately 2,800 ft (850 m) in the lower Tyonek Formation between −5,100 and −9,700 ft (−1,550 and −2,950 m). The entire sequence consists of sandstones, conglomeratic sandstones, conglomeratic shales, and coals deposited in a braided-stream environment. Up to 1,850 ft (565 m) of oil-productive sandstone and sandstone conglomerate is present over the crest of the structure. By using stratigraphic markers such as siltstones and coal beds located between sandstone groups, it is possible to correlate many intervals over the extent of the field. The State of Alaska Oil and Gas Committee has separated the gross oil-productive interval into seven pools designated "A" through "G" (see Fig. 8). Production is commingled in the B, C, and D pools and in the E, F, and G pools, respectively, but the A pool is treated separately. The characteristics and statistics for each pool are tabulated in Table 1.

In addition to these oil accumulations, gas is present in sandstones of the upper Tyonek Formation, commonly called the "SRS sands," between about −3,000 and −4,000 ft (−900 and −1,200 m). These have not been produced, but electric-log analysis (confirmed by a few drill-stem and wireline tests) indicates gas accumulations at the northern end of the field in the crestal position. Gas occurs in fairly thin sandstones or silty sandstones in a predominantly shale or siltstone interval.

The deepest well in the field, Amoco SMGS Unit No. 1, penetrated 900 ft (274 m) of basal

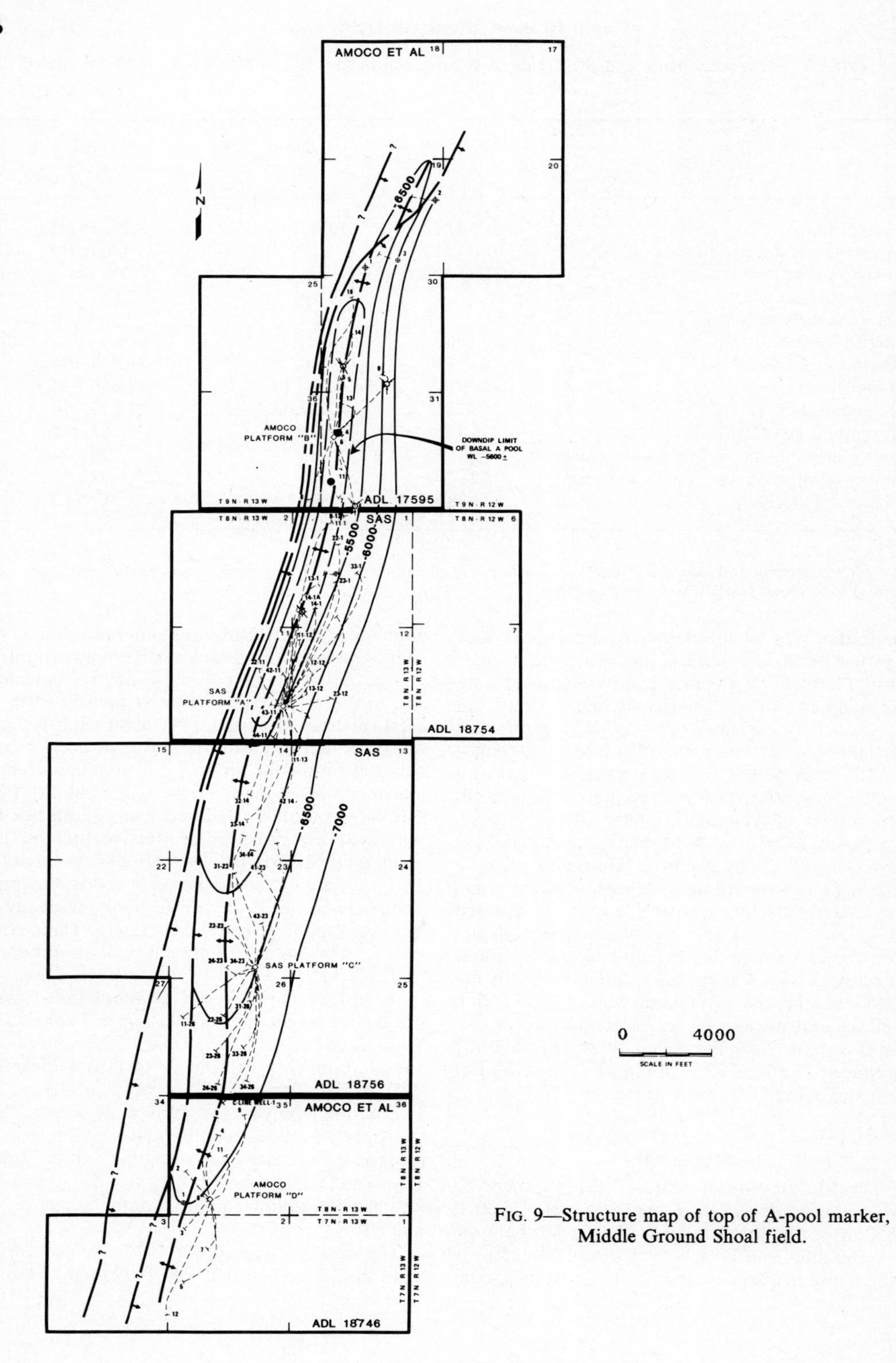

FIG. 9—Structure map of top of A-pool marker, Middle Ground Shoal field.

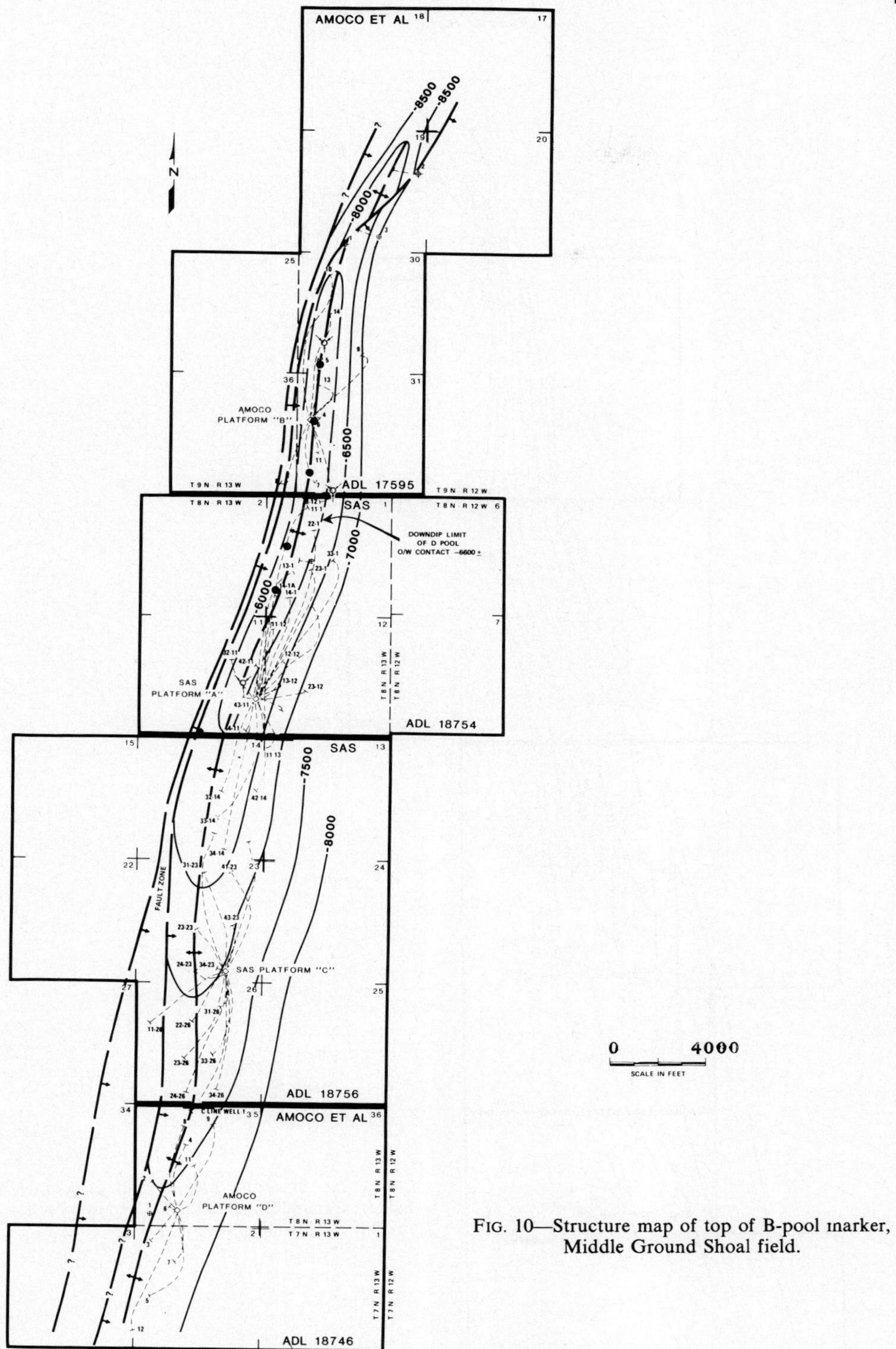

FIG. 10—Structure map of top of B-pool marker, Middle Ground Shoal field.

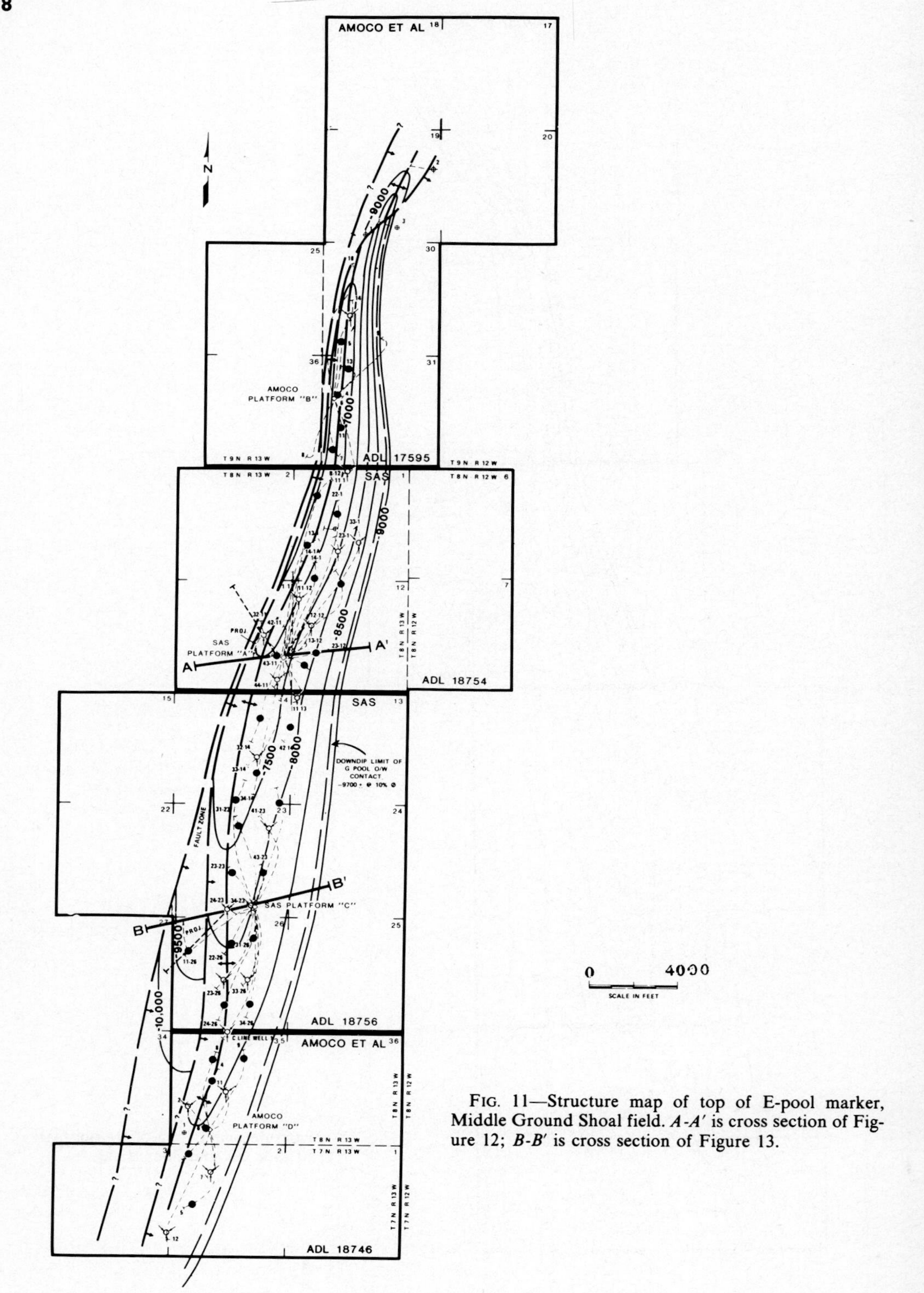

FIG. 11—Structure map of top of E-pool marker, Middle Ground Shoal field. *A-A′* is cross section of Figure 12; *B-B′* is cross section of Figure 13.

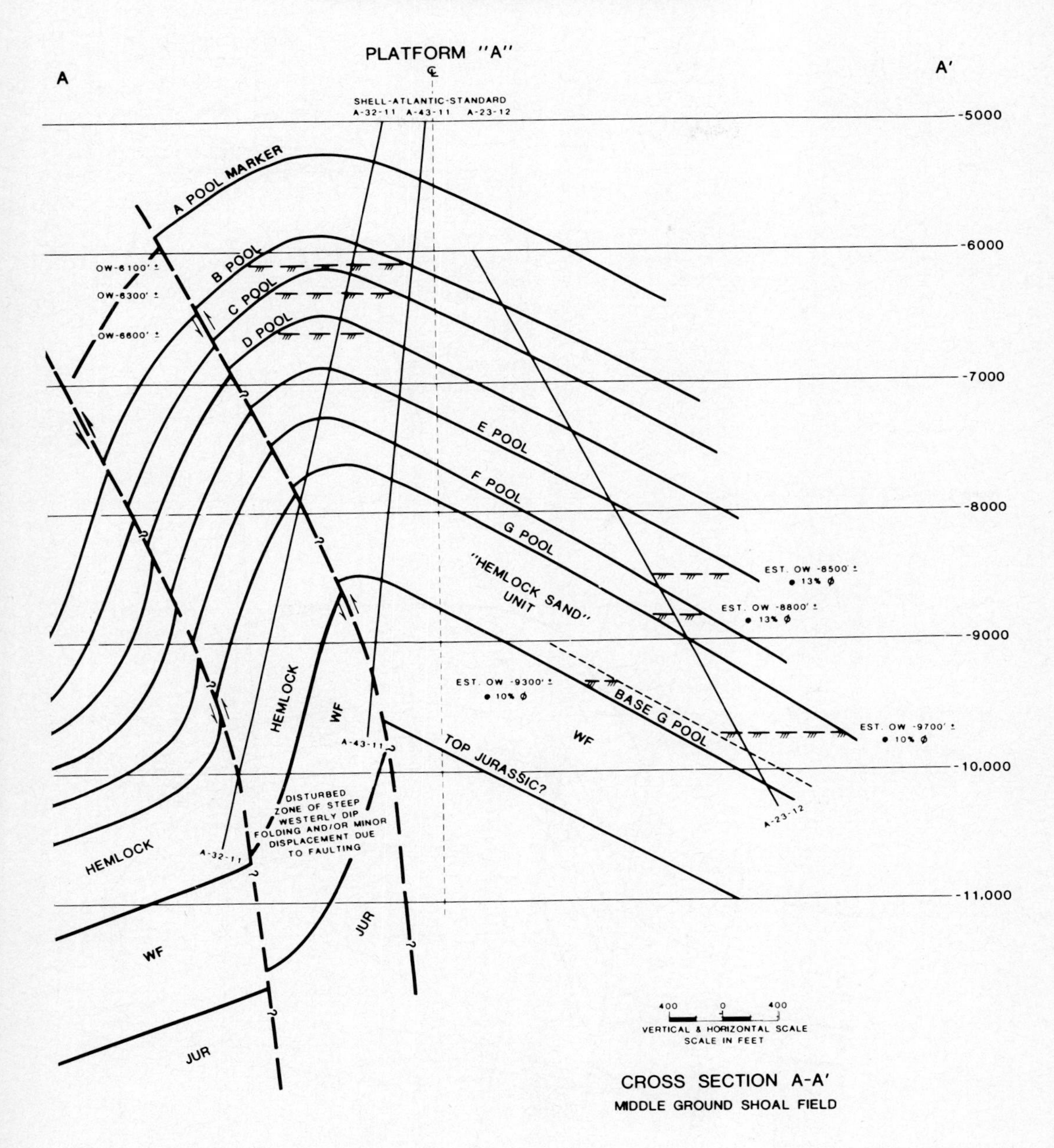

FIG. 12—Structural cross-section *A-A'*, Middle Ground Shoal field. Location is shown on Figure 11.

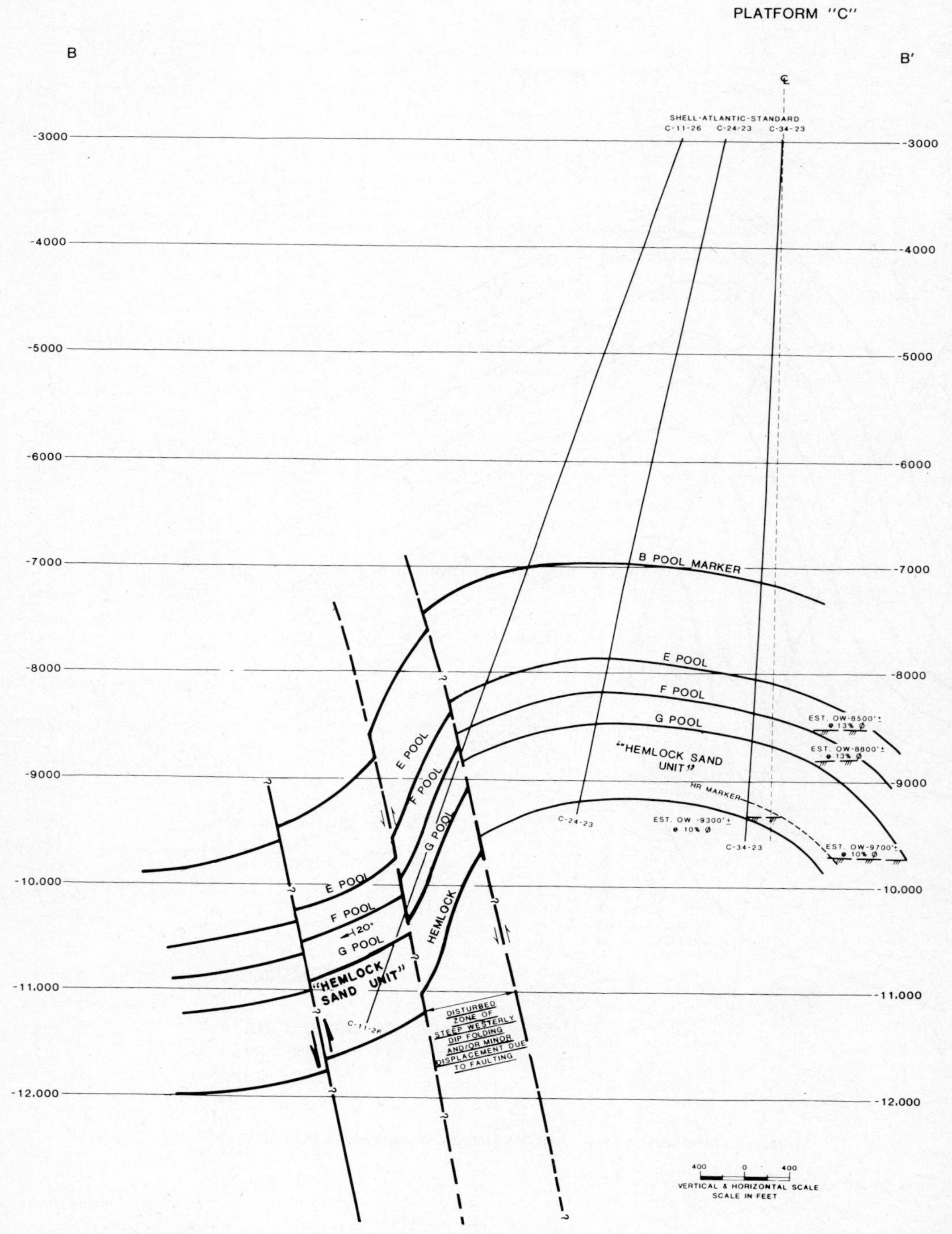

FIG. 13—Structural cross-section *B-B'*, Middle Ground Shoal field, Location is shown on Figure 11.

Tertiary West Foreland Formation below the Hemlock sandstone G pool; total depth is at 10,940 ft (3,335 m) in Upper Jurassic shale. Though not as deep as the Amoco well, the Shell MGS State No. 1 is in a more crestal position and penetrated deeper stratigraphically. It found about 900 ft (274 m) of West Foreland Formation and penetrated almost 600 ft (180 m) of Upper Jurassic shale. In both wells the West Foreland consisted primarily of coal and tightly cemented siltstone with a few very thin sandstones and conglomerates containing oil shows. Some tuffaceous material is present also.

Structure

The Middle Ground Shoal structure is a narrow anticlinal feature which strikes N10°E. The dip on the east flank ranges from 35 to 50° (see Figs. 9–13). At the south end it plunges gently to the south-southwest at about 5°. The structure dips more steeply (60–70°) on the west and may contain some easterly dipping reverse faults, one of which may transect the northern end of the structure. Structural contours on various markers in the lower Tyonek Formation are essentially parallel. Thus, little or no paleostructural growth is thought to have occurred during deposition of the oil-bearing sandstone sequence.

The actual structural picture on the west flank is not clear. As illustrated in Figures 12 and 13, wells on that flank have drilled strata with very steep westerly dips, underlain by strata with more gentle 15–20° dips in the same direction. It is not known if this change in dip is the result of (1) the pattern of folding on the very steeply dipping west flank (with a corresponding lower porosity in the more deformed sandstones), (2) a zone of easterly dipping reverse faults, or (3) a combination of the two. Regardless of the cause, the western limit of production for the lower intervals—namely, the E, F, and G pools—is the updip fault shown in Figures 11, 12, and 13.

Seismic data indicate that another large westerly dipping, steep reverse-fault zone roughly parallels the axis of the structure 0.5–2 mi (0.8–3.2 km) east of the axis. The fault has not been penetrated by any well, and its exact position and size are uncertain.

Productive Zones

A Pool

The A pool consists of approximately 500 ft (150 m) of section containing alluvial sandstones, coals, and floodplain shales. The channel origin of the sandstones is evidenced by a downward increase in grain size and by sharp basal contacts. The width of these channels is such that they are not penetrated by all the wells in the field; therefore, individual accumulations of hydrocarbons are discontinuous. As illustrated in Figure 9, hydrocarbon accumulations are limited to the crestal area in the northern end of the field. Although water levels differ in the individual sandstones, none of the sandstones appear to be productive below −5,600 ft (−1,700 m). These reservoirs produce with a high gas/oil ratio, and some contain only gas. The Shell MGS A-14A-1 tested two sandstones which produced only gas; at present, the well is shut in.

B, C, and D Pools

Directly below the A pool is a 1,000-ft (300 m) interval containing three state-designated pools (B, C, and D) which are commingled and thus are discussed here as one unit. The section consists of an interval of cut-and-fill sandstones deposited by braided streams. Alternating conditions of erosion and sedimentation are indicated by sharp basal sandstone contacts; abrupt changes from coarse-, medium-, and fine-grained layers of sandstone to conglomeratic sandstone are common. Grain-size variations can be followed on the resistivity log.

Medium- to coarse-grained sandstones average 15 percent porosity. Layers of finer grained sandstone, many near the top of individual sandstone bodies, have porosities as high as 20 percent, but pebbly sandstones have porosities of less than 12 percent. These two sandstone types are believed to have low permeabilities and to contribute little to the productivity of the pools. Thus, in general, the principal production is believed to be from thin permeable sandstone zones. None of the sandstones within the B, C, and D pools are productive below −6,600 ft (−2,012 m; see Fig. 12).

E, F, and G Pools

The 1,350-ft (410 m) lower portion of the oil-producing section is designated as the "E," "F," and "G" commingled pools. Approximately 90 percent (70,490,871 bbl) of the field's cumulative oil production through 1973 has come from these pools.

The upper 600± ft (183 m) of the section comprises the E and F pools, which have produced only about 10 percent of the combined pools' production. In the E and F pools, oil comes from a sandstone at the top of each interval. These sandstones are approximately 40 and 70 ft (12 and 21 m) thick, respectively; the remainder of the section is composed of floodplain shale and thin coal seams. Both sandstones have the characteristics of alluvial point-bar sandstone with a smooth or slightly "bell-shaped" SP curve, abrupt lower contact, upward decrease in grain size, and gradational upper contact. The water levels in the E and F pools are −8,500 and −8,800 ft (−2,590 and −2,680 m), respectively. These sandstones are oil bearing in all but the very southernmost wells in the field (see Figs. 11–13).

The state-designated "G" pool, in the Hemlock sandstone, is the main productive interval in the Middle Ground Shoal field. The gross section of the G pool is approximately 750 ft (230 m) thick and contains six correlatable sandstone and conglomeratic intervals which, in the crestal part of the structure, contain as much as 450 ft (137 m) of productive section. The bottom 150-ft (45 m) interval has an effective oil-water contact at approximately −9,350 ft (−2,850 m), which is 400 ft (120 m) higher than the oil-water contact of the main G pool at −9,750 ft (−2,970 m).

The productive interval in the G pool contains a group of moderate-energy, braided-stream deposits laid down in an area of low relief. The lower 450 ft (137 m) consists largely of sandstone and conglomeratic sandstone; only three shale intervals can be correlated from well to well. The predominant lithologic type is a very pebbly sandstone with well-rounded quartz, chert, and metamorphic and volcanic pebbles averaging 15 mm in size. Infilling and cementing the pebbles is a mixture of silt and sand with about 5 percent clay. This section is believed to have been deposited by a series of moderate-energy streams resulting in a widespread but complex cut-and-fill deposit.

The energy level appears to have been slightly lower during the deposition of the upper 300 ft (90 m) of the gross productive interval, producing a series of sandstones and pebbly sandstones separated by thicker and more correlatable shale intervals. The pebbly-sandstone layers are generally made up of alternate layers of moderately to poorly sorted, medium- to coarse-grained fragments with a mixture of pebbles ranging from 5 to 15 mm in diameter. The average sandstones are also medium to coarse grained, are moderately well sorted, and contain some medium-scale crossbeds.

References Cited

Burk, C. A., 1965, Geology of the Alaska Peninsula—island arc and continental margin: Geol. Soc. America Mem. 99, 250 p.

Calderwood, K. W., and W. C. Fackler, 1972, Proposed stratigraphic nomenclature for Kenai Group, Cook Inlet basin, Alaska: AAPG Bull., v. 56, no. 4, p. 739-754.

Crick, R. W., 1971, Potential petroleum reserves, Cook Inlet, Alaska, *in* I. H. Cram, ed., Future petroleum provinces of the United States—their geology and potential, v. 1: AAPG Mem. 15, p. 109-119.

Detterman, R. L., and J. K. Hartsock, 1966, Geology of the Iniskin-Tuxedni region, Alaska: U.S. Geol. Survey Prof. Paper 512, 78 p.

——— and D. L. Jones, 1974, Mesozoic fossils from Augustine Island, Cook Inlet, Alaska: AAPG Bull., v. 58, no. 5, p. 868-870.

Dickinson, W. R., 1971, Clastic sedimentary sequences deposited in shelf, slope, and trough settings between magmatic arcs and associated trenches: Pacific Geology, v. 3, p. 15-30.

Grantz, Arthur, and J. A. Wolfe, 1961, Age of the Arkose Ridge Formation, south-central Alaska: AAPG Bull., v. 45, no. 10, p. 1762-1765.

Imlay, R. N., and R. L. Detterman, 1973, Jurassic paleobiogeography of Alaska: U.S. Geol. Survey Prof. Paper 801, 34 p.

Karlstrom, T. N. V., 1964, Quaternary geology of the Kenai lowland and glacial history of the Cook Inlet region, Alaska: U.S. Geol. Survey Prof. Paper 443, 69 p.

Kirschner, C. E., and C. A. Lyon, 1973, Stratigraphic and tectonic development of Cook Inlet petroleum province, *in* Arctic Geology: AAPG Mem. 19, p. 396-407.

——— and D. L. Minard, 1948, Geology of the Iniskin Peninsula, Alaska: U.S. Geol. Survey Oil and Gas Inv. Prelim. Map 95, scale 1 in. = 4,000 ft.

Moore, J. C., 1972, Uplifted trench sediments: Southwestern Alaska–Bering shelf edge: Science, v. 1975, p. 1103-1105.

——— 1973, Cretaceous continental margin sedimentation, southwestern Alaska: Geol. Soc. America Bull., v. 84, p. 595-614.

Osment, F. C., R. M. Morrow, and R. W. Craig, 1967, Petroleum geology and development of the Cook Inlet basin of Alaska (with French abs.), *in* Origin of oil, geology and geophysics: 7th World Petroleum Cong. Proc., Mexico, 1967, London, Elsevier Pub. Co., v. 2, p. 141-150.

Permo-Triassic Reservoirs of Prudhoe Bay Field, North Slope, Alaska[1]

H. P. JONES[2] and R. G. SPEERS[3]

Abstract Production from the Prudhoe Bay field, which contains the largest single accumulation of oil ever discovered in North America, is expected to commence in 1977. Recent approval of the construction of an oil pipeline from Prudhoe Bay to Valdez has resulted in a marked increase in both drilling and construction activity in the field area. Already over 100 wells have been drilled as a result of almost 7 years of operation, and considerable subsurface data on the Permo-Triassic formations are now available. On the basis of these data, this paper describes geological and petrophysical aspects of the main Prudhoe Bay reservoirs.

Three reservoirs are contained within the Permo-Triassic. These are, in descending order, the Sag River Formation, the Shublik Formation, and the Ivishak Sandstone. The Ivishak Sandstone reservoir of the Sadlerochit Group contains most of the field's hydrocarbons.

Observations of the distribution of gas, oil, and water in the field have led to hypotheses regarding the origin and distribution of the hydrocarbons.

INTRODUCTION

Oil seeps near Cape Simpson, east of Point Barrow, attracted the first interest in the petroleum potential of the Arctic North Slope of Alaska. Mining claims were staked on these seeps in the early 1900s but, when Naval Petroleum Reserve No. 4 was established in 1923, private industrial interest temporarily ceased. During the period between 1923 and 1963, most of the exploration was conducted by the U.S. Navy and the U.S. Geological Survey. In the early 1960s, however, private industry resumed exploration of the North Slope, confining most of its activity to the area east of the Naval Petroleum Reserve and west of the Arctic Wildlife Range. Initial reconnaissance geologic surveys carried out in the foothills of the Brooks Range were augmented by geophysical surveys, and it was a combination of the results of this work, together with the incentive provided by the state's competitive leasing policy, that shifted industrial interest away from the foothills toward the coastal areas. By 1965, three major oil companies—Humble, Richfield, and British Petroleum—had acquired most of the leases covering the Prudhoe Bay structure, which by then had been delineated geophysically.

Two of the early wells drilled by British Petroleum on the North Slope, Colville No. 1 and Kookpuk No. 1, were extremely important. Although both were dry, they proved that the Permo-Triassic section and the Lisburne Group contained prospective reservoirs. However, this initial lack of success in finding significant hydrocarbon accumulations, plus the high costs and the remote and hostile environment, caused a lessening of interest in the area until the drilling of the Arco-Humble discovery well, Prudhoe Bay State No. 1, in 1968.

Following this discovery, several companies started drilling in the area in an effort to delineate the extent of the field prior to the September 1969 state lease sale. In 1969, British Petroleum joined with Standard Oil of Ohio (Sohio) for the purpose of development of their leases in Prudhoe Bay and its vicinity. BP Alaska Inc., a subsidiary of British Petroleum, is the operator for the joint companies.

After the 1969 lease sale, efforts were initiated to unitize the field. Negotiations toward the formation of the unit are continuing, with BP acting as operator for the western part of the field and Arco the operator for the eastern part.

Since the discovery well was drilled, over 100 wells have penetrated the Permo-Triassic reservoirs in and around the Prudhoe Bay field. Data from the wells show that the field can be divided into two structural areas (Fig. 1). To the west, several small, faulted structures are collectively known as the "Eileen" area. The major volume of hydrocarbons, however, occurs in the "Main" area, which is a faulted, generally south-dipping structure truncated on the east by an unconformity of Early Cretaceous age.

[1]Manuscript received, May 23, 1975; revised, September 5, 1975.

[2]BP Alaska Inc., San Francisco, California 94111.

[3]Oil Service Company of Iran (Ahwaz), P. O. Box 1095, Teheran, Iran.

The writers thank the management of BP Alaska, Inc. and Sohio Petroleum Company for allowing them to publish this paper. Although the opinions and interpretations expressed are those of the writers alone, the technical and editorial advice received from colleagues working in the Production Planning and Exploration Departments of BP Alaska and at the BP Research Center in Sunbury, England, were invaluable. The help and advice given by J. Schrak during preparation of the diagrams are especially appreciated.

The regional setting and general stratigraphy of the field have been described by Rickwood (1970) and by Morgridge and Smith (1972). This paper focuses on the stratigraphy, structure, petrophysics, and fluid content of the main reservoirs, which occur in the Permo-Triassic. The following brief description of the regional and stratigraphic setting is included (1) to relate the field to the broader structural aspects of the North Slope and (2) to illustrate the field's excellent position as a trap relative to the assumed source rocks.

Regional and Stratigraphic Setting

The Prudhoe Bay field lies on the north coast of Alaska approximately 250 mi (400 km) north of the Arctic Circle (Fig. 2). The structure is at the eastern end of the southeasterly plunging Barrow arch, a major regional high paralleling the coast east of Point Barrow. Dips are shallow south of Barrow arch, although as much as 30,000 ft (9,145 m) of sediments has accumulated in the deeper parts of the Colville trough (Fig. 2). This trough is an asymmetric syncline, and the Brooks Range forms the southern limit of its steeper south flank. The northern flank of the Barrow arch lies offshore under the Arctic Ocean. In the vicinity of Prudhoe Bay, this flank is represented by a series of down-to-north normal faults. North of the field, for example, depth to the Permo-Triassic increases more than 2,000 ft (610 m) in less than 5 mi (8 km).

Pre-Cretaceous sediments were derived from a source area north of the present coast. During the Early Cretaceous, erosion removed most of the sediments over the Barrow arch to the east of the field, producing a major unconformity. Subsequent to this erosion, the Brooks Range formed the source of the sediments. Most of the hydrocarbons of the Prudhoe Bay field and its environs occur in reservoirs which are truncated by the Lower Cretaceous unconformity and which directly underlie the Lower Cretaceous marine source rocks. The field is ideally situated for the entrapment of hydrocarbons: it is located updip from a large volume of potential source beds, and the reservoirs are sealed by unconformably overlying shales.

The stratigraphic column and the generalized stratigraphy of the Prudhoe Bay field are illustrated in Figures 3 and 4. Eroded, contorted, and slightly metamorphosed early Paleozoic rocks are overlain by interbedded sandstones and shales and some coal beds. These Early Mississippian strata are probably equivalent to the Kayak Shale. They are overlain by a thick sequence of carbonate rocks which are equivalent to the Lisburne Group of Mississippian to Pennsylvanian age. From west to east, this sequence thickens from less than 2,000 ft (610 m) to over 4,000 ft (1,220 m; Fig. 4). At least over the crestal part of the present structure, the Lisburne Group was eroded prior to deposition of the Permo-Triassic sediments. The Permo-Triassic strata thin toward the north-northeast and are overlain conformably by the Jurassic Kingak Formation, which comprises up to 2,000 ft (610 m) of shale and siltstone. West and north of the Prudhoe Bay field, the Kingak Formation is overlain conformably by the Lower Cretaceous Kuparuk River Formation. The shale, siltstone, and sandstone of this formation may have been deposited while the eastern area was being eroded in Early Cretaceous time. Progressively older beds were eroded from west to east across the field and, in the extreme east, all of the Jurassic and Permo-Triassic deposits were removed.

Subsequent to this erosion, deposition continued virtually without interruption. The sediments were derived from a southerly source (Brooks Range). These strata thicken consistently toward the east and northeast. Lower Cretaceous shales and mudstones are covered by approximately 2,000 ft (610 m) of interbedded sandstones and shales probably equivalent to the Seabee and Prince Creek–Schrader Bluff Formations. The upper part of this group of sandstones contains numerous coal seams, but the sandstones themselves are not as thick and clean as the lower sandstones. The Tertiary-Cretaceous boundary, at least in the field area, is thought to be near the upper limit of the coals, though it has not been precisely determined.

The Sagavanirktok Formation consists of a series of massive sandstones and conglomeratic sandstones interbedded with siltstones. It is overlain, possibly unconformably, by the Gubik Formation.

No description of the Prudhoe Bay field would be complete without some mention of the permafrost, an interval of permanently frozen ground commonly more than 2,000 ft (610 m) thick. Its presence complicates geophysical interpretations, and it is the subject of considerable study by engineers who are engaged in designing wells that will safely permit the production of hot oil. The permafrost at Prudhoe Bay has been described by Howitt (1971).

Within the stratigraphic column at Prudhoe Bay, no less than five different major stratigraphic intervals contain hydrocarbons (Fig. 3).

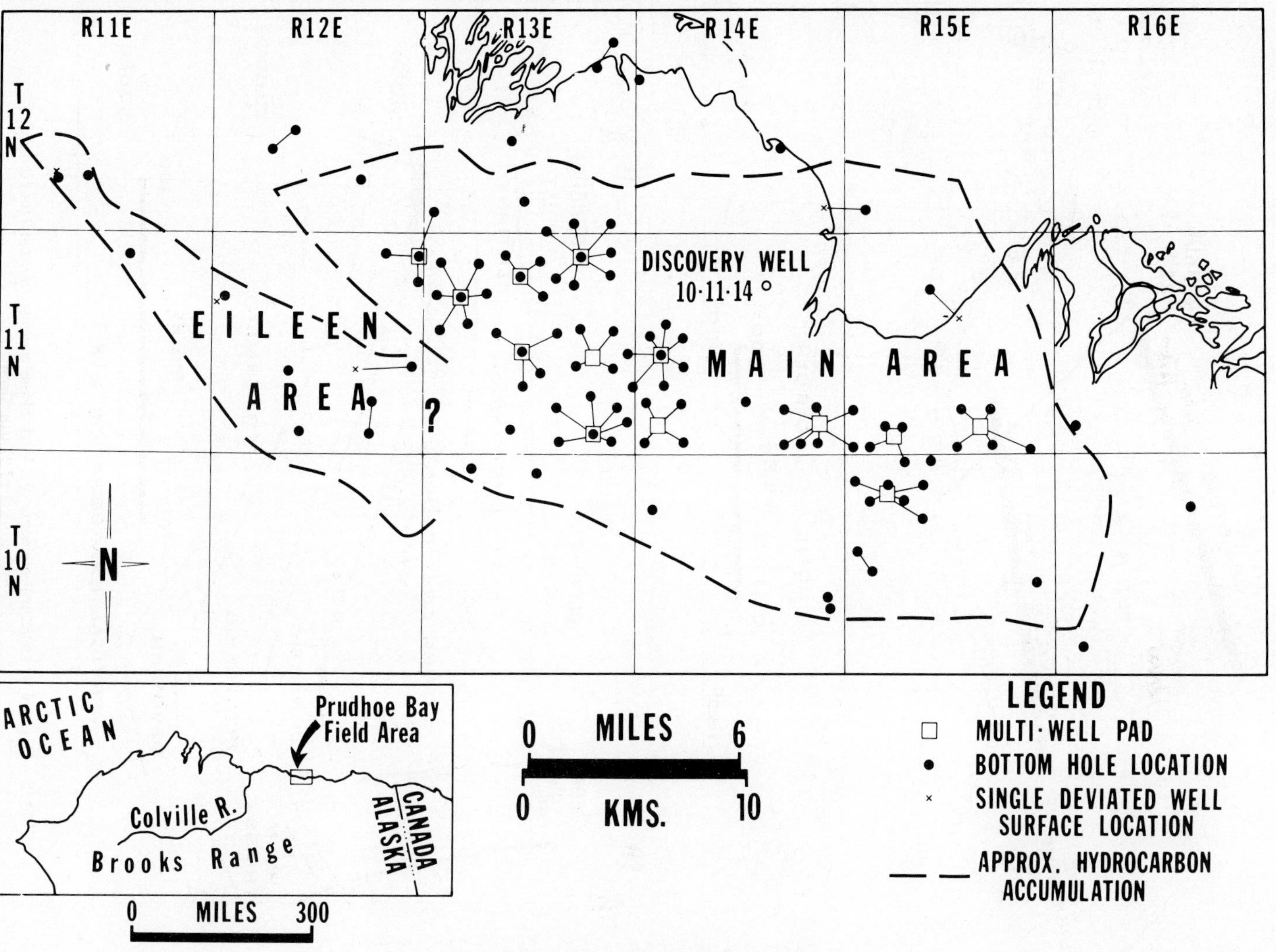

FIG. 1—Prudhoe Bay field, Alaska, location map.

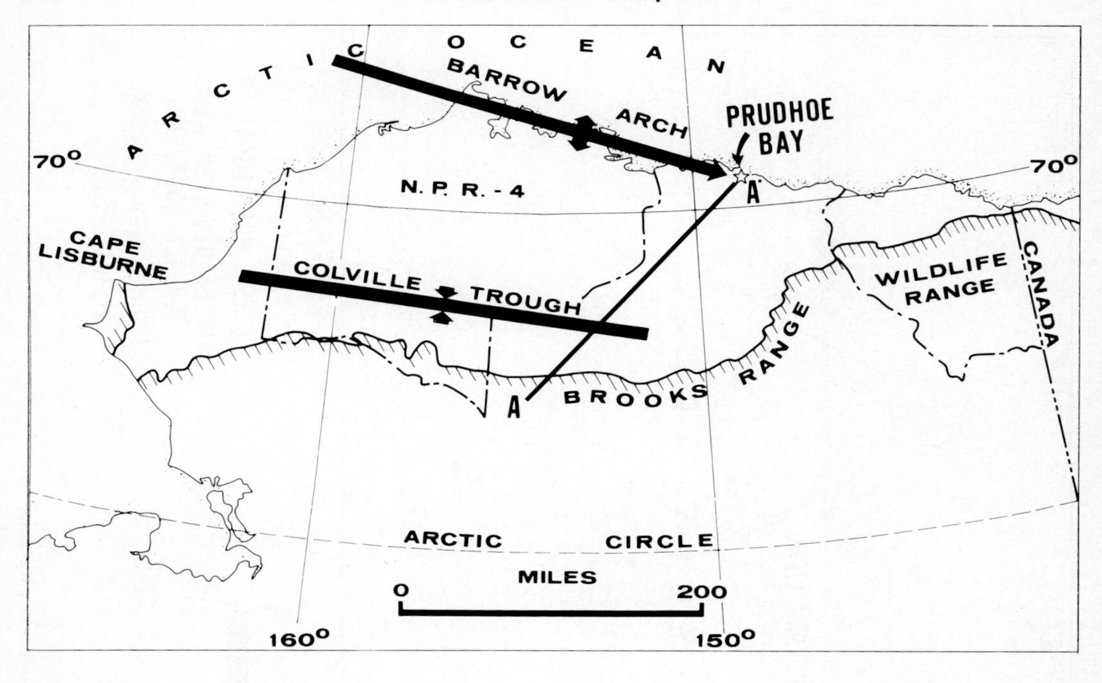

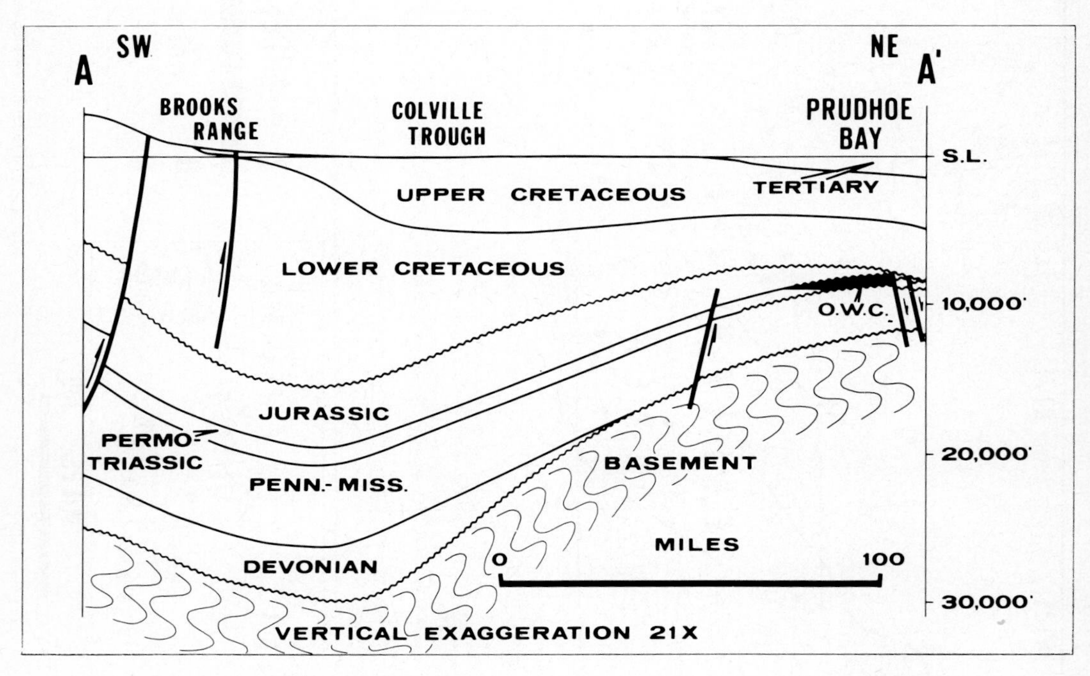

FIG. 2—*Top*: major structural elements of North Slope geology. *Bottom*: generalized cross section of North Slope.

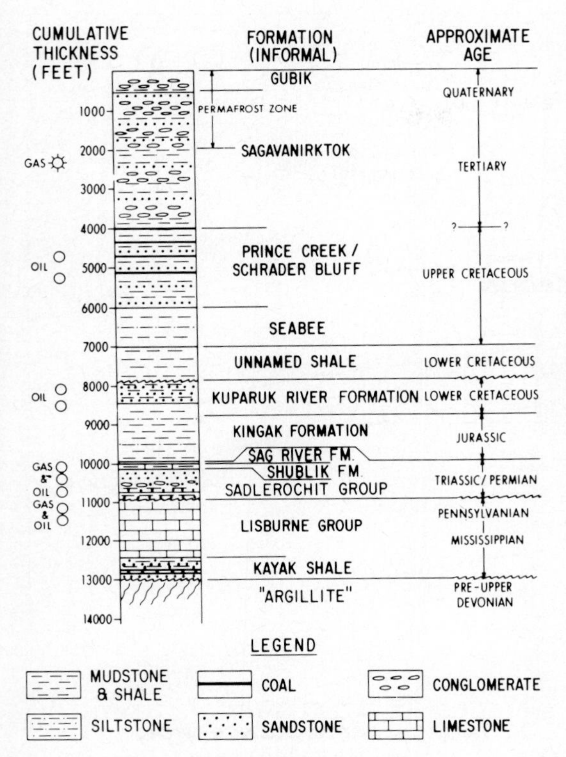

FIG. 3—Generalized stratigraphic column of Prudhoe Bay field.

The deepest of these is the Lisburne (Mississippian-Pennsylvanian) carbonate rocks, which are known to contain gas and oil in the area under and south of the bay itself. Accumulations in the Permo-Triassic rocks are the most significant. The Lower Cretaceous Kuparuk River Formation, which is limited to the area west and north of the field, contains oil at subsea depths in the order of 6,000 ft (1,830 m).

In the western part of the field, where they are shallowest, the Upper Cretaceous sandstones contain heavy, viscous oil. The shallowest hydrocarbons found in the field are believed to be in the form of relatively small gas pockets trapped under the base of the permafrost in the Sagavanirktok Formation.

PERMO-TRIASSIC STRATIGRAPHY

By far the most important reservoirs in the field are contained within the Permo-Triassic section. The largest of the three Permo-Triassic reservoirs is the Ivishak Sandstone of the Sadlerochit Group; it has an oil column of more than 450 ft (137 m). Published reserves of the field are 9.6 billion bbl of oil and 26 Tcf of gas.

The Permo-Triassic of the Prudhoe Bay field is currently subdivided into three formations. In descending order these are the Sag River Formation, the Shublik Formation, and the Sadlerochit Group or Formation. When these formations were described by Detterman in 1970 and by the North Slope Stratigraphic Committee of the Alaska Geological Society in 1970-1971, very few wells had been drilled and very few data were publicly available. Now that over 100 wells have been drilled, many of which have cored the complete Permo-Triassic sequence, the stratigraphy can be defined more accurately. The definitions and reference sections discussed below are those which are currently used informally. However, to standardize the nomenclature, the subdivisions are proposed for formal acceptance in the Prudhoe Bay field and its vicinity.

No individual well has sufficient core and log coverage to enable it to be used alone as the reference section for each of the subdivisions. Figures 5 and 6 show the logs of the BP/Sohio well in Sec. 19, T10N, R15E (19-10-15), and are used to illustrate the Permo-Triassic interval and its proposed subdivision. Well BP 21-11-13 (see Fig. 19) is suggested as the reference section for the Sag River Formation. As the well has complete core coverage of the interval, this section is preferable to the originally designated type section in the Arco/Exxon Sag River State No. 1 (04-10-15). Well BP 19-10-15 is, however, used as the reference section for the Shublik Formation and the proposed Sadlerochit Group.

Figure 7 illustrates the northeastward thinning of the whole Permo-Triassic interval. Its base lies unconformably on the eroded surface of the Lisburne Group, but its upper limit is transitional with the overlying Jurassic Kingak shales (Fig. 8).

Sadlerochit Group

In 1919, Leffingwell established the south flank of the Sadlerochit Mountains as the type locality for the Sadlerochit Formation. In the late 1940s and early 1950s, the U.S. Geological Survey (see Fig. 9), working in northeastern Alaska, subdivided the formation into two members, an Echooka Member of Late Permian age and an Ivishak Member of Early Triassic age.

In the Prudhoe Bay area, the Sadlerochit consists of three distinct lithologic types: a lower,

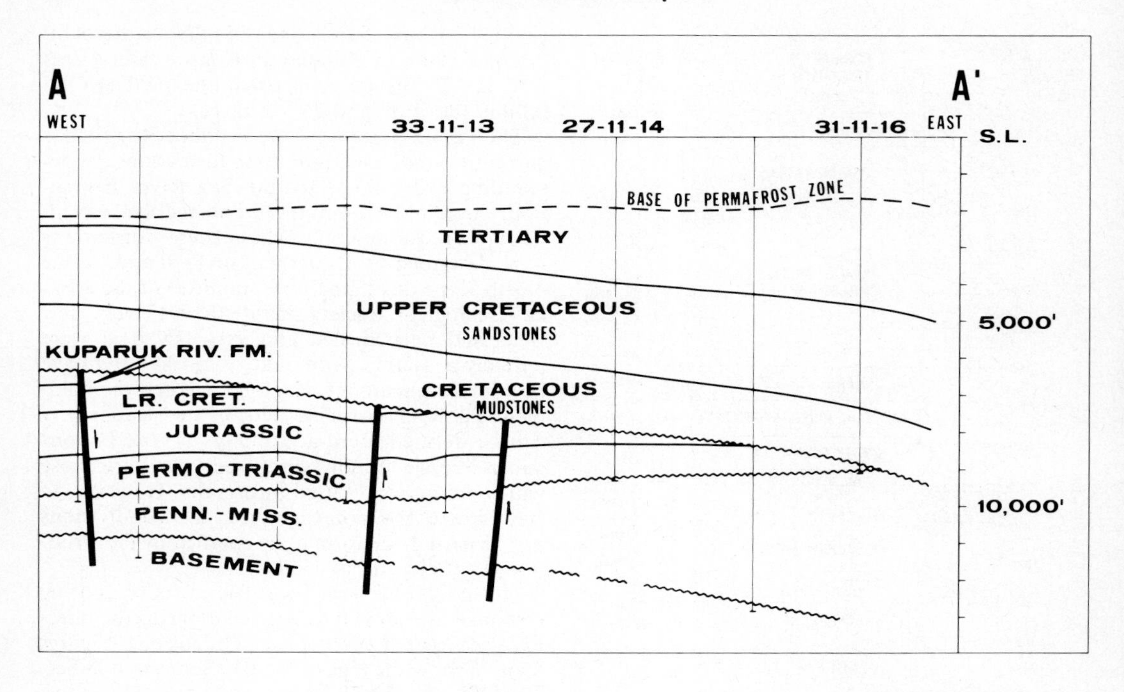

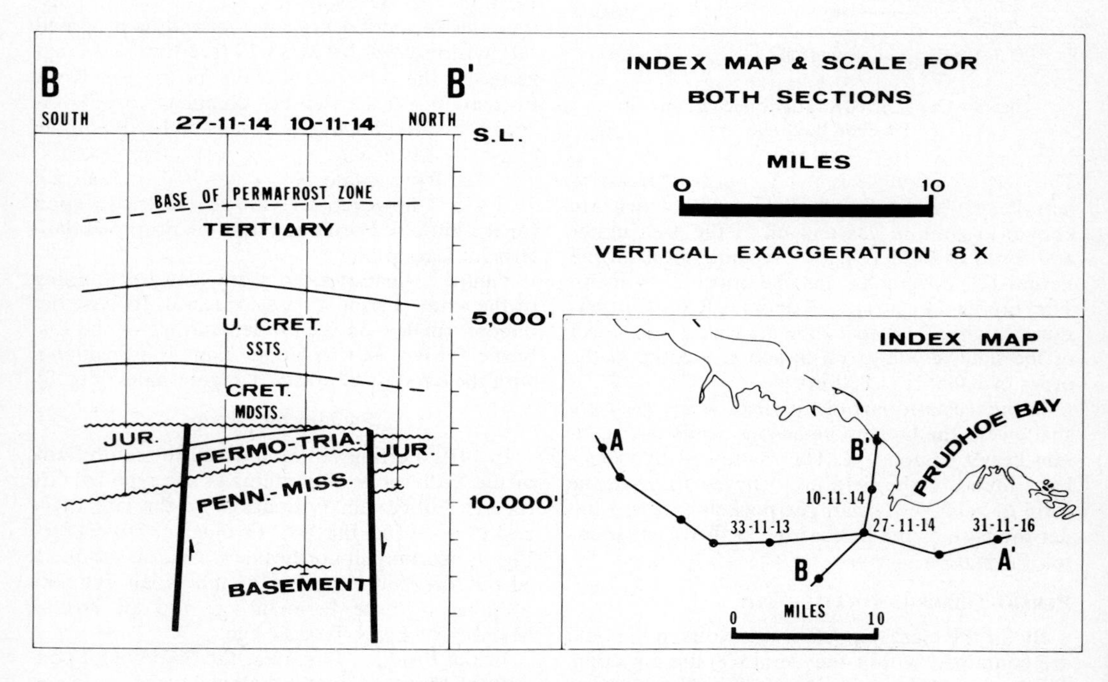

FIG. 4—Generalized cross sections of Prudhoe Bay field.

predominantly glauconitic sandstone interval; a middle shale and mudstone section; and an upper sequence of sandstones, conglomerates, and mudstones.

In view of the variation in lithologies, it is proposed that the Sadlerochit should be elevated to group status in the Prudhoe Bay area and subdivided into three formations. The subdivisions and nomenclature proposed in this paper are shown in Figure 9. The elevation to group status is supported by the U.S. Geological Survey in a paper to be published by Detterman et al (1975). Their proposed nomenclature is also shown in Figure 9.

The proposed Echooka Formation, as presently identified in Prudhoe Bay, is probably equivalent to the Ikiakpaurak Member of the Echooka Formation as defined by Detterman et al (1975). In the field area, the formation has a fairly consistent lithology and does not warrant further subdivision. Detterman et al's Kavik Member is elevated to formation status in the Prudhoe Bay area. The Ivishak Sandstone in the Prudhoe Bay area is probably equivalent to their Ledge Sandstone Member of the Ivishak Formation. The Fire Creek Siltstone, although probably absent over most if not all of the field area, could also be elevated to formation status.

Definition and subdivision—Well BP 19-10-15 (Figs. 5, 6) penetrated the entire Permo-Triassic section and, as the whole sequence is covered by one logging run, this well is proposed as the reference section for the Sadlerochit Group in the subsurface of the Prudhoe Bay field.

The *Sadlerochit Group* in the subsurface of the Prudhoe Bay field is defined as those strata occurring within and correlating with the interval 8,935–9,780 ft (2,723.3–2,980.8 m)—electric-log depths—in well BP 19-10-15 (Fig. 5). The group is subdivided into three formations. The deepest is the Echooka Formation, primarily a glauconitic sandstone unconformably overlying the Lisburne Group and conformably underlying the Kavik Shale.

The *Echooka Formation* is defined as those strata occurring within and correlating with the interval 9,729–9,780 ft (2,965.3–2,980.8 m)—electric-log depths—in well BP 19-10-15 (Fig. 5).

The Kavik Shale, consisting of dark gray shales, mudstones, and siltstones, conformably underlies the Ivishak Sandstone. The *Kavik Shale* is defined as those strata which occur within and correlate with the interval from 9,513 to 9,729 ft (2,809.4–2,965.3 m)—electric-log depths—in well BP 19-10-15 (Fig. 5).

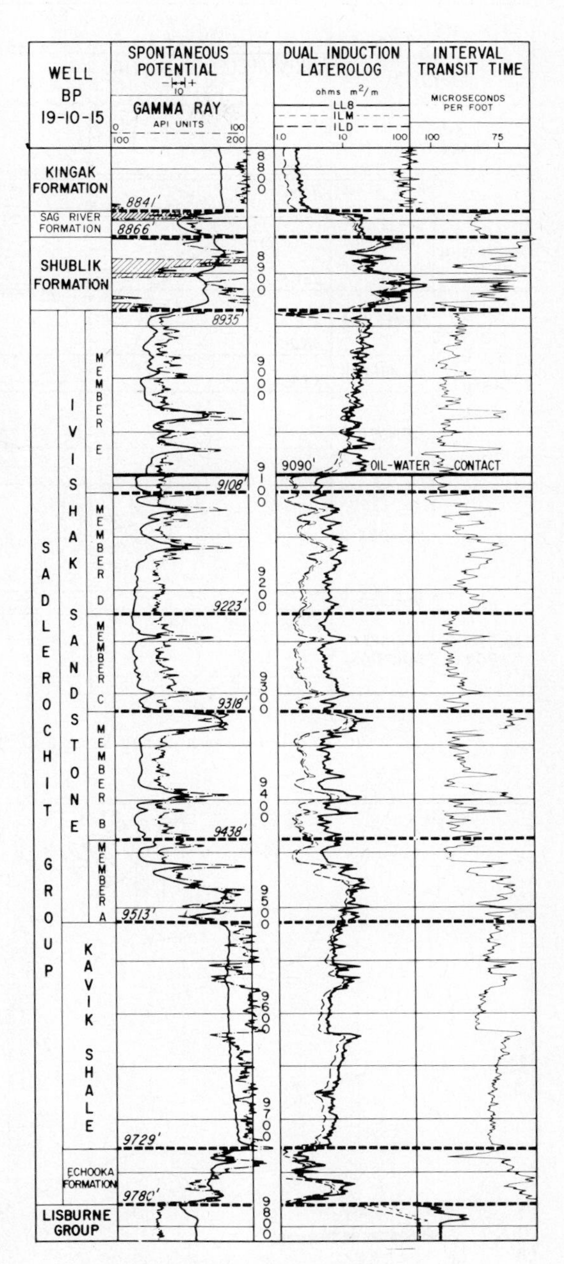

FIG. 5—Permo-Triassic reference section in BP well 19-10-15.

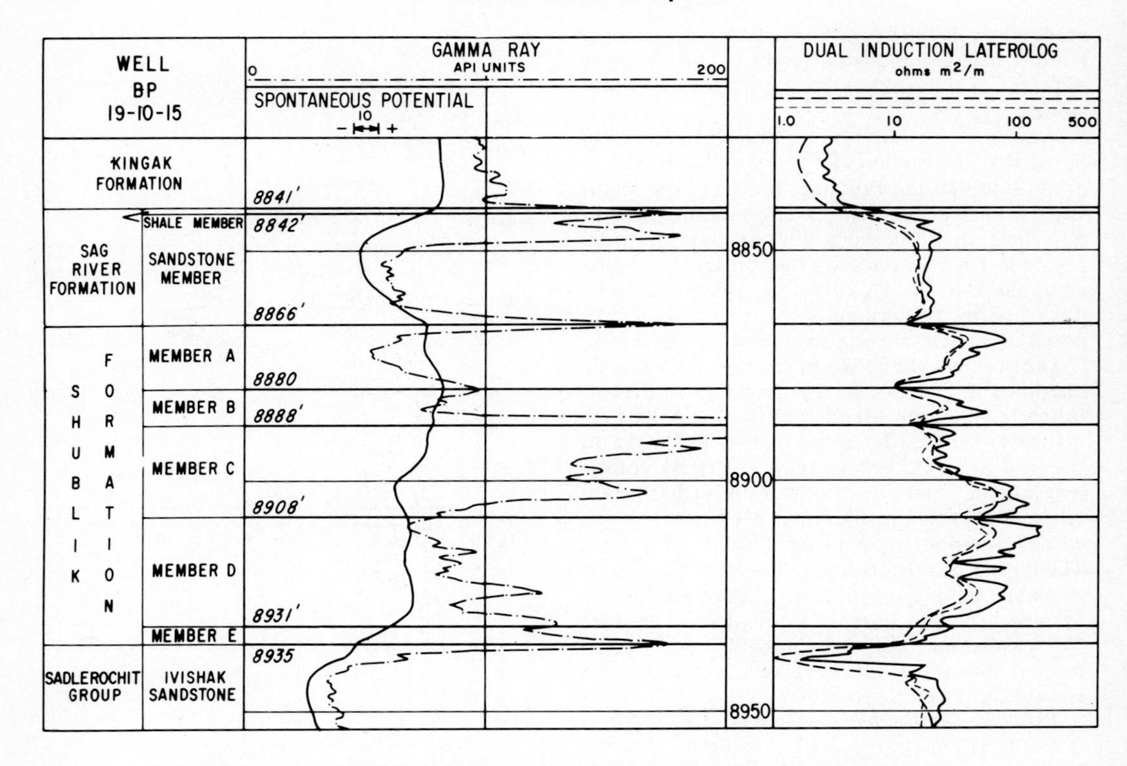

FIG. 6—Subdivisions of Sag River and Shublik Formations in BP well 19-10-15.

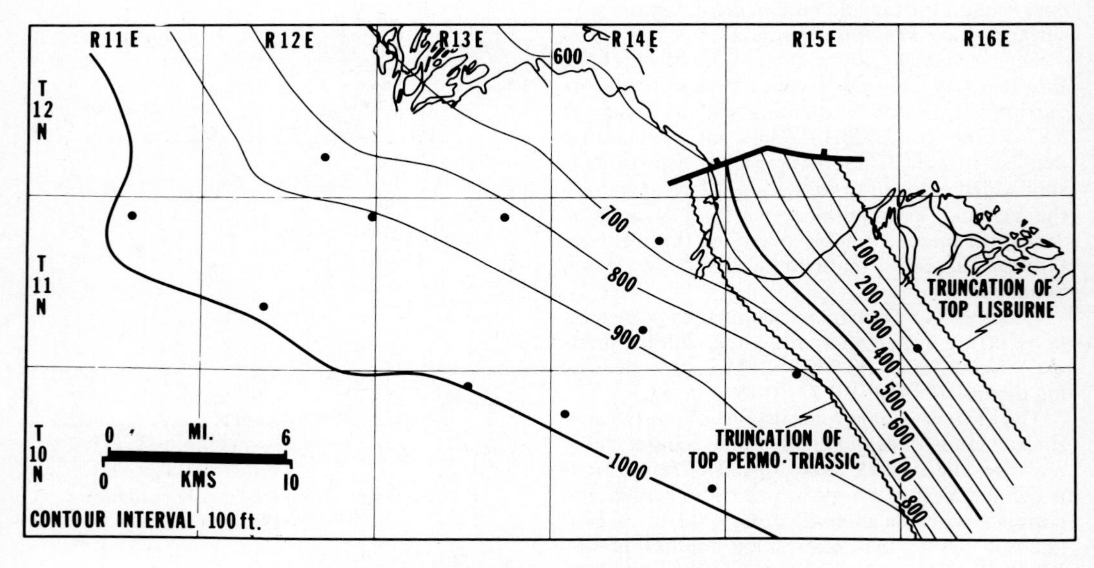

FIG. 7—Thickness variation of Permo-Triassic.

The shallowest formation is the Ivishak Sandstone, consisting of interbedded sandstones, conglomerates, and mudstones. South and west of the field area, the formation conformably underlies the Shublik Formation, but in the northeastern area the contact may be unconformable. The *Ivishak Sandstone* is defined as those strata occurring within and correlating with the interval 8,935–9,513 ft (2,723.3–2,899.4 m)—electric-log depths—in well BP 19-10-15 (Fig. 5). It is the main hydrocarbon-bearing reservoir in the Prudhoe Bay field.

The Sadlerochit Group thins to the north-northeast from ±900 ft (±275 m) in the southwest to less than 450 ft (137 m) in the north, as shown in Figure 10C. Each of the three formations thins in the same direction, though much of the overall thinning is due to the unconformable relation of the Sadlerochit Group with the underlying Lisburne Group and, in places, perhaps with the overlying Shublik Formation.

Echooka Formation—Within the field, the Echooka Formation thins from approximately 70 ft (21 m) in the southwest to zero north of an east-west line through the center of the field (Fig. 11C). The formation onlaps the eroded surface of the Lisburne Group and was probably deposited in a shallow, northward-transgressing sea.

In the field area it has been tentatively dated on the basis of palynomorphs as mid-Permian; the microfloral assemblages found in this sequence are too poor and scarce to permit a more precise dating. The mid-Permian age is based largely on the presence of *Vittatina*, a common constituent of mid-Permian assemblages, together with other multistriate pollen, and on the absence of *Lueckisporites virkkiae*. The latter, a characteristic Late Permian species, occurs in the overlying Kavik Shale.

Lithologically, the Echooka Formation consists of green and dark gray sandstones with thin laminae and stringers of clay and shale. The sand-

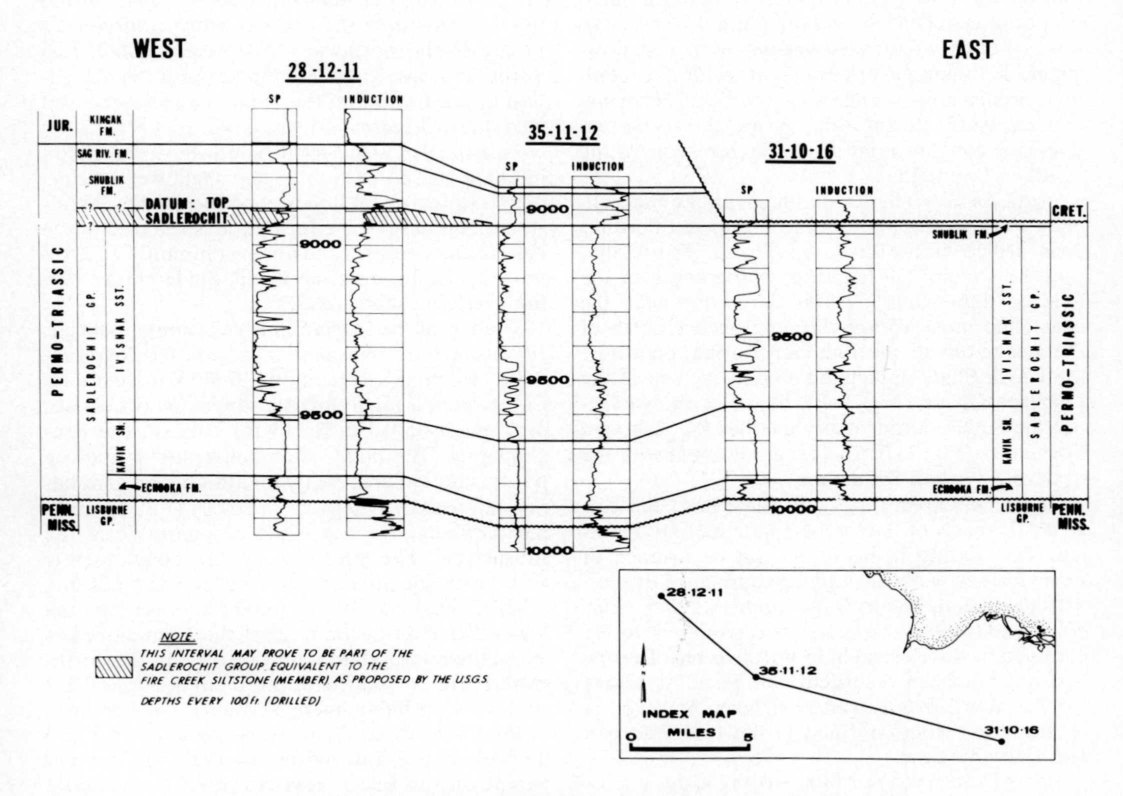

FIG. 8—Cross section illustrating Permo-Triassic subdivisions.

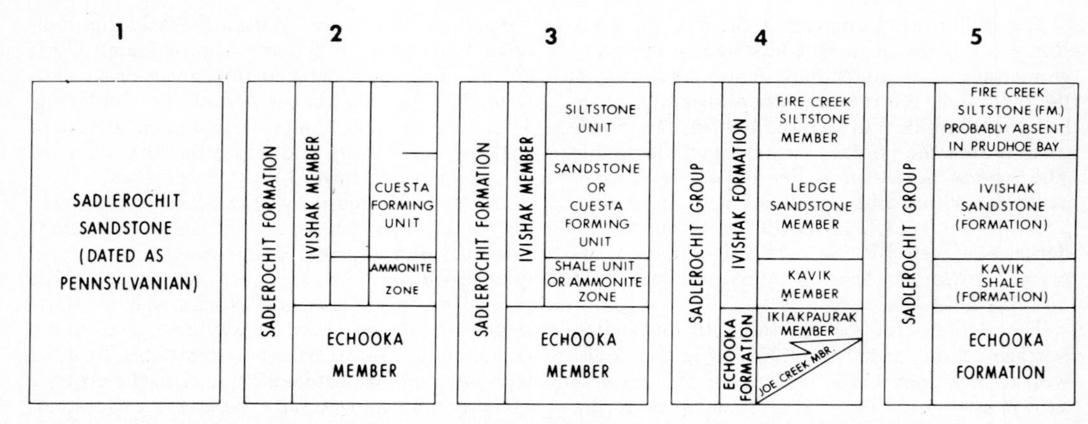

FIG. 9—Nomenclature used for Sadlerochit of North Slope. Column 1 from Leffingwell (1919); column 2, Keller et al (1961); column 3, Alaska Geological Society (1970-1971); column 4, Detterman et al (1975); column 5, proposed in this paper for Prudhoe Bay area.

stones are fine grained, argillaceous, glauconitic, phosphatic, and pyritic, and they contain small proportions of siderite and dolomite. Inasmuch as it was probably deposited near shore in a shallow-water, high-energy environment, with the common argillaceous stringers having been deposited at slack water during tidal cycles, the formation does not constitute an attractive reservoir in the Prudhoe Bay field.

Kavik Shale—The Kavik Shale, unlike the Echooka Formation, is present in every well that has been drilled to the Lisburne Group. The formation thins from 230 ft (70 m) in the south of the field to approximately 100 ft (30 m) near the coast; the most abrupt thinning occurs north of the wedge-out of the Echooka Formation where the Kavik Shale onlaps the eroded surface of the Lisburne Group (Fig. 11B). South of the wedge-out the Kavik conformably overlies the Echooka Formation (Fig. 11B, C). It is gradational with the overlying Ivishak Sandstone.

In the field area, the Kavik Shale has been dated on the basis of palynologic data as Late Permian. This dating is based on the occurrence of *Lueckisporites virkkiae* and an abundance of multistriate pollen, particularly *Striatoabietites richteri. Lueckisporites virkkiae* is considered to be confined to the Zechstein in northwestern Europe and has not been recorded with certainty above the Permian. *Striatoabietites richteri,* however, is regarded as being confined to the Late Permian and Early Triassic.

Apart from scattered thin, silty sandstone lenses which may be turbidites, the formation contains no reservoirs. It consists of fairly uniform, medium to dark gray, silty shales which are pyritic, noncalcareous, and micaceous. It was probably deposited in a shallow sea which transgressed progressively northward over the eroded Lisburne. However, at some time during the deposition of the formation this trend was reversed and a southward regression was begun.

Ivishak Sandstone—The southward regression initiated the deposition of the shallower marine, deltaic, and fluvial deposits of the Ivishak Sandstone. The base of the formation is selected at the base of the deepest sandstone, commonly an arbitrary depth, because the Kavik Shale grades into the overlying sandstones.

The top of the formation is commonly associated with a thin, phosphatic, radioactive conglomerate which is overlain by Shublik mudstones. Commonly, a highly pyritic sandstone occurs approximately 5–10 ft (1.5–3 m) beneath the conglomerate. The highly radioactive streak and the pyritic interval are readily identifiable on gamma-ray and resistivity logs, respectively (Fig. 6), making the formation top easily recognizable in the subsurface. The presence of the conglomerate and the pyritic interval, plus the fact that thinning of the Ivishak Sandstone apparently occurs at the top rather than at the base of the formation, has led to the interpretation that, at least in the northeastern area of the field, the Shublik-Sadlerochit contact is probably unconformable. Farther west, in the Eileen area, the contact is gradational and the radioactive and pyritic intervals are less apparent. In the Eileen area, an interval of slightly calcareous sandstones and shales occurs between the "typical" Shublik, as recognized in the Main

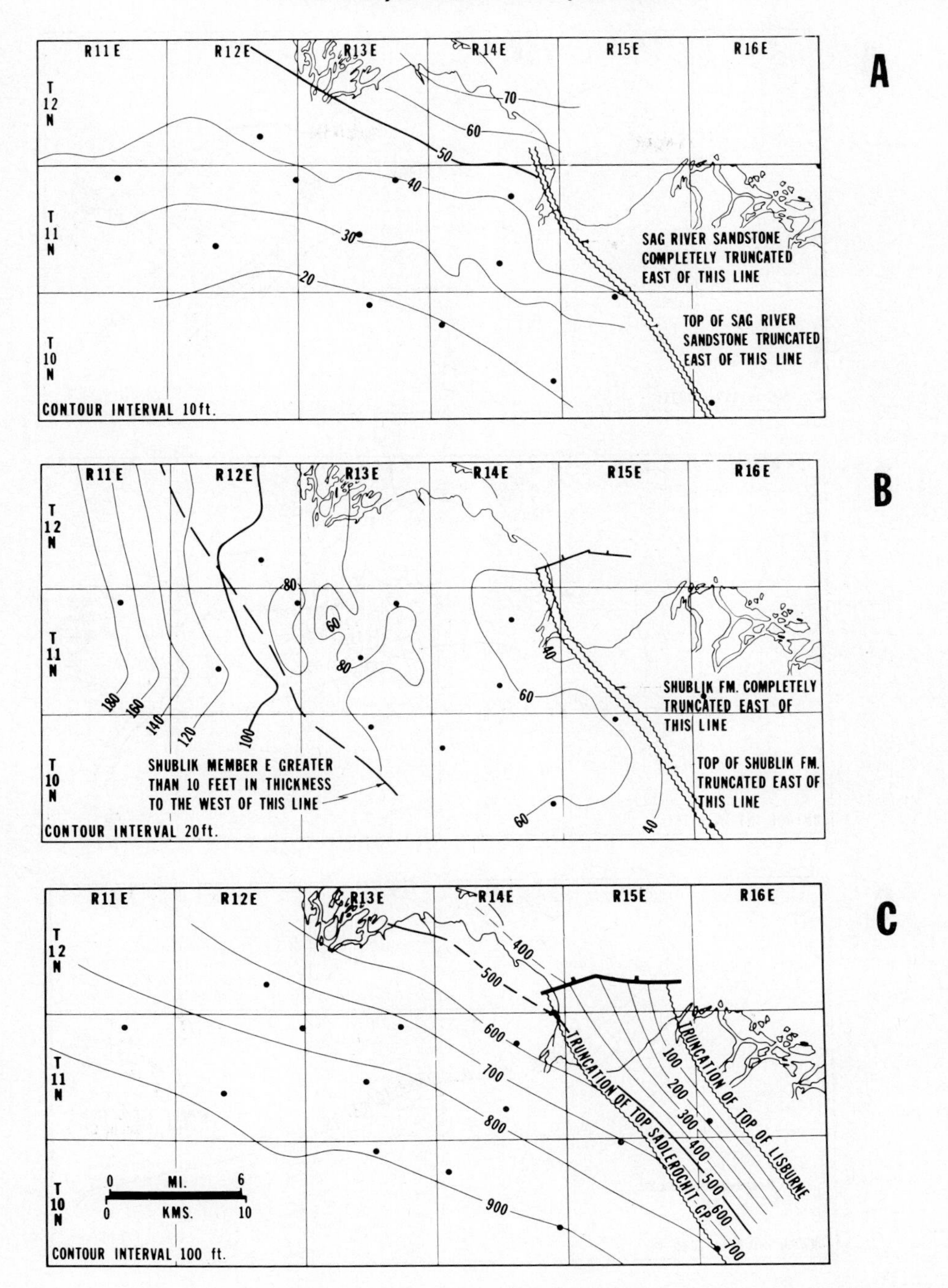

FIG. 10—Thickness variations within (A) Sag River Formation, (B) Shublik Formation, (C) Sadlerochit Group.

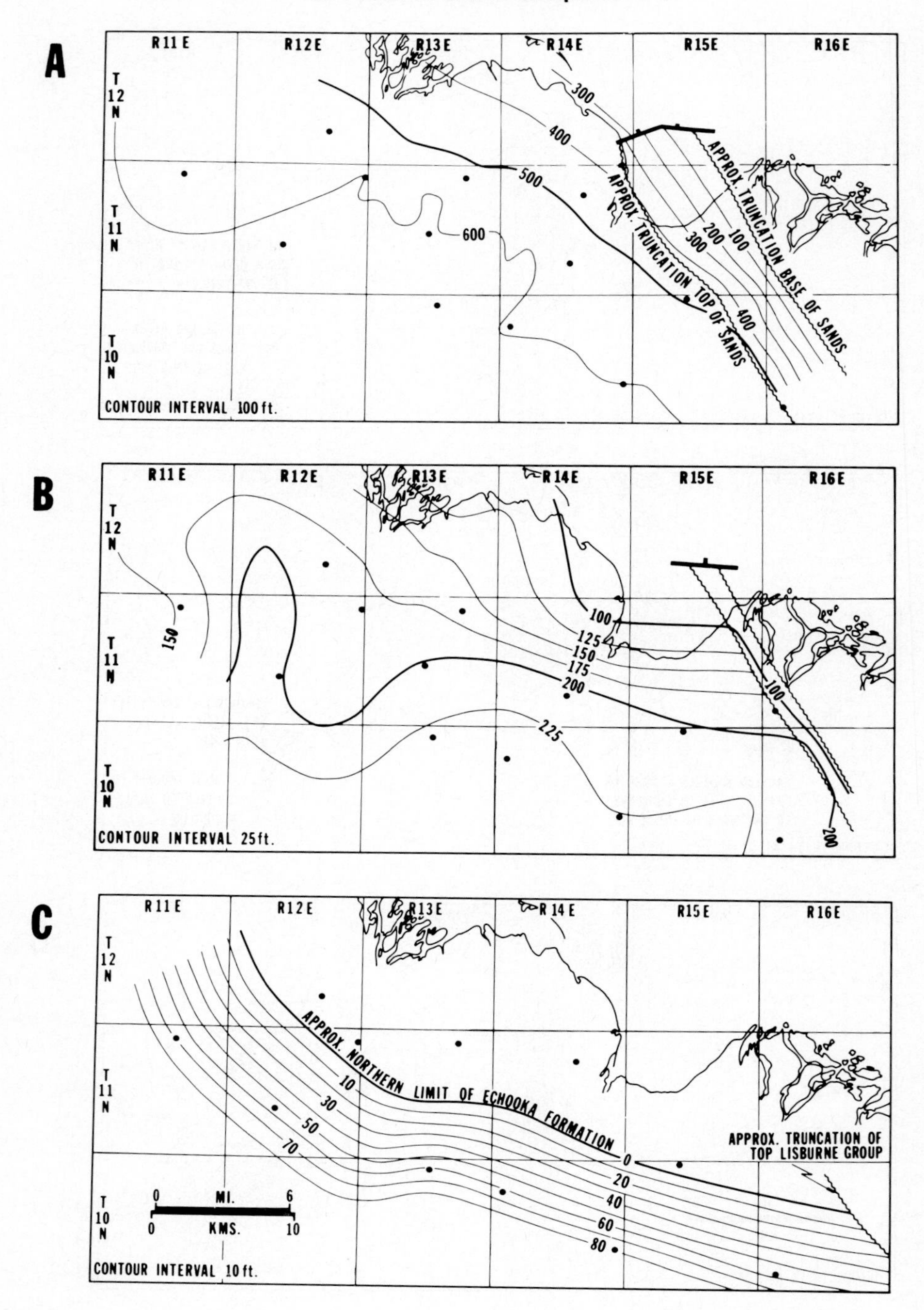

FIG. 11—Thickness variations within (**A**) Ivishak Sandstone, (**B**) Kavik Shale, (**C**) Echooka Formation.

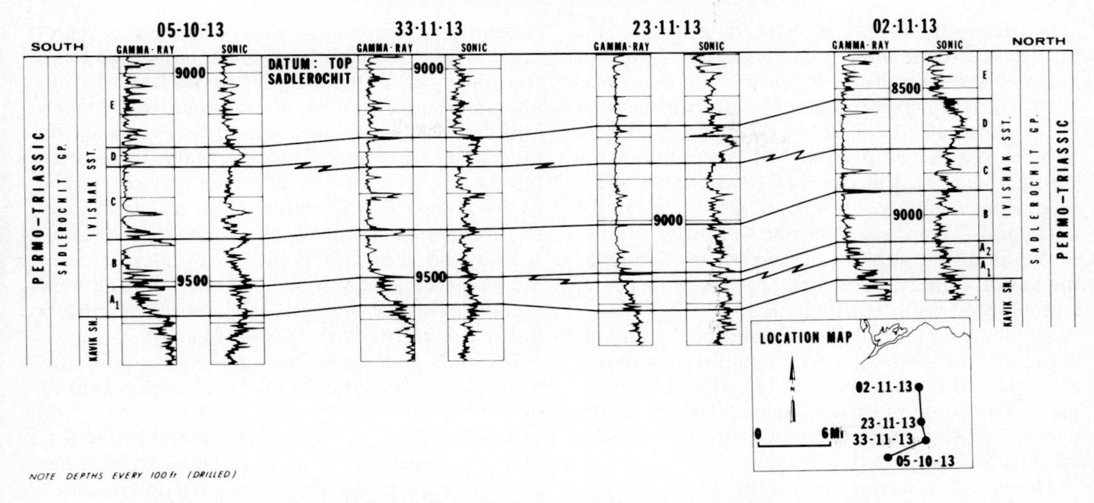

FIG. 12—South-north cross section illustrating subdivisions of Ivishak Sandstone.

area, and the Ivishak Sandstone (Fig. 8). This interval correlates with the "Siltstone Unit" of the "Ivishak Member" as recognized by the Alaska Geological Society (1970–1971) in the Colville No. 1. It is less arenaceous than the Ivishak Sandstone and less calcareous than the Shublik Formation, and probably represents a gradual transition between the depositional environments of these formations. We have included this interval in the Shublik Formation; however, when more data become available, the interval may prove to be equivalent to the Fire Creek Siltstone Member as recognized by Detterman et al (1975); thus, it may be part of the Sadlerochit Group.

The Ivishak Sandstone is virtually devoid of any macrofauna or microfauna, although several intervals containing rich microfloral assemblages have been recognized. The most abundant pollen type is *Striatoabietites richteri*, but other multistriate types such as species of *Lunatisporites, Protohaploxypinus, Striatopodocarpites,* and *Strotersporites,* as well as the nonsaccate gnetalean type *Gnetaceaepollenites steevesi,* have been identified. This assemblage is typical of the Middle to Late Permian and the Early Triassic (early to middle Scythian) of the northern hemisphere. The upper part of the formation lacks the distinctive Late Permian species, such as *Lueckisporites virkkiae,* which characterize the Kavik Shale. However, no types considered to be younger than Early Triassic have been recognized within the unit. The Ivishak Sandstone is therefore dated tentatively as Late Permian to Early Triassic.

The formation thickness ranges from approximately 650 ft (200 m) in the south to 300 ft (90 m) in the northeast (Fig. 11A). Most of the thinning appears to occur at the top of the formation, possibly in part the result of pre-Shublik erosion.

Figure 12 shows the subdivision of the formation. Because the Ivishak Sandstone contains such a varied suite of sediments, it is difficult to devise an acceptable subdivision. The informal subdivisons described here are used to facilitate the reservoir description. The formation has been divided into five members, lettered A through E from the base upward, because it was thought that other members eventually may be penetrated under the possible pre-Shublik unconformity. Initially, there were only three subdivisions—a lower sandstone, a middle conglomeratic interval, and an upper sandstone; however, owing to variations in the amount of conglomerate, in the porosity, and in the number of mudstones present, the conglomeratic interval was further subdivided into three members.

Member A is the lowermost interval, consisting of horizontally bedded siltstones, sandstones, and mudstones representing the initial influx of prodelta, coarser sediments into the southward-regressing sea. It is approximately equivalent to what Morgridge and Smith (1972, p. 498) described as a "stream-mouth bar" sequence in the Prudhoe Bay State No. 1. The member can be further divided into two submembers. The more silty and shaly, and generally older, submember A_1 occurs mainly to the southwest, and the cleaner, coarser submember A_2 is predominant in the northeast. Submember A_1 thins northeastward as it is replaced laterally by the nearer shore deposits of the upper submember. The whole member ranges from 60 to 140 ft (18–43 m) in thickness. Submember A_1 thins from 140 ft (43 m) in the west and is absent in the northeastern area of the field. Conversely, A_2 thins from 80 ft (24 m) in the north and disappears along the southwest margin of the field. The top of the member is selected as the top of a mudstone which appears to be correlatable throughout the field area.

Members B, C, and D compose the "conglomeratic section" of the formation and were deposited as part of a delta and braided-stream environment. The conglomerate content increases as the number of mudstones decreases, stratigraphically upward and laterally northward.

Member B contains the largest proportion of mudstones, particularly to the south, southeast, and west, and has been separated from the rest of the conglomeratic section on this basis. The member also contains sandstones, conglomeratic sandstones, and conglomerates which commonly have high sedimentary dips and eroded bed contacts. Depositional dips increase in an upward direction, suggesting a gradual change to a fluvial or perhaps distributary-type environment. The top of member B is picked at the top of a mudstone which is correlatable over much of the field. Over the Main field area, member B is between 100 and 140 ft (30–43 m) thick, but it thins toward the north and south.

Members C and D are readily recognized on gamma-ray logs, because they form a distinct "plateau" of low radioactivity caused by the lack of mudstones and the small proportion of argillaceous matrix in the sandstones and conglomerates.

Member C contains few mudstones in comparison with member B and has considerably higher porosity and less conglomerate than member D. The top of the member is arbitrarily selected from the sonic log at a depth where, because of decreasing porosity, the interval transit time decreases below 85 microseconds in an upward direction. There is no distinct lithologic change. Member C is thickest in the south (>160 ft or 49 m). It thins northward as its upper part is progressively replaced by the less porous conglomerates of member D.

Member D is primarily an interval of conglomerate and conglomeratic sandstone. The top of the member is taken at the top of a thin mudstone or group of mudstones which have been correlated across the field. In the east part of the field, the upper part of the member passes laterally into an interbedded sandstone and mudstone facies and eventually is almost completely replaced by mudstone. In the center of the Main field, member D contains numerous pyritic intervals. The pyrite affects the SP curve, causing positive deflections which, if not supplemented by a gamma-ray log and core data, could be misinterpreted as mudstone or a nonproductive interval. The conductive nature of the pyrite also results in anomalously low resistivity readings (Fig. 13).

Member D thins to the southeast, south, and southwest away from its thickest development of over 140 ft (43 m) in T12N, R13E.

Member E is the uppermost interval of the formation, consisting of homogeneous sandstones and thin mudstones. The lower part, comprising a series of "fining-upward" cycles, was probably deposited in a fluvial environment, but this was succeeded by a nearshore, possibly beach environment. Member E is over 200 ft (61 m) thick in the southwest part of the Main field, and thins fairly regularly toward the northeast. It is completely absent in the extreme northeast. It is not known how much of this thinning is due to original depositional thinning and how much, if any, is due to subsequent erosion.

Petrologically, the major constituents of the arenaceous deposit are chert and quartz, the former exhibiting varied degrees of weathering. The conglomerates contain pebbles of argillite, which forms the basement rock in this area of the North Slope. It is thought that both basement and Lisburne rocks were exposed and eroded to form the source of the Ivishak Sandstone. Pyrite and siderite are common as secondary minerals.

Figure 14, a map of the sandstone/shale ratio of the entire Ivishak Sandstone, shows that the proportion of "shales," or, more correctly, the mudstones and siltstones, generally increases away from the apparent northerly source area. It indicates a fan- or delta-shaped accumulation of sediments. The proportion of sandstone increases stratigraphically upward.

In general, grain size increases both upward and northward. Pebble size and the proportion of conglomerate increase in the same directions. These phenomena, together with changes in the sandstone/shale ratio, have led to the following paleogeographic interpretation (Figs. 15, 16). The Echooka Formation represents a period of transgression with fairly gradual rates of subsidence

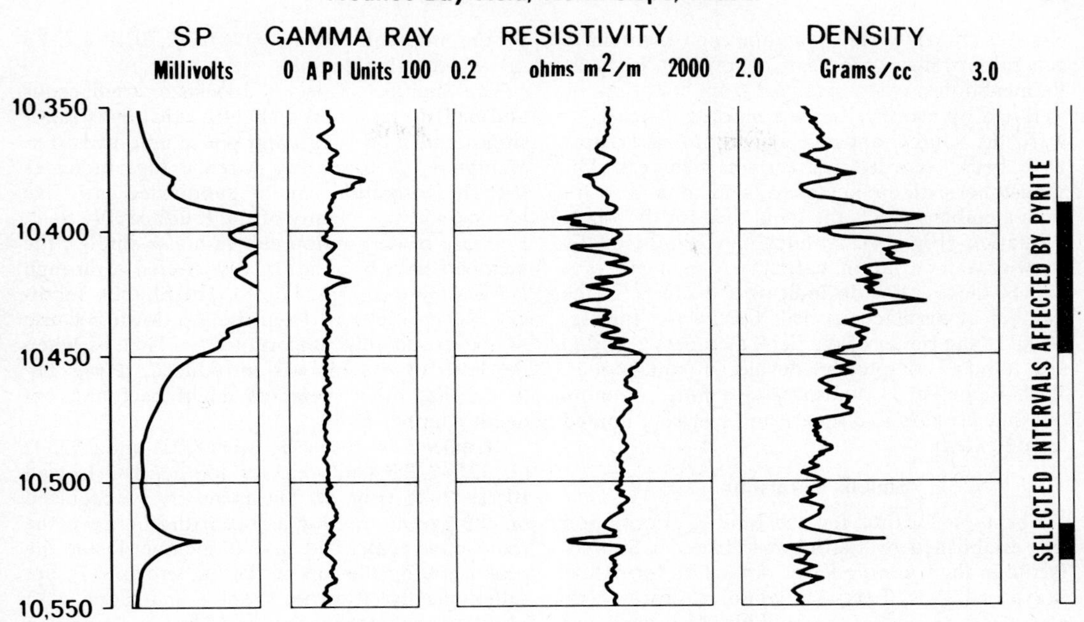

FIG. 13—Typical log response to pyritic intervals in BP well 13-11-13.

and sediment accumulation. A regressive phase began during the time of deposition of the Kavik Shale, and continued until the end of deposition of member D of the Ivishak Sandstone. In this period the rate of sediment supply increased as the stream velocities increased, possibly because of more rapid uplift of the source area, and the wedge of sediments built farther and farther south. Figure 16 shows a tentative and very generalized reconstruction of the prevailing paleogeography during the deposition of each member. As the members are not necessarily chronostratigraphic units, it is difficult to reconstruct their depositional history precisely. Members A through D probably represent a transition in deposition from prodelta through delta and distributary channels to meandering, high-energy rivers. Member E is thought to represent a period when the source area was almost completely denuded, when stream velocities decreased, and when the earlier deposits were reworked in a nearshore environment. The break between members D and E in the western part of the Main field is commonly fairly abrupt, suggesting that the environmental change may have been drastic.

The Ivishak Sandstone is a major reservoir partly because of its thickness and areal extent, but also because of its excellent petrophysical properties. The sandstones and conglomerates are generally poorly cemented, have a fairly low matrix content, are laterally continuous, and are not separated by numerous or thick mudstone intervals. The average sandstone/shale ratio within the reservoir is greater than 0.9 (Fig. 14).

Most of the sandstones and conglomerates probably have been winnowed and reworked during deposition by stream or current action. Apart from local secondary cementation by calcite, pyrite, or siderite, the sediments are clean and contain very little matrix. The postdepositional formation of quartz overgrowths may have slightly reduced the pore volume, but this phenomenon is largely confined to the uppermost member. Deformation of the chert grains in certain parts of the reservoir also reduces porosity. In general, the poorer reservoirs are in member A, because of its finer grained sandstones, and member D, because of its poorer sorting, pyritization, and greater proportion of pebbles.

The three major reservoir characteristics that have been examined all exhibit improvement laterally north-northeastward as well as stratigraphically upward, except in member D. Porosity commonly exceeds 30 percent; though it changes rather abruptly within each member, a general northward increase is present in each. Permeabilities, when averaged from core data by well and by member, show a marked increase toward the source, and average values of 1 darcy have been recorded for certain members. The sandstone/shale or net/gross sandstone ratio for each member reflects the trend seen for the whole formation (Fig. 14). Although they individually show greater variation, particularly in members A and B, these ratios do indicate a decrease in the number of argillaceous beds northward and upward. In the center of the field, members C and D are almost completely devoid of mudstones. Within member E, mudstones are more common, but they are thin and appear to be of very limited lateral extent.

Shublik Formation

The type locality for the Shublik Formation was established by Leffingwell (1919) as Shublik Island in the Canning River, where the formation is exposed. A well-exposed section on Fire Creek at the eastern end of the Shublik Mountains was described by Detterman (1970). A typical section of the Shublik Formation in the Prudhoe Bay field is seen in well BP 19-10-15 (Figs. 5, 6), within the interval 8,866–8,935 ft (2,702.2–2,723.3 m)—electric-log depths.

The Shublik consists of bioclastic argillaceous and pelletal limestones, slightly calcareous sandstones and mudstones, and phosphatic beds. Examination of cores and wireline logs indicates that the formation can be subdivided into five members in the vicinity of the Prudhoe Bay field. For ease of description and in-house studies, the members have been informally lettered A through E from the top (Figs. 17, 18). The Shublik subdivisions were lettered from the top down because of the pre-Shublik unconformity. That is, when the lettering system was introduced, it was believed that other members might be found beneath member E.

Member E is present between 8,931 and 8,935 ft (2,722.2–2,723.4 m)—electric-log depths—in well BP 19-10-15 (Fig. 6). The member is recognized on the gamma-ray log as occurring between the radioactive peak at the base of member D and the peak defining the top of the Sadlerochit Group. Lithologically, it varies from a mudstone with some siltstone partings in the Main field area to a sequence of interbedded, slightly calcareous sandstones, siltstones, and mudstones in the Eileen area. As mentioned, however, in the Eileen

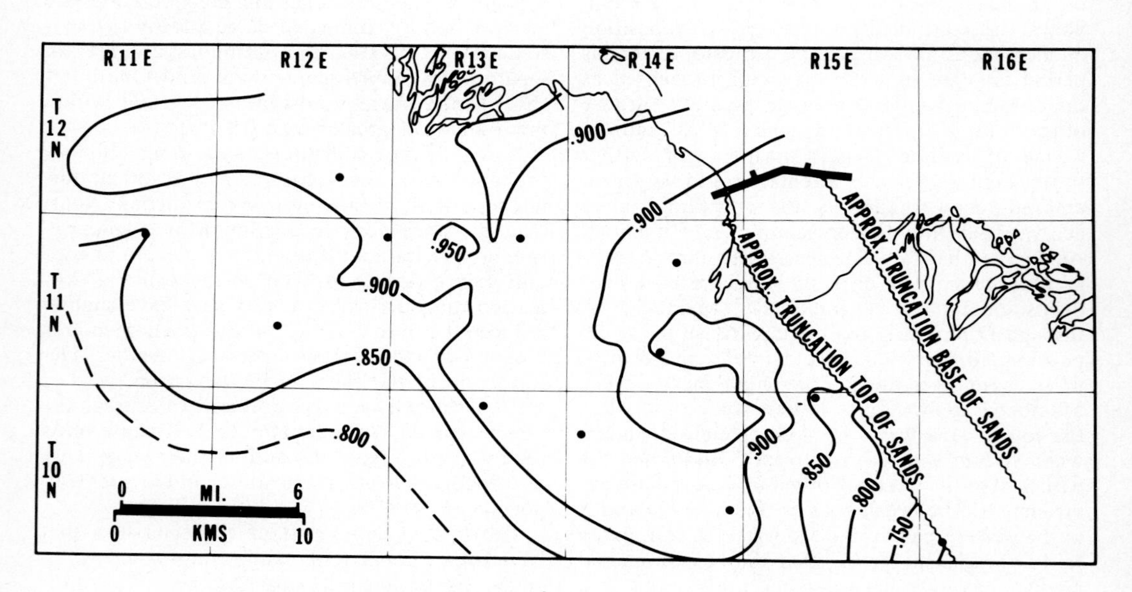

FIG. 14—Variation in sandstone/shale ratio in Ivishak Sandstone.

area this interval may be equivalent to the Fire Creek Siltstone Member of Detterman et al (1975).

Member D is present between 8,908 and 8,931 ft (2,715.2–2,722.2 m)—electric-log depths—in well BP 19-10-15. It consists of argillaceous, "tight" limestones interbedded with calcareous mudstones and shales. The top is picked at the lower limit of a zone of high radioactivity. The member is characterized by high resistivities, low sonic transit times, and low radioactivity.

Member C is present between 8,888 and 8,908 ft (2,709.2–2,715.2 m)—electric-log depths—in well BP 19-10-15. This member is lithologically complex, ranging from bioclastic pelletal and rubbly limestones to siltstones and shales; it is highly phosphatic and glauconitic in many places. The member is highly radioactive, and the top is picked at a correlatable resistivity low.

Member B is present between 8,880 and 8,888 ft (2,706.6–2,709.2 m)—electric-log depths—in well BP 19-10-15. It consists of shales and mudstones locally interbedded with siltstones, bioclastic limestones, and beds which contain pelletal phosphate. The top of the member is picked at a resistivity low which can be correlated across the field.

Member A is present between 8,866 and 8,880 ft (2,702.4–2,706.6 m)—electric-log depths—in well BP 19-10-15. In the field area the member is variable in lithology, ranging from bioclastic sandy limestones through calcareous siltstones to calcareous shales and silty mudstones. On the electric log the member appears as a resistive interval. The top is taken as the base of the Sag River Formation. Both members A and B appear to become increasingly arenaceous toward the northeast, where the boundary between member A and the base of the Sag River Formation would be indistinguishable in certain wells without a gamma-ray log.

Monotis ochotica and *Monotis obtusicosta* have been recognized in member A in the BP Put River 27-11-14. *Monotis* sp., *Halobia* cf. *zitteli*, and *Entolium* sp. have been recorded from member B. Members C and D contain *Halobia* cf. *zitteli* along with fragments of saurian bone.

The *Monotis* assemblage in members A and B suggests a late Norian age, whereas the abundant *Halobia* in members C and D indicate a late Karnian to early Norian age.

The Shublik in the BP/Arco Colville No. 1, which is located west of the Prudhoe Bay field, was found to contain the following faunal assem-

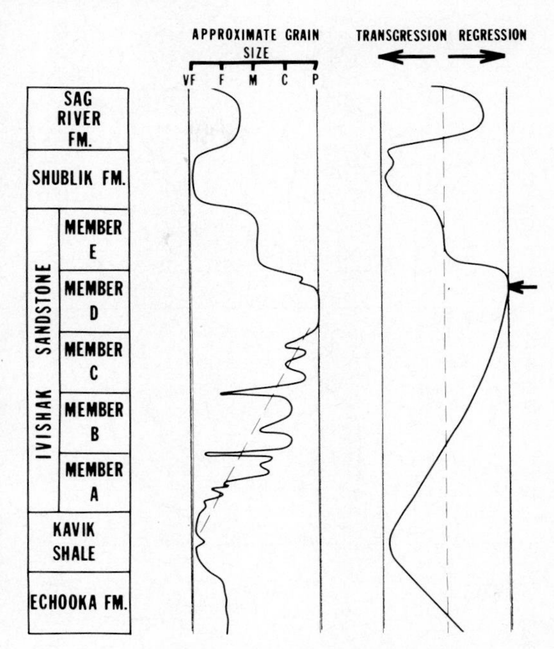

FIG. 15—Diagrammatic illustration of approximate grain size and transgressive/regressive movements in Permo-Triassic. Arrow at right center indicates maximum erosion of source area and maximum rate of sediment supply.

blage: *Variostoma* cf. *catilliforme* (= *Trochamina helicta* Tappan, 1951), *Astocolus connudatus, Vaginulinopsis acrulus, Nodosaria shublikensis,* and *Lingulina alaskensis.* The presence of *Variostoma* cf. *catilliforme* suggests a Norian age.

Thus the Shublik Formation appears to be Late Triassic, probably Karnian-Norian but possibly Rhaetian.

The Shublik was probably deposited in a low-energy, marine shelf environment with moderate water depths of between 200 and 1,000 ft (60–305 m; Detterman, 1970). The occurrence of highly phosphatic beds, which are in part "rubbly," suggests slow deposition and reworking of sediments. The abundant marine invertebrate fauna formed coquina beds. The occurrence of fossil bones in cores also indicates the presence of large vertebrates.

In the Main field area, the Shublik Formation ranges in thickness from 50 to 80 ft (15–24 m) and thins eastward (Fig. 10B). West of the Main field area, however, the thickness increases abruptly

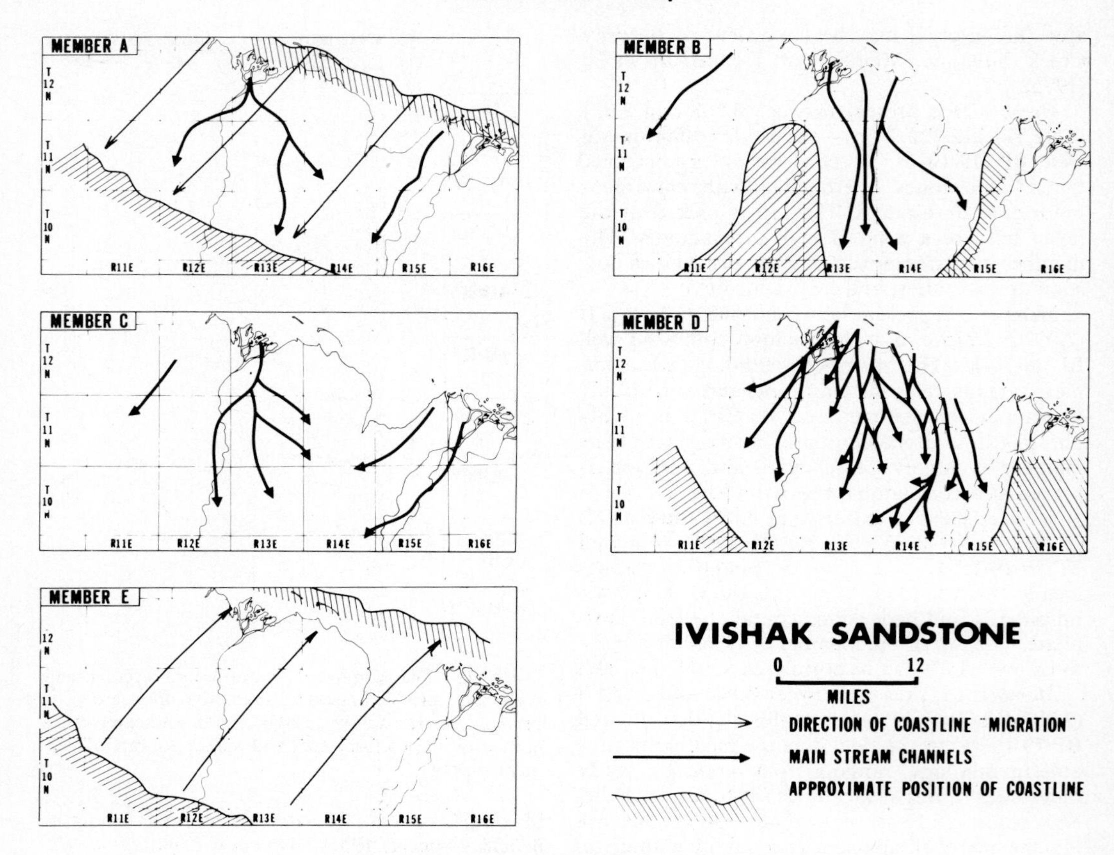

FIG. 16—Idealized paleogeography of Ivishak Sandstone.

and consistently, reaching a maximum of 192 ft (59 m) in well 29-12-11. Although all members become thicker, thickening is most pronounced in member E. It has a maximum thickness of 10 ft (3 m) in the Main field area but thickens to 47 ft (14.3 m) in well 29-12-11.

At present, the only potential Shublik reservoirs are thought to be in members C and E (Fig. 18). In member C, porosities of over 30 percent and, in a few instances, permeabilities of up to 400 md have been measured in intervals which are variously described as sandy, pelletal, and rubbly limestones; crumbly and porous calcarenites; and porous coquinas. It is possible that this porosity is developed only locally, as indicated by the sporadic occurrence of high sonic-log readings. The reasons for this erratic porosity distribution are presently being investigated; one possible explanation is that it may be related to slumping and faulting which occurred soon after deposition, causing brecciation of the partially consolidated sediments.

Member E forms a reservoir only in the Eileen area, where it occurs as interbedded sandstones and mudstones. Porosities range from 5 to 15 percent, but permeabilities are low.

Sag River Formation

The "Sag River Sandstone" was defined by the Alaska Geological Society (1970-1971) as ". . . the sandstone interval lying between the Jurassic, Kingak Shale above and the shales and limestones of the Triassic Shublik Formation below, in wells drilled in the Prudhoe Bay field, North Slope, Alaska." The type section was designated as the interval drilled in the Arco/Exxon Sag River State No. 1 (04-10-15). Inasmuch as the wireline logs of this well were confidential at the time of definition, the type section was described as being readily correlatable with the interval between 8,117 and 8,163 ft (2,474–2,488 m) in the Arco/Exxon Prudhoe Bay State No. 1 (10-11-14), which was designated as the reference section.

Over much of the Prudhoe Bay field, and particularly west of it, an interval of resistive silty shales and mudstones is present between the Kingak Shale and the Sag River Formation. The formation thus consists of an upper shale member and a lower sandstone member.

As the formation was not completely cored in either of the wells used in the Alaska Geological Society (1970-1971) definition, well BP 21-11-13 (Fig. 19) is proposed as a reference section.

The *Sag River Formation* in the subsurface of the Prudhoe Bay field is defined as those strata occurring within and correlating with the interval 8,555–8,591 ft (2,607.4–2,618.4 m)—electric-log depths—in well BP 21-11-13.

The top of the formation is chosen at a point where there is a downward increase in silt content; this is recognized as an increase in resistivity compared to that recorded in the overlying Kingak Shale. The base is identified on the basis of (1) a decrease in the proportion of sandstone, (2) an increase in calcareous content, and (3) the common presence of a thin bed of shell hash. Resistivity values are at a minimum, and the gamma-ray log indicates a peak of high radioactivity at this junction. By this definition, the base of the formation in well 10-11-14 is 5 ft (1.5 m) shallower than that selected by the Alaska Geological Society (1970-1971).

The "sandstone member" is defined as those strata correlating with the interval 8,560–8,591 ft (2,609.0–2,618.4 m)—electric-log depths—in well BP 21-11-13. It occurs between the two radioactive peaks which define the base of the "shale member" and the base of the Sag River Formation, respectively. High proportions of glauconite have been recorded opposite both peaks. In well 10-11-14, where the shale member is absent, the top of the sandstone member coincides with the top of the formation and with the top of the "Sag River Sandstone" as defined by the Alaska Geological Society (1970-1971).

The "sandstone member" is a uniform, well-sorted, fine-grained sandstone to siltstone. It is composed predominantly of quartz with variable amounts of glauconite and chert. Dolomite is the most common cement, but clay matrix occurs both as discrete laminae of brown hematitic clay and as thin sheaths around the clastic grains. The clay laminae are commonly distorted and broken, apparently as a result of bioturbation. Pyrite occurs both in disseminated form as dust and as

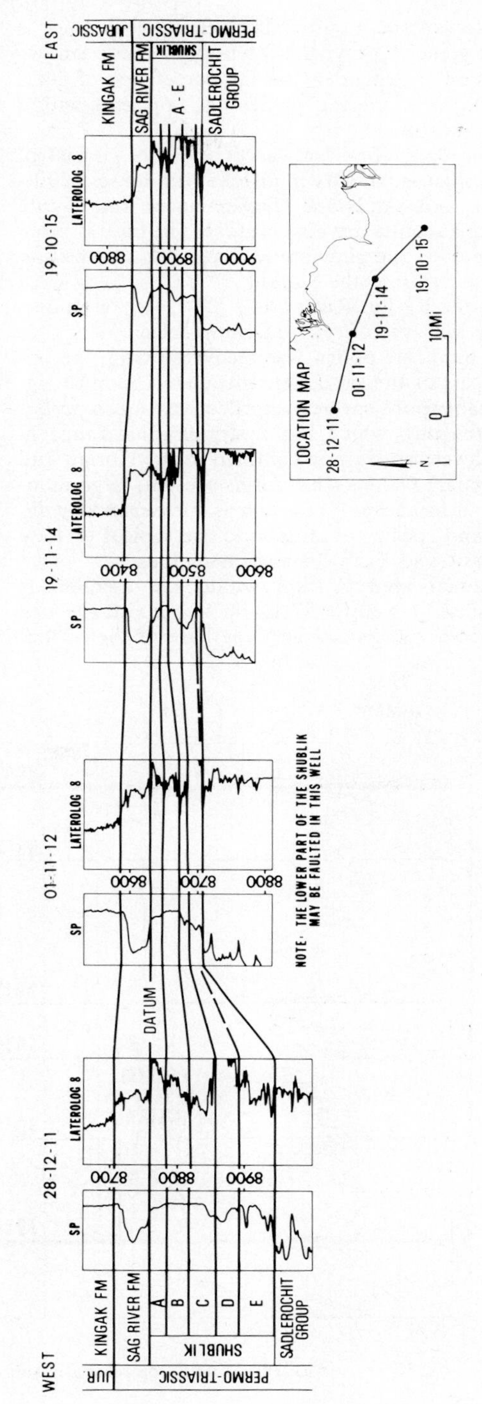

FIG. 17—Cross section illustrating Shublik subdivisions.

clusters of small cubes. The general rock color is light greenish-gray; the intensity of the green coloration is dependent on the percentage of glauconite. No bedding planes are apparent within the sandstone.

The upper few feet of the member consist of gray siltstone to silty mudstone with lenses of siltstone and sandstone. Disseminated fine pyrite and glauconite are also present. The base is more calcareous and glauconitic, and forms a transition downward into the Shublik.

In the BP Put River No. 1 (27-11-14), a moderately rich microflora consisting largely of spores and bisaccate pollen was recovered from the upper part of the sandstone member. The most significant spore species recorded was *Ricciisporites tuberculatus,* which has a stratigraphic range in northwestern Europe confined to the Norian and Rhaetian. Of the other forms present, large numbers of bisaccate types such as *Vitreisporites pallidus* and species of *Alisporites* are typical of Late Triassic and Early Jurassic assemblages.

Faunal identification within the member is minimal. In well BP 27-11-14, *Monotis* cf. *ochotica* has been recognized near the base of the formation. *Monotis* sensu stricto is confined to middle and late Norian and possibly Rhaetian strata. There does not appear to have been a documented positive identification of *Oxytoma* from the sandstone in the Prudhoe Bay area. Although the definition of the "Sag River Sandstone" by the Alaska Geological Society (1970-1971) cites the occurrence of *Oxytoma* sp., no reference is given for this identification. As the stratigraphic range of *Oxytoma* is from late Karnian to Late Cretaceous, it can occur in rocks of the same age as those containing *Monotis.* An identification of *Oxytoma* was applied to a lamellibranch from a siltstone occurring between the Kingak Shale and the Shublik Formation at an outcrop in Fire Creek in the Ignek Valley of the Brooks Range. Subsequently, the specimen was recognized as *Monotis ochotica* (Reiser, 1970). This siltstone is thought to be equivalent to the "sandstone member" of the Sag River.

The "sandstone member" may have been deposited as a barrier-beach complex which prograded over the shallow-marine sediments of the Shublik. The fossiliferous glauconitic and radioactive siltstones which occur near the base of the

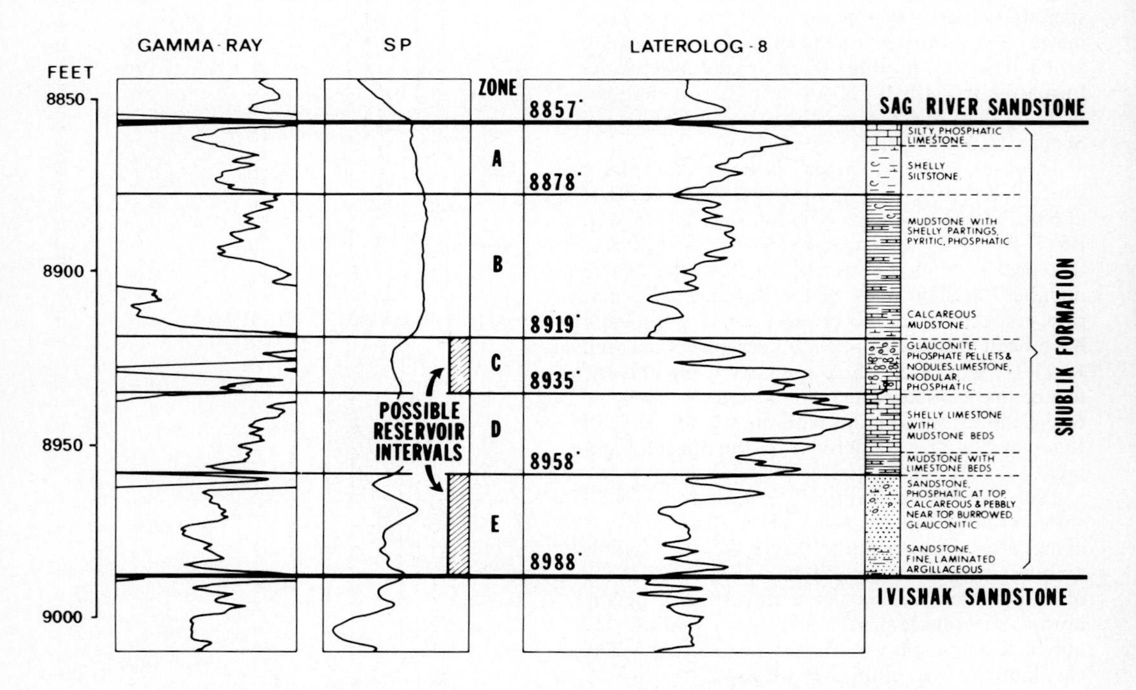

FIG. 18—Shublik log response, subdivisions, and summarized lithology.

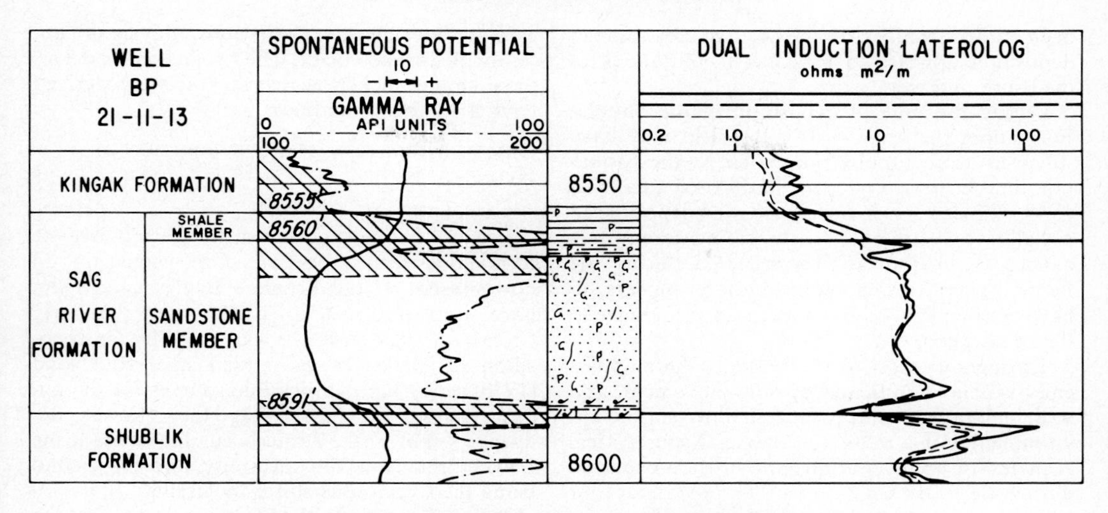

FIG. 19—Sag River Formation reference section, BP well 21-11-13.

formation, and also locally within the upper part of the Shublik, may represent time planes and also periods of slow, undisturbed deposition.

The "shale member" is defined as those strata correlating with the interval 8,555–8,560 ft (2,607.4–2,609.0 m)—electric-log depths—in well BP 21-11-13. The base of the member is recognized from the gamma-ray log and is normally coincident with the highest reading on the uppermost of the two radioactive peaks.

The Sag River shale consists of medium and dark gray to black glauconitic and pyritic shales and mudstones with varied amounts of silt. The member is usually less than 10 ft (3 m) thick in the field area, but it thickens westward to approximately 70 ft (21 m) in the Topagoruk No. 1.

No direct paleontologic dating is available for this interval. The lower part of the overlying unit, the Kingak Shale, has been dated as Early Jurassic on the basis of a microfloral assemblage dominated by the chlorophycean forms *Crassosphaera* and *Tasmanites*. Since the underlying Sag River sandstone has been dated as Late Triassic, the age of the "shale member" clearly is near the transition from Triassic to Jurassic time; however, this boundary has yet to be determined precisely. The writers believe that, in a rock-stratigraphic sense, the "shale member" is more correctly grouped with the Permo-Triassic than with the Jurassic strata.

All of the reservoir rocks within the Sag River Formation are in the "sandstone member." The sandstone forms a thin but continuous interval which thickens from less than 20 ft (6 m) in the south to about 70 ft (21 m) in the north (Fig. 10A). Nonreservoir rock is confined to thin intervals at the top and base of the member. Net-sandstone thickness ranges from 15 ft (4.6 m) in the south to over 60 ft (18 m) in the north.

South and west of the field, the lower part of the reservoir rock is more silty and fine grained than the upper part. Northeastward, however, all of the reservoir characteristics improve. Average porosities increase from 10 to 25 percent, and permeabilities, although low in comparison with those of the Ivishak Sandstone, increase to approximately 70 md. In areas where porosity and permeability are greatest, core descriptions indicate fairly common bioturbation, suggesting that the improved reservoir character is associated with this phenomenon.

POST-LISBURNE STRUCTURAL HISTORY

The Permo-Triassic of the Prudhoe Bay field was deposited over the eroded surface of the Lisburne Group. In the field area the Echooka Formation wedges out to the northeast as it onlaps this surface, indicating a structural high in that direction (Fig. 11C). The Kavik Shale and the Ivishak Sandstone isopachs show that this high persisted and may have controlled sedimentation until the end of deposition of the Sadlerochit Group (Fig. 11A, B). Prior to the deposition of the Shublik Formation, the Sadlerochit Group may have been eroded locally over the positive area, though deposition appears to have been continuous to the south and west.

During the time of deposition of the Shublik Formation, the area west of the field may have subsided more rapidly, inasmuch as the formation thins eastward over the Main field area (Fig. 10A). The Sag River Formation was deposited in a shallow sea with a northerly sedimentary provenance. No major faulting occurred during the Permo-Triassic, and sedimentation appears to have been controlled by more rapid subsidence to the south and west.

Throughout most of the Jurassic, there apparently was no significant tectonic movement, and sedimentation continued in a shallow-marine environment with a more northwesterly source area. A period of uplift, faulting, and erosion occurred during the Early Cretaceous. The pre-Cretaceous formations were tilted westward, and subsequently were completely truncated east of the field by the erosion, producing the Lower Cretaceous unconformity. Cretaceous marine mudstones which overlie this unconformity form the seal for the present hydrocarbon accumulation. Accompanying the tilting, major faulting occurred to the north and west of the field. The faults, which all appear to be normal tension faults, occur in two distinct trends. The most significant are aligned east-northeast to west-southwest (Fig. 20) and affect all pre–Early Cretaceous strata, causing them to be downthrown to the north. They are present only along the northern edge of the field, where throws up to 1,000 ft (305 m) have been recognized; south of the fault zone the top of the Sadlerochit Group is between 8,000 and 9,000 ft (2,440–2,740 m), but to the north it is generally below 10,000 ft (3,050 m).

Faults of the second group occur to the south and west and are mainly downthrown to the southwest. These faults control the structure of the Eileen area and the southwestern part of the Main field. The throws in a few places exceed 200 ft (60 m; Fig. 21), and many of the faults are thought to be complex zones of parallel and subparallel faults, rather than single fault planes. Approximately a fourth of the field wells have penetrated faults, the recognition of which is enhanced by the excellent correlations possible within the Jurassic shales and siltstones.

The Prudhoe Bay structure (Fig. 20) therefore formed during the Early Cretaceous. Since that time, it has been tilted toward the southeast and east and, apart from relatively minor periods of erosion later in the Cretaceous, no significant faulting has taken place. The Cretaceous and Tertiary strata thicken eastward and were derived from a southerly landmass.

Origin, Migration, and Distribution of Hydrocarbons

Origin

Because of the exposure of the Permo-Triassic formations during the Early Cretaceous, the oil now present in the Prudhoe Bay structure can have been emplaced no earlier than the Early Cretaceous. The structure was sealed by the deposition of shales of Barremian age. Rickwood (1970) suggested two possible sources for the hydrocarbons, the first being the "Lower Sequence," that is, the strata occurring beneath the Lower Cretaceous unconformity, and the second being the Cretaceous strata themselves.

Morgridge and Smith (1972) concluded that the Jurassic shales and the Cretaceous marine shales are both potential sources of hydrocarbons. Using two parameters which they considered to be diagnostic criteria for ranking the oil-generating potential of fine-grained sediments, they determined the total organic carbon and the C_{15+} hydrocarbon contents of several "shales" ranging in age from Mississippian to Cretaceous. The Jurassic shales proved to be rich and the Cretaceous marine shales, extremely rich, in both indices (e.g., the latter contained 3,000 ppm C_{15+} hydrocarbons and 5.4 weight-percent of organic matter). As only the Cretaceous marine shales are in contact with all of the field reservoirs, Morgridge and Smith concluded that the Cretaceous shales are the more likely source beds.

Crude oil samples from the Ivishak Sandstone, the Kuparuk River Formation, and the Upper Cretaceous sandstones have been examined using a variety of geochemical parameters including *n*-alkane variations, acyclic isoprenoid alkane "fingerprint" distributions, stable carbon isotope ratios, and porphyrin-type variations. A close genetic relation was indicated between the crudes from the Sadlerochit reservoir and those from the Kuparuk River sandstones; this supports the suggestion that the oils in the reservoirs adjacent to the unconformity are of common origin. Despite very marked differences in chemical composition, stable carbon isotope ratios for the heavy crudes from the Upper Cretaceous sandstones suggest that their origin was similar to that of the more deeply buried oils below the unconformity. Stable carbon isotope ratios also indicate that the oils

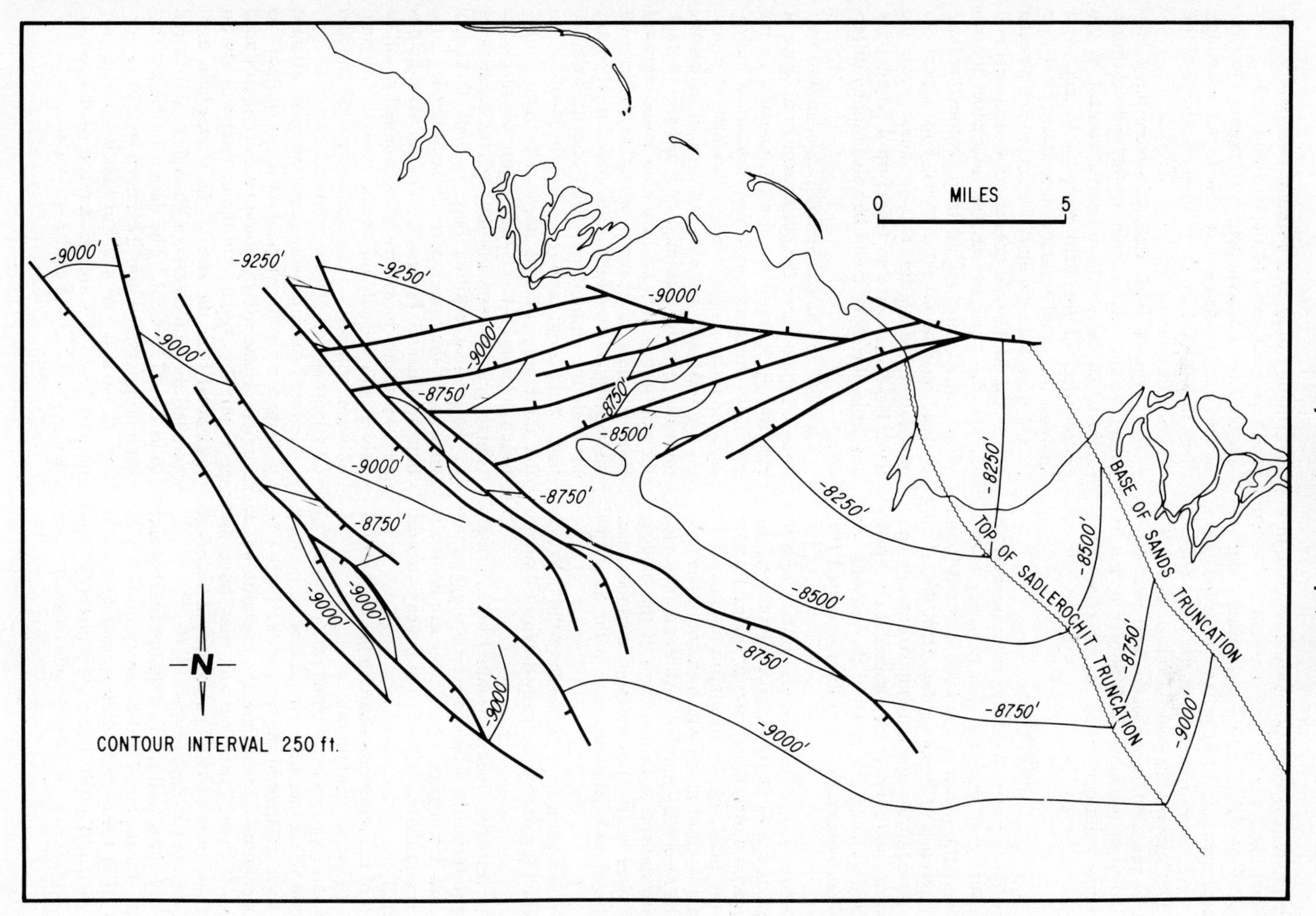

Fig. 20—Generalized structure map, top of Sadlerochit, Prudhoe Bay field.

from the three reservoirs probably originated in a rather restricted marine environment in which terrestrial material made up a significant part of the source material. Most of the Cretaceous sediments were deposited in such an environment. These data support the conclusions of earlier authors that the Cretaceous shales and mudstones appear to form the most likely source of the hydrocarbons contained in the Permo-Triassic reservoirs.

Migration

There is ample evidence to indicate that the Cretaceous, and possibly the Jurassic, marine shales were the source of the hydrocarbons found in the reservoirs of Prudhoe Bay. How the oil migrated structurally upward but stratigraphically downward, however, is not immediately apparent. The following are some of the reasons why the writers believe this migration occurred (Fig. 22).

1. The reservoirs are in direct contact with the Cretaceous shales along the truncation area, and they directly underlie the Jurassic shales. Furthermore, the plane of the unconformity may have formed a conduit along which the oil migrated.
2. The reservoirs may be in direct communication with the shales to the north of the field, by way of large faults. (For example, a similar reservoir–source rock relation exists in the Sarir field in Libya.)
3. Known faulting to the south of the field may have aided the stratigraphically downward migration.
4. The apparent lack of residual oil saturation in the aquifer to the west of the field suggests that the oil did not migrate through the sandstones in this area and that it entered from the east (truncation area).
5. All of the reservoirs abutting the unconformity are known to contain oil.
6. If the source was in older rocks underlying the Permo-Triassic, any oil generated in them would have escaped at the time of the Early Cretaceous erosion. At that time, the Lisburne was at least 3,000 ft (914 m) deep and may have had sufficient overburden and paleothermal history to have generated hydrocarbons.
7. The field is well situated on the Barrow arch to be the focal point for the migration of any oil generated in the deep basin surrounding it to the south, east, and north. There are probably at least 15,000 ft (4,575 m) of Lower Cretaceous strata in the Colville trough and a similar thickness of Upper Cretaceous rocks to the east and north.

If the hydrocarbons generated within the Cretaceous marine shales were able to move through them, these strata are the most likely source of the Prudhoe Bay oil.

The Permo-Triassic reservoirs were tilted subsequent to the emplacement of the hydrocarbons. Cores taken below the oil-water contact in the eastern part of the field were oil stained, and analyses indicated residual oil saturations. Those from below the oil-water contact in the western part of the Main field area contained no residual oil. On the basis of the few available control points, an attempt was made to contour the base of these residual oil saturations. This appears to be a surface which dips east-northeast across the field. However, since so few control points were available, the strike could be substantially different.

The field can be conveniently divided into four areas on the basis of the occurrence or absence of residual oil beneath the oil-water contact. To the northeast the sandstones have contained hydrocarbons since emplacement. This area is joined on the west and south by an area in which the oil-water contact has migrated upward through the sandstones; that is, the oil has moved upward from the base of the sandstones to the present oil-water contact. In this area, residual oil occurs between the present oil-water contact and the base of the sandstone (e.g., in well 24-10-14). In the third area, aligned northwest-southeast between Section 26-12-12 and Section 07-10-14, the contact moved vertically through the sandstones to the present position (e.g., in well 33-11-13). Here, only the upper part of the water leg contains residual oil. Finally, west of the Main field is an area where the sandstones contained no oil or only a thin oil column at the time the original oil-water contact formed, and there is no residual oil apparent beneath the contact.

Restored sections were prepared in an attempt to deduce when the original oil-water contact was formed. These are diagrammatically reproduced in Figure 23. Apparently the contact was almost horizontal near the end of the Cretaceous, as it parallels the shallowest correlatable coal, referred to as the "Top Coal." The data also suggest that oil had not spilled into the Eileen structures during the earliest Tertiary, but that it subsequently migrated into this area as a result of the continued eastward tilting of the structures.

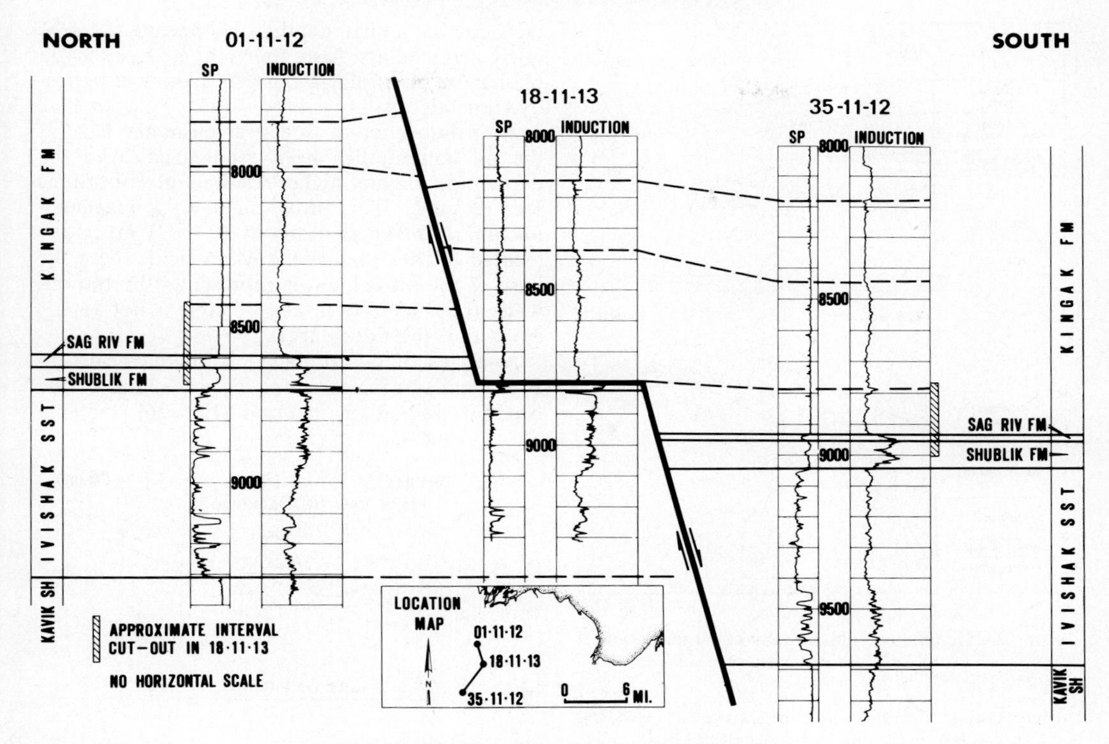

FIG. 21—Example of faulting in BP well 18-11-13.

Morgridge and Smith (1972) contend that the Colville structure contains no oil because the Jurassic strata separate the Permo-Triassic reservoirs from the Cretaceous source rocks. It is more likely that the absence of hydrocarbons in this western structure is due to its being much deeper than the Prudhoe structure at the time of oil migration. Subsequent easterly tilting placed the Prudhoe structure deeper than the Colville structure, but has not resulted in the spilling of oil into the latter, possibly because of the intervening syncline and its associated faults. The spill appears to have reached only the Eileen structures. Alternatively, oil may have spilled into and through the Colville structure. Evidence of oil at the top of the Ivishak Sandstone in the Colville No. 1 may support this possibility.

A tentative comparison was made between the volume of hydrocarbons contained in the present Ivishak Sandstone reservoir and the volume that might have been present in this reservoir at the end of the Cretaceous (Fig. 24). Allowances were made to account for the fact that the hydrocarbons were at a shallower depth at that time. If it is assumed that the Late Cretaceous accumulation had an average oil saturation of 65 percent, an average porosity of 22 percent, and an average net/gross sandstone ratio of 0.85, the volume of hydrocarbons in the reservoir at that time was approximately 25 percent greater than the present volume when adjusted to the shallower depth. If this was the case, a large volume of hydrocarbons is now missing and could have escaped through the Colville structure. However, if the average porosity and net/gross ratios are taken as 20 percent and 0.75, respectively—properties consistent with the reservoir's southeastward deterioration—the present and Late Cretaceous volumes are very similar.

Distribution

The gas-oil contact in the Ivishak Sandstone is at approximately −8,578 ft (−2,615 m) within the Main area of the field. Within the Eileen area,

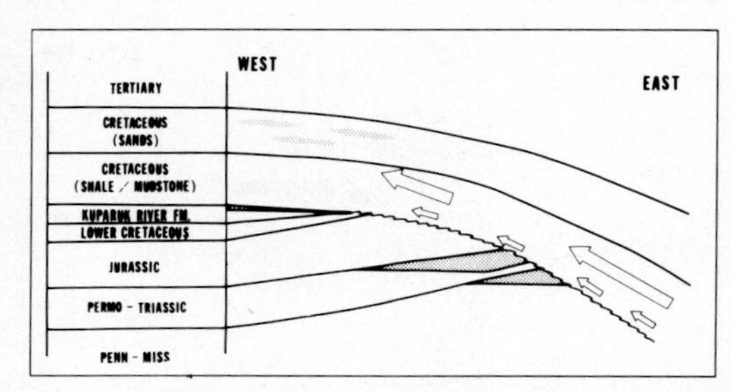

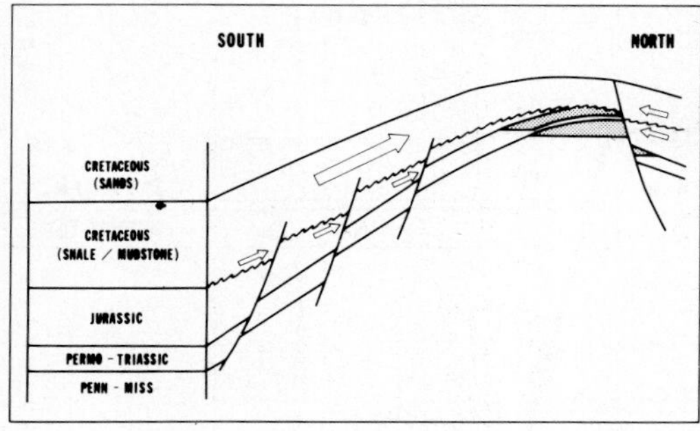

FIG. 22—Generalized illustrations of oil migration into Prudhoe Bay structure.

the contact is deeper—at approximately −8,780 ft (−2,676 m).

Properties of the oil contained in the Ivishak Sandstone reservoir vary vertically and possibly areally. Available data show that the gravity of the oil decreases with depth. Although the average is approximately 26.5° API, the gravity ranges from less than 15° near the base of the column to 28° at the gas-oil contact. Figure 25 illustrates the approximate gravity-versus-depth relationship based on numerous samples obtained in tests.

The contact between the oil accumulation and the underlying aquifer deepens eastward and northeastward from about −8,925 ft (−2,270 m) in the western part of the Eileen area to elevations of between −8,990 and −9,060 ft (−2,740–2,761 m) within the Main field. Within the Main field, the contact is deepest along an approximate northwest-southeast-trending line. This line generally coincides with the area where a zone of heavier oil at the base of the oil column is thickest. The variations in the elevation of the oil-water contacts in the Main field area have not been clearly explained. They may be related to the presence of the zone of heavier oil; they apparently are not due to faulting inasmuch as the contacts are at similar depths on opposite sides of fairly major faults. Also, they may be a reflection of changes in lithologic characteristics and hydrodynamic effects.

A distinct interval of denser oil at the base of the oil column has been recognized from the darker and slightly higher residual oil saturations seen in cores. This heavy-oil interval reaches a maximum thickness of about 70 ft (21 m) and is located in that part of the Main field where the base of the oil column is represented by the oil-water interface. The heavy-oil zone is not apparent within the Eileen area.

Analyses of oils obtained from the heavy-oil zone, both during tests and by extraction of the residuals remaining in cores (Fig. 26), indicate

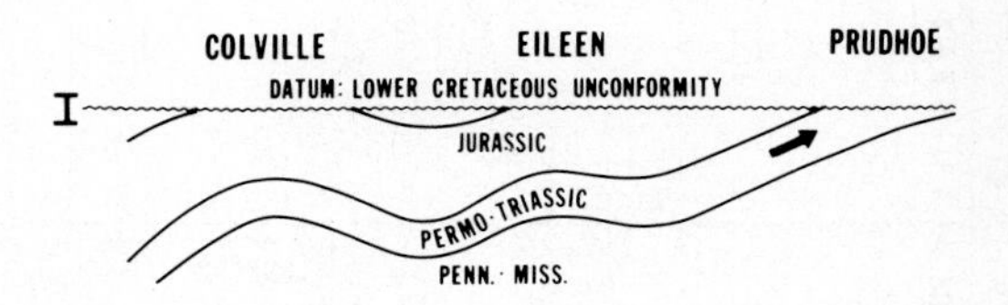

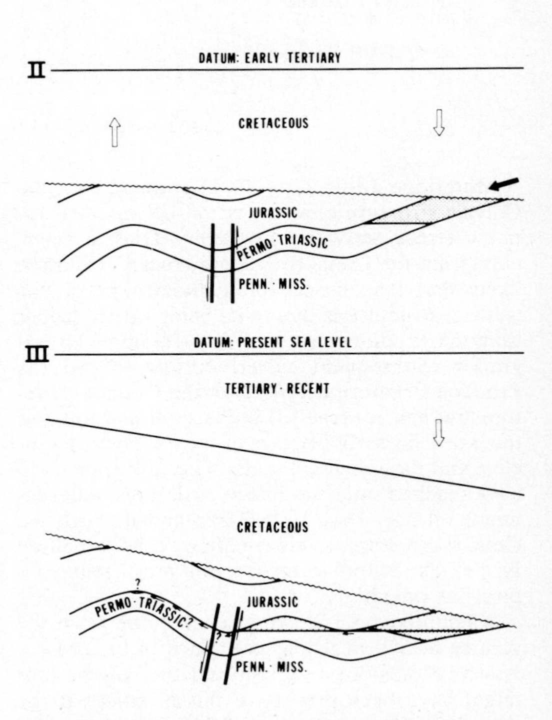

FIG. 23—Summary of oil migration in Permo-Triassic of Prudhoe Bay.

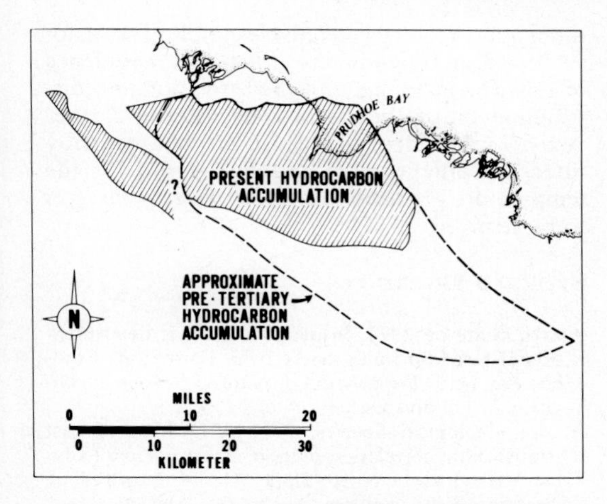

FIG. 24—Summary of redistribution of oil as result of post-Cretaceous tilting.

higher asphaltene contents than are found at shallower elevations in the oil column. The asphaltene contents of the extracted oils range from 13 to 36 weight-percent, whereas, higher within the oil column, the content is in the range of 1–5 weight-percent. It is suggested that the zone of heavy oil could have developed through a process whereby part of the asphaltene content of the crude had been precipitated and had accumulated at the base of the oil column as a result of gravity segregation. The precipitation of the asphaltenes may have resulted from a de-asphalting process initiated by the introduction of gas which was generated as a result of the gradual increase in the depth of the reservoir (Evans et al, 1971).

In well BP 21-11-13, admittedly in fairly low-porosity rocks, a possible "ghost" heavy-oil zone has been recognized within the water leg of the reservoir. Within this zone, residual oil saturations are greater than at shallower elevations. The base of this zone coincides with the base of the residual oil saturations and could represent a zone of heavy oil which formed above the original oil-water contact. This phenomenon has not yet been recognized in any other well, possibly because there are very few core data available from over the interval where it is likely to occur.

The absence of a heavy-oil zone in the Eileen area may be due to a combination of two things: (1) the accumulation formed later than the Main field structure, and (2) there is a thinner and less extensive column from which the heavy ends could precipitate. Analyses of formation water produced from the Ivishak Sandstone indicate an average salinity of 18,500 ppm NaCl equivalent and a total dissolved-solids content slightly in excess of 20,000 mg/l.

The distribution of hydrocarbons within the Sag River sandstone and the Shublik reservoirs is poorly defined. Within the field, no well has completely penetrated water-bearing reservoirs and the gas-oil contacts have not been firmly established, though in the Main field the gas-oil contacts may be the same as in the Ivishak Sandstone.

There is a marked temperature gradient from northeast to southwest within the Ivishak reservoir. Adjusted to a depth of −8,800 ft (−2,682 m), the range is from about 190°F (88°C) in the northeast to 230°F (110°C) in the west. The cause of this gradient is not known, but it may be associated with the increased thickness of Jurassic strata in the same direction. As they increase in thickness toward the west, these strata may act as a progressively more efficient insulator.

SUMMARY

The Prudhoe Bay field is an excellent example of a stratigraphic and structural trap. The main *reservoir,* the Ivishak Sandstone, is a highly porous and permeable sequence of sandstones and conglomerates with an excellent net/gross ratio. The *closure* is provided by structural dip, an un-

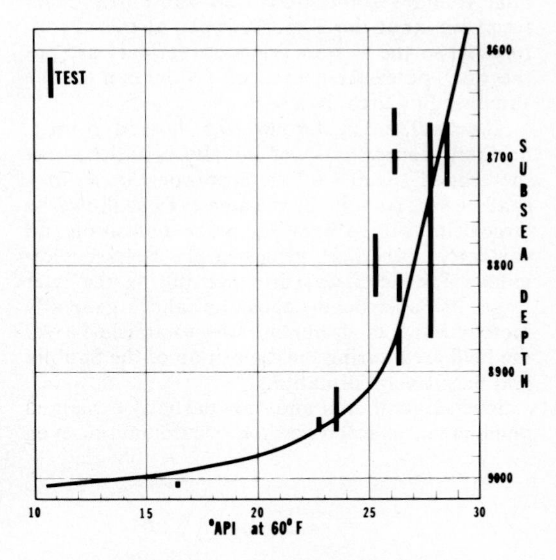

FIG. 25—Gravity-vs-depth plot, Sadlerochit crude oil.

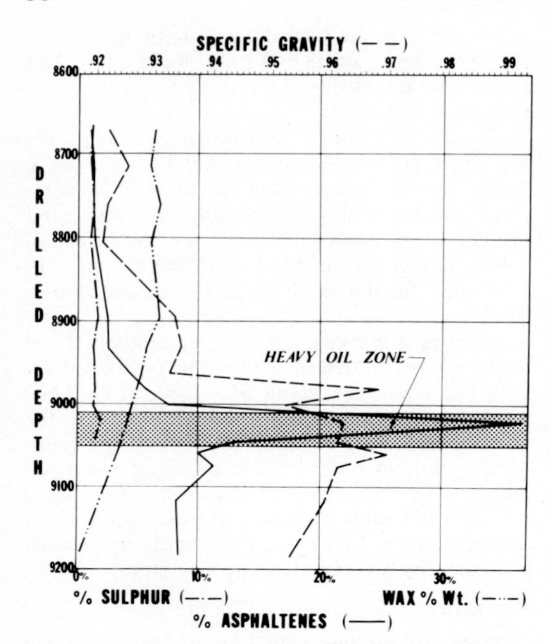

FIG. 26—Analyses of Sadlerochit crude oil extracted from cores in BP well 21-11-13.

conformity, and, to the north, faulting. The *source rocks* not only seal the reservoir, but also exist in huge volumes to the south, east, and north. Other reservoirs abut these source rocks at the unconformity, so the Permo-Triassic reservoirs are not the only potential source of production in the Prudhoe Bay area.

The Sadlerochit Group was derived from a northerly source and was initially deposited over the eroded surface of the Lisburne Group in a shallow sea. As this sea became even shallower, a large delta formed, resulting in the deposition and southerly spread of progressively coarser sediments. The delta was drowned during the later stages of Sadlerochit deposition, and a generally more marine environment was established over the field area during the deposition of the Shublik and Sag River Formations.

Several significant and only partially explained phenomena characterize the accumulation, even though the geology is relatively simple. The history of oil migration into the structure, the evidence of subsequent tilting, and the formation and distribution of the heavy-oil zone warrant further study. Furthermore, the cause of the apparently tilted oil-water contact and the reason for the temperature gradient within the reservoir are yet to be resolved.

SELECTED REFERENCES

Alaska, State of, 1974, In place volumetric determination of reservoir fluids, Sadlerochit Formation, Prudhoe Bay field: Department of Natural Resources, Division of Oil and Gas.

Alaska Geological Society, 1970-1971, West to east stratigraphic correlation section, Point Barrow to Ignek Valley, Arctic North Slope, Alaska: North Slope Stratigraphic Committee, Anchorage, Alaska.

Detterman, R. L., 1970, Sedimentary history of the Sadlerochit and Shublik Formations in northeastern Alaska, *in* Proceedings of the Geological Seminar on the North Slope of Alaska: Pacific Sec. AAPG, p. O-1 to O-13.

——— et al, 1975, Post-Carboniferous stratigraphy, N. E. Alaska: U.S. Geol. Survey Prof. Paper 886, in press.

Evans, C. R., M. A. Rogers, and N. J. L. Bailey, 1971, Evolution and alteration of petroleum in Western Canada, *in* Contribution to petroleum geochemistry: Chem. Geology, v. 8, p. 147-170.

Howitt, Frank, 1971, Permafrost geology at Prudhoe Bay: World Petroleum, v. 42, no. 8, p. 28-32, 37-38.

Keller, A. S., R. H. Morris, and R. L. Detterman, 1961, Geology of the Shaviovik and Sagavanirktok River region, Alaska, part 3: U.S. Geol. Survey Prof. Paper 303-D, p. 169-222.

Leffingwell, E. de K., 1919, The Canning River region, northern Alaska: U.S. Geol. Survey Prof. Paper 109, 251 p.

Morgridge, D. L., and W. B. Smith, Jr., 1972, Geology and discovery of Prudhoe Bay field, eastern Arctic Slope, Alaska, *in* R. E. King, ed., Stratigraphic oil and gas fields—classification, exploration methods, and case histories: AAPG Mem. 16, p. 489-501.

Reiser, H. N., 1970, Northeastern Brooks Range—a surface expression of the Prudhoe Bay section, *in* Proceedings of the Geological Seminar on the North Slope of Alaska: Pacific Sec. AAPG, p. K-1 to K-14.

Rickwood, F. K., 1970, The Prudhoe Bay field, *in* Proceedings of the Geological Seminar on the North Slope of Alaska: Pacific Sec. AAPG, p. L-1 to L-11.

Taglu Gas Field, Beaufort Basin, Northwest Territories[1]

T. J. HAWKINGS, W. G. HATLELID, J. N. BOWERMAN, and R. C. COFFMAN[2]

Abstract The Taglu gas field, discovered in 1971, is located near the outer fringe of the Mackenzie River delta, in the south-central part of the Beaufort Tertiary basin. Four wells and available seismic lines currently define about 12 sq mi (31 km^2) of productive area within a structural closure. The gas occurs in lower Tertiary sandstone reservoirs on the updip edge of a rotated down-to-basin fault block.

The basin is filled by a deltaic clastic sequence of sandstone and shale; deposition, which prograded northward, began in Late Cretaceous time and is continuing today at the shelf edge.

The southeast margin of the basin lies along the Tuk Peninsula. There, the Tertiary sediments prograded across a linear block-faulted hinge line formed of Lower Cretaceous transgressive sandstone and shale sequences overlying Paleozoic and older beds. Within the Lower Cretaceous beds, complex faulting and reservoir distribution control the hydrocarbon "play" that has yielded oil at Atkinson and gas at Parsons Lake.

The southwest margin of the basin lies along the locally designated "West coastal plain" (Yukon or Arctic coastal plain) where a thick Jurassic–Lower Cretaceous sedimentary section accumulated in front of a rising mountain belt. Only a small portion of the Tertiary-Cretaceous basin extends onto the onshore part of this southwest margin.

Offshore, the younger strata thicken abruptly into Mackenzie Bay and the Beaufort Sea.

During Tertiary time, the rapidly deposited, prograded sediments in the basin became involved in a partly gravity-induced, detached style of structural deformation. This deformation, which contrasts with the high-angle, basement-involved block faulting of the margins, has created diapir-like, shale-cored structures and growth faults. The association of these structures with prospective reservoir sequences has resulted in the style of hydrocarbon accumulation represented by the Taglu gas field.

Associated with sedimentation and deformation of this type is the phenomenon of overpressure, which has been encountered in most wells in the Mackenzie River delta. It is caused by undercompaction of fine-grained impermeable sediments, whereby pore waters are unable to escape and must support part of the weight of overburden.

Organic material within the Beaufort Tertiary strata has a predominantly terrestrial character. This fact, combined with low maturation levels, probably accounts for the predominance of gas and condensate discovered to date in the Beaufort Tertiary basin.

In the Taglu field, gas is produced from multiple sandstone reservoirs, probably of Eocene age, in a 1,700-ft (518 m) stratigraphic delta-front sequence below 7,700 ft (2,347 m). Gas columns of up to 1,700 ft (518 m) fill the closure almost to the spillpoint. The gas is usually wet and consists of 91–96 percent methane and 4–9 percent heavy hydrocarbons (C_{2+}), with no hydrogen sulfide. Wet-gas reserves are estimated to be in excess of 3 Tcf.

About 75 percent of the Taglu gas reservoirs consists of beach and stream-mouth bar sandstones; distributary-channel sandstones make up approximately 15 percent and shoreface sandstones about 10 percent. Net sandstone thicknesses in the wells range from about 450 to 600 ft (135–185 m).

The abundant, organic-rich, delta-front shales interbedded with the reservoir sandstones are the most likely source of the hydrocarbons. Therefore, a relatively small drainage area is required. An immature woody material, possibly with some biodegradation, is indicated as the hydrocarbon source by the relatively high aromatic content of the associated condensates.

The condensates have API gravities ranging from 33 to 48°, and a very low sulfur content. Oil gravities range from 17 to 32° API. Pour points range from +45°F (+7.22°C) in a paraffinic oil to −55°F (−48.33°C) in a naphthenic oil. The low gravity and pour point of some of the oils may be due to biodegradation which removed paraffins by bacterial action.

Regional Setting of Taglu Field

T. J. HAWKINGS and W. G. HATLELID

Location

The Taglu gas field is located in the Mackenzie River delta in the central part of the Beaufort basin (Fig. 1). Figure 2 shows the position of Taglu relative to the other hydrocarbon discoveries in the Beaufort basin. The basin is bounded on the southeast by the stable cratonic shelf of the Anderson Plain and on the southwest by the mobile belt of the British, Barn, and Richardson Mountains. To the north it extends out beyond the shelf edge to about the 600-ft (183 m) water depth.

Since 1965, about 90 wells have been drilled in the basin, north of lat. 68°, resulting in four oil, seven gas, and two gas/oil discoveries.

Structure and Stratigraphy

Several papers published by the Geological Survey of Canada and by oil companies give good accounts of various aspects of the Beaufort

[1]Manuscripts received, March 10, 1975.

Two papers presented at the Canadian Society of Petroleum Geologists' "International Symposium on Canada's Continental Margins and Offshore Petroleum Exploration," on September 29–October 2, 1974, Calgary, Alberta, Canada, have been combined for this publication. Permission to publish these papers has been granted by the Canadian Society of Petroleum Geologists.

[2]Imperial Oil Limited, Calgary, Alberta.

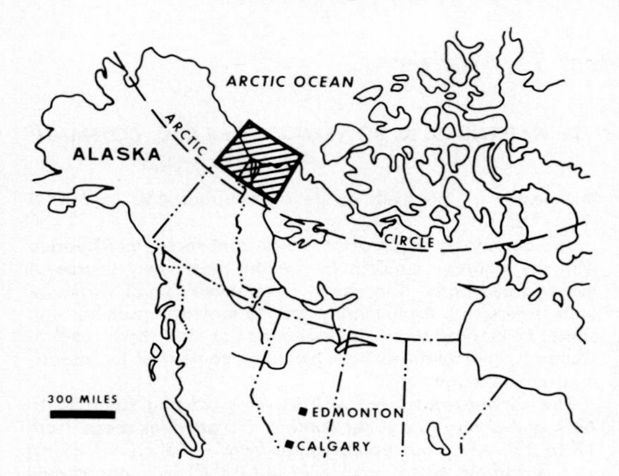

FIG. 1—Location map of Beaufort basin area.

basin. In particular, the paper by Lerand (1973) is a comprehensive account.

The main periods of deposition in the Beaufort basin occurred during Paleozoic, Jurassic–Early Cretaceous, and Late Cretaceous–Tertiary times (Fig. 3). Corresponding tectonic events that shaped the basin occurred in Late Devonian–Early Mississippian, early Mesozoic, and late Mesozoic–Tertiary times. All these depositional and tectonic events had some effect on the basin.

The late Paleozoic–early Mesozoic form of the basin is summarized in Figure 4, which shows the subcrop at the early Mesozoic unconformity. In the west, a northwest-southeast-trending arch of metamorphosed early Paleozoic rocks—the Neruokpuk Formation—was onlapped and transgressed by late Paleozoic and earliest Triassic clastic and carbonate sediments. The arch was eroded in the Early Triassic and then reactivated in pre-Jurassic time.

Eastward, the shallow-water late Paleozoic–Early Triassic sediments grade into basinal shales deposited in a geosyncline. The southeast margin of the geosyncline was formed by early Paleozoic shallow-water carbonate sediment that covered the cratonic shelf. In Late Devonian time, this shelf became covered by turbidites and shales of the Imperial Formation. The subcropping Proterozoic and early Paleozoic carbonate rocks along the Aklavik arch represent late Paleozoic structure eroded at the pre-Mesozoic unconformity.

Through Paleozoic and into early Mesozoic time, the area was dominated by a landmass on the north. In the early Mesozoic, a major reversal associated with the opening of the Beaufort Sea began the development of the basin that exists today.

The tectonic events that shaped or influenced the Beaufort basin are illustrated in Figure 5.

In the southwest, in the British and Barn Mountains, Mesozoic thrust faults were superposed on older Paleozoic thrust faults. In the Richardson Mountains, a complex structural zone had repeated normal and strike-slip faulting

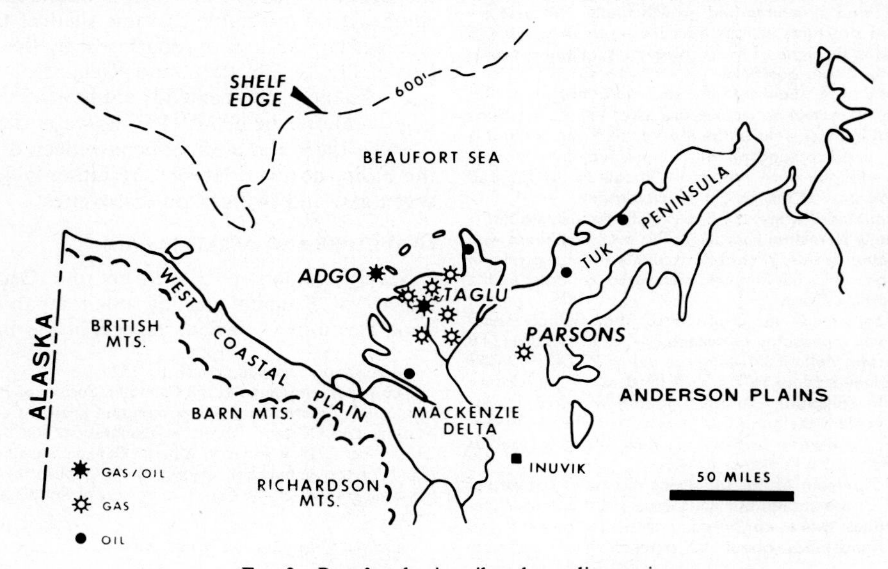

FIG. 2—Beaufort basin, oil and gas discoveries.

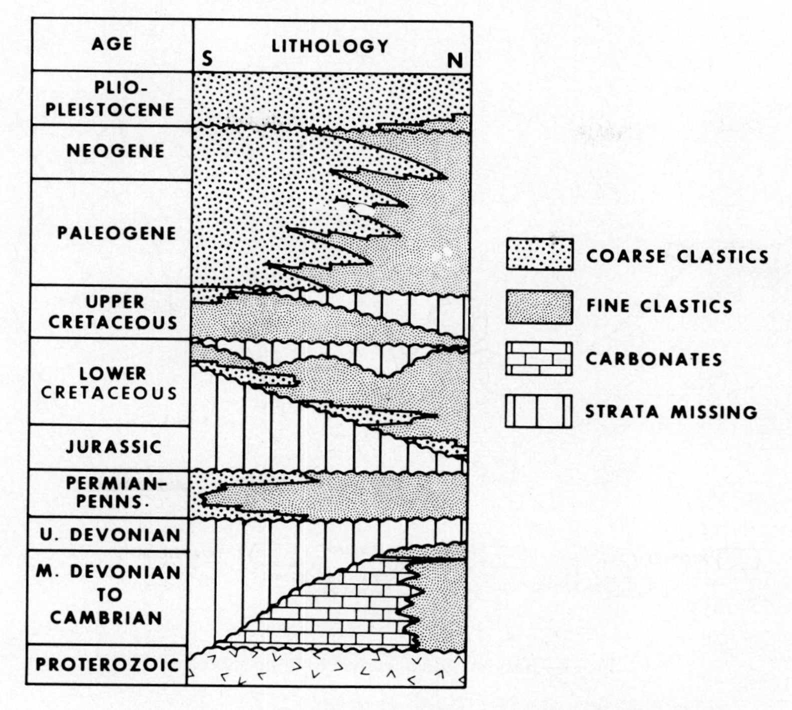

FIG. 3—Generalized stratigraphic column, Mackenzie delta area.

from early Tertiary to the present time. Deep strike-slip faults probably extend into the offshore, but they can only be postulated from indirect evidence because, if present, they are below the limit of present seismic resolution.

Along the West coastal plain (Yukon or Arctic coastal plain) and the Tuk Peninsula, normal block faults with horsts and grabens, typical of pull-apart margins, are present. The faulting along the West coastal plain apparently occurred in Triassic time, whereas that along the Tuk Peninsula occurred mainly during Early Cretaceous time.

In the thick Tertiary beds, the faults and structures are mostly detached rather than basement-involved. The deposition of thousands of feet of sediment set up gravitational forces, causing down-to-basin faulting and development of associated shale-cored anticlines and diapirlike structures with numerous adjustment faults. The faults indicated on Figure 5 are only a small fraction of those actually present.

Figure 6, showing the tectonic framework of the Beaufort basin, summarizes the structural and sedimentary styles of the Mesozoic and Tertiary basins. The mobile belt is shown at the lower left and the cratonic shelf of the Anderson Plain is in the area east of the Aklavik arch.

The Mesozoic basin was a Y-shaped continental-margin basin. The southeast margin was developed along the Tuk Peninsula (Fig. 7). It was formed by the shallow Aklavik arch and a deeper, downfaulted zone, the Tuk flexure zone. This margin was onlapped in Early Mesozoic time by a transgressive sequence; this sandstone-shale sequence includes a series of thick quartzose reservoir sandstones. Faulting, which continued through Early Cretaceous time, controlled the position of the transgressing shoreline and influenced sand deposition. The major faulting was over by the end of the Early Cretaceous, and in Late Cretaceous and Tertiary times sediments prograded out over the arch.

The southwest margin of the early Mesozoic basin was along the present West coastal plain. There, a thick Jurassic to Early Cretaceous sequence of medium-grained clastic material was derived from the mobile belt on the southwest

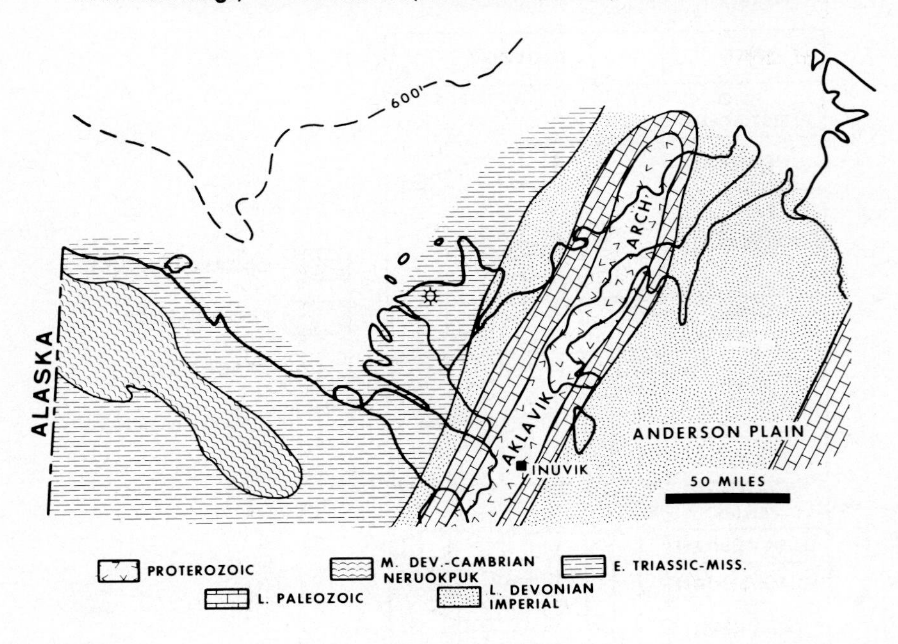

FIG. 4—Beaufort basin, early Mesozoic subcrop.

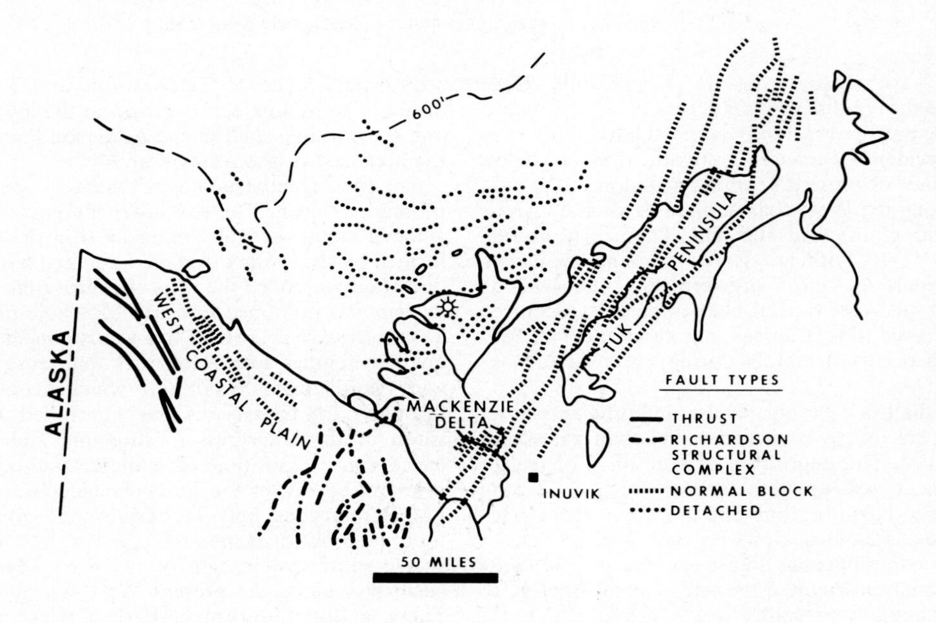

FIG. 5—Beaufort basin, types and distribution of faults.

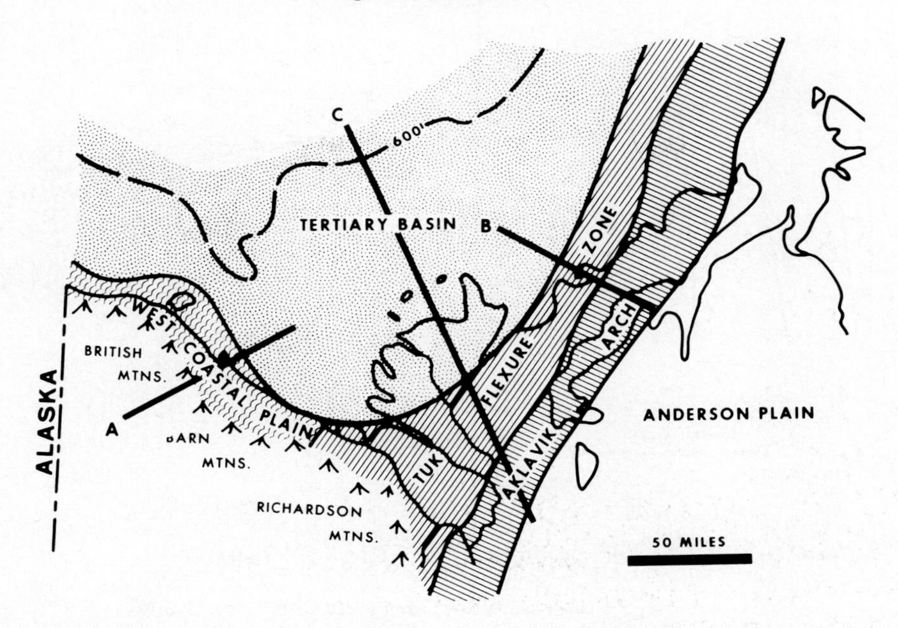

FIG. 6—Beaufort basin, tectonic framework. Line *A* is cross section of Figure 8; line *B* is cross section of Figure 7; line *C* is cross section of Figure 9.

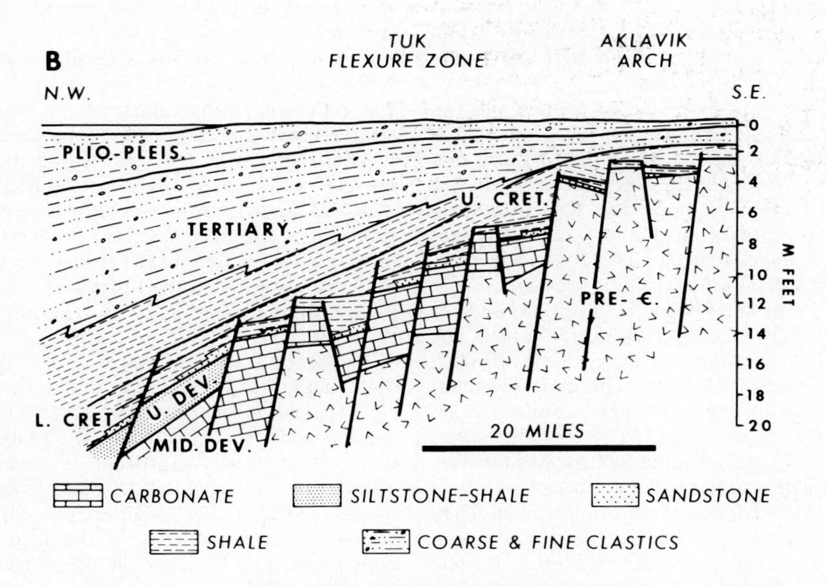

FIG. 7—Cross section of Beaufort basin, southeast margin. Line of cross section is shown on Figure 6.

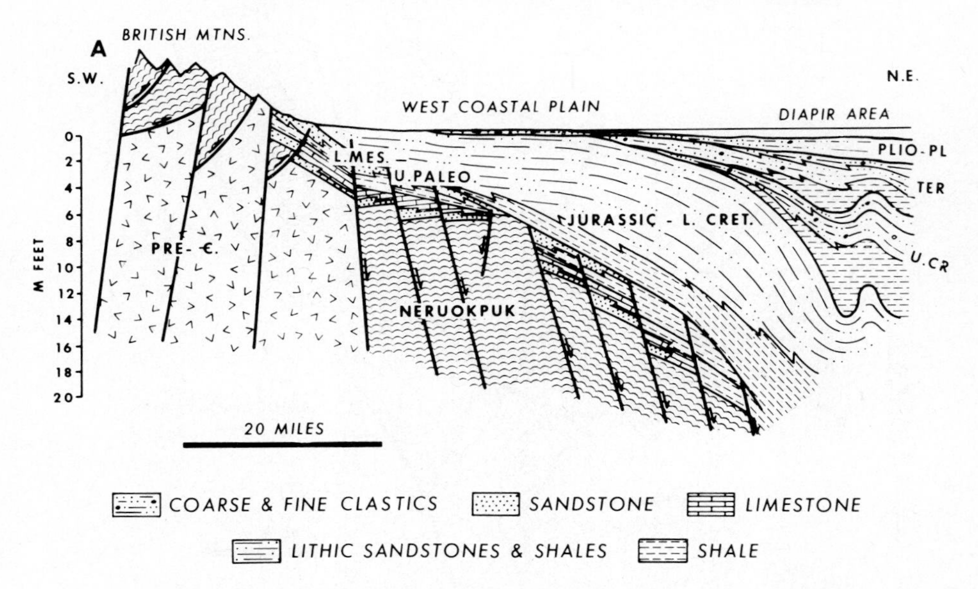

FIG. 8—Cross section of Beaufort basin, southwest margin. Line of cross section is shown on Figure 6.

(Fig. 8). The thrust-faulted area of the British Mountains passes eastward into the area of early Mesozoic continental-margin-type block faults. A thick, post–"pull-apart" Jurassic–Early Cretaceous sequence overlies block-faulted late Paleozoic and early Mesozoic beds.

The Jurassic–Early Cretaceous sequence is onlapped by a Late Cretaceous–Tertiary section, which thickens abruptly toward the offshore diapir area. At the end of the Early Cretaceous, the south end of the Y-shaped basin closed to form the U-shaped Cretaceous-Tertiary basin existent today. This basin is being filled by successively younger beds prograding farther seaward. From the margin to the shelf edge, the basin has an area of approximately 20,000 sq mi (52,000 km^2). The thickness of the Tertiary section exceeds 30,000 ft (9,100 m) at the thickest part. The main sediment source appears to have been the mobile belt on the southwest.

The cross section in Figure 9 illustrates the Tertiary basin. The southern block-faulted margin is formed of lower Mesozoic and older beds. They are overlain by Upper Cretaceous rocks which have been removed basinward by erosion. The overlying Tertiary section thickens abruptly basinward; the prograding Tertiary sequence includes interfingering shale facies. The Taglu gas field occurs on the high side of a rotated down-to-basin fault block. The gas reservoir is a series of prograding sandstones underlying a fairly thick transgressive shale.

Figures 10, 11, and 12 are seismic sections (exhibiting data from approximately along line *C* of Fig. 6) which further illustrate the structural and stratigraphic aspects of the Tertiary basin.

The first section (Fig. 10) shows the southeast margin of the basin. The block-faulted pre-Mesozoic and Lower Cretaceous beds are overlain by thin Upper Cretaceous strata; these strata are overlain by a prograding Tertiary section which thickens abruptly to the northwest. There is little structural deformation of the Tertiary strata in this area.

The second seismic section (Fig. 11), from the delta near the Taglu field, shows the structural style typical of that area. The down-to-basin fault block is similar to that at Taglu. The development of reservoir-quality sandstone in these blocks in association with early structural deformation makes them highly prospective. The markers above the fault are Tertiary. The deep part of the glide plane may be along the post-Cretaceous unconformity. The Taglu reservoir sandstones are just below the fourth marker from the top. The zone between the third and fourth markers is the

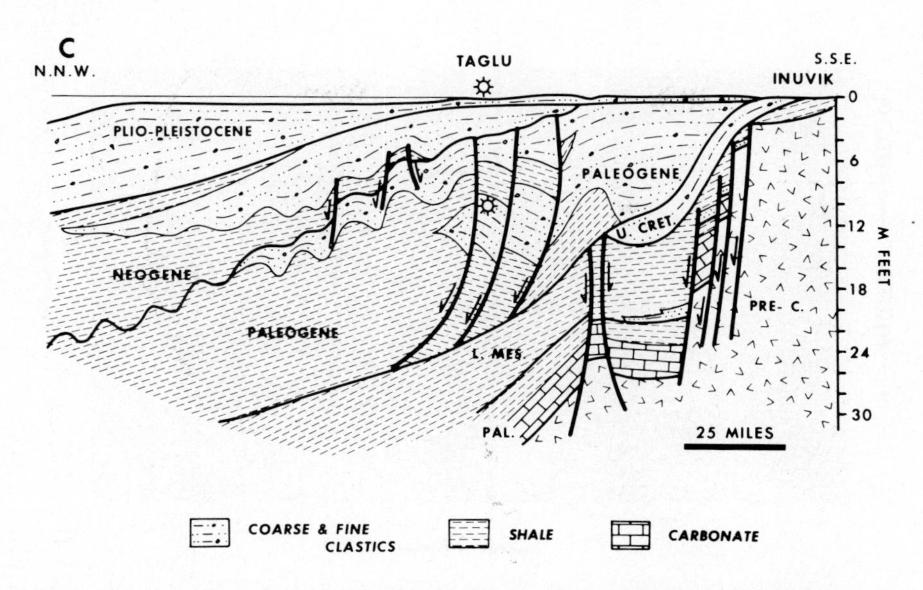

FIG. 9—Beaufort basin, Mackenzie delta section (facies shown schematically). Shows position of Taglu field in relation to basin. See Figure 6 for location of section.

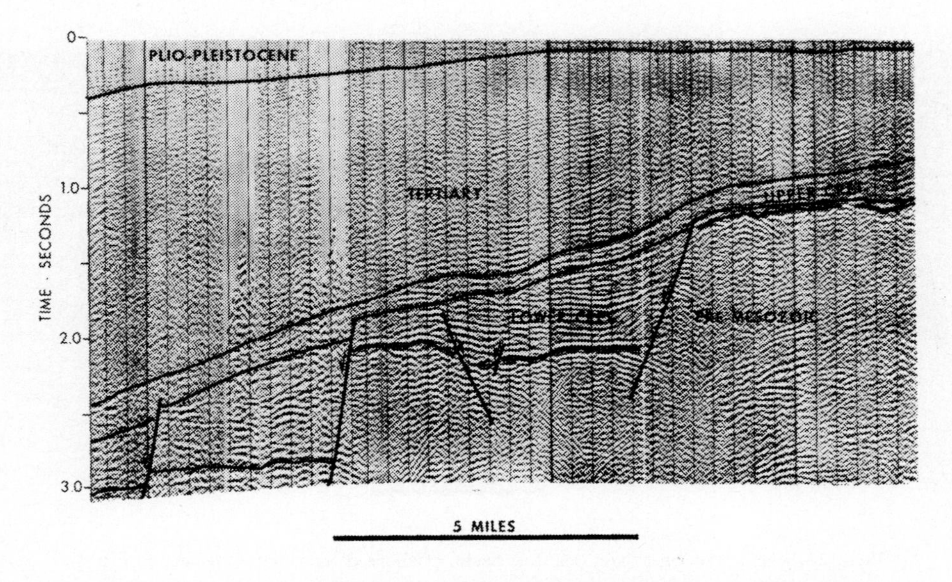

FIG. 10—Seismic section across Beaufort basin, southeast margin (along line *C*, Fig. 6).

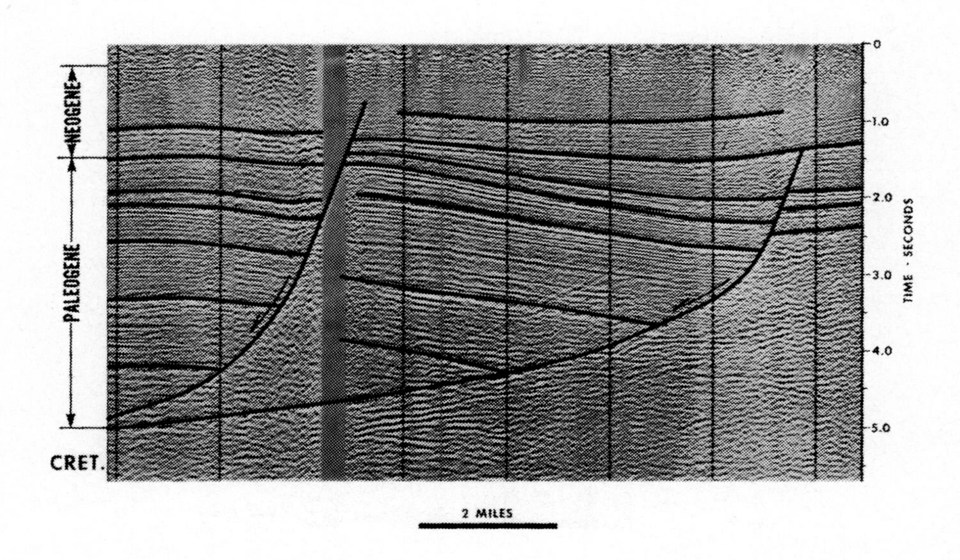

FIG. 11—Seismic section across Beaufort basin, delta area (along line *C*, Fig. 6).

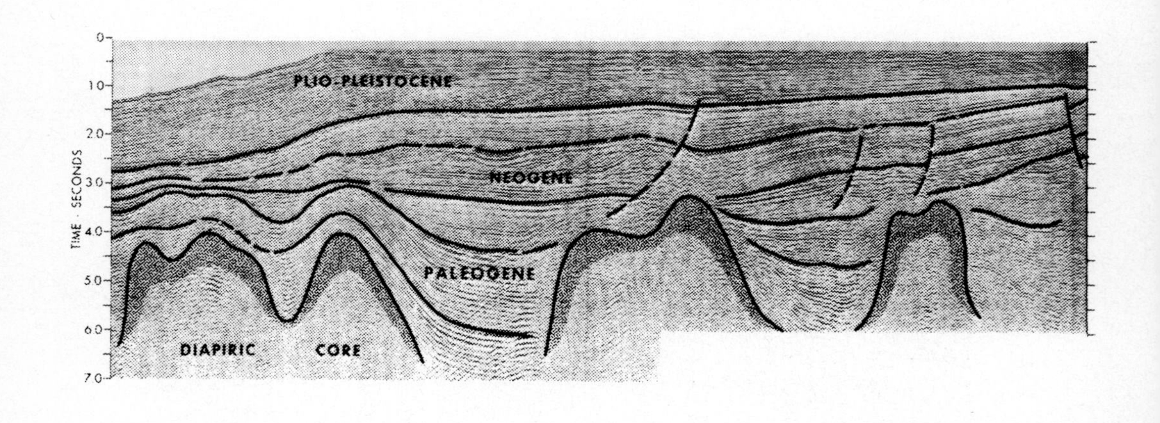

FIG. 12—Seismic section across Beaufort basin, offshore delta area (along line *C*, Fig. 6).

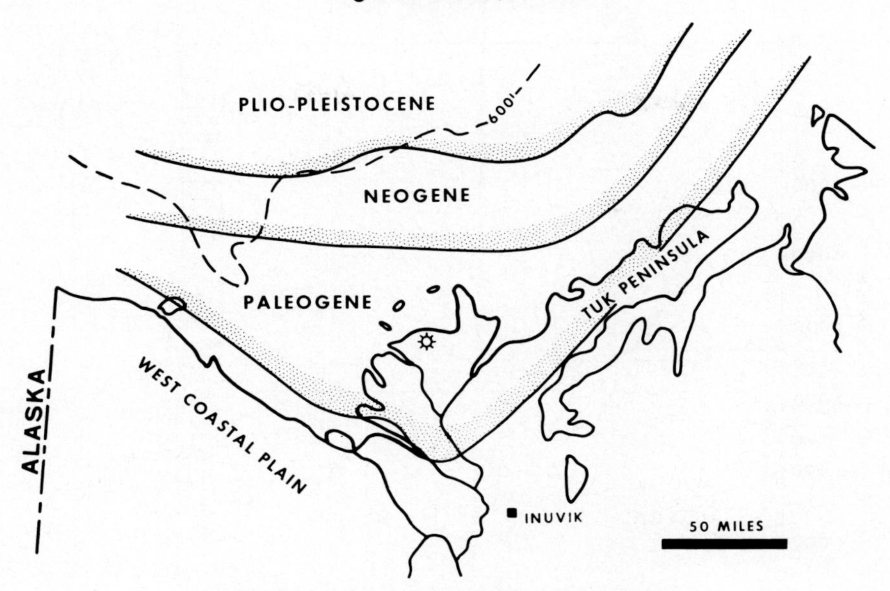

FIG. 13—Beaufort basin, Tertiary deltaic trends.

transgressive shale. The age of structural movement is indicated by the growth zone between the second and fourth markers.

The third seismic section (Fig. 12) is from the offshore area and extends to the shelf edge. The obvious anticlinal or diapirlike structures contain no evidence of salt; therefore, these structures are believed to be entirely shale-cored. They are best developed in the depocenter of the basin, where they have relief of up to 15,000 ft (4,600 m). Their potential for hydrocarbon accumulation depends on development of reservoir-quality sandstones on the flanks and crests. Other prospective structural features have been created by the adjustment faulting around these structures.

Figure 12 illustrates particularly well the sedimentary trends that are developed through the offshore area. Notice the clinoforming sequences in the Neogene and Plio-Pleistocene strata.

Figure 13 shows the Tertiary deltaic trends. Well control in the delta and the geometry of seismic reflection patterns offshore indicate that successively younger strata prograde farther into the basin and Plio-Pleistocene beds are present out to and beyond the shelf edge.

In the Tertiary beds of the southern Beaufort basin, the correlations are based on spores and pollen which are adequate for local correlations. Detailed integration of the Beaufort Tertiary sequence with the worldwide correlations has not been made. At present, the age designation is generalized to Plio-Pleistocene, Neogene (representing early Pliocene and Miocene), and Paleogene (representing Oligocene, Eocene, and Paleocene). Undoubtedly, continued exploration will result in a more refined time framework.

OVERPRESSURE

A feature of the Beaufort Tertiary rocks and, indeed, of Tertiary deltas in general, is the phenomenon of overpressure. Overpressure, caused by undercompaction of fine-grained impermeable units, results when trapped pore waters support part of the weight of the overburden. Most wells in the Mackenzie River delta have encountered some degree of overpressure. Pore-pressure profiles of shale from two of these wells illustrate some of the aspects of overpressure (Figs. 14, 15).

Figure 14, depicting the Imperial Ivik J-26 well, shows a lithologic column on the right and a pore-pressure profile on the left. The pressure is expressed in terms of mud-weight equivalent in pounds per gallon. By Imperial Oil Ltd.'s definition, normal pressure ranges up to 9.5 lb/gal, overpressure is between 9.5 and 12.5 lb/gal, and ab-

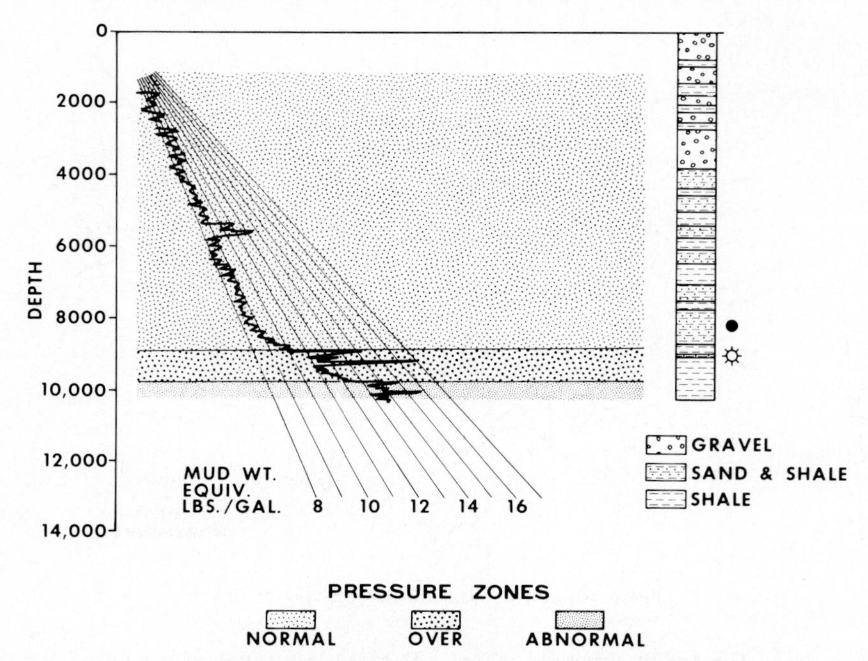

FIG. 14—Pore-pressure profile of shale from Imperial IVIK J-26.

normal pressure is over 12.5 lb/gal. The pressure profile shown in Figure 14 is derived from a sonic log. Because overpressure is a function of compaction, the pore pressure can be derived from sonic travel times through the shale zones.

In this well the pressure is normal down through the sandy section to about 8,500 ft (2,590 m). At the zone of transition to shale, the pressure increases fairly abruptly to abnormal. In this simple pressure profile, a thin overpressured zone marks the transition between delta-front sandstones and the abnormally pressured prodelta shales.

Figure 15 shows a more complicated profile from the Taglu field. In this profile, pressures are normal down to almost 7,000 ft (2,135 m). The absence of permeable sandstones in the interval from 7,000 to 8,200 ft (2,135–2,590 m) results in an increase in pressure to just over 12 lb/gal. However, below 8,200 ft (2,590 m) the massive Taglu sandstones have allowed pore waters to escape and the pressure is nearly normal. Below 10,000 ft (3,048 m), pressure increases to abnormal in a shaly, nonporous zone. Below 12,000 ft (3,658 m), a series of permeable sandstones has allowed pore waters to escape from the shale, and the pressure is 10 lb/gal. Finally, below 14,000 ft (4,267 m), the pressure increases to abnormal. The productive sandstones here are found beneath overpressured shales. Overpressuring no doubt has added to the sealing effectiveness of the overlying shale.

Hydrocarbons in Beaufort Basin

Figure 16 illustrates the stratigraphic range of Beaufort discoveries and the variation in oil and gas properties. These hydrocarbons all occur in Tertiary strata except for one oil sample from the Lower Cretaceous.

The gas is usually wet, with 91–96 percent methane and 4–9 percent heavy hydrocarbons (C_{2+}). One sample with 99 percent methane comes from a gas hydrate zone. The gas is sweet and has no hydrogen sulfide. The associated condensates are clear to honey colored, have API gravities ranging from 33 to 48°, and have very low sulfur content. The fact that the condensates are naphthenic in composition accounts for the low gravities.

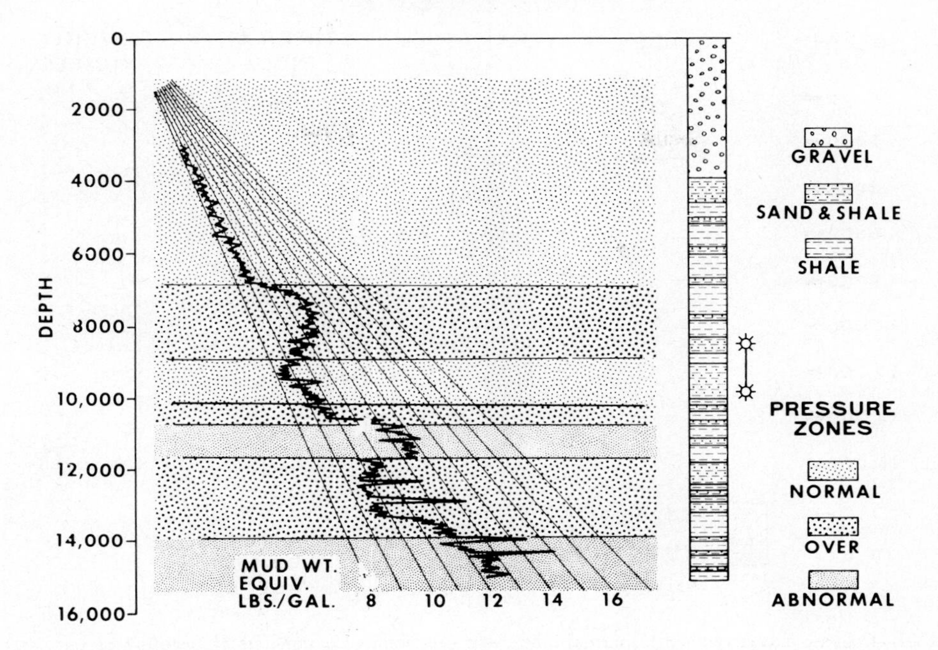

FIG. 15—Pore-pressure profile from Taglu shale.

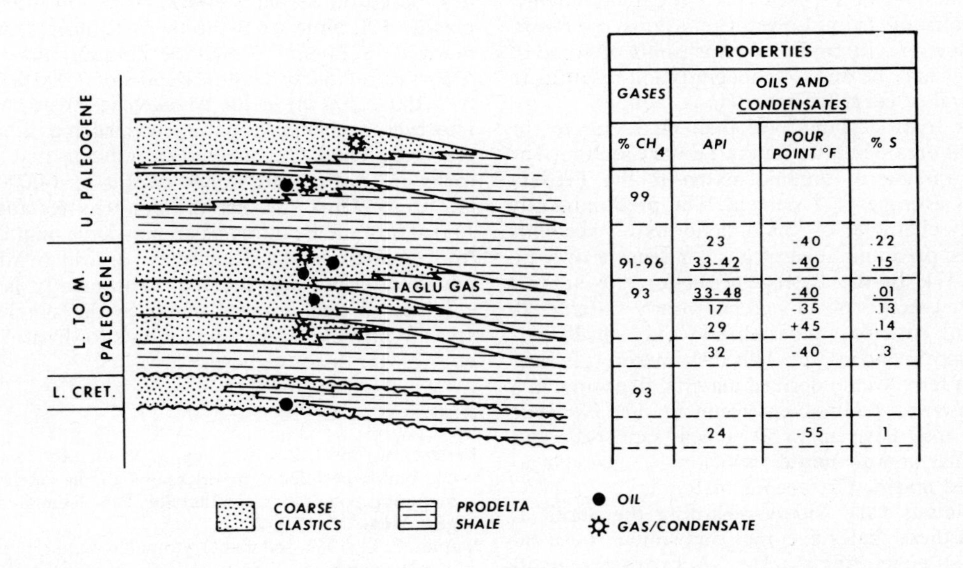

PROPERTIES			
GASES	OILS AND CONDENSATES		
% CH_4	API	POUR POINT °F	% S
99			
91-96	23 33-42	-40 -40	.22 .15
93	33-48 17 29	-40 -35 +45	.01 .13 .14
	32	-40	.3
93			
	24	-55	1

FIG. 16—Beaufort basin, hydrocarbon description.

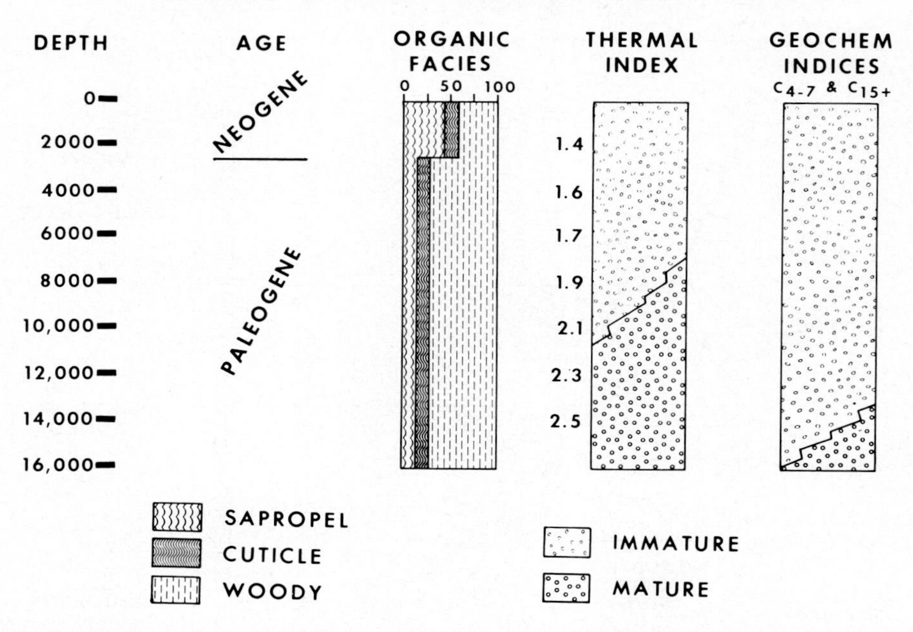

FIG. 17—Organic matter (percent), thermal index, and geochemical parameters of Beaufort Tertiary section.

Oil gravities range from 17 to 32° API. Pour points range from +45°F (+7.22°C) in a paraffinic oil to −55°F (−48.33°C) in the naphthenic-aromatic oil from Lower Cretaceous reservoirs. The low gravity and low pour point of some of the oils may be due to biodegradation resulting in removal of paraffins by bacterial action.

The hydrocarbons owe their character to the type of organic matter in the Tertiary section. The total amount of organic matter in the Tertiary shales averages 1–2 percent. The predominantly woody character of this organic matter accounts for the predominance of gas condensate over oil (Fig. 17). In the Paleogene section the organic matter averages 60–80 percent woody material, 20 percent cuticular material, and only 10–20 percent sapropelic matter. This distribution is indicative of terrestrially derived material deposited in a shallow-water deltaic environment. The Neogene strata may have up to 50 percent sapropelic and cuticular organic matter, which may represent reworked marine Cretaceous material.

Opinions vary widely regarding the depth at which these shales become good source rocks capable of generating mature, oily hydrocarbons. It has been suggested, using data from other parts of the world, that enough heat has been applied to the rocks to generate oil when the thermal index, based on Staplin's (1969) alteration indices, reaches 1.8. Note from the thermal indices summarized in Figure 17 (middle column) that 1.8 occurs at relatively shallow depths of 7,000–8,000 ft (2,100–2,400 m) in the Mackenzie River delta. However, geochemical parameters, such as analyses of C_{4-7} and C_{15+} (Fig. 17), indicate that the rocks are not mature above depths of 14,000 ft (4,250 m). Thus, the oils and condensates above 14,000 ft in the Beaufort basin may have migrated from greater depths. Regardless of which interpretation is correct, there are sufficiently large volumes of rich, mature shales in the synclinal areas to provide an excellent source for hydrocarbons in the Beaufort basin.

REFERENCES CITED

Lerand, M., 1973, Beaufort Sea, *in* R. G. McCrossan, ed., Future petroleum provinces of Canada—their geology and potential: Canadian Soc. Petroleum Geologists Mem. 1, p. 315-386.

Staplin, F. L., 1969, Sedimentary organic matter, organic metamorphism, and oil and gas occurrence: Bull. Canadian Petroleum Geology, v. 17, p. 47-66.

Geology of Taglu Field

J. N. BOWERMAN and R. C. COFFMAN

INTRODUCTION

The Taglu gas field, which produces from Tertiary rocks of the Beaufort basin, is located on the outer edge of the Mackenzie River delta 6 mi (9.7 km) from the Arctic Ocean (Fig. 2). The field is 190 mi (305 km) north of the Arctic Circle and 75 mi (121 km) northwest of Inuvik, N.W.T., the main transportation center for the delta.

The average elevation of the surface is 3–5 ft (1–1.5 m) above sea level. Scattered small pingos (conical ice-cored mounds) and an esker which reaches an elevation of 40 ft (12 m) provide the only vertical relief. Distributary channels of the Mackenzie River cut the area (see Fig. 18), and small thaw pools, shallow lakes, and marshy areas abound in the brief summer. Minor flooding is common during spring ice breakup. The flora consists of a low (3–4 ft; 1–1.2 m) willow, sedge, and herb community. Holocene sediments are clay, silt, and fine-grained sand deposited on the river floodplain.

Taglu was discovered in June 1971. The discovery well, Imperial IOE Taglu G-33, flowed gas at the rate of 28 MMcf/day on an open-hole drill-stem test of an Eocene sandstone at a depth of 8,170 ft (2,490 m). The discovery well tested gas from seven sandstone beds within the interval 8,150–9,800 ft (2,484–2,987 m); an additional 13 sandstones are interpreted to contain gas. Of the approximately 450 ft (137 m) of net porous sandstone in this interval, none is believed to be water-bearing.

STRUCTURE

The Taglu structure is believed to have been formed along the upthrown side of a major, detached growth fault that is characteristic of early Tertiary structural development in the Beaufort basin. The structural configuration appears to have been modified by deep-seated shale movement during a later period of structural growth. Strike-slip faulting at depth also may have been involved and may have triggered the shale movement.

There appears to be about 2,100 ft (640 m) of structural closure on the gas reservoirs and about 1,700 ft (518 m) of gas column in the uppermost reservoir sandstone (Fig. 19). Since the discovery, three confirmation wells have tested gas reservoirs in the same stratigraphic interval as the discovery well. A fourth well, north of the field, tested high-pressure water from equivalent sandstones. The indicated productive area of the field is about 12 sq mi (31 km^2).

FIG. 18—Rig at Taglu discovery location adjacent to Mackenzie River distributary channel. Photo courtesy of Tom Watmore.

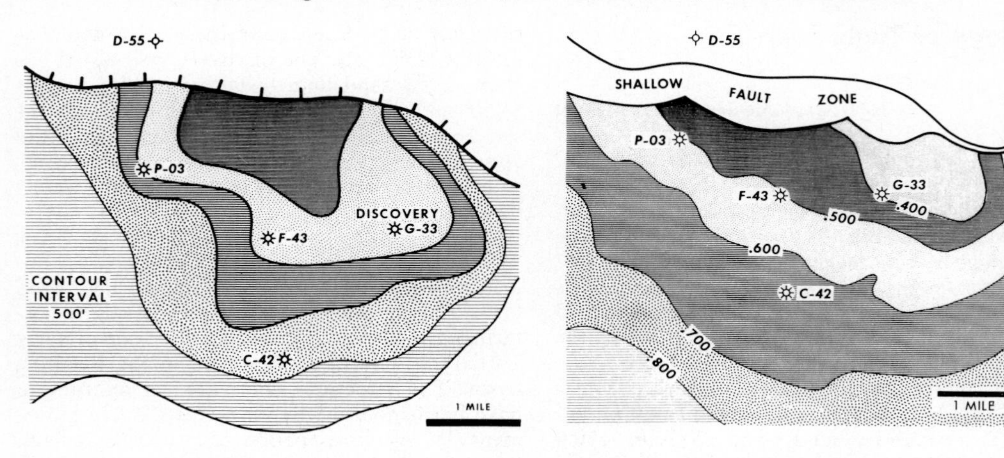

FIG. 19—Taglu structure map of a Paleogene horizon.

FIG. 20—Isochron map of interval directly above Taglu gas-bearing sandstones.

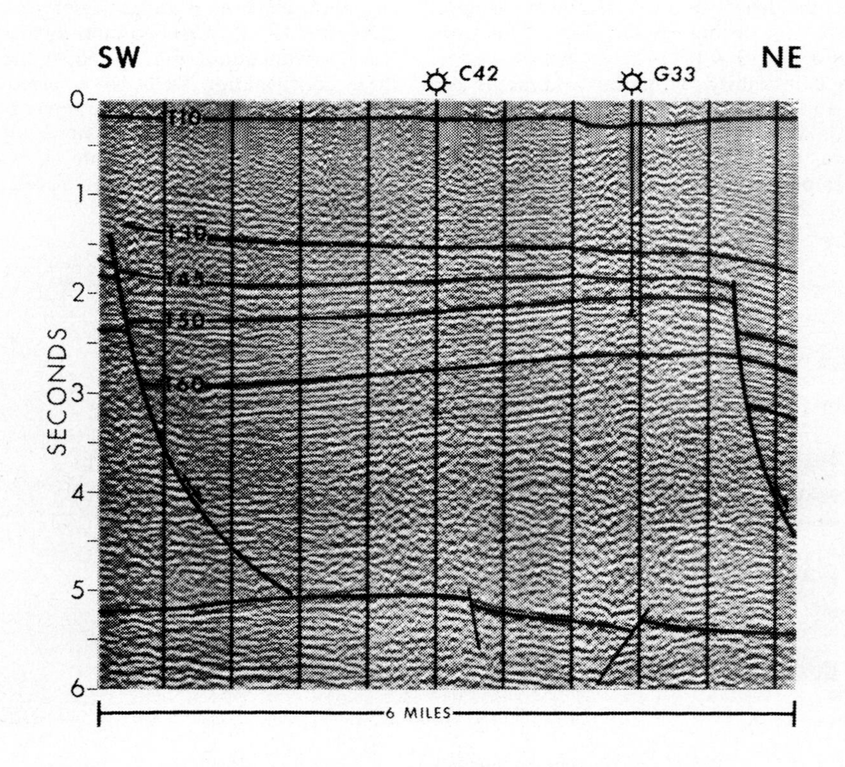

FIG. 21—Seismic section across Taglu field. Seismic horizons are identified by prefix "T."

An isochron map of the prodelta shale interval directly above the gas reservoirs (Fig. 20) illustrates a primary period of structural development at Taglu. The presence of the thinnest interval northeast of the discovery well (G-33) indicates that at one time the high was centered near this well. The crest of the structure has shifted about 3 mi (4.8 km) westward and now is located north of well F-43 (see Fig. 19). The wide fault zone at the north (Fig. 20) is due to shallow faults which terminate at a sole plane within this interval.

A typical seismic section across the Taglu field (Fig. 21) illustrates the seismic data quality and reflection continuity. The two main faults cutting the Tertiary section terminate at a sole plane in basal Tertiary shales directly overlying a probably Early Cretaceous surface.

Permafrost

In mapping structures in the Beaufort basin, the effects of changes in the permafrost distribution must be considered. These changes occur quite rapidly in many areas of the land portion of the delta, particularly in the Taglu area, and are related to the presence of lakes and channels. A sag, most noticeable on the T10 marker (Fig. 21), is an example of such a permafrost change. It is attributed to a reduction in the amount of frozen material in the vicinity of the G-33 well relative to that in the surrounding area. This causes slower travel time through the sediment, resulting in delay of reflections from underlying strata. The abrupt changes also cause shifts in ray-path geometry, which makes proper stacking of these multifold data difficult. The deterioration in record quality associated with this sag area is evident on Figure 21.

Permafrost variations in Taglu wells are shown in Figure 22. Basic data are interval velocities collected at 25-ft (7.6 m) spacings from velocity surveys run to approximately 2,000 ft (600 m) in four Taglu wells. Velocities range from less than 5,000 ft/sec (1,524 m/sec) in well D-55 at 400 ft (122 m) to slightly over 12,000 ft/sec (3,658 m/sec) between 800 and 900 ft (244–274 m) in all of the wells. The base of permafrost in all wells is about 1,700 ft (518 m), and this is where interval velocities drop to about 6,000 ft/sec (1,829 m/sec).

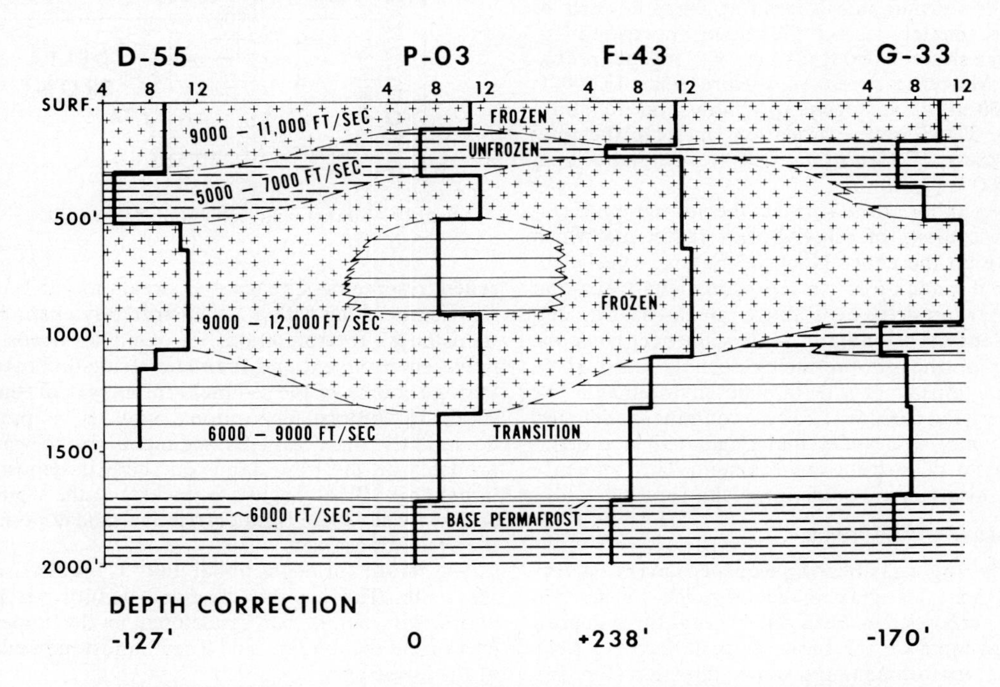

FIG. 22—Permafrost zone in Taglu wells. Heavy line below each well indicates interval velocity. Scale below each well name is in thousands of feet per second.

Although each of the four wells has a unique permafrost configuration, it is possible to distinguish several common zones in the wells. Next to the surface is a "cap" of frozen material up to 300 ft (91 m) thick. This zone is underlain by an interval of unfrozen sediment ranging in thickness from 75 to 200 ft (23–61 m). Water was produced from a depth of 220 ft (67 m) in a shallow hole drilled near the G-33 well. The third zone is mostly massive permafrost; in the F-43 well it is 850 ft (259 m) thick. Below this zone is a "transition zone" of complexly interlayered, frozen and unfrozen material. The cumulative effect of these varying velocities is given at the bottom of Figure 22. The depth correction between wells can be as much as 400 ft (122 m). Thus the need for detailed velocity information is readily apparent.

Stratigraphy

In the Taglu field, Tertiary strata have been penetrated to 16,000 ft (4,877 m). However, detailed age determinations have been hindered by sparsity of floral and faunal assemblages as well as by the appearance of many new species.

The stratigraphic column at Taglu consists of approximately 1,000 ft (305 m) of Pliocene to Holocene strata, 2,000 ft (610 m) of Neogene (probably Miocene) strata, and more than 13,000 ft (3,960 m) of Paleogene (probably Eocene) strata (Fig. 23). Seismic correlations indicate the total thickness of Cenozoic rocks to be more than 25,000 ft (7,600 m).

Beginning at the surface, the stratigraphic column consists of alluvial-plain sediments which make up the entire Neogene section. The Paleogene is made up of alluvial-plain, delta-front, and shallow prodelta beds above the gas-bearing delta-front sandstone and shale sequence. Near the base of the gas-producing section, units show characteristics of a delta-plain environment. The underlying 6,000 ft (1,830 m) contains sandstones, siltstones, and shales that originated in a delta-plain to delta-front environment. This sequence may also contain some interdelta shales.

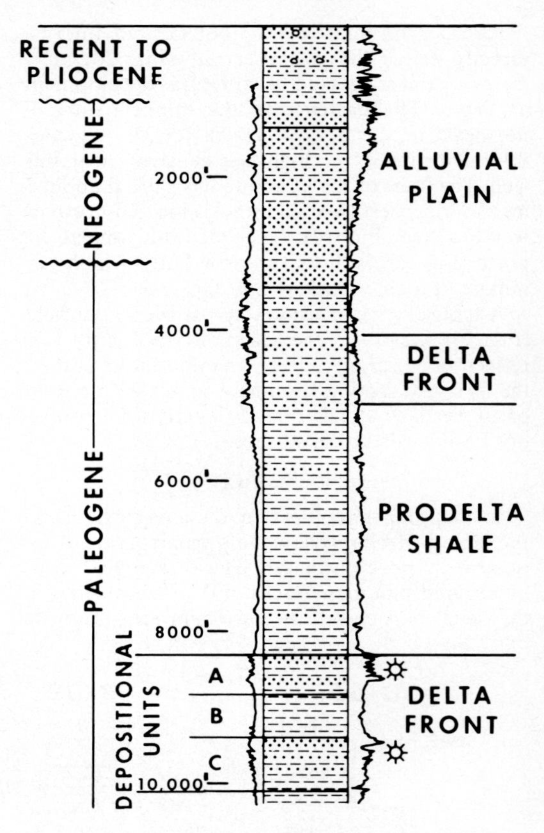

FIG. 23—Stratigraphic column of Taglu field.

Reservoir Sequence

The Taglu gas-bearing sequence covers a 1,700-ft (518 m) stratigraphic section which, for descriptive purposes, has been divided into three depositional units on the basis of similarity of depositional environment and source direction (Fig. 24). The upper unit, depositional unit "A," is about 500 ft (150 m) thick and contains some excellent reservoirs of beach and stream-mouth bar sandstones along with a few distributary-channel sandstones. Interpretations of regional depositional environment and dipmeter results indicate that the source of the sediment influx was on the south. In contrast, depositional unit "B" is predominantly shale, and its sediment source was probably on the west. Sandstone beds in depositional unit "B" at Taglu are thickest in the westernmost well; there is no correlative sandstone in the easternmost well.

The source for depositional unit "C" is also on the south. The unit has reservoir-quality beach and stream-mouth bar sandstones in the upper part of the section but only a few sandstone beds in the lower part.

Total net sandstone in the three depositional units ranges from 450 to 600 ft (137–183 m).

More detailed sections for the three depositional units as penetrated in the Imperial IOE Taglu West P-03 well are given in Figures 25–27. Well P-03 is typical for depositional unit "A" within the field (Fig. 25), and the section is correlated easily with the other wells. Excellent reservoir sandstones are present at the top of the section and at 8,800 ft (2,682 m). In this well, three intervals were perforated in the "A" unit and were tested by the closed-chamber drill-stem-test method. Actual flow rates into the well bore on the three tests were between 7 and 9 MMcf/day. For environmental and safety reasons, much of the testing at Taglu has been done by the closed-chamber method in which the drill pipe is closed at the surface during the time the bottom valve is open for flow from the formation.

All of the gas-bearing "A" sandstones tested in the four Taglu field wells have indicated a common gas-water contact; therefore, the sandstones are believed to form one interconnected gas reservoir.

Depositional unit "B" is much shalier than the "A" unit. The Taglu West P-03 well (Fig. 26) has the most sandstone in this unit and is the only well in which it was tested. Gas was recovered from thin beach and shoreface sandstones just below 9,200 ft (2,804 m), and water was tested from a sandstone about 100 ft (30 m) lower in the section.

In depositional unit "C" (Fig. 27), there is more sandstone in the upper part of the unit than in the lower part. The reservoirs in the upper 400 ft (122 m) are interpreted to be beach, stream-mouth bar, and shoreface sandstones. Tests in three wells indicate one interconnected gas reservoir within these upper sandstones. The sandstone just above 10,100 ft (3,080 m) is interpreted as a distributary-channel sandstone (Fig. 27). It has a reservoir pressure different from that of the other sandstones, which indicates an isolated gas reservoir. Sandstones in the lower several hundred feet of the "C" unit appear to represent a delta-plain environment; however, nearly all the other Taglu reservoir sandstones probably represent a delta-front environment.

Reservoir-Sandstone Types

Based on core descriptions and core-to-log comparisons, the approximate proportions of various types of delta-front sandstones are: beach and stream-mouth bar sandstones, 75 percent;

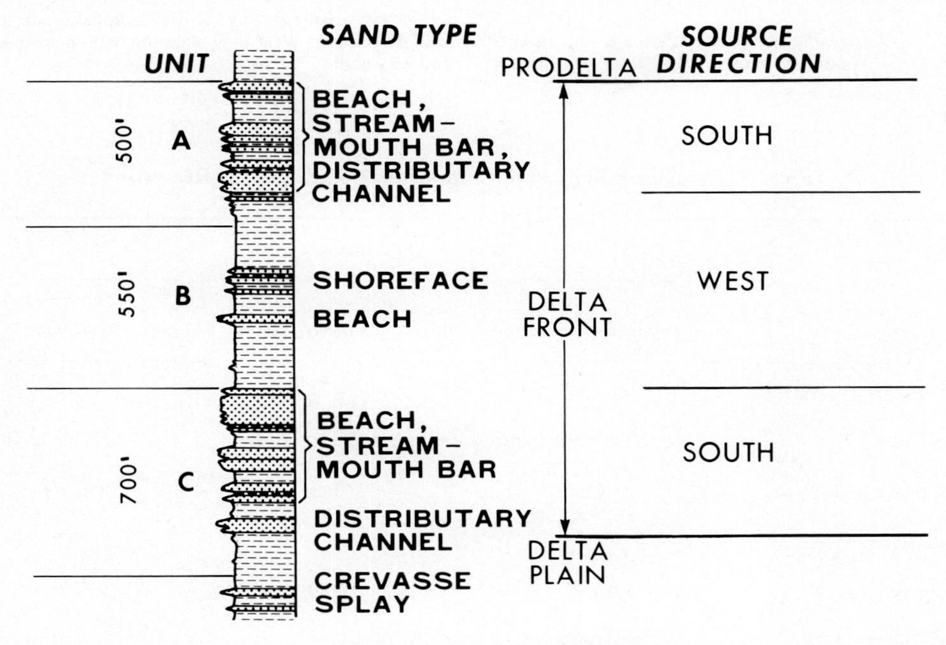

FIG. 24—Reservoir sequence in Taglu gas field.

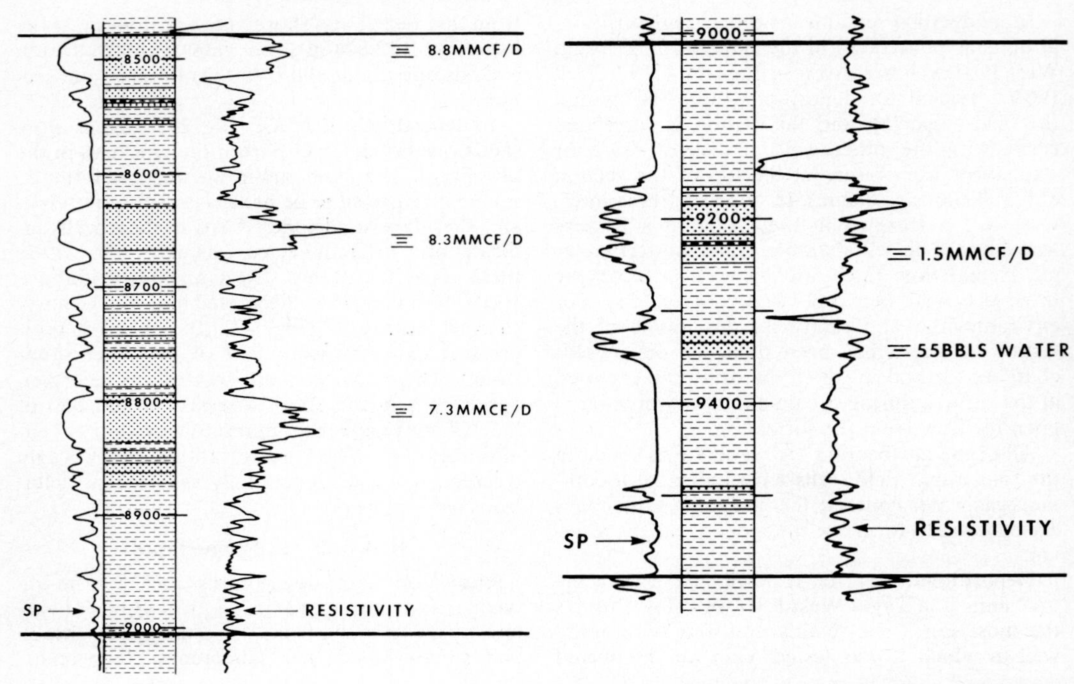

FIG. 25—Stratigraphy of "A" depositional unit in IMP IOE Taglu West P-03 showing drill-stem-test flows.

FIG. 26—Stratigraphy of "B" depositional unit in IMP IOE Taglu West P-03 showing drill-stem-test flows and recoveries.

Table 1. Sandstone Types and Composition of Taglu Gas Reservoirs

	Sandstone Type		
Parameter	*Distributary Channel*	*Beach and Stream-mouth Bar*	*Shoreface*
Percentage of reservoir	15	75	10
Grain size	Medium to coarse	Medium to fine	Very fine to fine
Dominant grain type	Chert	Chert	Quartz
Mud matrix (%)	4	7	15
Porosity (%)	14-23	12-19	12-17
Permeability (md)	10-900	5-260	2-30

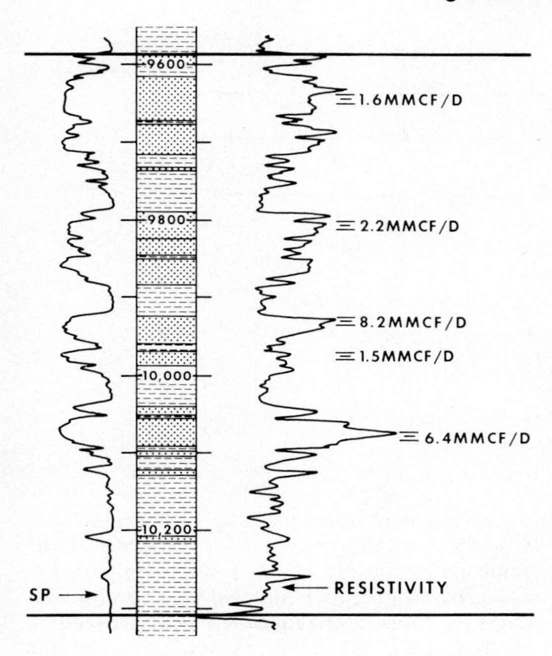

FIG. 27—Stratigraphy of "C" depositional unit in IMP IOE Taglu West P-03, showing drill-stem-test flows.

distributary-channel sandstones, 15 percent; and shoreface sandstones, 10 percent. These sandstones have different characteristics (Table 1). The average grain size is largest—medium to coarse—in the distributary-channel sandstones; in comparison, shoreface sandstones are very fine to fine grained.

One of the features of Taglu sandstone is the large proportion of nonquartz grains; this can lead to problems in the quantitative interpretation on electric logs. Chert grains are dominant in the coarser grained rocks. Mud matrix is lowest and porosity and permeability are highest in the distributary-channel sandstones.

Thin sections of two Taglu reservoir sandstones are shown in Figure 28. The upper photograph shows a medium-grained beach sandstone. The gray grains are chert, the light-colored ones are quartz, and the dark gray area is plastic-filled pore space. Very little clay matrix is present. In contrast, the lower photograph is a thin section of a fine-grained shoreface sandstone with a greater number of white quartz grains and a higher proportion of mud matrix.

Reservoir Properties

Ranges of reservoir parameters have been compiled for the three depositional units in the Taglu field (Table 2). The porosity is greatest in the structurally highest and stratigraphically youngest "A" unit. Poorer overall reservoir quality in the "B" unit is due to the presence of a larger proportion of shoreface sandstone than in the other units.

Calculated water saturations appear to be relatively high, but the fairly long gas columns and resultant high capillary pressures may actually cause the average water saturation to be toward the low side of the indicated range.

A pressure-versus-depth plot (Fig. 29) summarizes reservoir-pressure data collected at Taglu. Pressures from nine separate gas tests of "A" unit sandstones in four wells plot in a single gas-pressure envelope. The common gas-water contact is at −9,470 ft (−2,886 m). Similarly, six gas tests of sandstones in the upper 400 ft (120 m) of depositional unit "C" indicate a common gas reservoir with a gas-water contact at −10,190 ft (−3,106 m).

FIG. 28—Thin-section photographs of two Taglu reservoir sandstones. *Top*: beach sandstone. *Bottom*: shoreface sandstone. Photo courtesy of Hans Nelson.

Table 2. Reservoir Parameters of Taglu Depositional Units

Parameter	*Depositional Unit*		
	A	B	C
Porosity (%)	13-22	12-16	12-18
Water saturation (%)	25-55	High	25-55
Gas-water contact ft	-9,470		-10,190
(m)	(-2,886)		(-3,106)
Gas column ft	1,700		1,400
(m)	(518)		(427)
Pressure (psia)	4,240	4,370	4,530
Temperature in °F	146	152	159
(°C)	(63)	(67)	(71)

The regional water gradient (Fig. 29) for the "A" and "C" sandstones is from pressures obtained in well IOE Taglu C-42, which was deliberately drilled far down the flank of the structure in order to obtain information on the hydrocarbon-water contacts. The well substantiated within a few feet the earlier predictions of structure and gas-water contacts. Normal water pressures in the "A" and "C" depositional units are present beneath the gas reservoirs.

Stratigraphically below the Taglu gas-reservoir section, pressures increase abruptly, reaching abnormal conditions in water-bearing sandstones below 12,000 ft (3,660 m) in well F-43.

An unusual occurrence of high-pressure water was encountered in well IOE Taglu D-55, located north of the field on the downthrown side of a large fault (Fig. 30). Sandstone beds stratigraphically equivalent to the upper gas-bearing sandstones in the field have pressures 3,400 psi higher than the aquifer system for the field at the same depth. In well D-55, sandstones equivalent to the lower gas-bearing sandstones in the field have pressures 3,800 psi higher than the field aquifer system.

In the Taglu area, condensate recovered with the gas contains a relatively large proportion of aromatics. This property and an average gravity of 47° API suggest an immature nonmarine source or, possibly, some biodegradation.

Oil has been found only in Taglu C-42, where it was recovered from a sandstone several hundred feet below the gas-water contact of the "C" unit. A drill-stem test yielded 15 bbl of 29° API gravity oil and 45 bbl of water.

The sediment below 5,000 ft (1,525 m) includes a fairly constant amount—1-2 percent—of organic matter, mostly woody or coaly. Suitable hydrocarbon source-rock material is present in adequate amounts throughout the Paleogene section. Consequently, the likely source beds are the downdip delta-front shales interbedded with the reservoir sandstones. The reasoning for this postulation is that the source-rock material in the downdip beds is more mature because of the

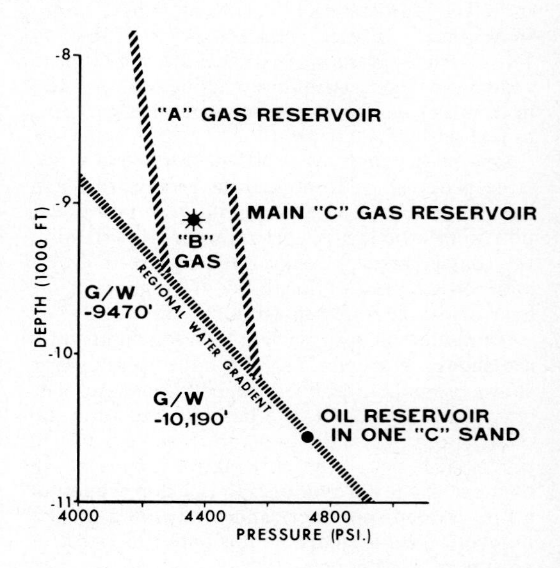

FIG. 29— Pressure-versus-depth plot for Taglu reservoirs.

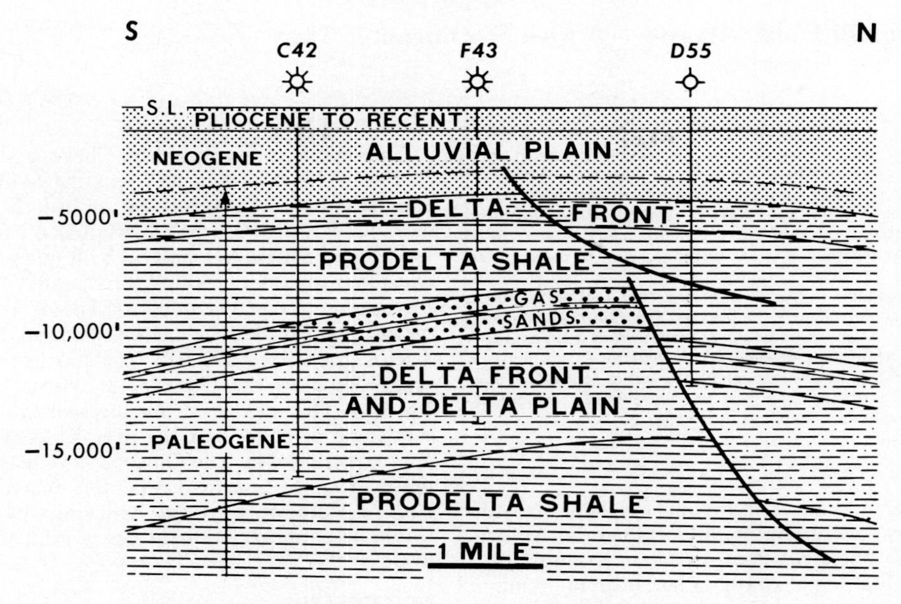

FIG. 30—South-north structural cross section through Taglu field.

greater depth of burial. The net amount of normally compacted, organic-rich shale and mudstone interbedded with the Paleogene reservoirs is quite large—approximately 1,400 ft (425 m). With conservative assumptions of 1 percent organic matter and a yield factor of less than 100 Mcf of gas per acre-foot of source rock, an area of less than 10 mi by 5 mi (16×8 km) could account for Taglu's long gas columns and large reserves.

Taglu gas consists of about 94 percent methane, 4 percent ethane, 1 percent propane, and 1 percent butane and heavier hydrocarbons. The carbon dioxide content is about 0.2 percent. No hydrogen sulfide has been detected. The gas density is 0.6, compared to air at standard conditions. Between 5 and 25 bbl of 47° API condensate has been recovered per million cubic feet of gas.

An unusual property of the reservoir sequence in the Taglu area is that the formation water is predominantly sodium bicarbonate. An approximate average composition of the brackish reservoir water is as follows:

Sodium	2,600 mg/l
Calcium	11 mg/l
Chloride	1,550 mg/l
Bicarbonate	4,170 mg/l
Total solids	8,400 mg/l
R_w at 70°F (21°C)	1.1 ohm-meters

SUMMARY

The Taglu field combines thick, good-quality reservoir sandstones with abundant suitable source rocks in an ideal structure. This combination has resulted in thick pay zones and long gas columns filling a large proportion of the available reservoir. A calculation based on the available seismic and well data, using somewhat conservative water saturations and a recovery of 85 percent of the gas in place, yields probable reserves in excess of 3 Tcf.

Mitsue Oil Field, Alberta—A Rich Stratigraphic Trap[1]

HAL H. CHRISTIE[2]

Abstract The Mitsue field, in north-central Alberta, produces large quantities of oil and minor gas from the Middle Devonian Gilwood Sandstone Member of the Watt Mountain Formation. This sandstone is generally clean, well sorted, fine to medium grained, and quartzose. The average oil column at the field is 4 m, and porosity averages 15–20 percent.

In retrospect, it seems surprising that for more than a decade explorationists condemned the Gilwood as being either "tight" or water bearing. The Gilwood is an excellent example of the type of widespread sandstone units common in sedimentary basins throughout the world. Many such units have not been explored fully and undoubtedly contain petroleum reservoirs awaiting discovery.

Introduction

Mitsue oil field is approximately 240 km north-northwest of Edmonton, in north-central Alberta, Canada. The field is an excellent example of a typical lithologic stratigraphic trap in a major sedimentary basin. The area of study of this paper includes the immediate area around the Mitsue oil field (Fig. 1). Production is from the Gilwood Sandstone Member of the late Middle Devonian Watt Mountain Formation.

The field trends north-northwest to south-southeast and is about 55 km long and ranges in width from a few kilometers to approximately 13 km. On trend in a northwest direction about 30 km away are the Nipisi and Utikuma oil fields. The Nipisi field also produces oil from the Gilwood and from an older Devonian sandstone, the "Granite Wash," which overlies the Precambrian; the Utikuma field produces oil from the "Granite Wash."

The accumulation of hydrocarbons in the Gilwood Sandstone Member at Mitsue field is due to an almost ideal example of a lithologic stratigraphic trap. Because of the nature of the sandstone and the regional geologic setting, the history of Mitsue field is an excellent illustration of the principles of exploring for this type of stratigraphic trap.

Mitsue Oil Field

The Mitsue field consists of approximately 200 wells, most of which are either presently producing oil or have been completed as water-disposal and injection wells.

The field was discovered by Chevron Standard in 1964. The discovery well was the SOBC Calstan Hondo 2-1-71-4, located in Lsd. 2, Sec. 1, T71, R4W5. The estimated original oil in place for Mitsue field is 671 million STB (87×10^6 metric tons) or 611 bbl/acre-ft. Estimated primary recoverable oil is 134 million STB (17.3 million metric tons) or 122 bbl/acre-ft (Century, 1966). The oil gravity is 42.9° API. The approximate average depth of the Gilwood at Mitsue field is 1,800 m. The maximum thickness of oil-column sandstone is 11 m, and the average thickness is approximately 4 m. The Gilwood is water bearing updip from Mitsue field, and this fact discouraged explorationists from believing that there would be commercial quantities of oil in the Gilwood in this area.

Stratigraphy

The Watt Mountain Formation typically consists of 6–24 m of gray-green dolomitic shale containing minor amounts of siltstone and thin limestone breccias (Fig. 2). Near the Peace River uplands, the Watt Mountain includes fine- to medium-grained sandstone interbeds and lenses named the "Gilwood Sandstone Member" (Guthrie, 1956).

The Gilwood Sandstone Member of the Watt Mountain Formation is readily mappable over most of the area, because it has distinct lithologic contacts with the overlying and underlying units.

Geologic Physiography

In the area of the Mitsue field, the Gilwood Member is present on the east side of the Peace River arch (or uplands), which was a terrestrial upland during the time of Gilwood deposition (Fig. 3). Surrounding the Peace River uplands on the north, south, and east is the Western Canada sedimentary basin. Northeast of the study area is the Athabasca high, which may have been the

[1]Manuscript received, November 25, 1974. Modified slightly from a paper presented in Proceedings of the 8th World Petroleum Congress, 1971, v. 2, p. 269-274. Published with permission of World Petroleum Congresses.

[2]Christie and Company, Calgary, Alberta T2S 1W7.

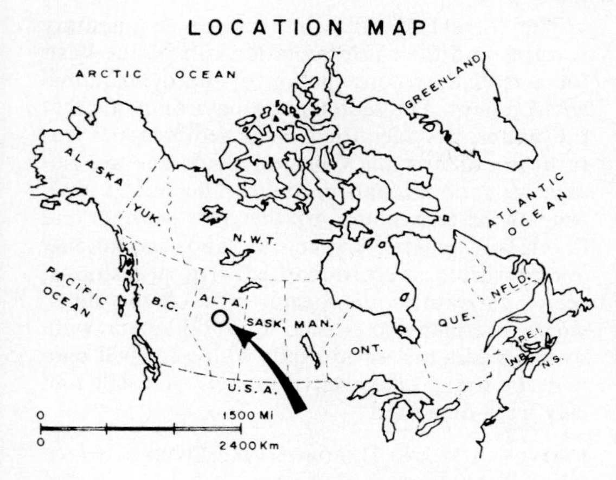

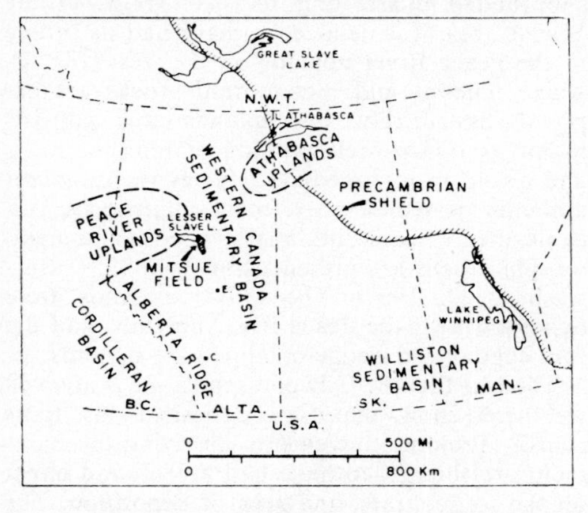

FIG. 1—Location map and geologic setting of Mitsue field, Alberta.

source of some of the clastic sediments that constitute the Gilwood. However, the main source area of Gilwood sandstone is believed to have been the Peace River uplands; this conclusion is based primarily on the degree of weathering, sorting, grain size, sand type, and sand-dispersal patterns.

In the study area, the Gilwood sandstone occupies a wide belt south and west of the Mitsue field. The sandstone belt narrows around the eastern nose of the Peace River uplands and then widens abruptly. At the north of the uplands, the edge of the sandstone is irregular.

Gilwood Sandstone Characteristics

Quality and Porosity

The Gilwood sandstone is mainly feldspathic but includes subsidiary arkose. In the Nipisi area, north of Mitsue field, arkoses are dominant.

A study by Kramers and Lerbekmo (1967) showed that the cementing materials in the Gilwood of the Mitsue area are of three principal types: (1) anhydrite, (2) carbonate, and (3) silica. It is concluded that the anhydrite appears to have replaced both carbonate cement and matrix (consisting of clay, mica, feldspar, and quartz), and probably was the last cement to be precipitated.

Porosity in the Gilwood is variable and may be as much as 30 percent where conditions are favorable. Although a comprehensive study of the porosity throughout the field and surrounding area has not been undertaken, where the sandstone is well developed and has no excessive infilling of cementing material, average porosities of 15–20 percent are found. The lower end of this scale is more common.

Permeabilities change abruptly as a result of difference in cementing materials and interrela-

EUROPEAN	AMERICAN		PERIOD	MITSUE AREA (THIS PAPER)	NORTHERN ALBERTA (LAW, 1955)		WILLISTON BASIN (SASK.)
FRASNIAN	FINGERLAKESIAN		UPPER DEVONIAN	SLAVE POINT FM. / FT. VERMILION	SLAVE POINT FM.		SOURIS RIVER FM.
GIVETIAN	ERIAN	TAGHANIC TIOUGHNIOGAN	MIDDLE DEVONIAN	WATT MTN FM. / GILWOOD	WATT MOUNTAIN FM.		FIRST RED BEDS
				MUSKEG FM.	MUSKEG FM.	PRESQU'ILE FM.	DAWSON BAY FM.
							SECOND RED BEDS
							PRAIRIE FM. (EVAPORITE)

FIG. 2—Stratigraphic relations of Middle and Upper Devonian of Mitsue area and equivalents elsewhere. (Right column is from McCrossan and Glaister, 1964.)

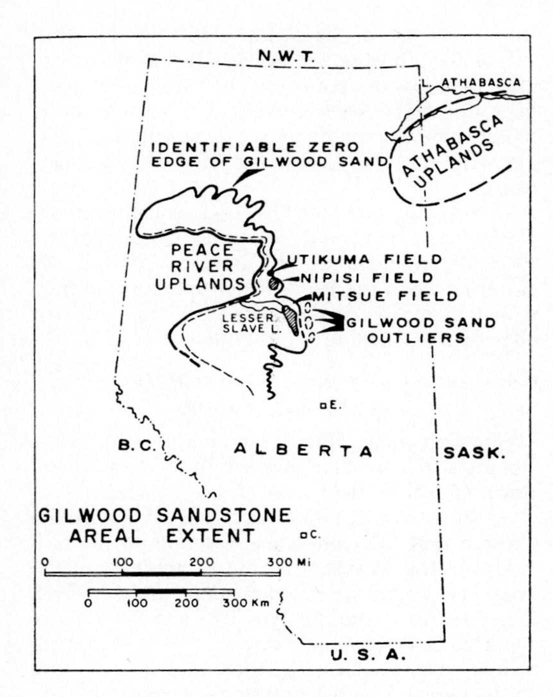

FIG. 3—Areal extent of Gilwood sandstone.

tions with argillaceous material. The Gilwood can be subdivided into two mappable units over part of the area included within the Mitsue field; the upper unit has an average permeability of approximately 140 md, and the lower unit has an average permeability of 250 md.

Texture

The Gilwood is a unimodal or, in places, bimodal sandstone which consists of grains that are classed as subrounded to well rounded. The shape of the sand grains commonly is nearly spherical to very slightly elongate. The feldspar grains are subrounded and generally partly decomposed, resulting in poorly formed clay particles.

Sedimentary Features

The more common sedimentary features which have been recognized in the Gilwood sandstone are: (1) crossbedding, (2) graded bedding, (3) ripple-drift cross-laminations, (4) convolute laminations, (5) rip-up clasts, and (6) organic reworking (Kramers and Lerbekmo, 1967; Shawa, 1969).

The correct identification of the sedimentary features and their interpretation can be the basis for a realistic reconstruction of the depositional environment. The sedimentary environments that the writer has identified from sedimentary features in cores of the Gilwood sandstone are fluvial, transitional, and prodelta. The fluvial environment is represented by the river channel and floodplain, where very coarse sands, commonly poorly sorted, were deposited. The transitional realm constitutes the beach part of the delta, where sediments consisted of moderately well-sorted medium-grained sands with a low silt content. In the prodelta environment, fine silt and clay were deposited.

Provenance and Depositional Environment

Because the Gilwood sandstone consists of subrounded quartz and feldspar grains in the Mitsue area, it is deduced to have had its source in the Peace River uplands on the west (Fig. 4), where igneous and metamorphic rocks are exposed. Specifically, the uplands area consists mainly of quartzo-feldspathic metamorphic rocks and acidic to intermediate igneous rocks; minor amounts of sedimentary, metasedimentary, volcanic, metavolcanic, and mafic igneous and metamorphic rocks are present (Burwash, 1957; Burwash et al, 1964). The Gilwood sands were deposited near the Peace River uplands, and the thin edge of the wedge onlapped the uplands.

The fact that the Gilwood sandstone is not well weathered shows that it was deposited close to its source. Tectonic movements involving the basement are thought to have had a profound effect on the source, rate, and area of deposition. The sediments are believed to have traveled in an

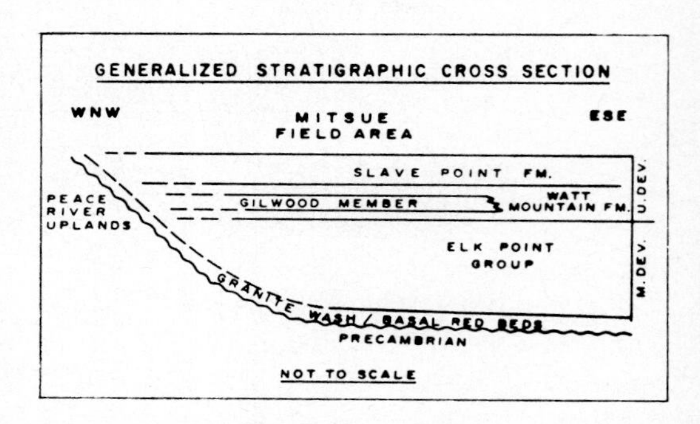

FIG. 4—Generalized stratigraphic cross section, Mitsue field.

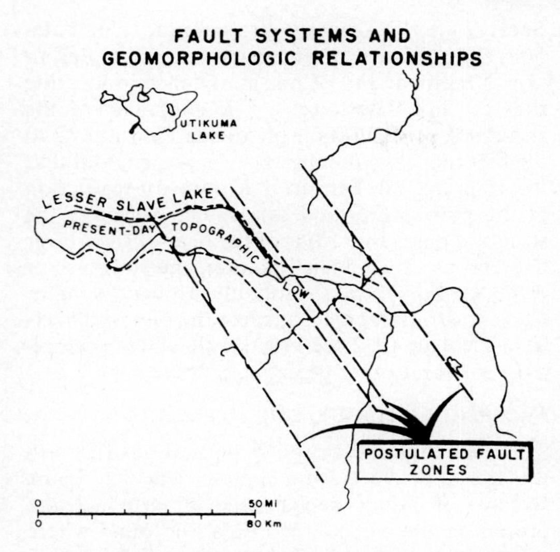

FIG. 5—Fault systems and geomorphologic relations, Mitsue field area.

easterly direction across the Lesser Slave Lake area, and, near the southeast extremity of the lake where the Gilwood Member formed, the sediments formed a birdfoot-delta dispersal pattern. The Gilwood here is a typical deltaic deposit having characteristics similar to those of other ancient and modern deltas of the world.

The coincidence in position and direction of the main Middle Devonian fluvial channel with the present-day topographic low occupied by Lesser Slave Lake is not due to chance but to the reactivation of tectonic movements along persistent trends.

Faulting (Fig. 5) is known to have been common during the Middle Devonian, and some of the faults affected the source area as well as the location of sand deposition. The dispersal pattern as interpreted by Kramers and Lerbekmo (1967) and the superposition of faults or fault zones inferred by the writer from the present-day geomorphology have been superimposed in Figure 5.

Historical Comments and Exploration Techniques

Of special significance to the explorationist is the historical background that led to the eventual discovery of the Mitsue oil field in an area that previous exploration had appeared to condemn. Probably the main reason the true potential of the Gilwood was overlooked for more than a decade was that geologists thought the prime area for exploration was near the updip sandstone limit. There, water had been tested from a porous sandstone section, and the area between this water test and the updip edge was considered to be too small to warrant further exploration. The error that most explorationists made was to assume that the porous and wet sandstone near the updip edge was the extremity of a blanket sandstone present on the west and south. Although well control was sparse (Fig. 6), enough control was available to reveal that the Gilwood is not a blanket sandstone but is a complex of bars, channels, and fluvial deposits. Moreover, although no commercial hydrocarbons of any significance had been found in the Gilwood prior to the discovery of the Mitsue field, the fact that several marginal oil wells as well as gas shows were present downdip to the southwest attested to the presence of hydrocarbons. The presence of these hydrocarbons downdip provided the significant clue that exploration should be concentrated between these shows and a "tight" facies, as interpreted from logs, that was present downdip from the water-bearing wells near the zero edge of sandstone development.

The examination of cores and their environmental interpretation can be the most effective means for determining the environmental condi-

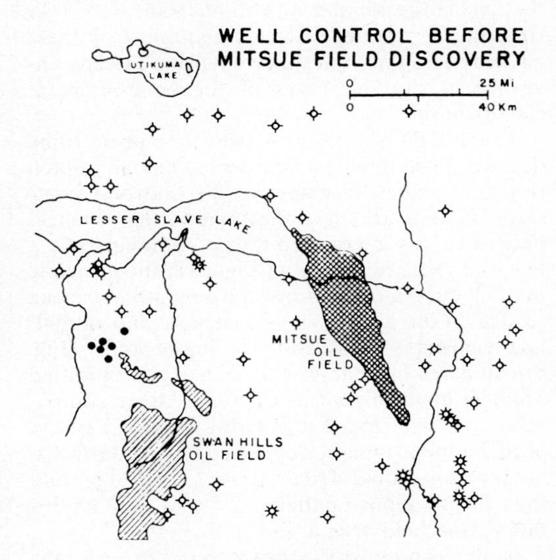

FIG. 6—Well control before Mitsue field discovery.

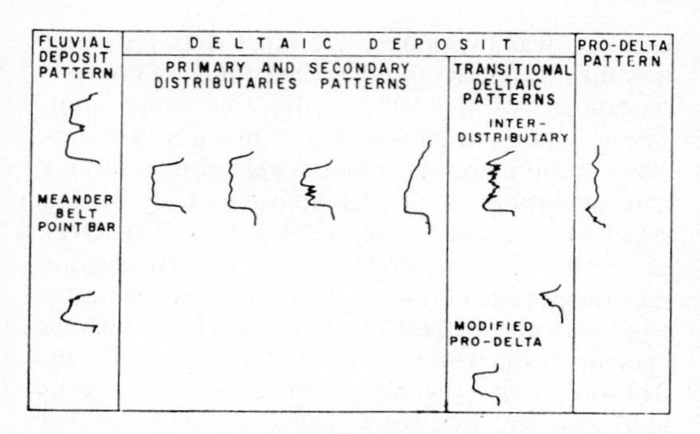

FIG. 7—Flow diagram of SP-log shapes. From Saitta-B. and Visher (1968).

tions that existed during a period of sand deposition. However, in wildcat areas such as the Mitsue area prior to the field discovery, scarcity of core information is common and the explorationist is forced to rely on other available data.

Several authors have documented the use of SP curves as a useful tool in environmental studies (e.g. Saitta-B. and Visher, 1968; see Fig. 7). Saitta-B. and Visher stated that the following characteristics of the SP curve were found to be useful: (1) whether the upper contact of the sandstone with shale is abrupt, transitional, or transitional serrated; (2) whether the central part of the curve is smooth or serrated; and (3) whether the base of the sandstone section is abrupt, transitional, or transitional serrated. The combination of these patterns and their constant repetition in each environment are the bases of the environmental classification.

The well density in the Mitsue area prior to the discovery was low, but the writer has attempted to relate several SP curves of the Gilwood sandstone of wells drilled at that time to the dispersal pattern as it is interpreted today. Although only a few well logs are shown on Figure 8, the principle that SP curves of the Gilwood sandstone can be related to the known fluvial, deltaic, and prodeltaic deposits is significant. The log of well 1 (Fig. 8) compares favorably with that of the modified prodelta environment in Figure 7. Other comparable logs are: log 2 and Saitta-B. and Visher's prodelta environment, log 3 and their interdistributary transitional deltaic pattern, and log 4 and their fluvial deposit pattern. A typical well log for the Mitsue field area is shown in Figure 9.

In the exploratory process that preceded the discovery of Mitsue field, the lack of a positive seismic anomaly almost condemned the prospect. Because this is a purely stratigraphic accumulation, the presence of any reliable seismic definition is minimal and of no significance in locating sites for initial test wells. The simplicity of the structural configuration of Mitsue field attests to the fact that no structural nose is present and that the trapping mechanism is a porosity restriction of the porous Gilwood sandstone. Although the structure map (Fig. 10) is of the base of the Lower Cretaceous "Fish Scale" marker, the structure of the Devonian strata probably is very similar. Clues to stratigraphic traps like the one which exists at Mitsue field are very subtle where preliminary well and other geologic data are scarce.

PROSPECTING FOR DELTAIC DEPOSITS

Some of the most prolific oil and gas fields in the world produce from deltaic deposits. Thick sections of deltaic deposits like those which are present in the Mitsue field area are found where coastal sediments subsided rapidly and the supply of sediment was sufficient to keep the subsiding area filled.

Deltaic deposits have all the primary requirements for oil and gas accumulations: (1) porous

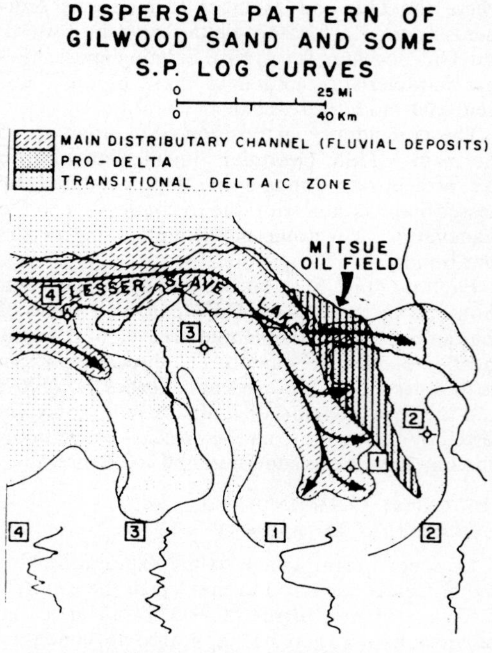

DISPERSAL PATTERN OF GILWOOD SAND AND SOME S.P. LOG CURVES

0 25 Mi
0 40 Km

MAIN DISTRIBUTARY CHANNEL (FLUVIAL DEPOSITS)
PRO-DELTA
TRANSITIONAL DELTAIC ZONE

MITSUE OIL FIELD
LESSER SLAVE LAKE

FIG. 8—Dispersal pattern of Gilwood Sandstone Member and some SP-log curves.

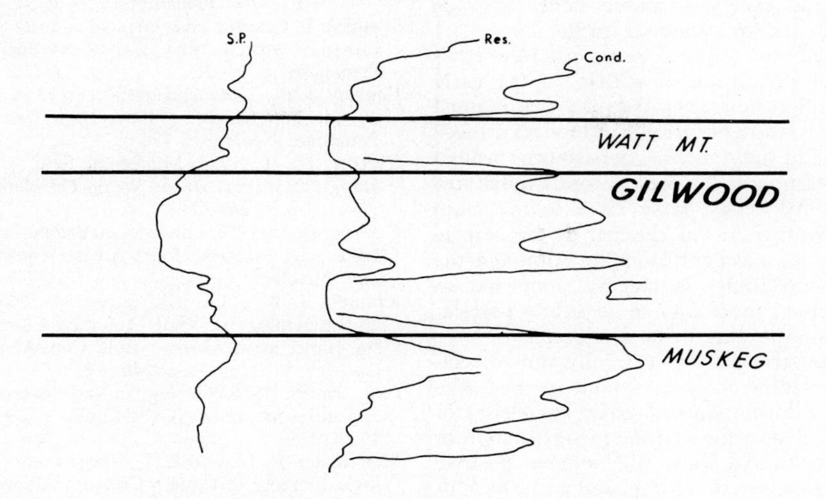

FIG. 9—Typical induction electrical log, Mitsue field area.

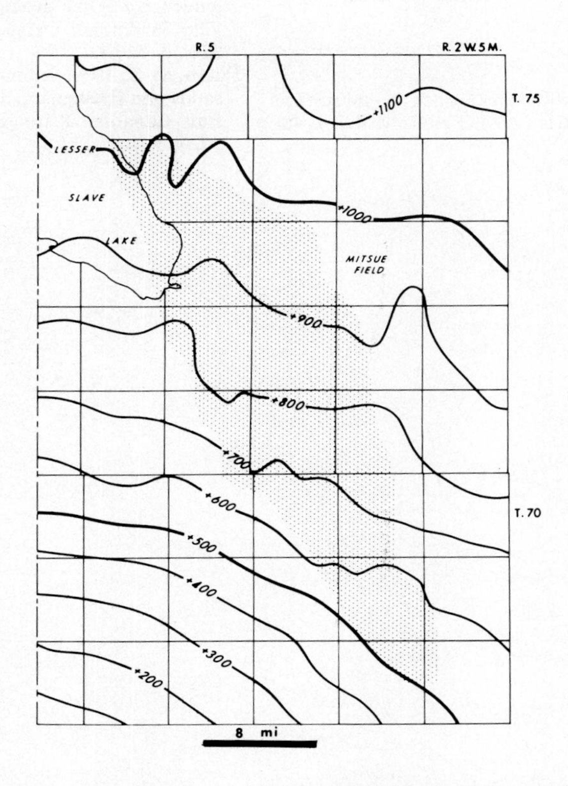

FIG. 10—Structural contour map of base of "Fish Scales" marker, Mitsue field, Alberta.

reservoirs, (2) excellent source beds, and (3) stratigraphic and/or structural traps.

A geologist attempting to reconstruct the environments of deposition of a field in the early stages of exploration when data are scarce must be constantly aware of tectonic influences on sedimentation and their possible persistence through time. The relation of the present topographic low represented by Lesser Slave Lake to the main Middle Devonian fluvial channel in the Mitsue field area is an excellent example. Although the study of neotectonics is lagging somewhat in North America, more and more subtle correlations of this nature are being disclosed.

The astute geologist who carefully and correctly identifies sedimentary environments, and who applies this information to a basic knowledge of sedimentary deposition and the reconstruction of environmental conditions, will achieve positive results in the exploration for oil and gas. The Mitsue field provides an excellent example of how basic exploratory techniques can be used to lead to the discovery of a very prolific oil field.

References Cited

Burwash, R. A., 1957, Reconnaissance of subsurface Precambrian of Alberta: AAPG Bull., v. 41, p. 70-103.

——— et al, 1964, Precambrian, *in* R. G. McCrossan and R. P. Glaister, eds., Geological history of western Canada: Alberta (now Canadian) Soc. Petroleum Geologists, p. 14-19.

Century, J. R., 1966, Mitsue field, p. 73 *in* Oil fields of Alberta supplement: Alberta (now Canadian) Soc. Petroleum Geologists, 136 p.

Christie, H. H., 1971, Mitsue oil field: a rich stratigraphic trap (Part I): 8th World Petroleum Congress Proc., v. 2, p. 269-274.

Guthrie, D. S., 1956, Gilwood sandstone in the Giroux Lake area, Alberta: Jour. Alberta (now Canadian) Soc. Petroleum Geologists, v. 4, p. 227-231.

Kramers, J. W., and J. E. Lerbekmo, 1967, Petrology and mineralogy of Watt Mountain Formation, Mitsue-Nipisi area, Alberta: Bull. Canadian Petroleum Geology, v. 15, no. 3, p. 346-378.

Law, James, 1955, Geology of northwestern Alberta and adjacent areas: AAPG Bull., v. 39, no. 10, p. 1927-1975.

McCrossan, R. G., and R. P. Glaister, eds., 1964, Geological history of western Canada: Alberta (now Canadian) Soc. Petroleum Geologists, 232 p.

Saitta-B., S., and G. S. Visher, 1968, Subsurface study of the southern portion of the Bluejacket delta, *in* A guidebook to the geology of the Bluejacket-Bartlesville Sandstone, Oklahoma: Oklahoma City Geol. Soc., p. 52-68.

Shawa, M. S., 1969, Sedimentary history of the Gilwood sandstone (Devonian), Utikuma Lake area, Alberta: Bull. Canadian Petroleum Geology, v. 17, no. 4, p. 392-409.

Kaybob Oil Field, Alberta, Canada[1]

N. H. SCHULTHEIS[2]

Abstract The Kaybob Beaverhill Lake oil field is located in the Swan Hills area of Alberta, approximately 150 mi (241 km) northwest of Edmonton. Production is from the upper Swan Hills Formation of the Beaverhill Lake Group, at an approximate depth of 9,800 ft (2,987 m). The reservoir is a limestone reef complex of Middle Devonian to early Late Devonian age. Maximum reef thickness is 257 ft (78.3 m). The reef complex is divisible into four major facies: organic-reef, backreef, forereef, and offreef facies. The thickest "pay" sections and best reservoir characteristics occur in the organic-reef facies. Original oil in place is estimated to have been 300 million bbl. There is no gas cap or water table in the pool.

INTRODUCTION

The Kaybob oil field of northwestern Alberta is located in T63–65, R19, west of the Fifth Meridian (Fig. 1). The discovery well, Phillips Kaybob 7–22–64–19 W5M, was completed in June 1957 for an initial potential flow of 1,560 BOPD through a 32/64-in. choke. The well was one of three discoveries made during 1957 in Devonian oil-bearing reefs in the sparsely explored Swan Hills area. It was drilled by Phillips Petroleum on farmout lands obtained from Chevron Standard and Gulf Oil. The well was located on a questionable seismic reflection anomaly.

In the general Swan Hills area, oil production is obtained both from the upper Swan Hills reef member and the lower Swan Hills platform member (Fig. 2). Only the upper reef member is productive at Kaybob. This study is limited to a discussion of the productive upper reef complex.

Previous Work

The first detailed facies analysis of a Swan Hills reef complex was made by Edie (1961). Since then, the great economic importance of these reefs has provided the incentive for several detailed petrographic, paleoecologic, and facies-analysis studies. In recent years, major studies on various Swan Hills reefs have been published by Fischbuch (1960, 1962, 1968), Murray (1966), Jenik and Lerbekmo (1968), Leavitt (1968), and Hemphill et al (1970).

GEOLOGIC SETTING AND NOMENCLATURE

The Kaybob reef complex is elongate in shape and measures about 10 by 3 mi (16 by 4.8 km). The reef axis trends approximately north-south, and the thickest part of the reef occurs near the south end of the field. The buildup thins toward the north. Regional dip of the Beaverhill Lake Group in the Kaybob area is approximately 50 ft/mi (9.4 m/km) in a S60°W direction.

The reef developed on the western margin of a broad and extensive organic carbonate platform (Fig. 2) on which the facies changes north and west into argillaceous limestones and shales of the Waterways Formation. The Waterways Formation entirely encloses the isolated reef and forms the seal (Fig. 3).

In the Swan Hills area, the Beaverhill Lake Group consists, from oldest to youngest, of the Fort Vermillion Formation, the Swan Hills Formation, and the Waterways Formation. The Swan Hills Formation can be subdivided into a lower "Dark Brown" platform member and an upper "Light Brown" reef member (Fong, 1960). Only the upper reef member is productive at Kaybob (Fig. 4). For a more detailed discussion of the geologic history of this area, see Hemphill et al (1970).

Reserves

There are 16 known hydrocarbon-bearing Beaverhill Lake reservoirs in the Swan Hills area (Fig. 2). Total oil in place as of December 1972 was approximately 6 billion bbl and total gas in place was 8.4 Tcf (Energy Resources Conservation Board, 1973).

Estimated original oil in place at Kaybob is 300 million bbl. Estimated recoverable oil, both primary and secondary, is 126 million bbl. The productive area of the field is approximately 18,000 acres. A total of 93 oil wells has been drilled in the field on 160-acre spacing. The field has been unitized and is operated by Chevron Stan-

1Manuscript received, November 7, 1974.

2Chevron Standard Limited, Calgary, Alberta T2P 0L7.

The writer is grateful to Chevron Standard Limited for permission to publish this paper. The writer owes much to the work of many present and former Chevron employees. Special thanks are extended to D. A. Pounder and E. W. Mountjoy, who offered valuable suggestions and critically read the manuscript.

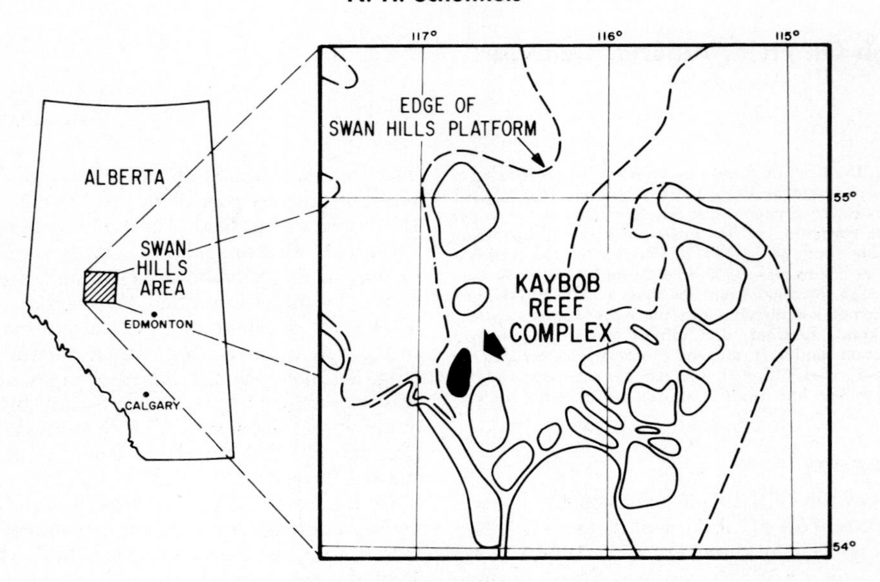

FIG. 1—Location map, Kaybob oil field, Alberta.

dard Limited. Cumulative production to August 1974 was 52.5 million bbl. Net "pay" thicknesses are varied, reaching a maximum of 115 ft (68.6 m) and averaging 56 ft (17 m). Average calculated weighted porosity for the pool is 7.4 percent and average permeability is 24 md. The oil is sweet and has a 42.5° API gravity.

Reservoir Performance and Secondary Recovery

The reservoir has no water table or natural water drive. The high-gravity oil was originally undersaturated by about 1,500 psi. The original reservoir pressure at a subsea elevation of 7,240 ft (2,206.7 m) was 4,630 psi.

All completed wells have penetrated the entire porous reef section. About 60 percent of the wells are cased through the "pay" zone; the rest are open-hole completions. About 30 percent of the wells were stimulated by acid treatment.

Primary recovery was estimated to be 16 percent of the original oil in place. This prediction was based on the expansion of the reservoir rock and fluids to the saturation pressure, followed by solution-gas drive. Waterflooding was found to be the most suitable method of secondary recovery. It is estimated that the ultimate recovery efficiency will be equal to 42 percent of the oil in place. Pressure maintenance was begun in 1964 with the initiation of a line-drive waterflood. Fresh water for the flood is obtained from nearby Iosegun Lake.

Reef Terminology

Terms used in this study are defined as follows:

Reef—"A mound or lens of skeletal material formed by, and derived from, organisms that were capable of constructing wave-resistant structures. A reef has significantly greater vertical extent than that of the contemporaneous flanking deposits" (Fischbuch, 1968).

Reef complex—"An aggregate of reef limestones and genetically related carbonate rocks" (Henson, 1950).

Organic reef—"That portion of the reef which is or was built directly by organisms, and is responsible for the reef's wave-resistant character" (Klovan, 1964).

Geometry and Reef Morphology

The well control in the Kaybob field is sufficient to give a fairly detailed picture of the reef geometry and paleotopography. Maximum relief of the reef above the surrounding platform is 257 ft (78.3 m). The most abrupt decrease of reef thickness at the reef flank is 210 ft (64 m) over a distance of approximately 0.5 mi (0.8 km; Fig. 5). This decrease indicates a gentle slope averaging 4–6°. No steeply dipping reef contacts, forereef contacts, or forereef megabreccias have been observed in cores. Because of the regional dip, the structurally highest part of the reef is in the north (Fig. 6). The gentle depositional dip of the forereef facies was controlled partly by the shallow depth of water in front of the growing reef. It is concluded that the relief of the reef above the seafloor during reef formation was much less than its present maximum relief of 257 ft (78.3 m). Individual reef stages rarely exceed a thickness of 50 ft (15.2 m). It seems likely, therefore, that maximum relief at any one time between the reef and the offreef and interreef area was never more than 40–50 ft (12.2–15.2 m) in the immediate Kaybob reef area.

In other Swan Hills reefs, the presence of considerable relief has been interpreted between the reef complex and the offreef seafloor (Murray, 1966; Leavitt, 1968).

Reef growth by stages is apparent in the Kaybob reef complex; a platform stage and five informal reef stages (stages II–VI, Fig. 4) are recognized. Criteria for recognizing stages in the reef are: (1) shale breaks which are widespread in the backreef area and are easily recognized and correlated on gamma ray logs (see Fig. 9); (2) backstepping of the reef stages resulting in reef "terraces"; (3) overstepping of reef stages onto forereef and interreef sediments; and (4) recognizable disconformities in core samples.

Backstepping occurred on the windward (east) side of the reef, and overstepping on the leeward side. The overstepping onto interfingering basinal and forereef sediments further suggests that offreef sedimentation, especially in proximity to the reef, did not lag far behind reef growth, inasmuch as a suitable hard substrate was required for growth by the reef-building organisms.

The shape of the reef probably was controlled by wind, wave, and current action; postdepositional erosion did not alter the final reef configuration significantly. A general absence of fine bioclastic forereef sediments on the east side of the reef suggests that prevailing winds and currents were from the northeast (Fig. 7).

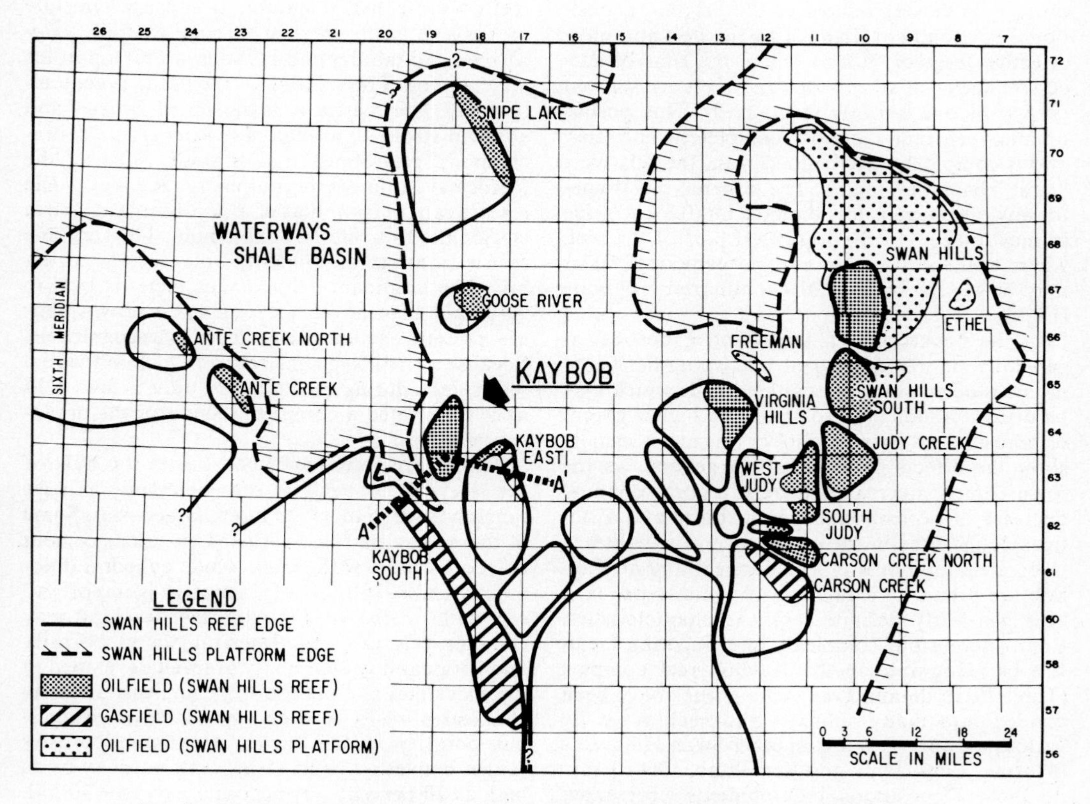

FIG. 2—Distribution of hydrocarbon accumulations in Devonian Swan Hills Formation.

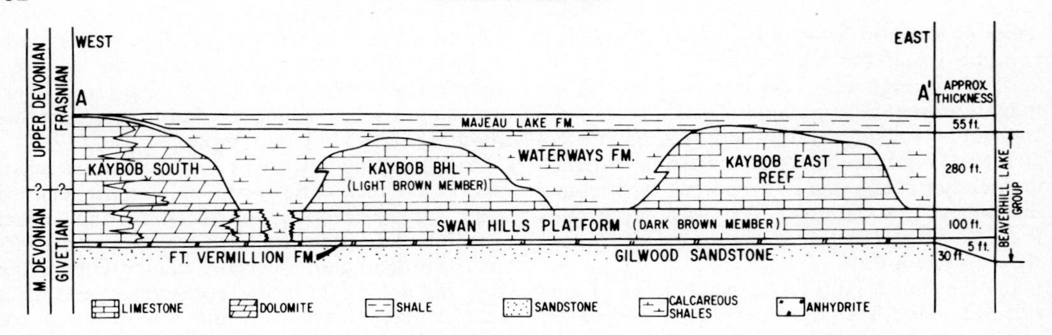

FIG. 3—Diagrammatic stratigraphic cross section illustrating relations of Kaybob Swan Hills reef complexes and enclosing Waterways Formation.

Facies and Porosity Relation

Porosity and permeability in the Kaybob Swan Hills reservoir are largely dependent on the number and types of colonial fossils present. Thus, porosity is closely related to the original depositional environment. In general, the best and most effective reservoir occurs where the rims of successive stages of stromatoporoid reefs are stacked on top of one another (Figs. 8, 9). This porous organic-reef facies almost completely encircles, and is complexly interbedded with, the relatively "tight" backreef facies. The organic-reef facies for any one stage is less than 0.5 mi (0.8 km) wide in most places. In the southwest part of the reef, where well control is poor, the organic-reef facies may be absent or only a few hundred feet wide (Fig. 7).

In the backreef area the effective porosity is generally poorly developed and sporadic. Commonly it occurs in thin, scattered beds which may be discontinuous and isolated. The higher grades of porosity and permeability in this facies usually show the effects of leaching. In general, as the fossil content decreases, there is a corresponding decrease in porosity and permeability and a noticeable increase in the occurrence of stylolites.

Because the porosity and permeability at Kaybob are primarily a function of facies type, each type has fairly definite reservoir characteristics. Four discrete but complexly interfingering facies can be recognized in the Kaybob reef complex (Table 1). In detail, these facies groups have been divided into many subunits and rock types by various authors working on other Swan Hills reefs (Murray, 1966; Jenik and Lerbekmo, 1968; Leavitt, 1968). Depositional environments represented by these facies range from open marine to restricted quiet-water lagoon.

Organic-Reef Facies

The organic-reef facies was responsible for the reef's wave-resistant nature. It consists typically of massive hemispherical stromatoporoids. *Stachyodes* and tabular and bulbous stromatoporoids are common. Preservation of the fauna is generally good. The matrix is made up of broken and abraded stromatoporoids, *Amphipora*, corals, brachiopods, gastropods, and crinoids. Some of the fossils have thin coatings of encrusting algae. The texture ranges from that of a poorly sorted coarse calcarenite to a fine carbonate mud. The very fine matrix consists of unidentifiable carbonate debris. The dominant color of this facies is buff to very light brown, but a few gray to brown beds are present. Bedding is usually unrecognizable. Because of shifting and retreating of individual reef stages during reef growth, only a few wells have penetrated a complete section of the organic-reef facies.

The organic-reef facies constitutes the bulk of the reservoir. Porosity is present mainly as vugs ranging from pinpoint to much larger sizes. Some of these vugs are filled with sparry calcite cement or, more rarely, with small, white, euhedral dolomite rhombs. Intraorganic porosity is important, especially in the stromatoporoids; leaching may enhance this porosity. Interfragmental porosity and permeability commonly are well developed in the biocalcarenite matrix. Core analysis and mechanical porosity logs commonly show continuous porosity in this facies. Maximum porosity ranges between 12 and 15 percent, but may be as high as 20 percent. Permeability averages 30 md.

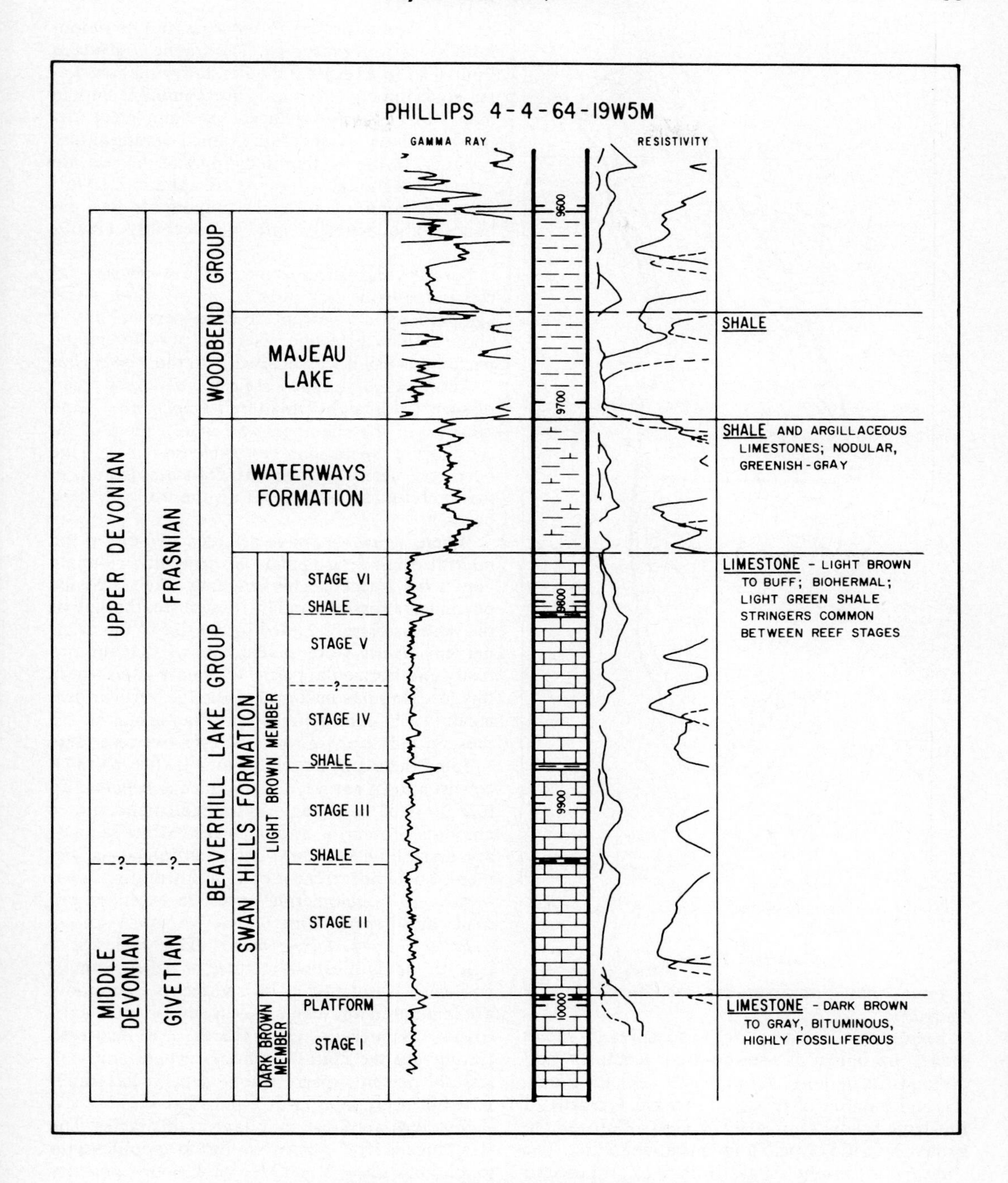

FIG. 4—Stratigraphic nomenclature, Kaybob reef complex, Alberta.

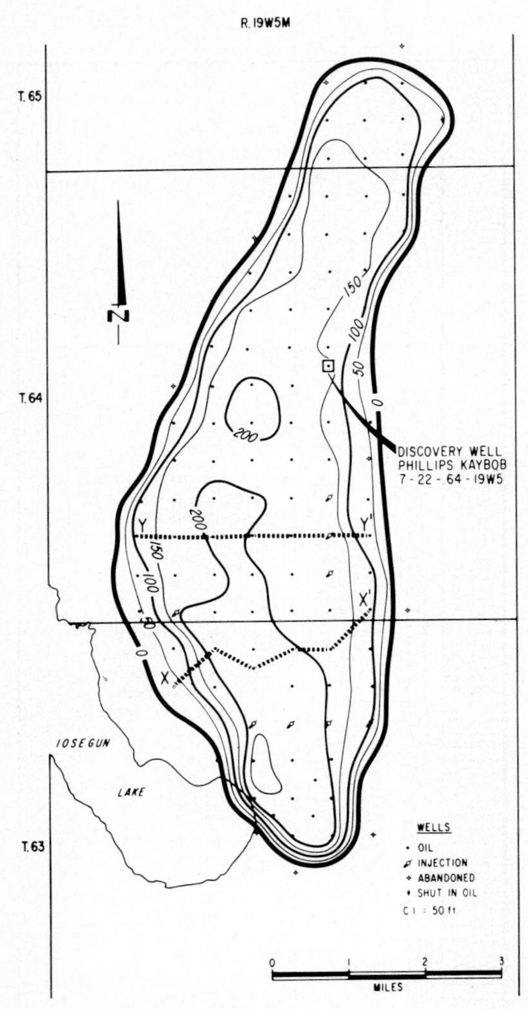

FIG. 5—Isopach map of reef complex, Kaybob field.

Backreef Facies

Four main subfacies are recognized in the backreef.

Amphipora facies—This widespread facies makes up most of the backreef complex and abounds in detailed variations of texture, porosity, and number of fossils. It consists typically of flat-lying branches of *Amphipora* in a buff to light brown, micritic or pelletoid limestone matrix. The *Amphipora* content ranges from 5 to 50 percent; few intervals are without some trace of *Amphipora*. The state of preservation varies, but commonly they are well preserved. The fragile *Amphipora* flourished in a restricted, quiet, fairly shallow-water environment. Stylolites are common, and in many places have reduced or completely destroyed much of the fine-grained intergranular porosity. In the northernmost part of the reef, no extensive lagoonal sediments are apparent. In this area, leaching of skeletal components has enhanced the porosity and permeability significantly.

Porosity and permeability in the *Amphipora* facies range from very poor to fair or good. Average porosity is 5 percent; average permeability is about 8 md. Effective reservoir porosity in the backreef is usually associated with the *Amphipora*.

All stages of leached *Amphipora* porosity can be seen, from "tight" unaltered *Amphipora*, to fine porosity in the chambers and axial canals of recrystallized specimens, to skeletons that have been completely leached out. Oil staining and, in places, bitumen are present on the walls of these pores.

Where *Amphipora* have been leached out of the micritic matrix, irregular, horizontally elongate vugs have formed. This leaching rarely extends beyond the fossil boundary; therefore, the size of the pores is closely related to the size of the original *Amphipora*. Core analysis shows that the porosity and horizontal permeability are good where this leaching has occurred, although vertical permeability is commonly poor. Fracturing, where present, makes these rocks an effective reservoir.

Some beds and lenses up to 3 ft (0.9 m) thick consist almost entirely of entangled *Amphipora* up to 2 in. (5.08 cm) long. Crystalline calcite, minor amounts of micrite, and, in places, white secondary crystalline dolomite occur as interstitial cement. Local occurrences of this "intraformational *Amphipora* conglomerate" provide excellent porosity and permeability.

Pellet-microcrystalline facies—This facies is a light to medium brown mixture of pellets, lumps, and carbonate mud. The sediments commonly are laminated and contain very few or no macrofossils. Numerous calcispheres are scattered throughout the unit. Commonly, where only pellets are present, sparry calcite cement has eliminated primary interparticle porosity.

Stylolites are most abundant in this facies. The size range is from microstylolites to stylolites with amplitudes up to 3 in. (7.6 cm). Usually, porosity and permeability are completely destroyed. Fractures are commonly cemented with calcite or dolomite.

Fragmental-limestone facies—This facies is most common on the outer edge of the lagoon adjacent to the reef wall. Sorting is usually poor, and the bioclastic fragments range in size from very fine to coarse sand. The fragments are subangular to subrounded, and most can be recognized as brachiopod, crinoid, stromatoporoid, and *Amphipora* debris. Much of the rock has an appearance similar to that of the matrix in the reef-wall facies and represents a transition between the reef-wall facies and the fine-grained lagoonal facies. Intergranular porosity and permeability are good wherever they are present; in many places they have been obliterated by sparry calcite cement.

Green shale facies—This facies consists of light green to gray, slightly calcareous, laminated pyritic shale. The shales are common between reef stages in the backreef area. They are volumetrically insignificant and by themselves do not constitute permeability barriers. They range in thickness from thin, irregular stringers of less than 1/8 in. (3 mm) to beds up to 2 ft (0.6 m) thick. They are good correlation markers within the reef and are easily recognized on gamma ray logs. The shale beds are most widespread near the top of stages II and III in the backreef area (Fig. 9), and generally are absent above the organic-reef and forereef facies. The more prominent shale beds have been recognized in other Swan Hills reefs and are regionally correlatable. Some angular limestone clasts from underlying rocks are present at the base of the thicker shale beds. These clasts decrease in size and number toward the top.

The origin of the green shale units is unknown. They are interpreted to be the results of subaerial erosion, each shale marker being a reflection of an interruption in the slow, regular, regional subsidence of the Beaverhill Lake basin.

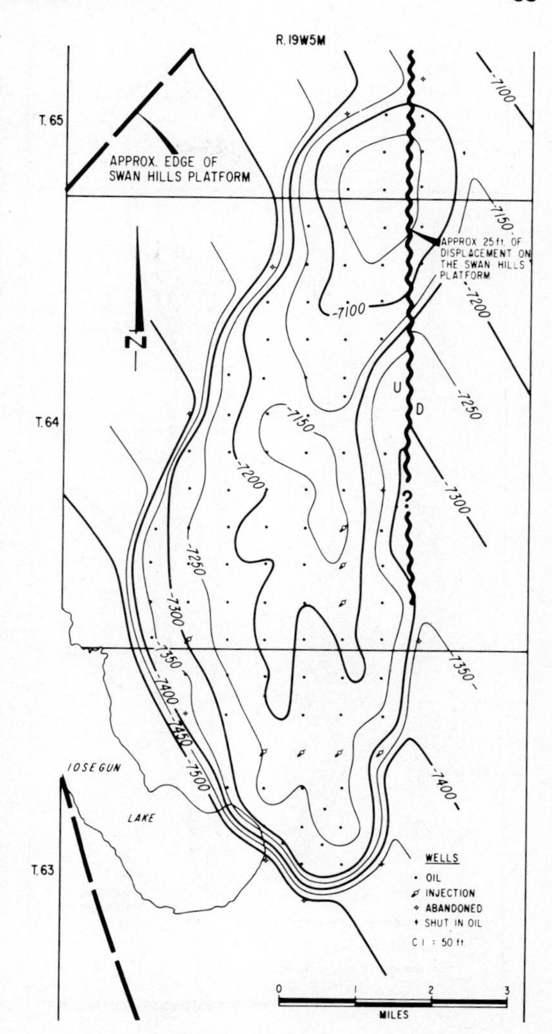

FIG. 6—Structural contours of Swan Hills reef, or of Swan Hills platform where reef is absent.

Forereef Facies

The forereef facies lies seaward of the organic reef and is characterized by fine- to coarse-grained, in places conglomeratic, gray to dark gray, argillaceous limestone. It is mainly organic-reef-derived rubble, and contains minor amounts of backreef rocks and organisms indigenous to the reef flanks. The facies interfingers with the organic-reef facies and the offreef and interreef Waterways Formation. The depositional slope of this talus averages 4–6°.

Between Kaybob and Kaybob East, the unit forms a narrow band on the windward, east side of the reef, where currents did not allow much accumulation of detritus. A broader band of skeletal detritus accumulated on the leeward side. Accumulation of this detritus is at least partly responsible for the slight east-to-west asymmetry of the reef. The most common fossils are crinoids and brachiopods. Gastropods, ostracods, oncolites, pellets, intraclasts, and broken and abraded fragments of stromatoporoids and *Amphipora* are

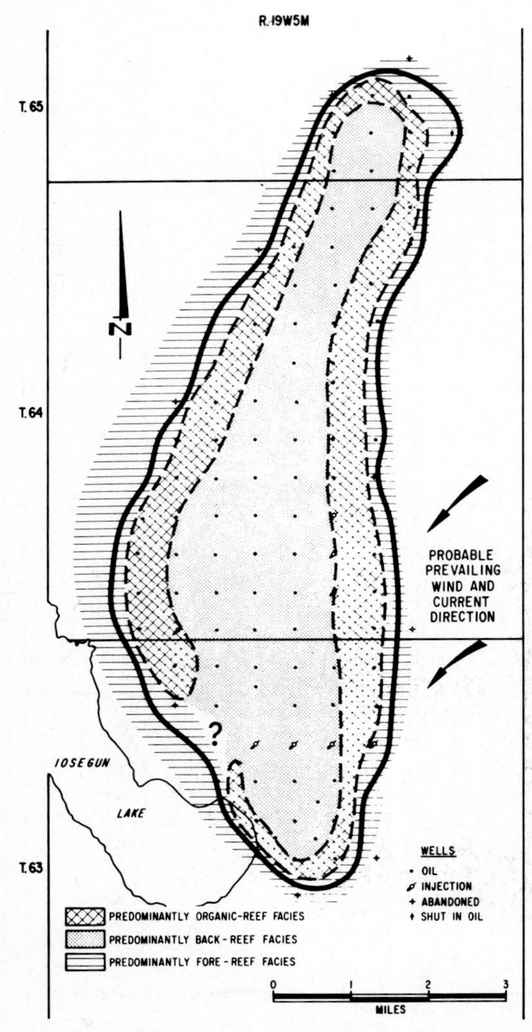

FIG. 7—Generalized facies distribution of Kaybob reef complex.

common. This facies probably was deposited in zones of moderate but persistent currents.

There is almost no effective porosity or permeability in this facies. What once was good intergranular porosity has been destroyed almost completely by calcite cementation of the micritic matrix.

Offreef Facies

The offreef and interreef Waterways Formation surrounds and overlies the entire reef complex. The basinal Waterways Formation consists typically of laminated to nodular, dark brown to dark gray, calcareous shales and argillaceous limestones. Blebs and thin bands of dark gray chert and siliceous argillite are common. A sparse fauna of *Styliolina* and *Tentaculites* and locally abundant crinoids, brachiopods, gastropods, and bryozans are present. In proximity to the reef, and in the interreef area, the carbonate-mud content tends to be much greater than that of the typical basinal Waterways sediments. On gamma ray logs, these limestones appear very clean and commonly are indistinguishable from reef limestones.

The Waterways Formation is most likely the source rock for the oil at Kaybob. Deroo et al (1973) determined that hydrocarbon generation in the Beaverhill Lake Group occurred at a depth of burial ranging from 7,500 to 12,000 ft (2,286 to 3,658 m), and that the principal phase of hydrocarbon genesis could not have begun until the Late Cretaceous.

SECONDARY ALTERATIONS

The Kaybob reef limestones have been affected by four main types of secondary alteration: solution, cementation, dolomitization, and pyritization.

Postdepositional solution and cementation have been the most significant factors in modifying the reservoir characteristics. Pressure solution has been an especially destructive agent in the lagoonal sediments. Stylolites and microstylolites are ubiquitous in the micrites and very common in the *Amphipora* limestones, where they cut into and across fossils. Although stylolitization greatly reduced porosity, leaching significantly enhanced it in many parts of the reservoir. Selective solution in places removed many fossils, leaving the matrix intact and leaving only an external mold. The skeletal molds commonly have a drusy lining of calcite crystals.

Calcite cementation is widespread in the backreef areas and is especially noticeable in the forereef areas. The most abundant form of calcite cement is clear crystalline calcite that infills vuggy, intergranular, intraorganic, and fracture porosity. Much of the calcite cement in the forereef probably originated in backreef areas. Where pellets were observed in a spar matrix, the matrix is believed to represent neomorphosed carbonate mud.

Dolomitization is a widespread but minor process in all reef facies. Dolomite rarely is present in

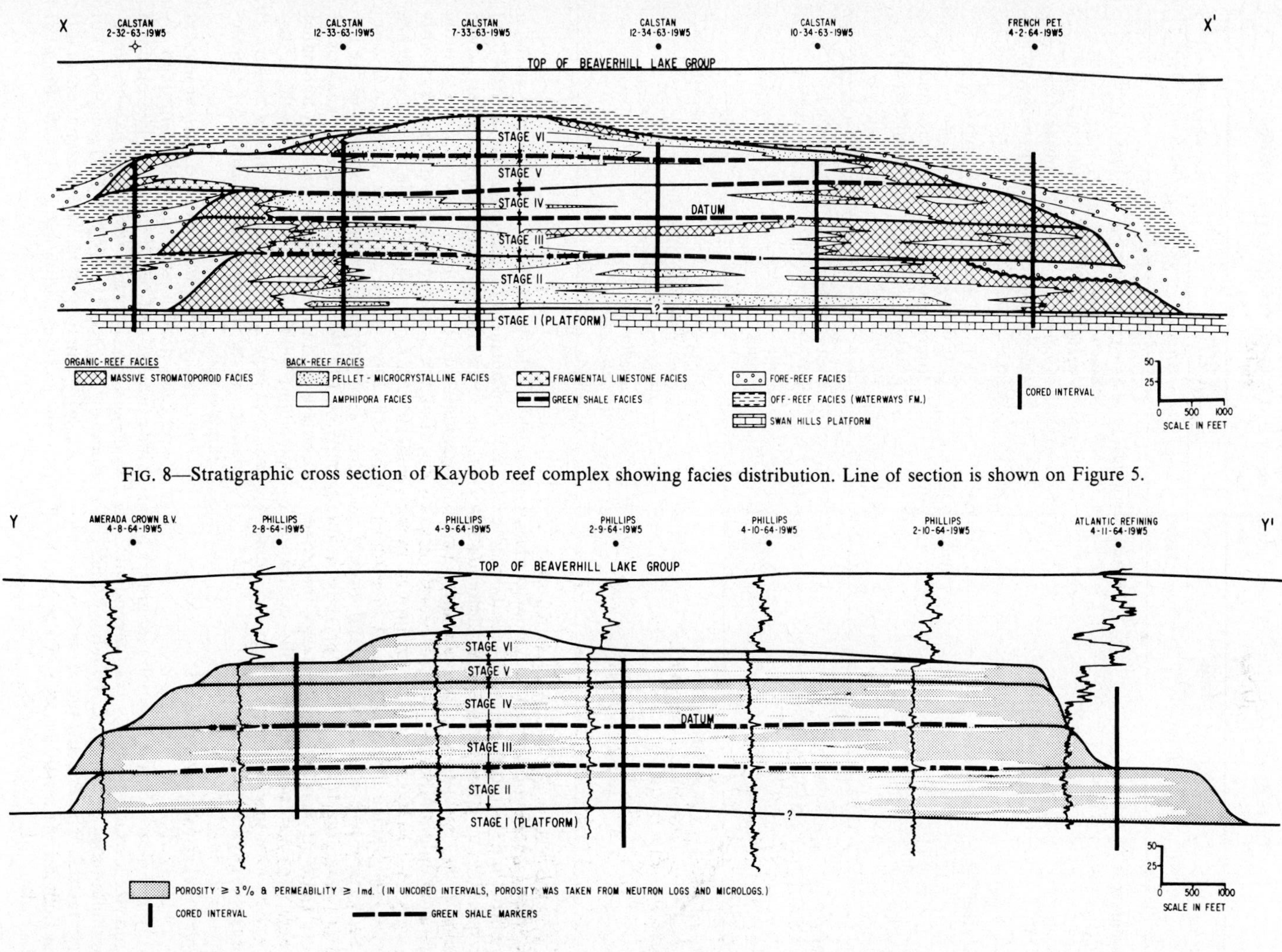

FIG. 8—Stratigraphic cross section of Kaybob reef complex showing facies distribution. Line of section is shown on Figure 5.

FIG. 9—Stratigraphic cross section of Kaybob reef showing porosity distribution and gamma ray–log correlations. Line of section is shown on Figure 5.

Table 1. Kaybob Field, Major Reef Facies and Associated Reservoir Grades

Facies	*Dominant Lithology*	*Fossils*		*Av. Reservoir Grade and Type*
		Major	*Minor*	
Organic reef	Limestone	Stromatoporoids	*Stachyodes* Brachiopods Corals Crinoids Algae	Good-excellent Vuggy Interorganic Intraorganic Interfragmental
Backreef	Limestone, minor shale	*Amphipora* Calcispheres	Algae Gastropods Stromatoporoids Corals Brachiopods	Poor-fair-good Vuggy Intraorganic Interfragmental
Forereef	Limestone Argillaceous limestone	Crinoids Brachiopods	Gastropods Ostracods *Amphipora* Corals	Nil-poor Nonreservoir
Offreef	Calcareous shale Argillaceous limestone		Crinoids Brachiopods Gastropods *Styliolina* *Tentaculites*	Nonreservoir

amounts exceeding 2 percent, and nowhere has it enhanced the reservoir quality. Where it occurs as crystals in vugs and fractures, the dolomite has reduced the porosity and permeability. Small, light gray dolomite rhombs or, more rarely, small patches of dolomite are scattered through much of the fine-grained limestones.

Secondary pyrite is common and occurs as finely disseminated blebs and crystals in the green shale markers and the adjacent limestones. Where the Waterways–Swan Hills contact has been observed in cores, the top 6 in. (15.2 cm) of the reef is highly pyritized.

Reef Development

It is not clear why the Kaybob reef grew where it did. Its location probably was dependent on a combination of geomorphic, structural, and ecologic factors. There is seismic evidence of a normal fault on the northeast side of the reef. Well control in the northern part of the reef shows a fault displacement of approximately 25 ft (7.6 m) on the Swan Hills platform (Fig. 6). Reef growth was initiated on the upthrown west side, and the linear nature of the east side of the reef probably reflects this fault. There is also some evidence that shoaling on the carbonate platform is a reflection of irregularities on the surface of the underlying Elk Point rocks. A regional analysis of structure on this surface shows small anomalies coincident with Swan Hills reef growth. Abundant reef organisms in the dark brown biohermal platform made it a suitable substrate for reef development.

Once reef growth was initiated, the subsequent shape and growth were largely controlled by winds, waves, and ocean currents. The platform stage and five informal reef stages can be recognized at Kaybob (Fig. 4). A fairly distinct, but complexly interfingering, lateral zonation of organisms is developed in stages II to V. Stage VI consists primarily of calcarenites, and has no apparent lateral facies zonation or organic-reef rim.

At the top of stage II, a sharp break overlain by forereef sediments is interpreted as evidence of erosion on the east flank of the reef. North of Township 63, subsequent reef stages were not able to reestablish themselves on this surface of calcarenite rubble. As a result, the front of the organic reef retreated approximately 0.5 mi (0.8 km) west (from approximately the 50-ft contour to the 100-ft contour; Fig. 5) and reestablished itself on the backreef facies of stage II (Fig. 9). Where backstepping has occurred and where only stage II is present, clean oil still is produced, but, on the west in adjacent wells that have a complete reef buildup, injection water has entered the upper part of the reef (Fig. 10).

Shale breaks and truncation of fossil colonies at the termination of most reef stages are evidence of minor erosion. Reef organisms were reestablished quickly as reef-building was renewed.

The contact of the reef with the overlying sedimentary units is sharp and irregular. There is evidence of truncation and channeling of lithified limestones; however, there is no evidence of extensive erosion. Rapid subsidence and the inability of reef organisms to grow on the calcarenites of stage VI most likely resulted in reef extinction. Rapid burial preserved most of the reef.

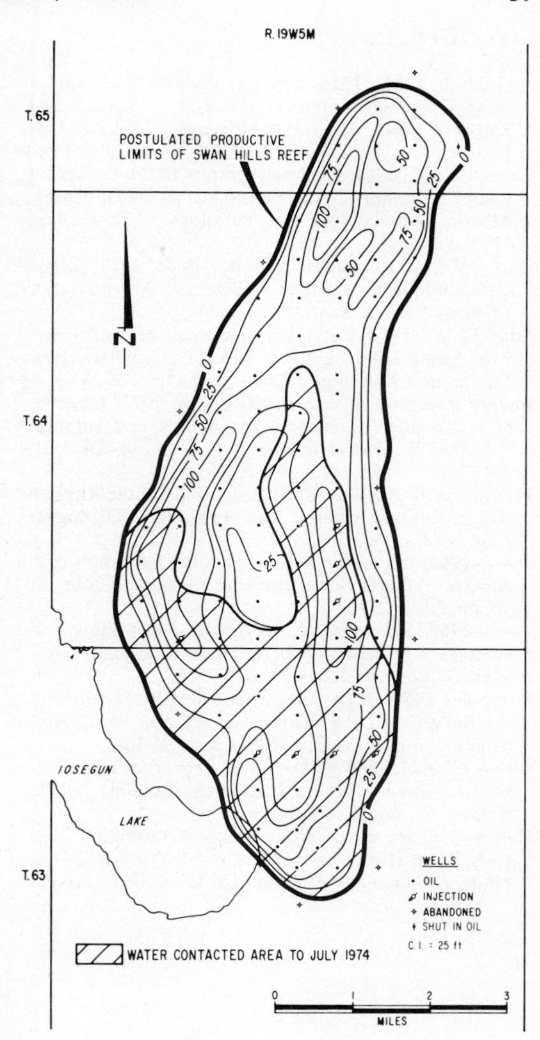

FIG. 10—Isopach map of Swan Hills reef net "pay," showing area that has been contacted by injected water.

CONCLUSIONS

1. Four major facies are recognized in the Kaybob complex: organic-reef, backreef, forereef, and offreef facies. The reef is a buildup of successively smaller reef stages.
2. Massive, hemispherical stromatoporoids were the most important reef builders at Kaybob.
3. Five informal reef stages (stages II–VI) can be recognized readily. These stages probably reflect interruptions in the slow regional subsidence of the Beaverhill Lake basin.
4. The initiation and areal extent of the lowest reef stage (stage II) were controlled by local block faulting, shoaling on the platform, and wind and current activity. In many places, subsequent reef stages retreated on the windward side over backreef deposits of the lowest stage; on the leeward west side, they advanced basinward over their own forereef detritus or over offreef sediments.
5. The thickest "pay" sections and best reservoir characteristics are in the organic-reef facies. The backreef facies has fair to good porosities, but the forereef facies is "tight" in most places. The distribution of porosity in the reservoir generally reflects the pattern of reef growth.
6. Minor erosion, followed by rapid burial by clay and carbonate mud, preserved most of the reef.

Selected References

Andrichuk, J. M., 1958, Stratigraphy and facies analysis of Upper Devonian reefs in Leduc, Stettler, and Redwater areas, Alberta: AAPG Bull., v. 42, no. 1, p. 1-93.

Carozzi, A. V., 1961, Reef petrography in the Beaverhill Lake Formation, Upper Devonian, Swan Hills area, Alberta, Canada: Jour. Sed. Petrology, v. 31, no. 4, p. 497-513.

Deroo, G., J. Roucache, and B. Tissot, 1973, Etude Geochimique du Canada Occidental, Alberta: Inst. Français Pétrole, Ref. 21270, 230 p.

Edie, E. W., 1961, Devonian limestone reef and reservoir, Swan Hills oil field, Alberta: Canadian Inst. Mining and Metallurgy Trans., v. 64, p. 278-285.

Energy Resources Conservation Board, 1973, Reserves of crude oil, gas, natural gas liquids and sulphur, Province of Alberta (Canada): Edmonton, Dec. 31, 1973.

Fischbuch, N. R., 1960, Stromatoporoids of the Kaybob reef, Alberta: Alberta Soc. Petroleum Geologists Jour., v. 8, p. 113-131.

———1962, Stromatoporoid zones of the Kaybob reef, Alberta: Alberta Soc. Petroleum Geologists Jour., v. 10, no. 2, p. 62-72.

———1968, Stratigraphy, Devonian Swan Hills reef complexes of central Alberta: Bull. Canadian Petroleum Geology, v. 16, no. 4, p. 446-587.

Fong, G., 1959, Type section, Swan Hills Member of the Beaverhill Lake Formation: Alberta Soc. Petroleum Geologists Jour., v. 7, no. 5, p. 95-108.

———1960, Geology of Devonian Beaverhill Lake Formation, Swan Hills area, Alberta, Canada: AAPG Bull., v. 44, no. 2, p. 195-209.

Hemphill, C. R., et al, 1970, Geology of Beaverhill Lake reefs, Swan Hills area, Alberta, p. 50-90 *in* M. T. Halbouty, ed., Geology of giant petroleum fields: AAPG Mem. 14, 575 p.

Henson, F. R. S., 1950, Cretaceous and Tertiary reef formations and associated sediments: AAPG Bull., v. 34, no. 2, p. 215-238.

Jenik, A. J., and J. F. Lerbekmo, 1968, Facies and geometry of Swan Hills Reef Member of Beaverhill Lake Formation (Upper Devonian), Goose River field, Alberta, Canada: AAPG Bull., v. 52, no. 1, p. 21-56.

Klovan, J. E., 1964, Facies analysis of the Redwater reef complex, Alberta, Canada: Bull. Canadian Petroleum Geology, v. 12, no. 1, p. 1-100.

Langton, J. R., and G. E. Chin, 1968, Rainbow Member facies and related reservoir properties, Rainbow Lake, Alberta: Bull. Canadian Petroleum Geology, v. 16, no. 1, p. 104-143.

Leavitt, E. M., 1968, Petrology and paleontology of Carson Creek North reef complex, Alberta: Bull. Canadian Petroleum Geology, v. 16, no. 3, p. 298-413.

———and N. R. Fischbuch, 1968, Devonian nomenclature changes, Swan Hills area, Alberta, Canada: Bull. Canadian Petroleum Geology, v. 16, no. 3, p. 288-297.

Martin, R., 1967, Morphology of some Devonian reefs in Alberta: a palaeogeomorphological study, *in* International symposium on the Devonian System, v. 2: Alberta Soc. Petroleum Geologists, p. 365-385.

Murray, J. W., 1966, An oil producing reef-fringed carbonate bank in the Upper Devonian Swan Hills Member, Judy Creek, Alberta: Bull. Canadian Petroleum Geology, v. 14, no. 1, p. 1-103.

Stearn, C. W., 1963, Some stromatoporoids from the Beaverhill Lake Formation (Devonian) of the Swan Hills area, Alberta: Jour. Paleontology, v. 37, no. 3, p. 651-668.

Thomas, G. E., and H. S. Rhodes, 1961, Devonian limestone bank–atoll reservoirs of the Swan Hills area, Alberta: Alberta Soc. Petroleum Geologists Jour., v. 9, no. 2, p. 29-38.

Big Piney–La Barge Producing Complex, Sublette and Lincoln Counties, Wyoming[1]

ROBERT E. McDONALD[2]

Abstract The Big Piney–La Barge complex, with cumulative gas production of 1.2 Tcf, has been the leading gas-producing area in Wyoming since 1956. It also contributes significant oil production, which has amounted to about 65 million bbl. Ultimate reserves are estimated to be 2.5 Tcf of gas and 75 million bbl of oil. In addition, condensate production averages about 3 bbl/MMcf of gas.

Production is from rocks as young as Paleocene and as old as Triassic. The major gas reservoir in the area is the Cretaceous Frontier Formation. Structural, stratigraphic, and combination traps are all common in the Big Piney–La Barge area. Probably significant to trapping the hydrocarbons in the Paleocene strata was the transgression of Paleocene units onto the Big Piney–La Barge platform or anticline. The area was anticlinally folded during Late Cretaceous and early Paleocene times, and Mesaverde units reflect the influence of the Moxa arch to the south and the Monument Buttes–Blue Rim arch to the southeast. Accumulations in the Frontier are essentially structurally controlled west of the La Barge thrust; however, east of that thrust, production is mainly from stratigraphic traps.

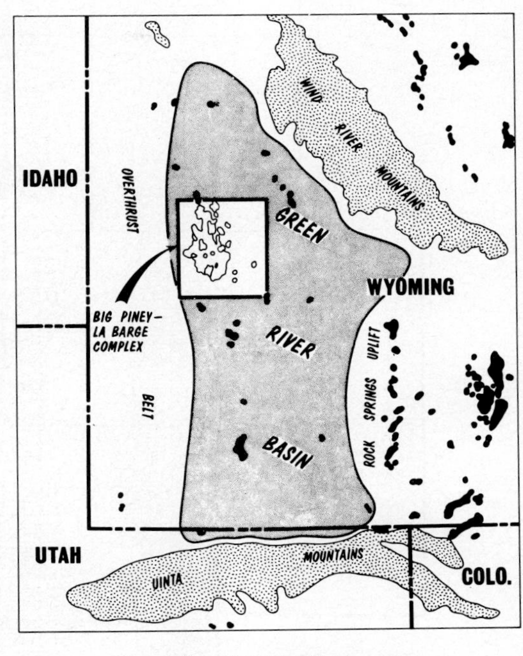

FIG. 1—Location map, Green River basin, Wyoming.

INTRODUCTION AND HISTORY

The Big Piney–La Barge complex (Fig. 1) has been the leading gas-producing area in Wyoming since 1956 and a major producer of oil since 1960. Cumulative gas production is over 1.2 Tcf, and total oil production is about 65 million bbl. Probable ultimate reserves are 2.5 Tcf of gas and 75 million bbl of oil. In addition, condensate averages 3 bbl/MMcf of gas.

The first development phase began with the discovery of the La Barge oil field (T26 and 27N, R113W; Fig. 2) in 1924. The discovery was located on a gently folded surface anticline, and multiple oil-bearing Tertiary sandstones were penetrated between 600 and 1,200 ft (183–366 m). This spurred a 4-year cable-tool boom at Tulsa, Wyoming, later named "La Barge."

Sporadic wildcatting in the 1930s and 1940s yielded occasional reports of gas in very shallow Tertiary strata, and General Petroleum found apparently subcommercial gas in low-permeability Frontier sandstones and a three-well oil field in the Triassic-Jurassic Nugget Sandstone on Tip Top anticline. However, these discoveries did not create significant industry interest. In 1952, Ar-

[1]Manuscript received, February 25, 1975; revised from *25th Field Conference—1973, Wyoming Geological Association Guidebook*. Published with permission of Wyoming Geological Association.

[2]Exploration Manager, Wolf Energy Company, Denver, Colorado 80201.

Dick Welch, consultant, was of great assistance in reviewing maps, sections, and interpretations, particularly of Paleocene and Mesaverde sediments. John Dunnewald, Belco Petroleum, provided data helpful in compiling Nugget structure, and Milton O. Childers, Power Resources Corp., assisted in both structural and stratigraphic reviews. Barbara Childers deserves much thanks for editorial review and revision, and Marvin Wolf, Wolf Energy Co., generously provided drafting time and reproduction expense for illustrations.

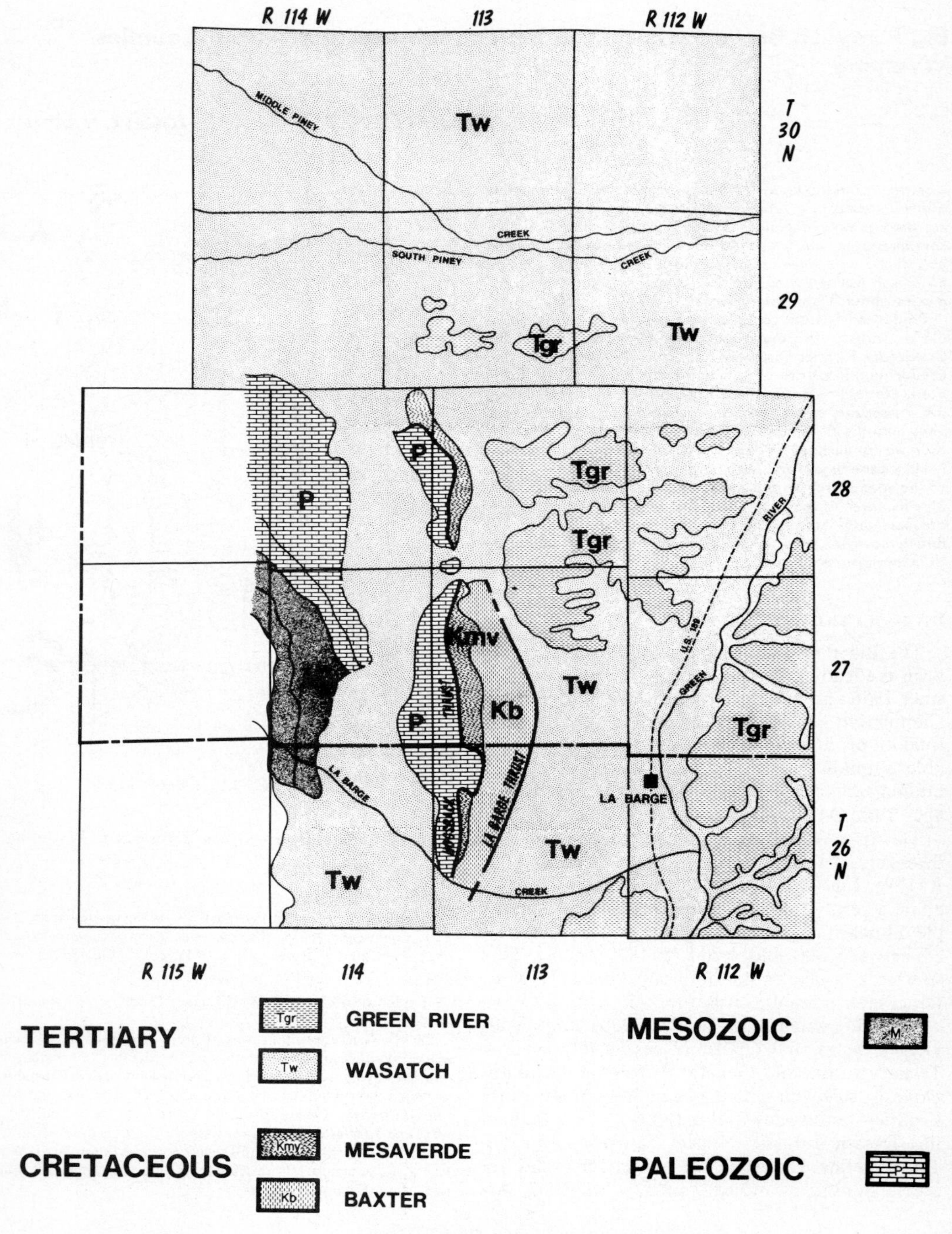

FIG. 2—Areal geologic map, La Barge area, Wyoming.

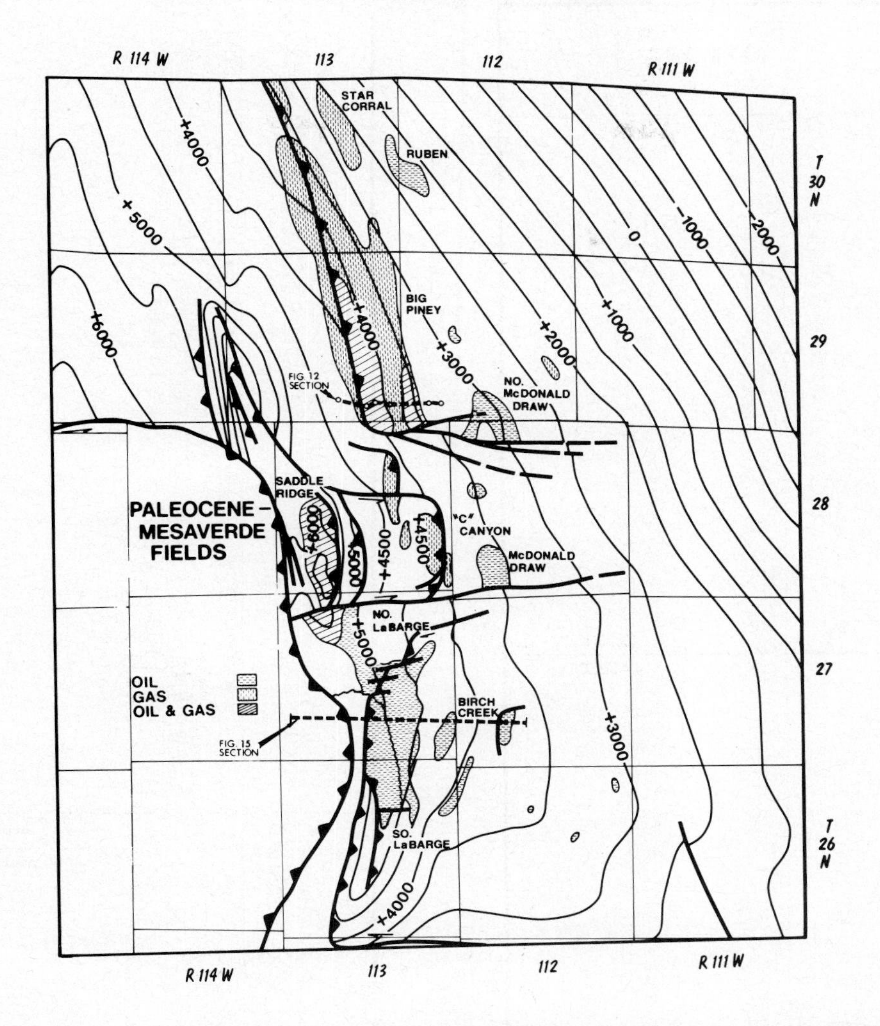

FIG. 3—Structural contour map of Tertiary-Mesaverde unconformity and location of Paleocene-Mesaverde fields. Cumulative production to date is 430 Bcf of gas and 54 million bbl of oil. Lines are shown for cross sections of Figures 12 and 15.

thur Belfer (later Belco Petroleum Corp.) of New York began serious development of gas reserves in the Tertiary strata. Later, both Mobil and Belco began development of gas reserves in the Frontier using the then-new sand-fracturing techniques. Other operators subsequently entered the area, and a natural gas pipeline outlet was provided by construction of the Pacific Northwest pipeline in 1956.

SURFACE GEOLOGY

The Big Piney–La Barge producing complex is situated on and around the flanks of a Laramide anticlinal fold that was formed subjacent to a salient along the "Disturbed Belt." On the surface map (Fig. 2), this salient is apparent from the exposures on the Hogsback thrust, which carries Cambrian rocks in its sole. This thrust was long

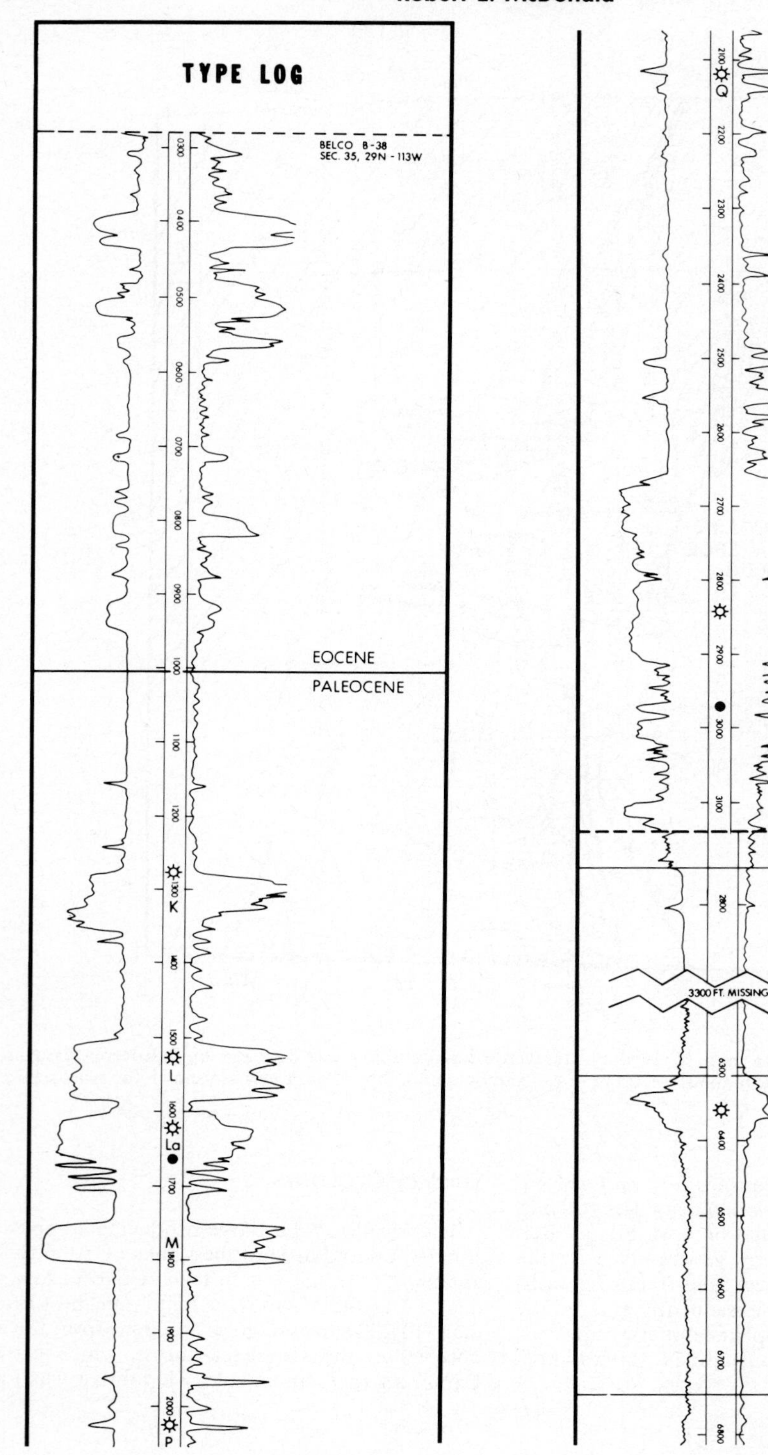
TYPE LOG
BELCO B-38
SEC. 35, 29N - 113W
EOCENE
PALEOCENE
K
L
La
M
P
Q
TRANSITION SAND
MESAVERDE
BELCO BNG 39-9
SEC 9, 27N-113W
BAXTER SHALE
3300 FT. MISSING
1st FRONTIER
1st BENCH 2nd
FRONTIER

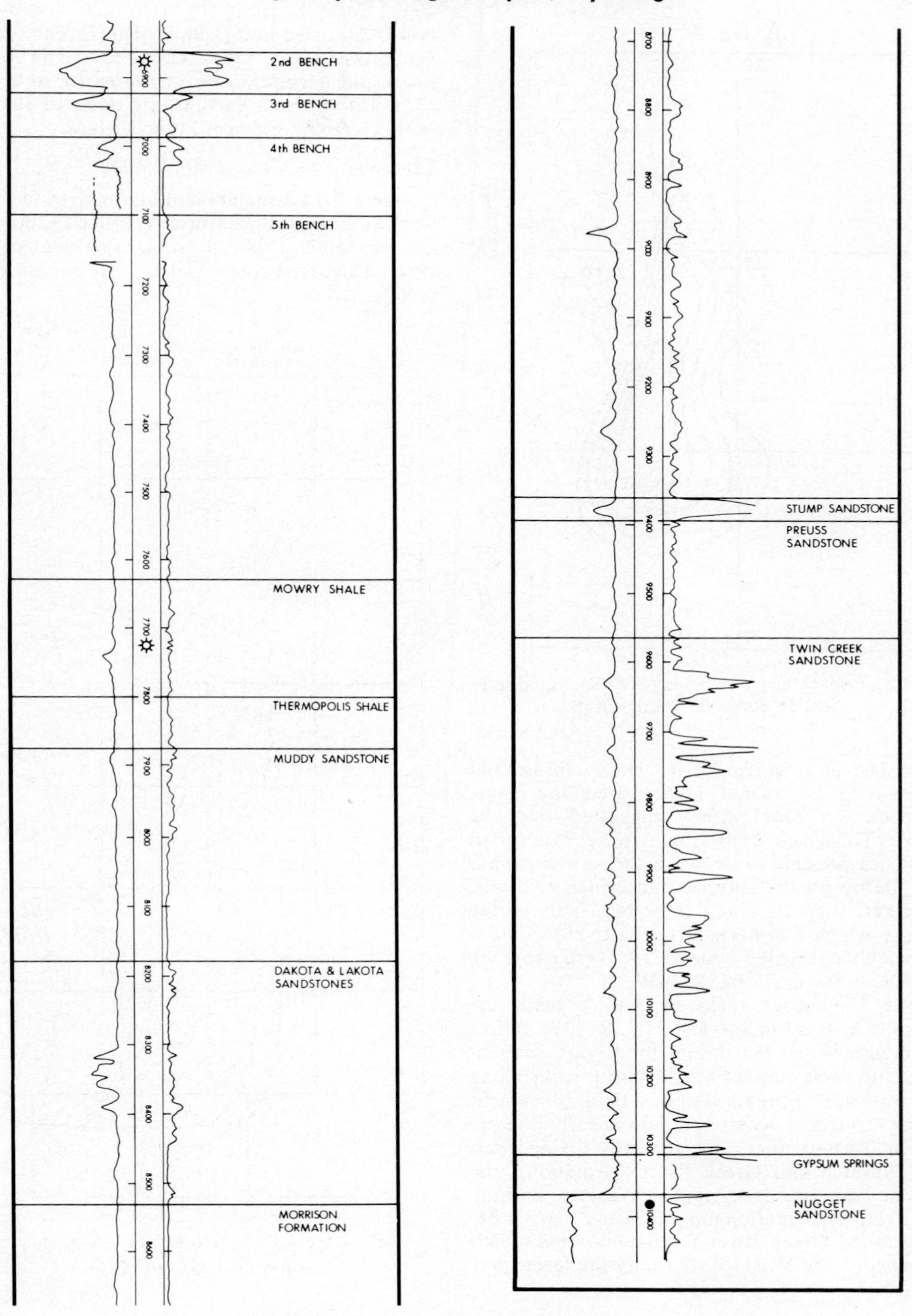

FIG. 4—Composite type log, Big Piney–La Barge area.

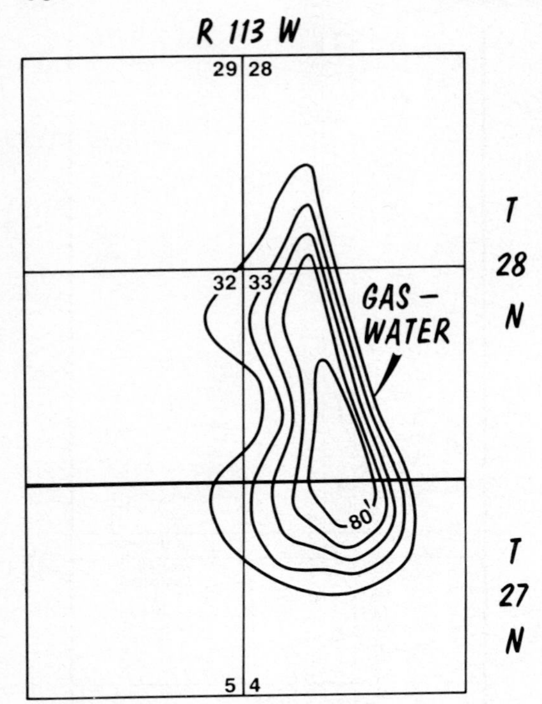

FIG. 5—Isopach map of net "pay" of "Kb" sandstone, Shallow Ridge field. C.I. = 20 ft.

described as a salient of the Darby thrust, but Armstrong and Oriel (1965) used the name "Hogsback." Oriel (1969) suggested that the name "Hogsback thrust" be used because it had not been possible to trace this thrust system into the Darby on the surface. Subsurface evidence, however, suggests that it may be a part of the Darby which has been thrust eastward along a tear fault concealed beneath the Tertiary along the south line of T29N, R114W.

The Cretaceous rocks exposed beneath the Hogsback thrust in T26 and 27N, R113W, reflect a younger thrust that brings the Cretaceous section into fault contact with Tertiary rocks along the west side of the La Barge oil field. Movement along this thrust occurred as late as early Eocene.

The Tertiary rocks are generally divided into the Wasatch and Green River Formations; the fluvial strata are designated as Wasatch, and the overlying and intertonguing lacustrine strata belong to the Green River Formation. Oriel (1969) subdivided the Wasatch, but only the lower portion is discussed in this paper. The Tertiary cover unconformably overlies Cretaceous and older rocks and generally dips gently eastward about 2°. Local anticlinal flexures are reflected slightly in the Tertiary in places at the surface.

TERTIARY AND MESAVERDE FIELDS

Figure 3 is a structural contour map of the Tertiary-Cretaceous unconformity, which for most of the map area is the unconformable contact between Paleocene rocks and the Mesaverde Formation.

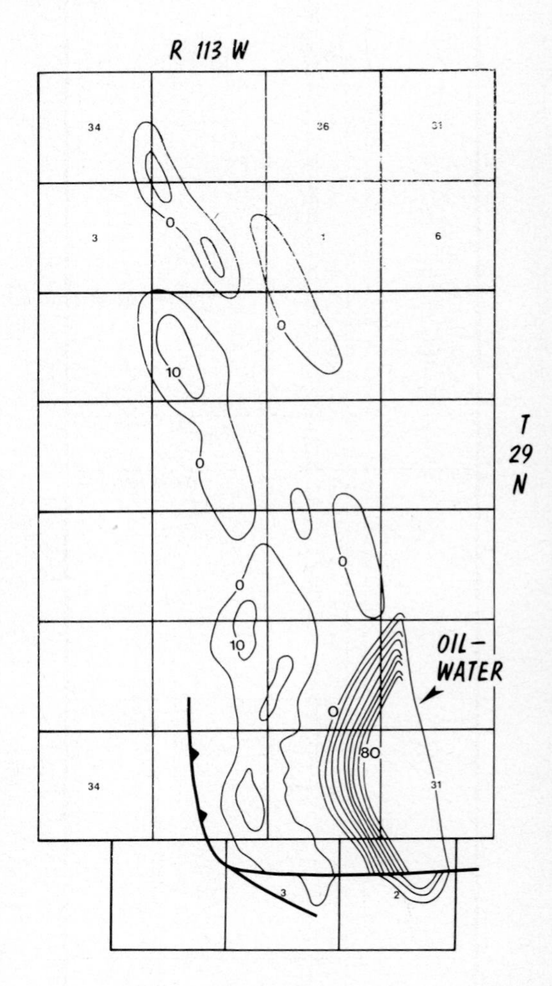

FIG. 6—Isopach map of "P" sandstone, Big Piney field. C.I. = 10 ft.

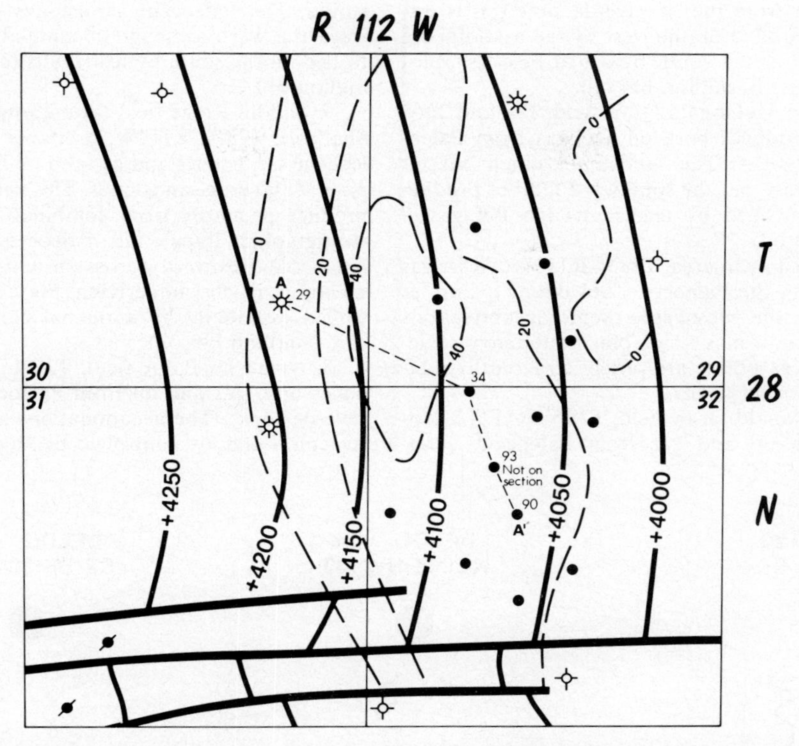

FIG. 7—Structure map of top of "Q" sandstone (solid lines), McDonald Draw field; C.I. = 50 ft. Isopachs of "Q" sandstone (dashed lines); C.I. = 20 ft. Location for cross-section *A-A'* (Fig. 8) is shown.

Formerly these Paleocene strata were inappropriately called "Almy." Subsequently, many have called them "Fort Union," which has become a catch-all for Paleocene rocks in the Rocky Mountains. Oriel (1969) suggested using the name "Hoback Formation" for the Paleocene strata (Dorr, 1952), but this name has not been generally accepted by petroleum geologists and presents a few problems. Thus, in this paper, I shall refer simply to the "Paleocene strata."

The Tertiary-Cretaceous unconformity is the first marker beneath the surface that is suitable for regional mapping. Stratigraphic markers within the fluvial Paleocene strata are suitable for local field mapping, but only zonal correlations can be carried over wide areas.

The oil and gas fields delineated in Figure 3 produce only from the Tertiary and the Upper Cretaceous Mesaverde; the accumulations occur in purely stratigraphic, purely structural, and combination structural-stratigraphic traps. Listed generally from north to south, the important fields and their trapping mechanisms and probable ultimate reserves (given in parentheses) are as follows:

1. Star Corral field, T30N, R113W, is a gas field producing from stratigraphically controlled accumulations in Paleocene strata (4.5 Bcf gas).

2. Ruben field, T30N, R112 and 113W, is an oil field producing from stratigraphically trapped accumulations in Paleocene sandstone (7 Bcf gas, 5 million bbl oil).

3. Big Piney field, T29 and 30N, R113W, produces oil and gas from several Paleocene sandstones and subcropping Mesaverde sandstones. The Paleocene sandstones pinch out or grade into impermeable siltstones and conglomerates westward. Some are wholly stratigraphically controlled, but the most significant accumulations are also terminated on the south by tear faults.

Production from the Mesaverde subcrops is primarily controlled on the west by the east-dipping thrust and on the south by a tear-fault complex (120 Bcf gas, 10 million bbl oil).

4. North McDonald Draw field, T28 and 29N, R112W, produces both oil and gas from Paleocene sandstones. The sandstones pinch out toward the west, but the southward limit of production is controlled by tear faults (45 Bcf gas, 7 million bbl oil).

5. "C" Canyon area, T28N, R113W, yields gas mainly from the Paleocene, but minor quantities come from the Mesaverde. Some accumulations occur in sandstones which pinch out across structure, whereas others are purely structurally controlled (15.5 Bcf gas).

6. McDonald Draw field, T28N, R112W, produces both oil and gas from Paleocene sandstones. The Paleocene sandstones pinch out toward the west, although accumulations are controlled on the south by tear faults (12 Bcf gas, 1.2 million bbl oil).

7. Saddle Ridge field (also known as Tip Top Shallow), T28N, R113W, produces gas from Paleocene sandstones and gas and oil from subcropping Mesaverde sandstones. The Paleocene strata produce primarily from combination structural-stratigraphic traps; the Paleocene sandstones pinch out westward across structure. The accumulation in the underlying Mesaverde is controlled essentially by anticlinal closure (8.5 Bcf gas, 3 million bbl oil).

8. North La Barge field, T27N, R113W, produces both gas and oil from Paleocene and Mesaverde rocks. The accumulations are essentially structural and are controlled by an east-west tear

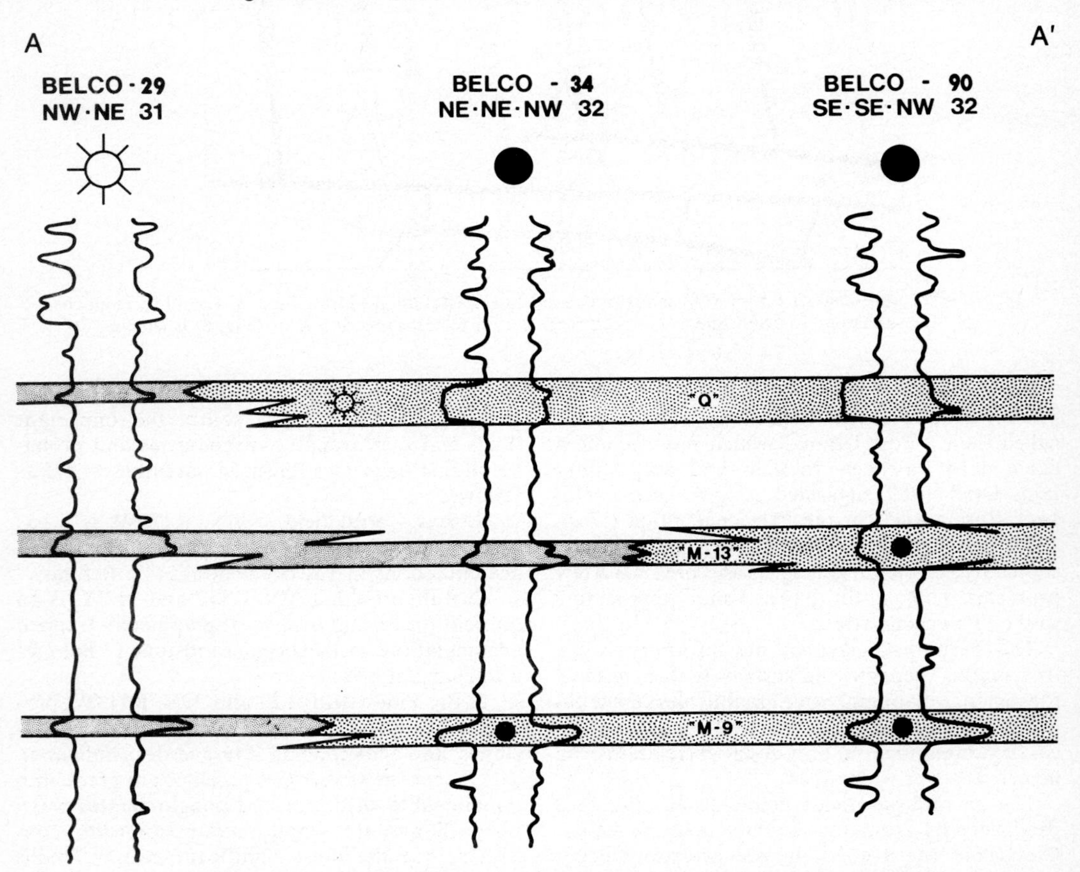

FIG. 8—Section *A-A'* across McDonald Draw field. See Figure 7 for location. Well 93 on section line is not included.

fault across the south plunge of the Saddle Ridge anticline. The south side of the fault is downthrown. Stratigraphic accumulations occur in some Paleocene sandstones which pinch out westward across the anticlinal nose (2 Bcf gas, 1 million bbl oil).

9. South La Barge field, T26 and 27N, R113W, produces oil from Paleocene and Mesaverde sandstones. Production from the Paleocene is controlled largely by a fold just east of the La Barge thrust. The Mesaverde has been removed over the fold by truncation but, where the sandstones subcrop east and north of the La Barge anticline, their upturned edges produce from traps at the Paleocene-Cretaceous unconformity (20 million bbl Paleocene oil, 12 million bbl Mesaverde oil).

10. Birch Creek field, T27N, R113W, produces oil from Paleocene sandstones. This accumulation is purely stratigraphic; the sandstones loop slightly updip and westward across a gently plunging anticlinal nose (7.5 Bcf gas, 9 million bbl oil).

Figure 4 is a composite type log for fields in the area.

Stratigraphy and Trapping Mechanisms

Tertiary

Figures 5–13 depict in more detail the geometry of Paleocene sandstones and trapping mechanisms involved in Paleocene accumulations. The transgression of the Paleocene units onto the Big Piney–La Barge platform or anticline is a significant factor. While earlier Paleocene units were being deposited in basinal areas to the east and northeast, erosion was still taking place across the platform area. The oldest Paleocene units within the area of discussion are middle Paleocene, and they are present only in the subsurface; at the surface, only upper Paleocene units are present. An unconformity within the Tertiary, which appears to be early Eocene in age, represents a period of significant local tectonic activity—probably the last pulse of thrusting.

Figure 5 is an isopach map of the net "pay" of the "Kb" sandstone in the Saddle Ridge (Tip Top Shallow) area. The accumulation in this sandstone is controlled on the west by pinchout across structure and, on the east, by a gas-water contact. This is one of several Paleocene "pays"; it occurs at a depth of about 800 ft (244 m). This "Kb" sandstone grades westward into "tight" sandstones, siltstones, and drab variegated shales, which persist for about 1 mi (1.6 km) then grade

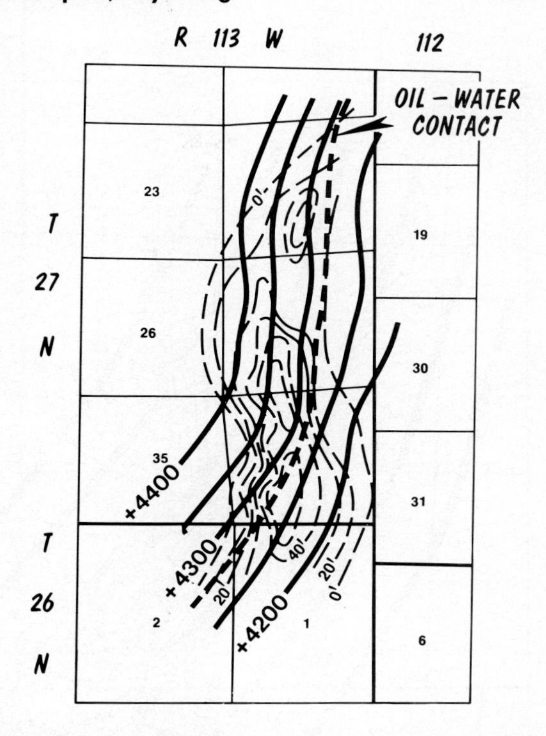

FIG. 9—Structure map of top of Birch Creek sandstone (solid lines), Birch Creek field; C.I = 100 ft. Isopachs of Birch Creek are light dashed lines; C.I. = 20 ft. Oil-water contact is shown by heavier dashed line.

into conglomerates subjacent to the Hogsback thrust.

Figure 6 is an isopach map of the "P" sandstone series across the Big Piney field. The "P" series is lower in the Paleocene than the "Kb" sandstone. The discovery history of this field is typical of such reservoirs. First, a few pods of thin, rather impermeable, hydrocarbon-bearing sandstones were discovered on the west. Later, the main sandstone was encountered on the east, where it develops in a short distance into a thick, porous sandstone with excellent reservoir characteristics. Accumulation in the "P" pods is strictly stratigraphic; productive limits of the main sandstone body are controlled by pinchout on the west, the oil-water contact on the east, and tear faults on the south. The equivalent stratigraphic interval west of the development of the "P" series is typified by drab, gray to variegated claystones, siltstones, and clay-filled sandstones for a dis-

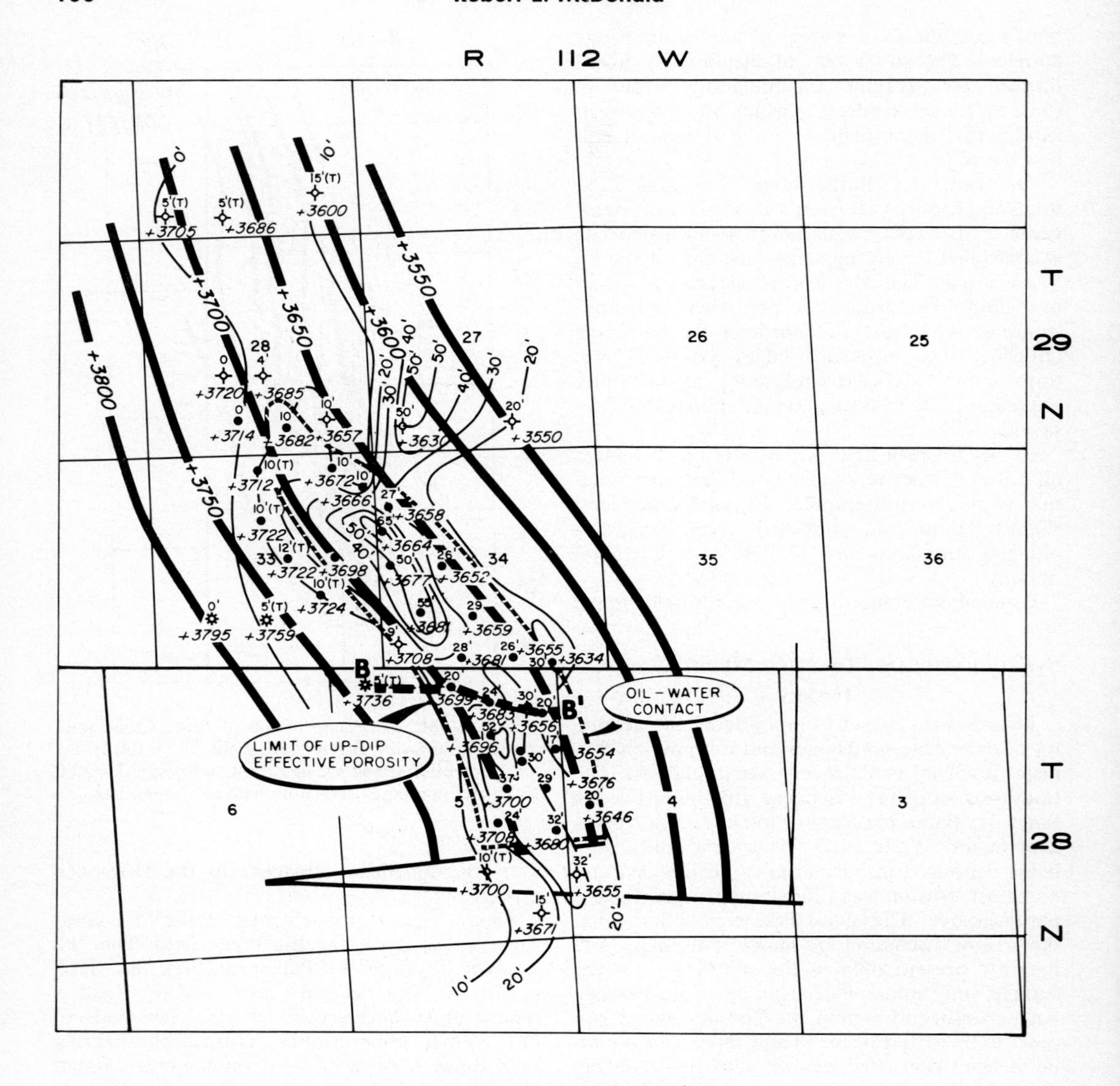

FIG. 10—Structural contour and isopach map of "M-20" sandstone, North McDonald Draw field. Location is shown for cross-section *B-B'* (Fig. 11).

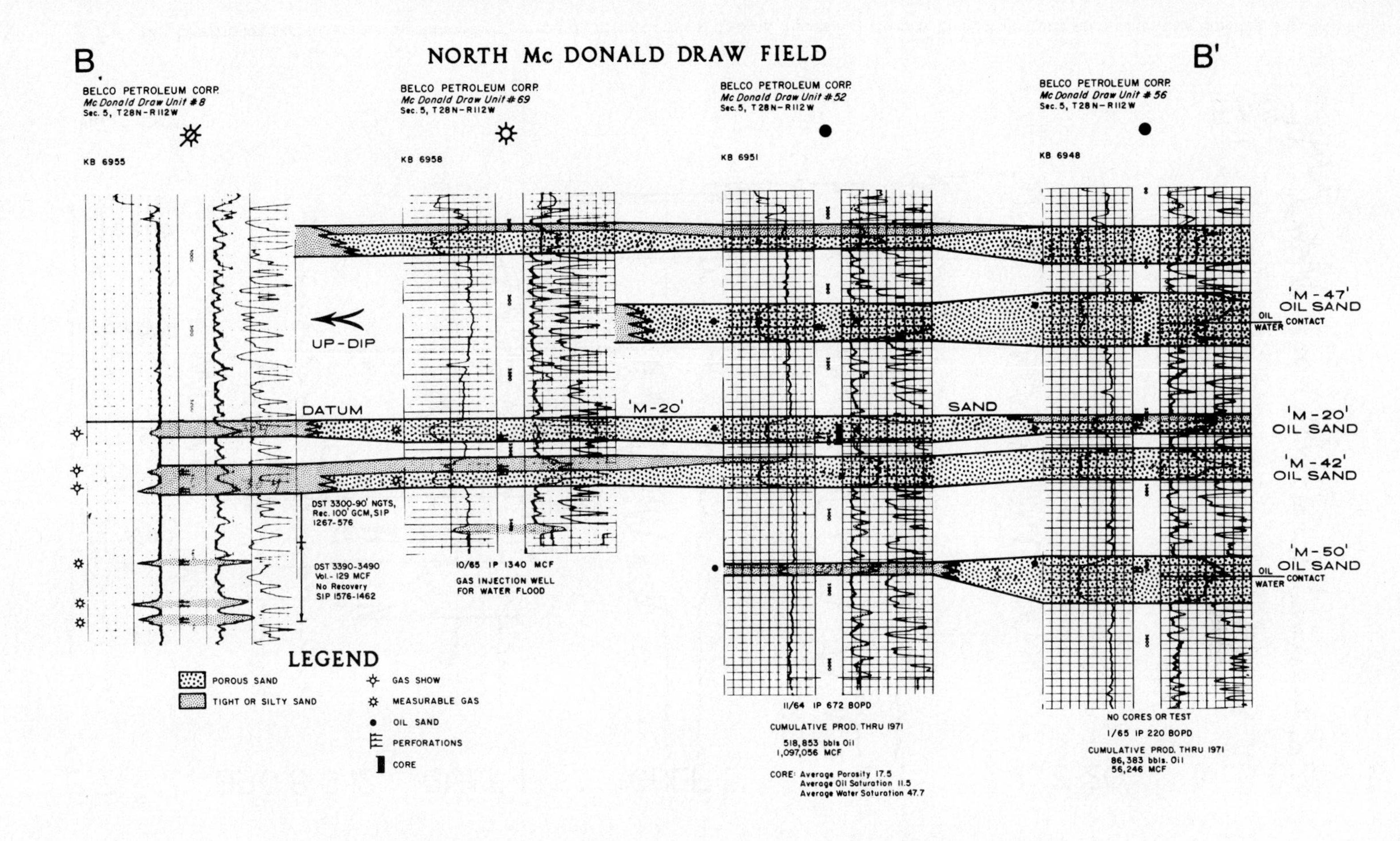

FIG. 11—Cross-section *B-B'*, North McDonald Draw field. Location is shown on Figure 10.

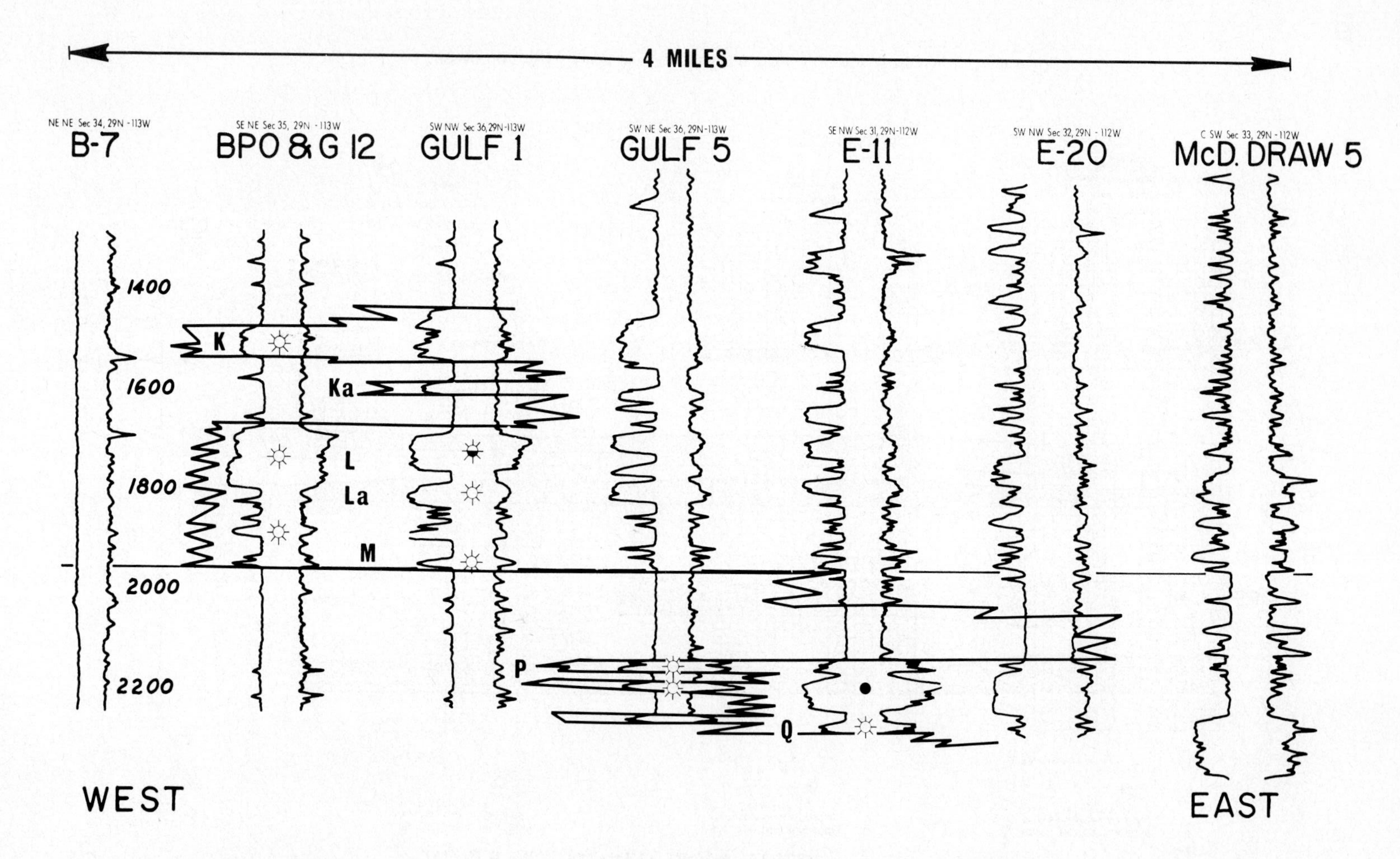

FIG. 12—Electric-log section across Big Piney field. Location of section is shown on Figure 3. Letters K-Q are Paleocene sandstones (see type log, Fig. 4).

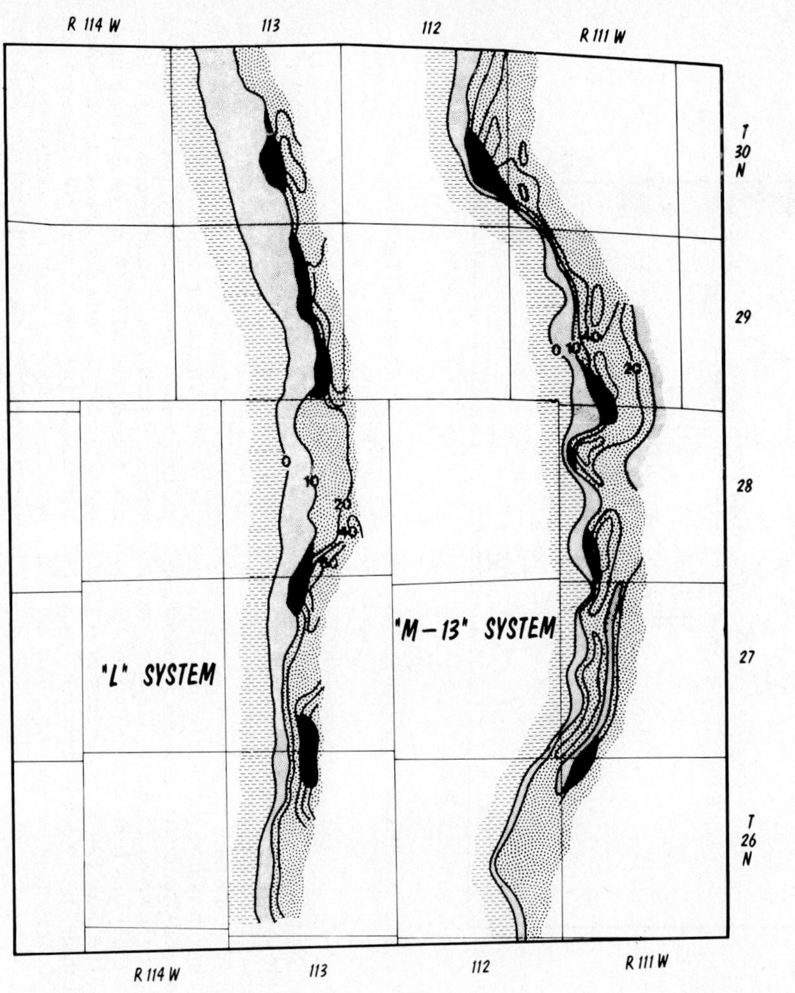

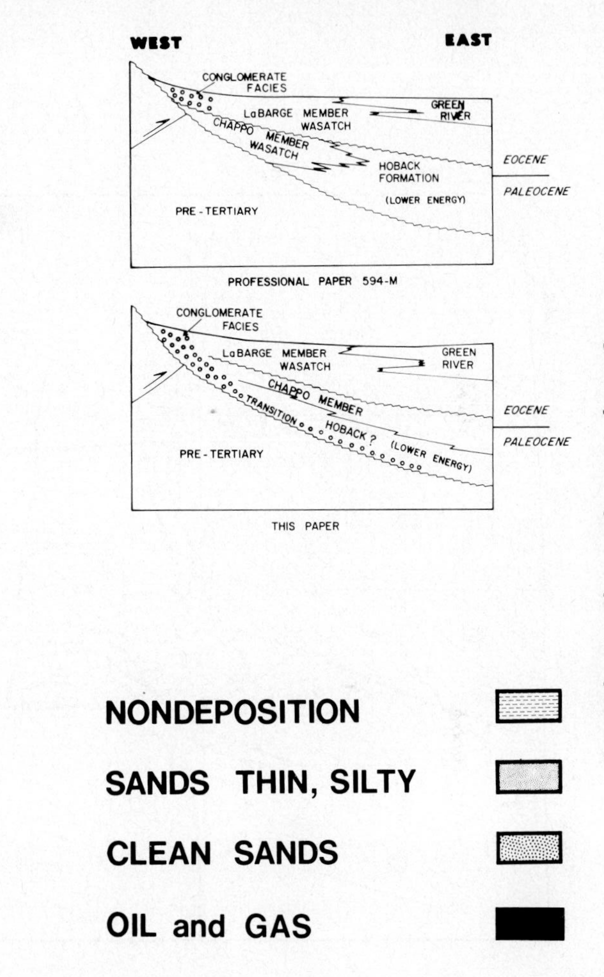

FIG. 13—Paleocene channel systems. Cross sections show interpretations of present stratigraphy: *upper*, Oriel (1969); *lower*, this paper.

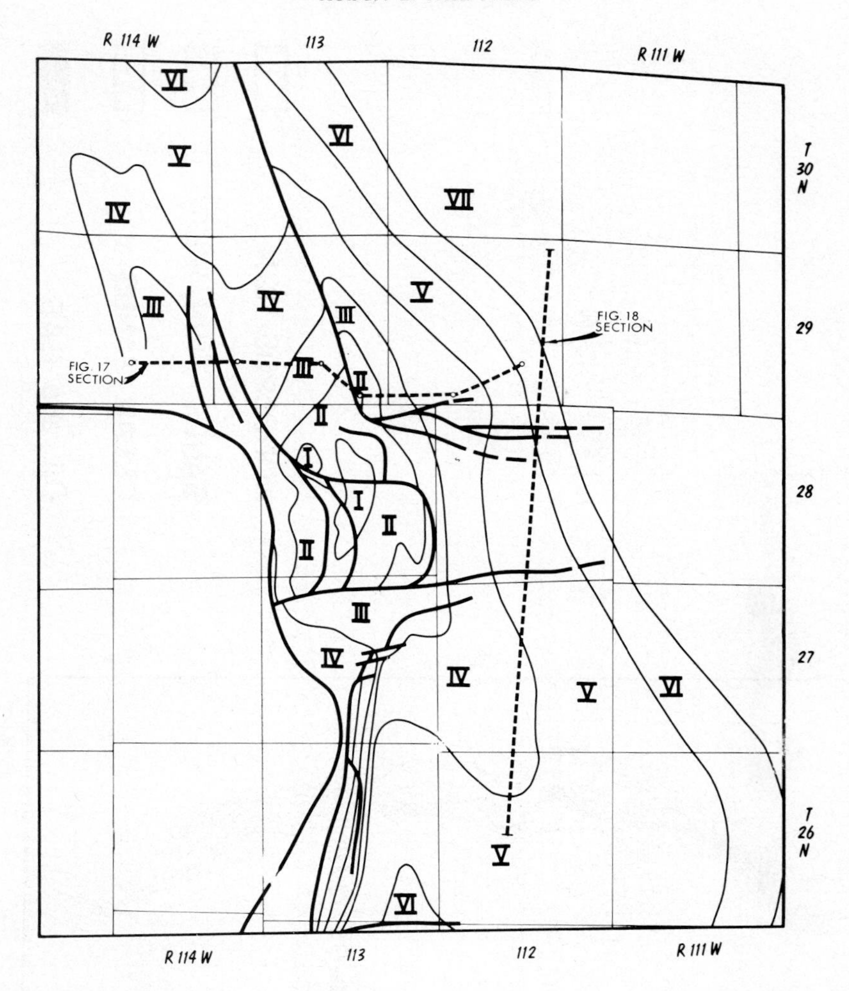

FIG. 14—Mesaverde subcrop map, Big Piney–La Barge area. Locations of cross sections of Figures 17 and 18 are shown.

tance of about 2 mi (3.2 km). Then these lithologies grade into conglomerates.

Figure 7 is a structure map of the McDonald Draw field; it includes isopachs of the "Q" sandstone, which produces gas in this field. There are also two oil-productive Paleocene sandstones, which are shown in Figure 8. Trapping mechanisms are obvious from examination of the two figures. The three Paleocene sandstones all grade into siltstone in a westerly direction; the south limits of accumulation are controlled by tear faults.

Figure 9 is a structure and isopach map of the Birch Creek field. The stratigraphically controlled accumulations occur in Paleocene sandstones, and the reservoir has been referred to as an example of an offshore bar deposited in a lacustrine environment. This is an excellent oil field in which the main "pay" sandstone has a net effective thickness of 60 ft (18 m). Its downdip limits are defined by an oil-water contact, but the effective sandstone grades to zero both eastward and westward. To the west the sandstone grades abruptly into low-energy siltstone and shale deposited in a chiefly reducing environment; to the east the interval grades abruptly into high-energy fluvial rocks. Local examination of the Birch Creek sandstone leads logically to its interpretation as a lacustrine bar sandstone. Regionally, however, such an interpretation does not hold. The interval equivalent to the Birch Creek sandstone thickens eastward and thins westward; within a few miles the homolog grades into the Paleocene "transition zone" at the Paleocene-Cretaceous unconformity.

Figures 10 and 11 show the structure, sandstone thickness, and stratigraphic relations in McDonald Draw field. Four Paleocene sandstones produce oil and/or gas in this field; Figure 10 is a structure and isopach map of one of these—the "M-20" sandstone. The figures indicate the similarity between the North McDonald Draw and McDonald Draw fields, both of which are typified by Paleocene pinchouts to the west and tear faults to the south.

Figure 12 is an electric-log section across the Big Piney field, showing the manner in which the high-energy fluvial environment on the east transgresses the small-stream fluvial, alluvial, and piedmont environments on the west. The line of section extends east from Sec. 34, T29N, R113W, to Sec. 33, T29N, R112W. A low-energy fluvial environment on the east is indicated by the logs. Without sample examination, the apparently lower energy environment suggested by the B-7 log could be interpreted as indicative of a lacustrine siltstone and shale section. In reality, the section recorded by this characterless log is in part conglomeratic, and, as samples and bit records reveal, nearly the entire section is conglomerate a short distance farther west.

Each transgressive Paleocene high-energy sequence or "sandstone series" may be traced from north to south along the platform. On the west in Figure 13 is an example of the "L" series of sandstones, showing (a) area of nondeposition where the unit is unrecognizable; (b) area of siltstones and "tight" clay-filled silty sandstones; (c) area of clean porous sandstones representing maximum westward extent of the meander system and buildup of the channel, bar, and point-bar sandstones; and (d) areas of production from the "L" series, whether stratigraphically or structurally controlled. A similar depiction of the strati-

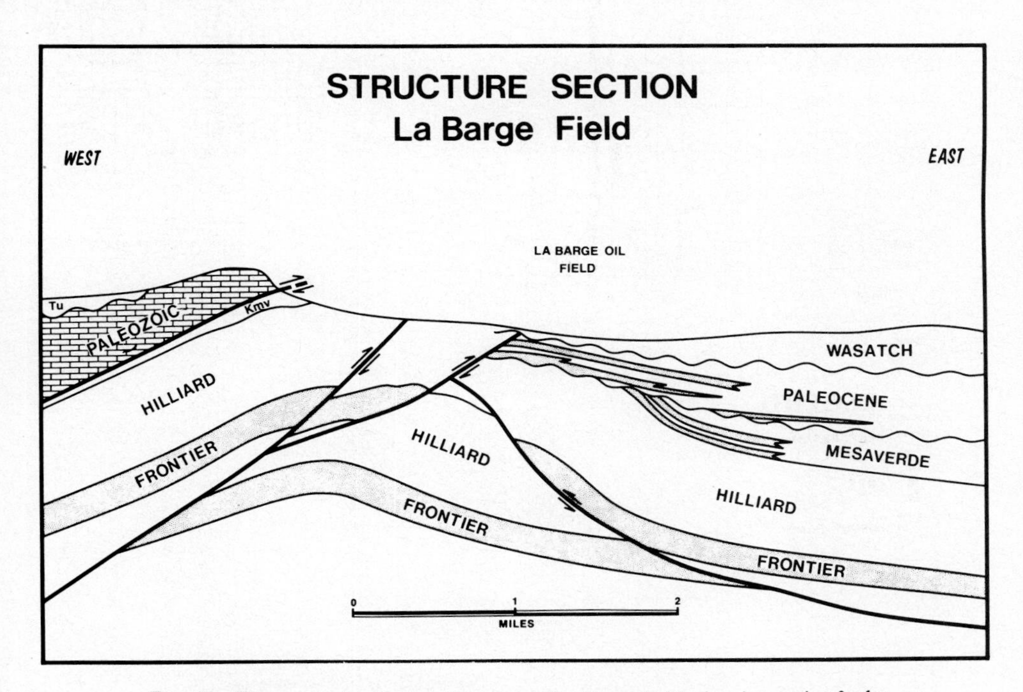

FIG. 15—West-east structure section across La Barge field, showing major faults. See Figure 3 for location.

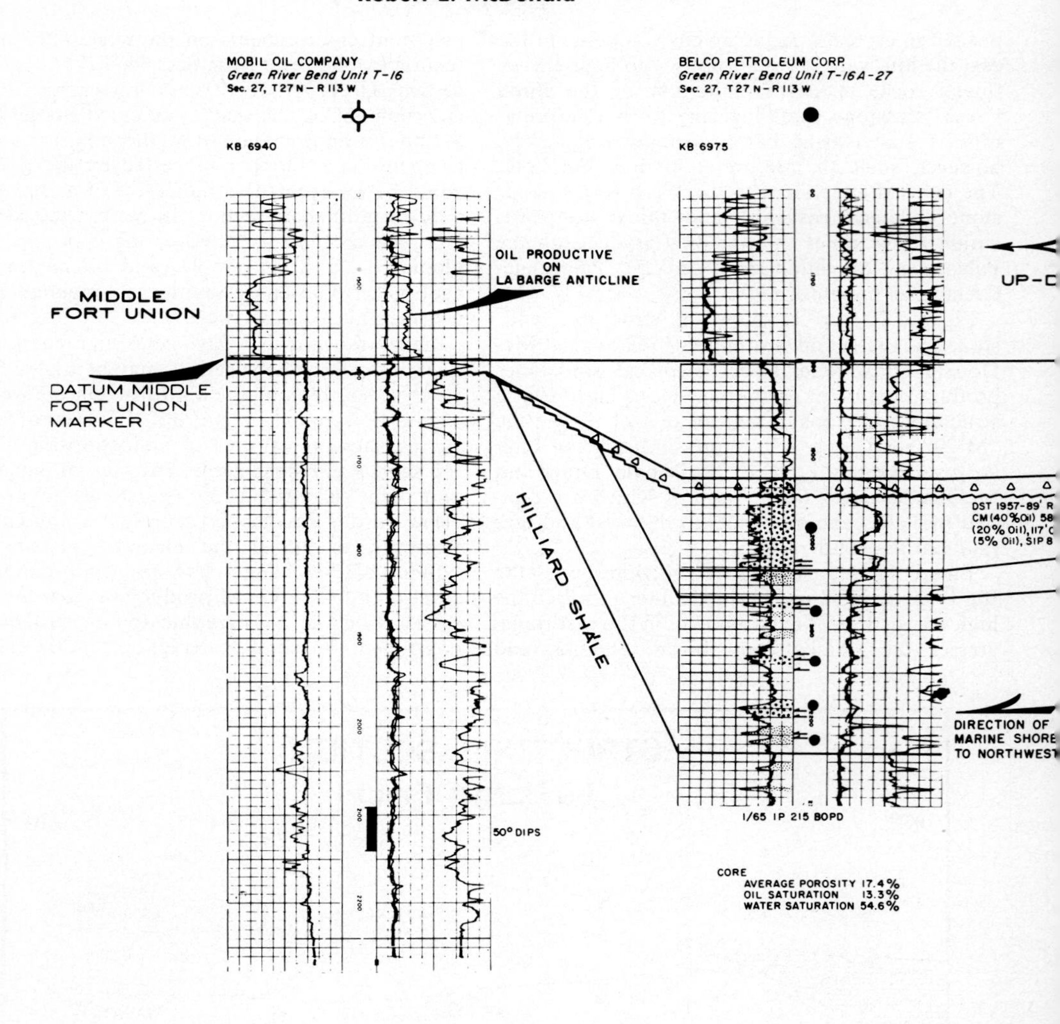

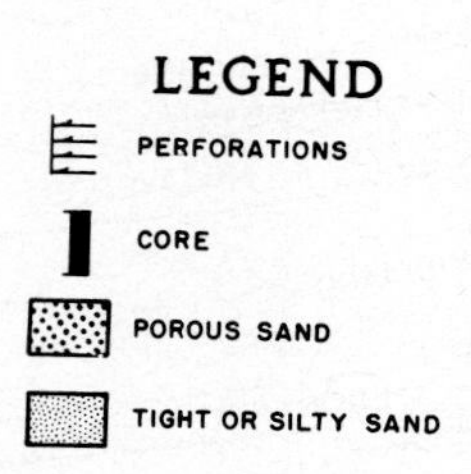

FIG. 16—Cross section of South La Barge field. Line of sec

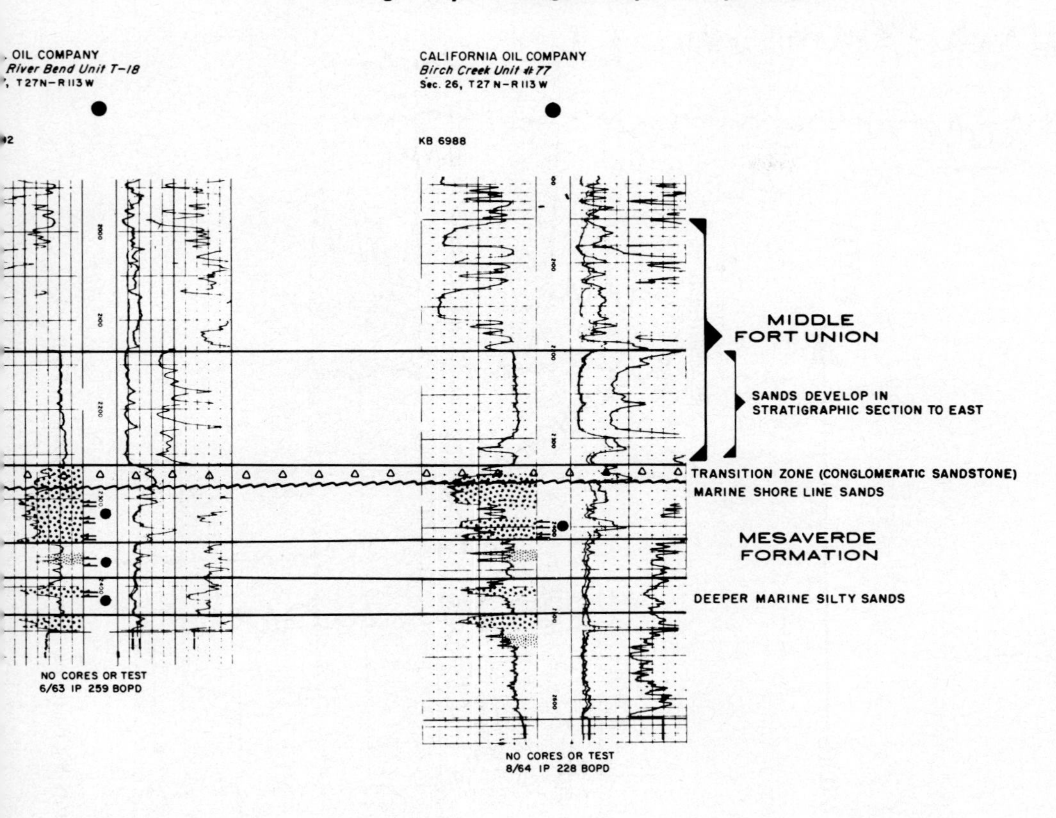

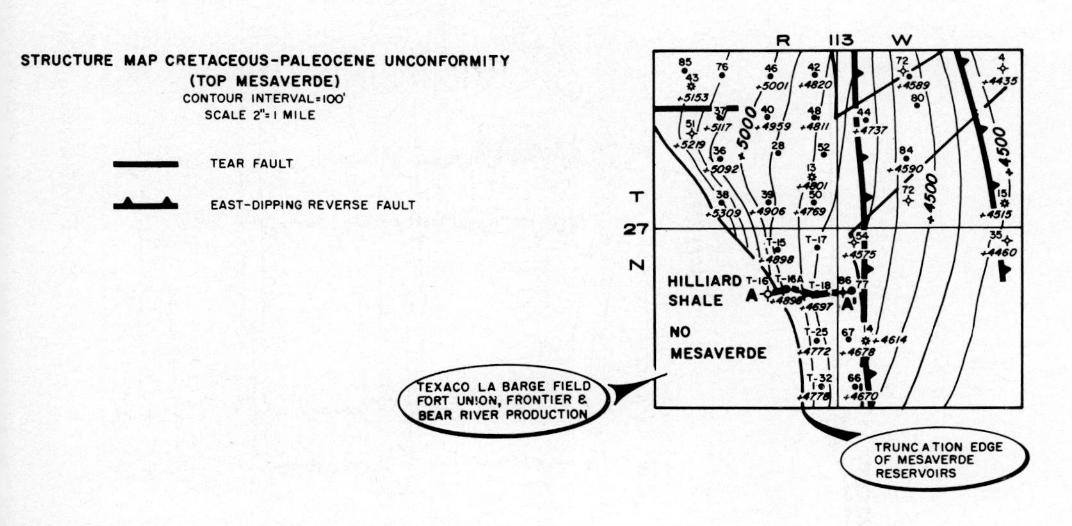

′ on structure map of Cretaceous-Paleocene unconformity.

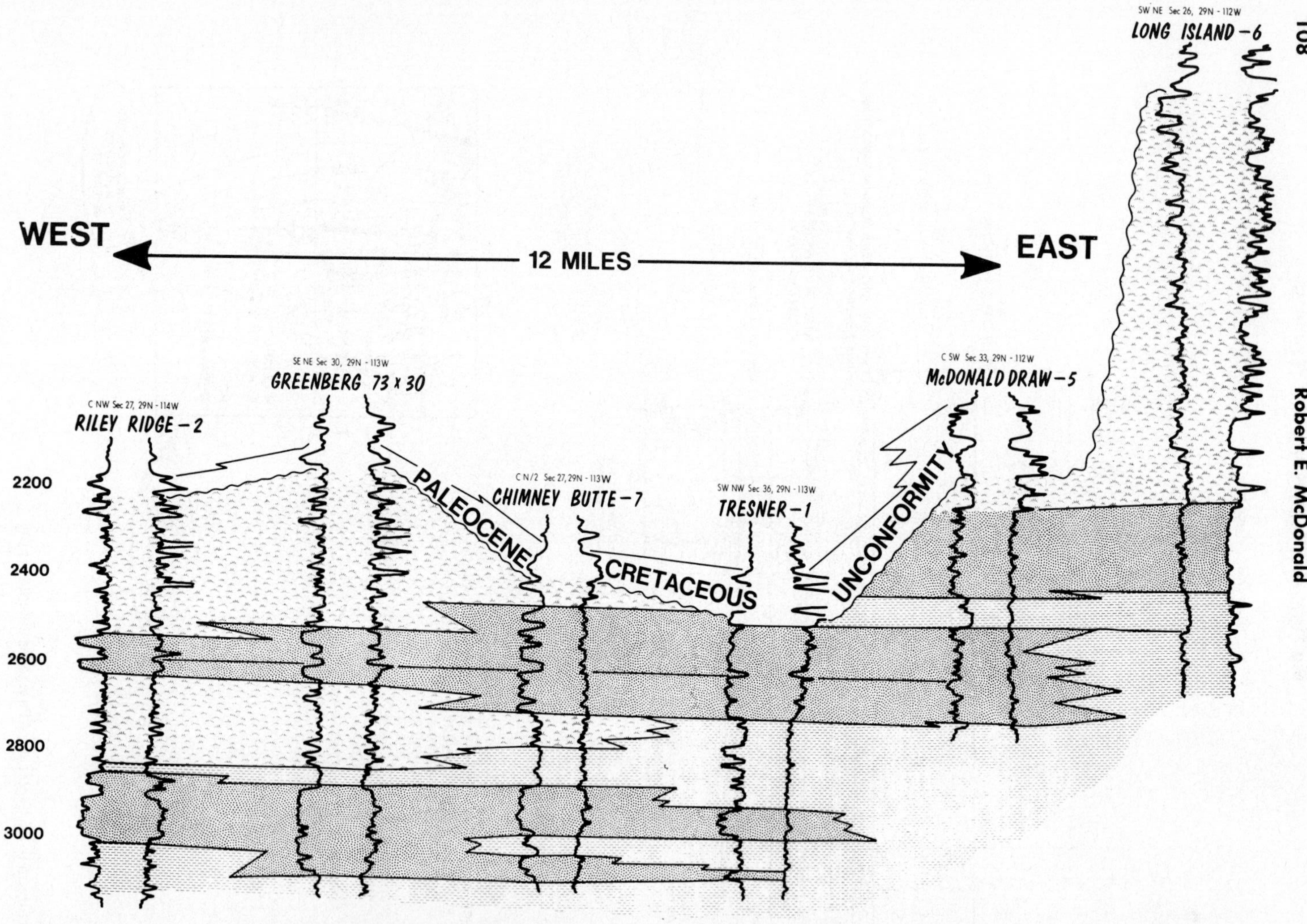

FIG. 17—Mesaverde correlations across Big Piney field. Line of section is shown on Figure 14.

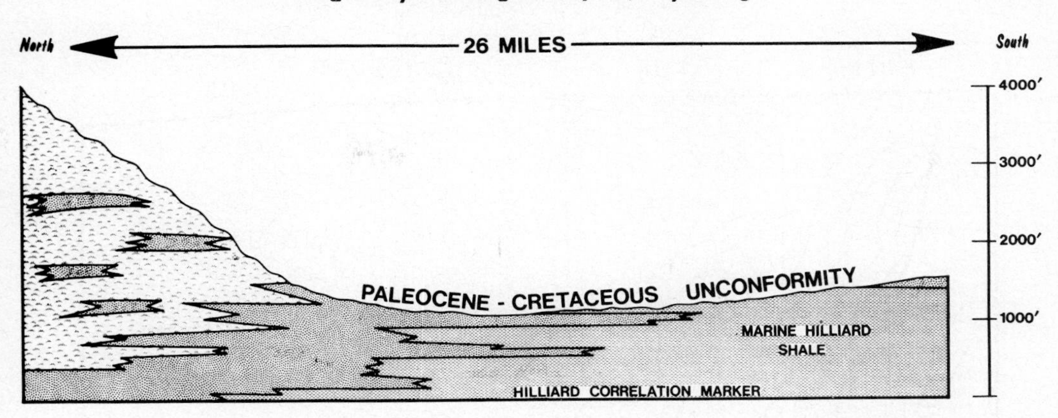

FIG. 18—Diagrammatic section across Big Piney field. Line of section is shown on Figure 14.

graphically lower "M-13" series is shown in Figure 13. Although the area of porous sandstone does not represent a continuous sandstone body, reservoir performance indicates that a single sandstone unit commonly extends 3–4 mi (5–6 km) along depositional strike. It is obvious that some movement along folds and faults occurred during Paleocene deposition, creating slight topographic barriers that limited lateral extent of the floodplain of the major drainage system. That movement occurred concurrent with Paleocene deposition is particularly apparent in T29N, R113W, where several producing Paleocene sandstones, representing over 700 ft (213 m) of stratigraphic interval, all pinch out within an east-west zone of about 3 mi (5 km; see Figs. 3, 12). These pinchouts occur over and directly east of an east-dipping thrust fault that cuts Paleocene strata.

A brief statement is necessary to tie the foregoing interpretations into Oriel's breakdown of the Tertiary into the Hoback Formation and the Chappo and La Barge Members of the Wasatch Formation. According to Oriel (1969), a conglomerate facies on the west—in many places a diamictite—grades eastward into the early to middle Eocene Wasatch and Green River Formations (Fig. 13, upper cross section). I believe that the conglomerate facies also extends downward into the Paleocene in the subsurface and is time-transgressive from east to west (Fig. 13, lower cross section).

Oriel (1969) recognized the early Eocene unconformity east of the conglomerate facies. He named the late Paleocene to early Eocene fluvial sequence beneath the unconformity the "Chappo Member of the Wasatch Formation," and named the overlying fluvial, early to middle Eocene strata the "La Barge Member of the Wasatch Formation." He described both as manifesting similar fluvial redbed lithologies on the surface, but recognized that a drab, lower energy Paleocene unit existed in the subsurface farther eastward. He suggested that the Chappo graded into the lower energy Paleocene unit, which he called the "Hoback Formation."

I believe that Oriel's high-energy fluvial Chappo Member continues into the subsurface (Fig. 13), thickening eastward, and is the upper high-energy portion of the subsurface Paleocene. The lower energy Hoback intertongues with the Chappo and is time-transgressive from east to west.

Cretaceous

Mesaverde Formation—There are three major and two minor areas of accumulation in the Mesaverde on the La Barge platform. Major production is from the Big Piney field in T29N, R113W, the Saddle Ridge field in T28N, R113W, and the South La Barge field in T27N, R113W. The trapping mechanisms were described briefly in the section on "Tertiary and Mesaverde Fields."

Figure 14 is a paleogeologic map of the Mesaverde prior to Tertiary deposition. It is apparent that the area was anticlinally folded during Late Cretaceous and early Paleocene times, and that the broader areas of outcrop (unit V) reflect the influence of the Moxa arch to the south and the Monument Buttes–Blue Rim arch to the southeast.

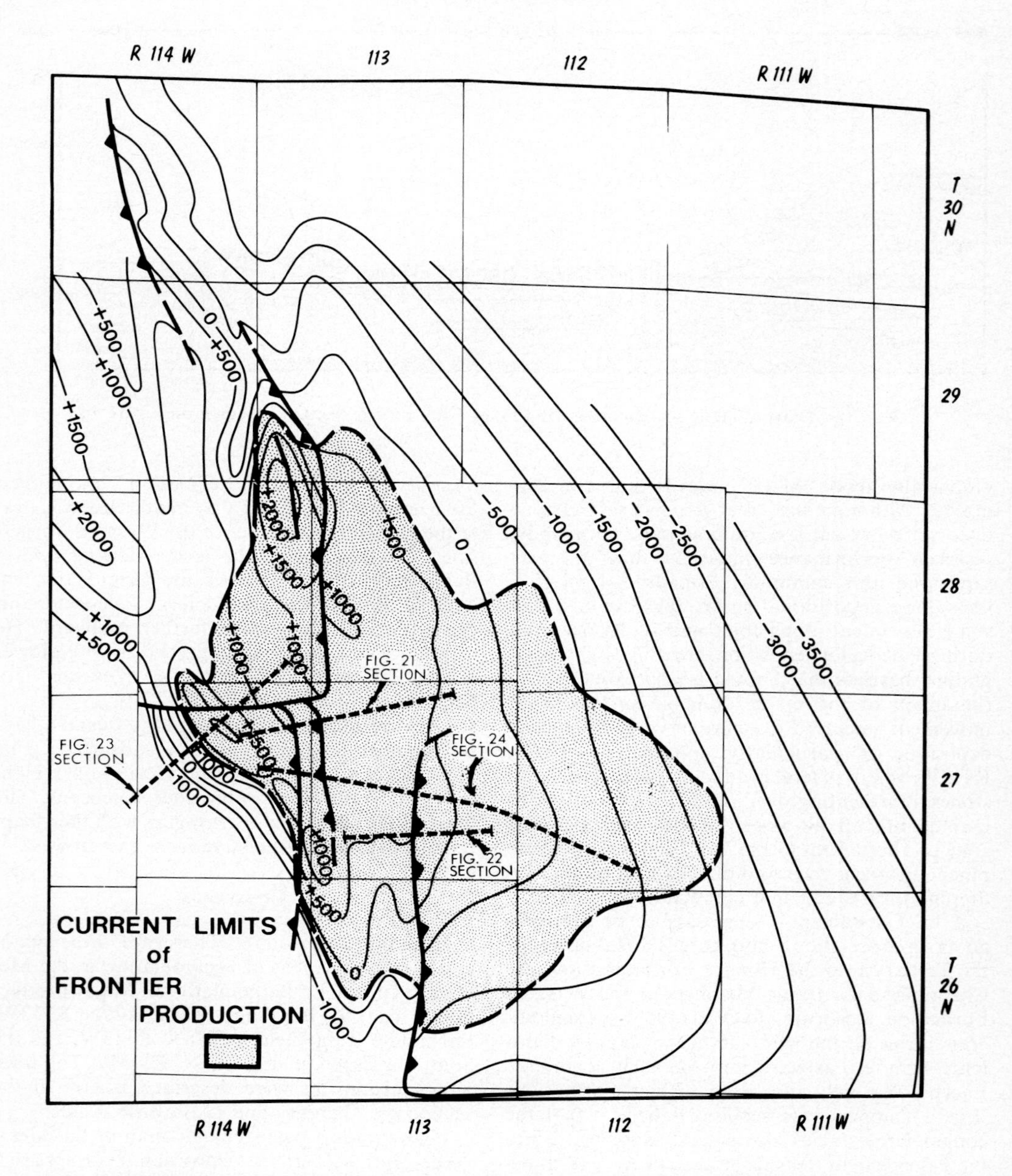

FIG. 19—Structural contour map, Big Piney–La Barge area. Datum is bentonite marker bed between first and second Frontier sandstones. Lines of sections for Figures 21-24 are shown. Cumulative production from Lower Cretaceous reservoirs has been 780 Bcf of gas and 9 million bbl of condensate.

These subcropping units (I–VII) are time units, not lithologic units; unit I is the oldest and VII is the youngest. The Mesaverde was deposited during a Late Cretaceous regression that progressed, in general, from northwest to southeast; some local exceptions probably were caused by shoaling over the Big Piney anticline. Within any given unit, therefore, lithologies may be expected to grade from paludal to littoral-paralic to marine in a northwest-southeast direction.

Figure 15 is a diagrammatic section across the South La Barge field running east-west through the S½ of T27N, R113W. The Mesaverde is entirely eroded from the crestal portions of the La Barge structure, and Mesaverde units dip steeply eastward along the eastern flank of this thrust structure.

Figure 16 includes a structure map of a part of the South La Barge field and an electric-log cross section illustrating the Paleocene-Mesaverde relationships. Datum for the section is a Paleocene correlation marker.

Figure 17 is an electric-log cross section prepared with an Upper Cretaceous correlation marker as datum. The section illustrates both the stratigraphic downcutting by erosion across the old Big Piney anticlinal fold and the facies changes that occur within time-equivalent stratigraphic units. This is a west-east section across the southern part of T29N, R112–114W. It plainly reflects the northern plunge of the anticlinal fold. Figure 18 is a north-south section, also based on an Upper Cretaceous correlation marker, from about the northeastern corner of T29N, R112W, southward to the center of T26N, R112W.

Frontier Formation—Between the Mesaverde Formation and the gas-bearing Frontier sand-

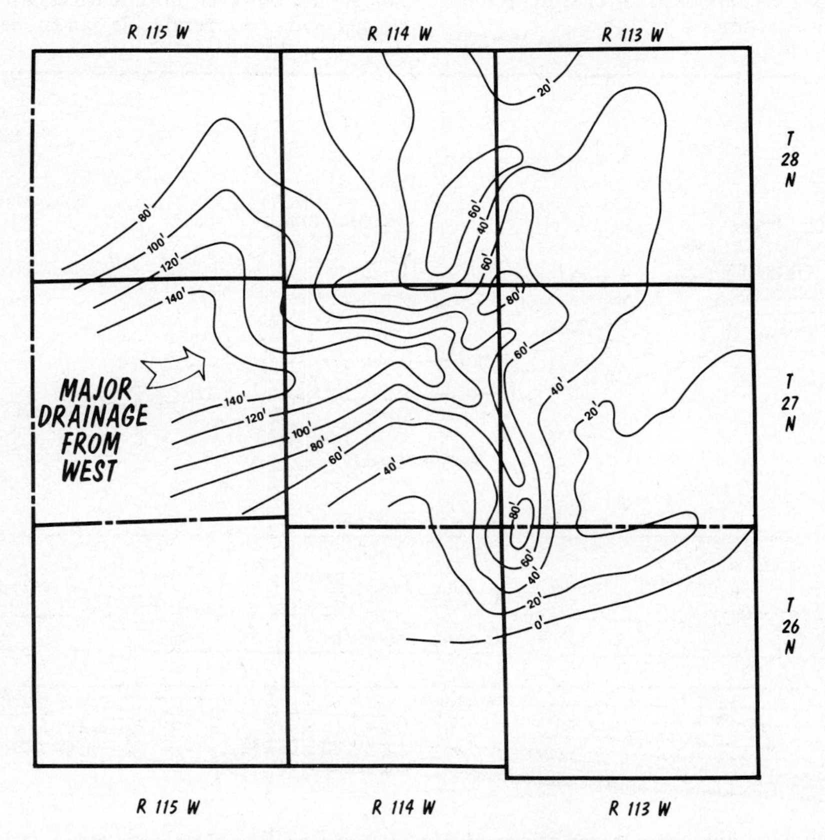

FIG. 20—Isopach of net sandstone, first Frontier sandstone. C.I. = 20 ft.

stones, a monotonous sequence of marine shale is present. This shale is usually designated as the "Hilliard Shale" in the western Green River basin. Average thickness within the mapped area is about 3,500 ft (1,070 m). The Frontier Formation is the major natural gas reservoir in the Big Piney area. Figure 19 is a structural contour map of a bentonite correlation marker between the first and second Frontier sandstones.

West of the La Barge thrust—or Hilliard thrust of Oriel (1969)—accumulation in the Frontier is essentially controlled by structure but, to the east, northeast, and southeast, Frontier production is most commonly obtained from stratigraphic traps. Some exceptions and further information are given by Shipp and Dunnewald (1962). Exploration of the Frontier has ceased basinward from the La Barge platform area because of deteriorating sandstone conditions and subcommercial flow rates—not because of its lower structural position in that direction.

The first Frontier sandstone is the most prolific producing zone on the Dry Piney structure in T27N, R114W (Fig. 19), but reservoir quality and sandstone thicknesses diminish basinward. Figure 20 is an isopach map of the first Frontier sandstone, and Figure 21 is a diagrammatic cross section. The first Frontier appears to represent the eastern edge of a major delta built into the Coloradan seas from the west. Petrographic work reveals the presence of an unusual low-temperature feldspar peculiar to the Idaho batholith; thus this area may be a source area.

The second Frontier (Fig. 21), which represents a period of intermediate depositional environments, has been subdivided roughly into five benches or sandy units across the platform area. Again the source appears to have been to the west, because the sandstones are generally thicker and more numerous in this direction; thicker sandstones, however, do not necessarily indicate cleaner and more permeable sandstones. Perme-

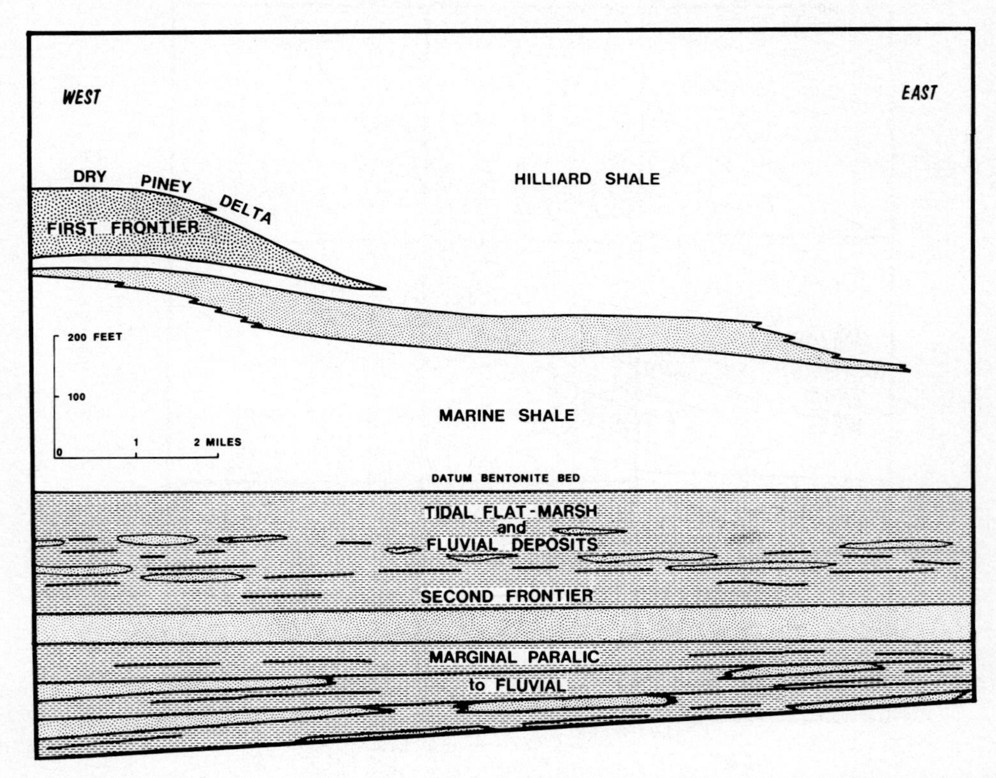

FIG. 21—Diagrammatic west-east cross section of Frontier sandstone. Line of section is shown on Figure 19.

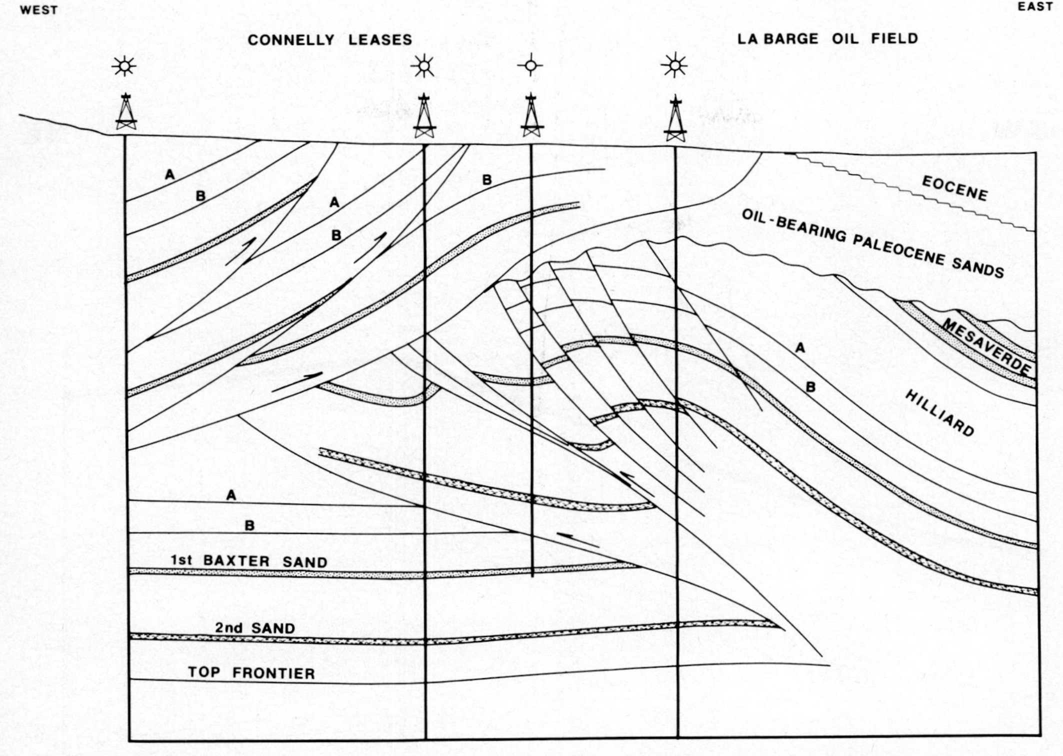

FIG. 22—West-east cross section through La Barge area. Line of section is shown on Figure 19.

ability trends are more commonly oriented NNE-SSW.

The first bench or sandy interval—the top of the second Frontier—is typified by sporadic pods of sandstone, siltstone, coal, and mudstone, suggesting widespread tidal flats and marshes with a distributary system from the west. Although the first bench sandstones are generally thin and discontinuous, many are extremely porous and have excellent gas deliverabilities through the central platform area.

The second bench of the second Frontier represents a very widespread transgressive marine unit which can be traced extensively to the east and south off the La Barge platform. Where this unit is not clay filled, as in the North La Barge area, deliverability may be very prolific, and some wells produce 6–7 MMcf/day against line pressure with no stimulation. Complete geologic treatment of the depositional patterns of the second bench could occupy a full paper. Any explorationist working the Green River basin should devote much more study to this unit.

The third, fourth, and fifth benches of the second Frontier are all lenticular sandstone bodies in intermediate paralic, tidal-flat, marsh, and distributary depositional environments. Although any one of these sandstones may be a good reservoir over relatively small areas, they are either clay filled or absent over large areas. These benches are not readily traceable beyond the immediate Big Piney–La Barge producing complex.

Fault relations are extremely interesting and are critical to development across the La Barge anticline in T27N, R113W (see Fig. 15). The east-dipping thrust that subcrops beneath the west-dipping La Barge or Hilliard thrust was not generally recognized in earlier literature, although relations are actually very clear-cut. Explanation of fluid contacts requires a more reasonable solution

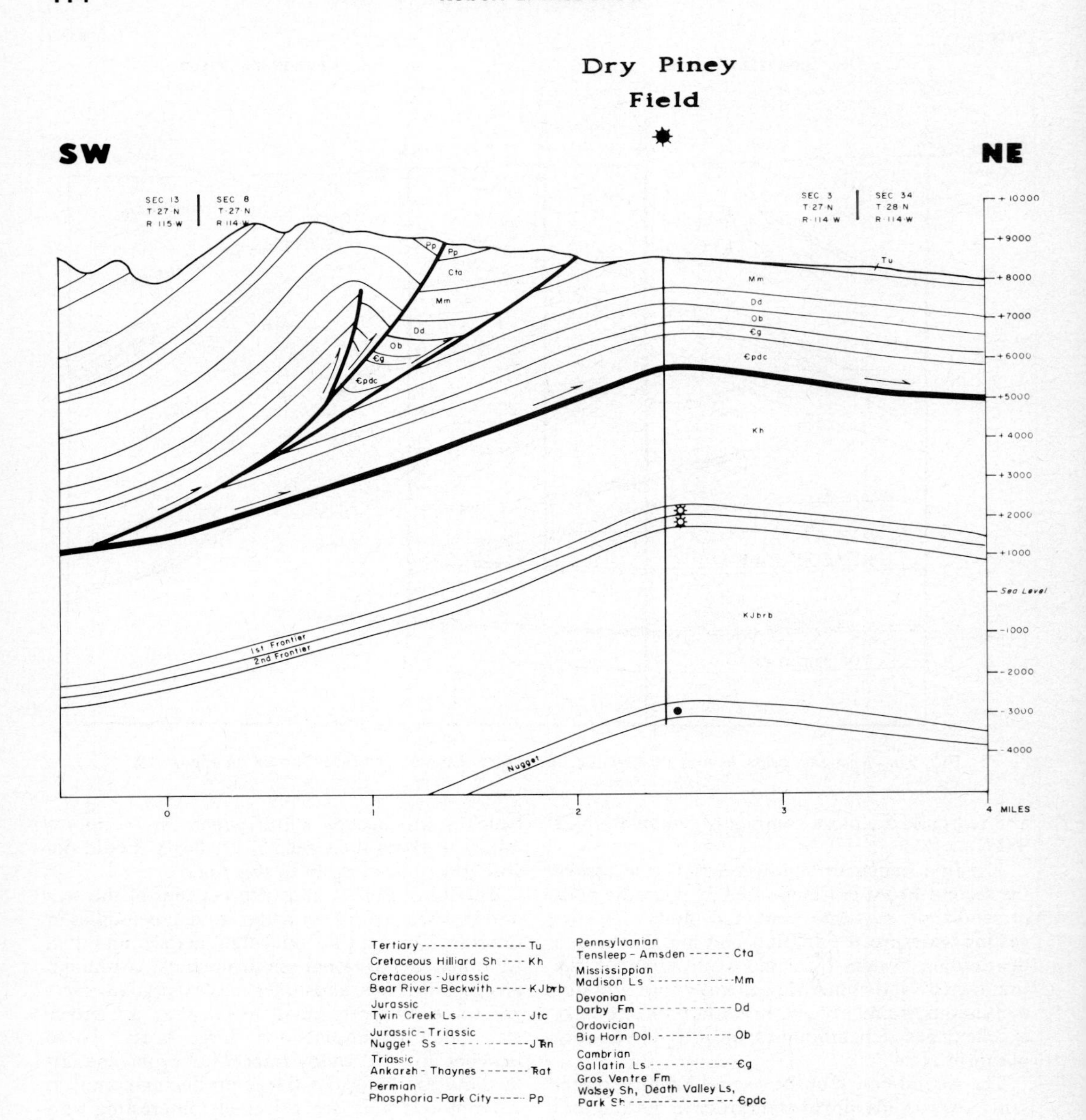

FIG. 23—Southwest-northeast cross section of Dry Piney field. Line of section is shown on Figure 19.

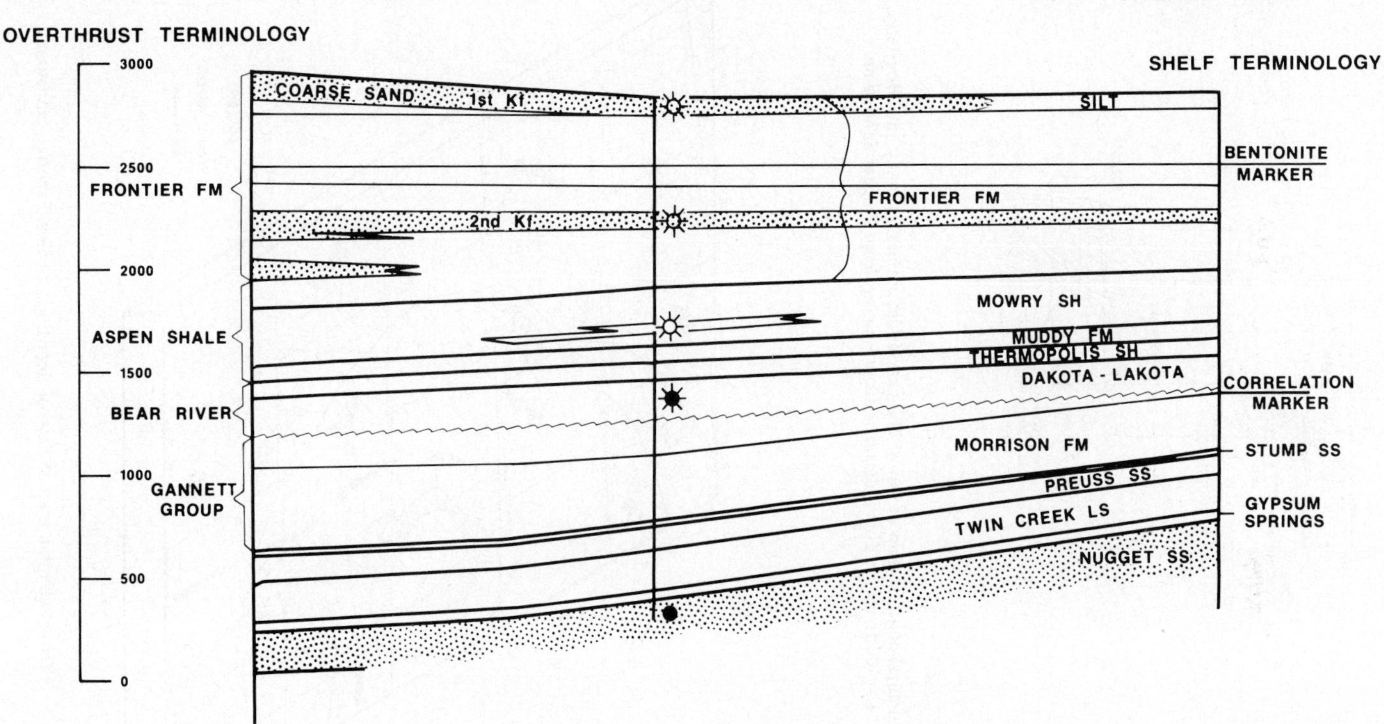

FIG. 24—Diagrammatic section of Jurassic–Lower Cretaceous from Dry Piney to East La Barge area. Line of section is shown on Figure 19.

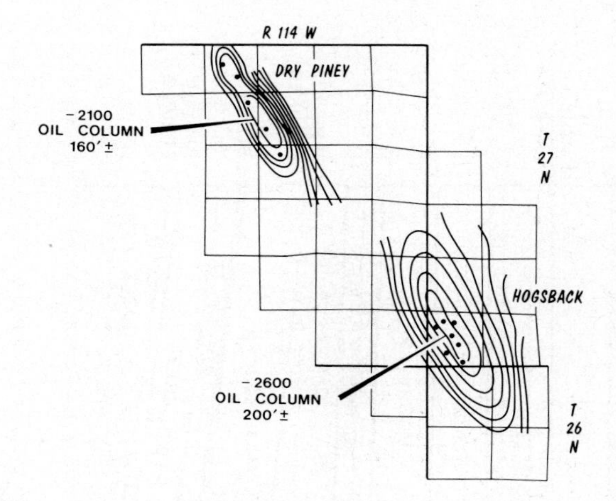

FIG. 25—Structural contour map of top of Nugget Sandstone in Dry Piney and Hogsback fields. C.I. = 100 ft. Black dots indicate wells producing from Nugget Sandstone.

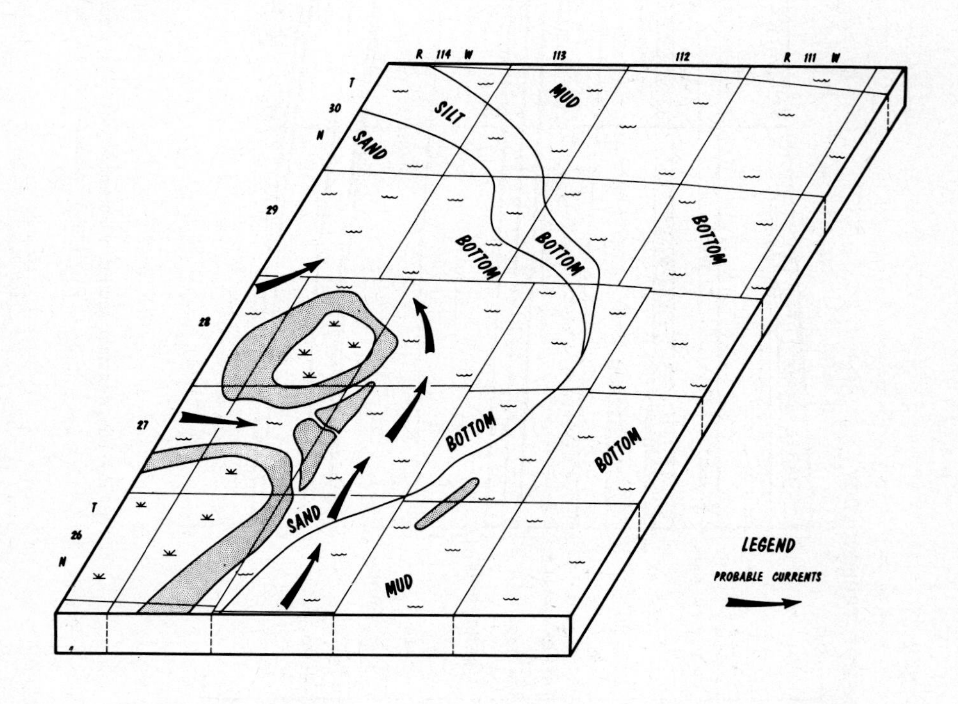

FIG. 26—Paleogeographic map at time of deposition of first Frontier sandstone.

than simply the imbrication of west-dipping thrusts. Figure 22 is another cross section extending west-east about 5 mi (8 km) through the southern part of T27N, R113W. Well data show in detail the west-dipping La Barge or Hilliard thrust with its attendant imbrication, as well as the imbricated east-dipping thrust fault that is truncated by the La Barge thrust.

Figure 23 illustrates another interesting and significant facet of the Dry Piney field, located in the northern part of T27N, R114W. This field, in which accumulation is structurally controlled, is situated within the "Disturbed Belt." Wells are spudded in overthrust Paleozoic rocks—most commonly Mississippian—and are drilled through the Darby or Hogsback thrust into underlying Cretaceous rocks. The crestal portion of the fold in the Cretaceous is nearly identical in position to the top of an anticlinal feature in the overlying Cambrian or any other overthrust unit. This fold developed in both the overthrust and subthrust during or subsequent to the final stages of thrusting.

Triassic to Lower Cretaceous

Figure 24 is a scaled diagrammatic section of an area extending about 12 mi (19 km) from Sec. 10, T27N, R114W, to Sec. 33, T27N, R112W. It includes the Triassic-Jurassic Nugget Sandstone at the base and the first Frontier sandstone at the top. In addtion to the Frontier, productive units include a sandy facies of the Mowry Shale, the Dakota Sandstone, and the Nugget Sandstone.

The Mowry sandstone facies locally is called "Muddy" or "Bear River," both of which I believe to be incorrect. As indicated on Figure 24, the thin sandstones are clearly Mowry equivalents and lie above both the Muddy and Bear River units. The production is largely controlled stratigraphically and, on parts of the platform, is fairly significant.

The Dakota-Lakota interval is relatively thick overall, but the sandstones are thin and discontinuous. It produces some oil over the Hogsback anticline but is not an important productive zone. The thick sandstones typical of the Church Buttes

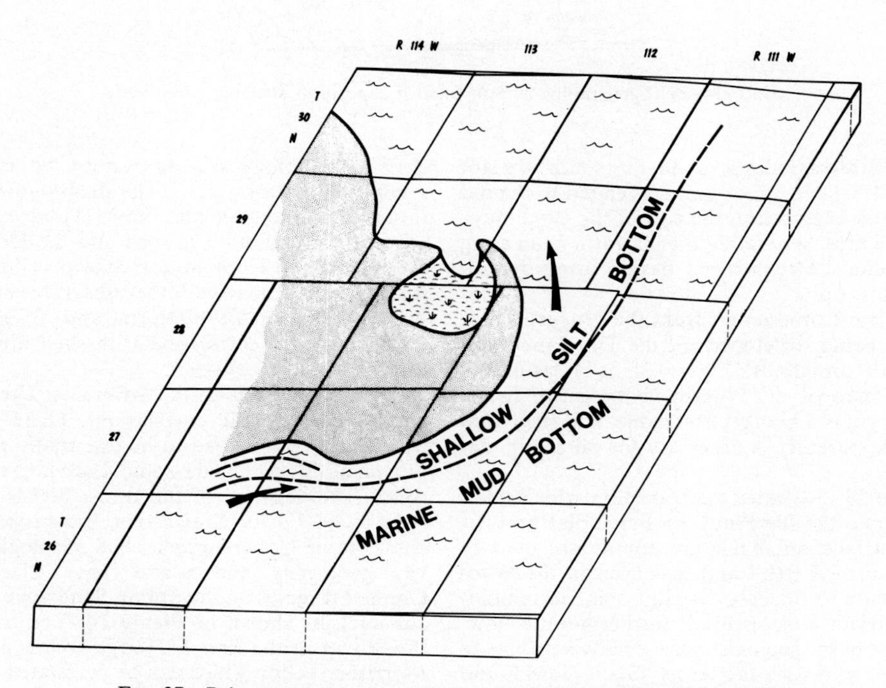

FIG. 27—Paleogeographic map at time of deposition of basal Mesaverde.

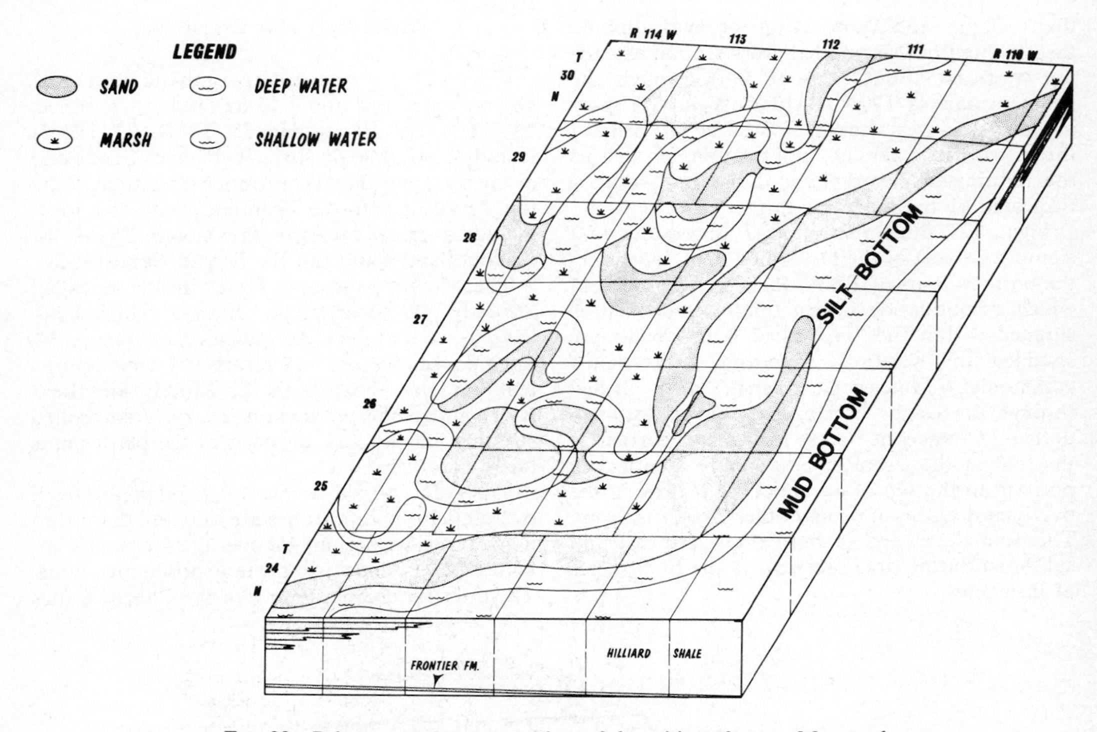

FIG. 28—Paleogeographic map at time of deposition of upper Mesaverde.

and the Bridger Lake areas to the south are not present. I believe minor arching related to the old Moxa arch occurred to the east of the Big Piney–La Barge area, separating the platform from a sag in the central Green River basin during deposition of this unit.

Significant production from the Nugget Sandstone is being developed on the Dry Piney and Hogsback anticlines. Figure 25 is a structural contour map of the Nugget in these two fields. The Nugget is a blanket sandstone, and structural closure apparently is necessary for oil accumulation.

Figure 24 illustrates the transition which takes place across the Big Piney–La Barge platform between surface nomenclature commonly used in the "Disturbed Belt" and that used in the Wyoming craton to the east. Usually, some terminology from both areas is used; furthermore, a Powder River basin geologist occasionally ventures to Big Piney and uses the terms "Skull Creek" and "Fall River," which further confuse the situation. By careful use of both samples and electric logs, shelf terminology can no doubt be extended across the platform area. The diagrammatic section is drawn to scale and shows (1) the extension of shelf correlations and tops into the Dry Piney area, (2) an attempt to correlate overthrust-belt surface terminology with the subsurface strata on the west side of the platform, and (3) an intermeshing of this correlation with shelf nomenclature.

A problem exists in the lowermost Cretaceous and Upper Jurassic correlations. Generally the Morrison is considered to be essentially Jurassic, but possibly to include some Lower Cretaceous rocks. The Gannett of the thrust belt is considered to be Lower Cretaceous, but probably includes some Upper Jurassic. U.S. Geological Survey geologists and others have placed the Gannett directly on the Stump Sandstone (Upper Jurassic), as shown in Figure 24. Yet, overlying the Stump in the East La Barge area is a typical Morrison section which can be correlated into the so-called "Gannett" section within the thrust belt. There is a good correlation marker lying within the Morrison, only 60 ft (18 m) below the unconformity at East La Barge. At Dry Piney, however, this marker lies 208 ft (63 m) below the Lakota unconformity, and this interval perhaps represents the lowermost Cretaceous Gannett.

Paleogeographic Maps

Figure 26 is a paleogeographic map at the time of deposition of the first Frontier sandstone. It depicts the postulated deltaic buildup into the Coloradan sea. Sand, silt, and marsh areas are based on lithologies interpreted from samples and electric logs.

Figure 27 is a paleogeographic map at the time of deposition of the basal Mesaverde. Although the Mesaverde is clearly Montanan farther eastward, this basal Mesaverde in the Big Piney area may be late Coloradan in age, because it is unquestionably older than the Mesaverde of the Rock Springs uplift. Lithologic units are interpreted from actual well penetrations. Buildup of good littoral sandstones digresses somewhat from the usual northeast-southwest strandlines over the old Big Piney anticline. Perhaps slight flexing had commenced, causing shoaling and high-energy littoral and paralic depositional environments.

Figure 28 represents the paleogeography at the time of deposition of the upper Mesaverde, which probably extended into early Montanan time. Some deviation of the strandline from the basic northeasterly trend is noted again. This trend is based on the probable stratigraphic position of a massive littoral sandstone that occurs in the subsurface east and south of the La Barge platform, although well density is not sufficient to establish definite continuity.

Figure 29 depicts the paleogeography of about middle to late Paleocene time. A major drainage system flowed southward from the northwest, working southeastward through the northwestern

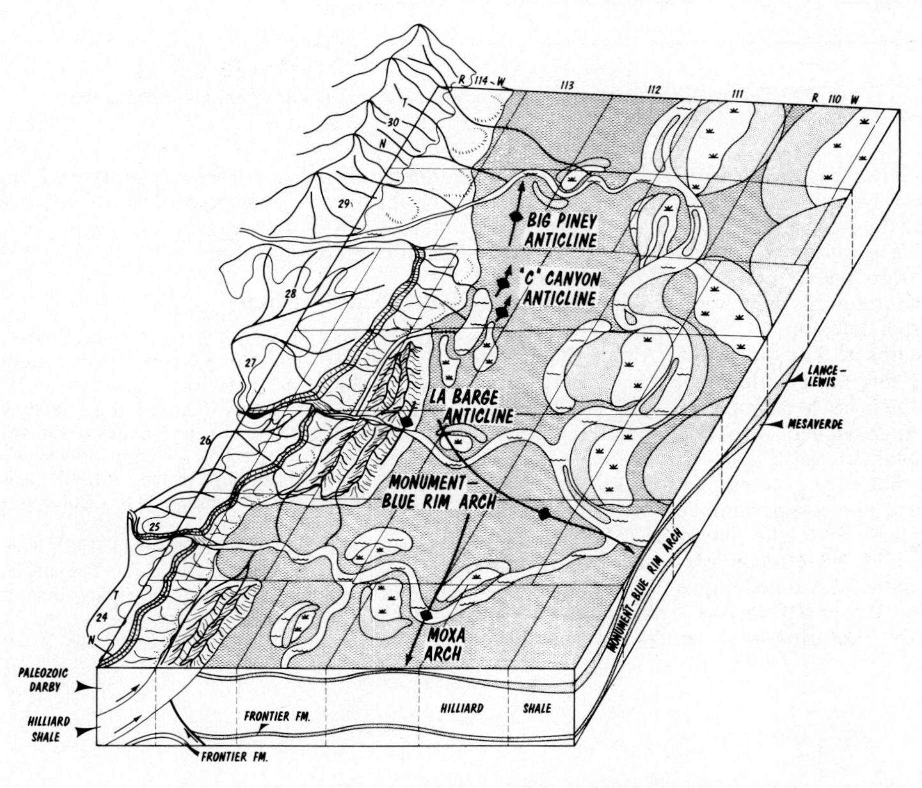

FIG. 29—Paleogeographic map of middle to late Paleocene time.

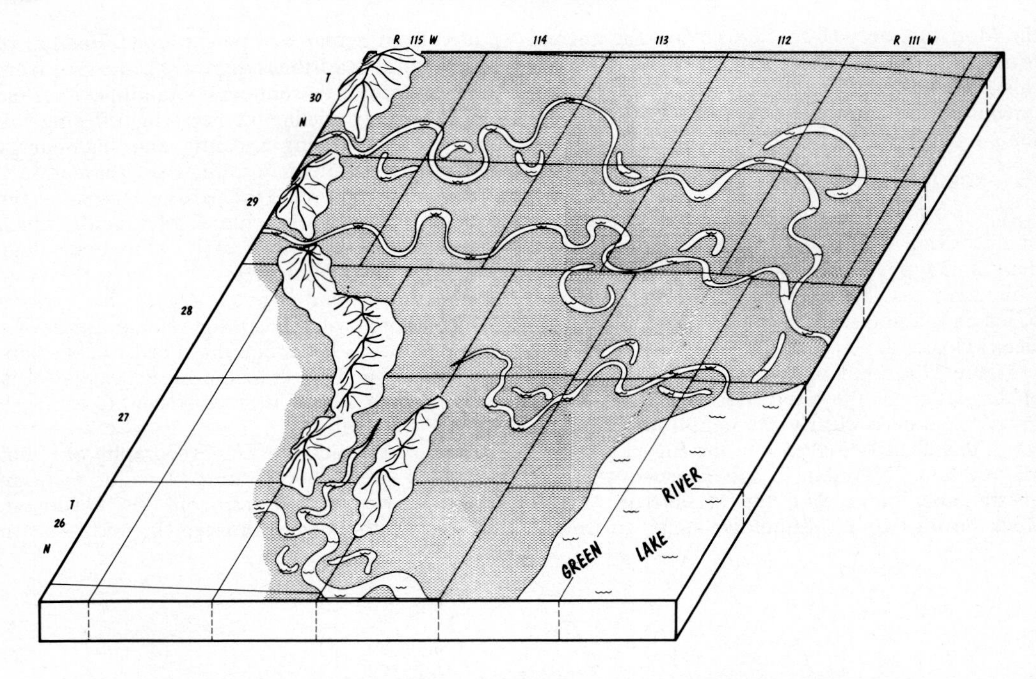

FIG. 30—Paleogeographic map of Big Piney–La Barge area during middle Eocene time.

Green River basin. In earlier Paleocene time, the drainage system was limited to the deeper basinal area to the northeast. However, continued basin subsidence caused the drainage system to broaden its floodplain, and it was forced southwestward continuously by the rapid alluviation resulting from erosion of the rising Wind River Mountains to the northeast. Along the northeastern and eastern flanks of the La Barge platform, the massive channel, bar, and point-bar systems intertongued with the piedmont, alluvial, and small-stream fluvial depositional environments that were predominant on the west.

Figure 30 is a representation of middle Eocene paleogeography. The mountains of the "Disturbed Belt" were much less pronounced, and alluvial material reached high onto the slopes. Lake Gosiute (Green River Lake) encroached onto the La Barge platform and eventually left tongues of the lacustrine Green River Formation lying nearly against the Paleozoic rocks of the Hogsback thrust before withdrawing as a result of regional uplift and desiccation.

REFERENCES CITED

Armstrong, F. C., and S. S. Oriel, 1965, Tectonic development of Idaho-Wyoming thrust belt: AAPG Bull., v. 49, no. 11, p. 1847-1866.

Dorr, J. A., 1952, Early Cenozoic stratigraphy and vertebrate paleontology of the Hoback basin, Wyoming: Geol. Soc. America Bull., v. 63, p. 59-93.

Oriel, S. S., 1969, Geology of the Fort Hill Quadrangle, Lincoln County, Wyoming: U.S. Geol. Survey Prof. Paper 594-M, 40 p.

Shipp, B. G., and J. B. Dunnewald, 1962, The Big Piney–La Barge Frontier gas field, Sublette and Lincoln Counties, Wyoming, *in* Symposium on Early Cretaceous rocks of Wyoming: Wyoming Geol. Assoc. 17th Ann. Field Conf. Guidebook, p. 273-279.

Altamont-Bluebell—A Major, Naturally Fractured Stratigraphic Trap, Uinta Basin, Utah[1]

PETER T. LUCAS and JAMES M. DREXLER[2]

Abstract The Altamont-Bluebell trend is composed of a highly overpressured series of oil accumulations in naturally fractured, low-porosity, Tertiary lacustrine sandstones. It now covers more than 350 sq mi (907 km^2) located across the deeper part of the Uinta basin of northeastern Utah. Postdepositional shift of the structural axis of the basin in late Tertiary time produced a regional updip pinchout of northerly derived sandstones into a lacustrine "oil-shale" sequence. Facies shifts during the deposition of more than 15,000 ft (4,570 m) of lacustrine sediments have resulted in a changing pattern of reservoir distribution and hydrocarbon charge at various stratigraphic levels. About 8,000 ft (2,440 m) of stratigraphic section is oil bearing, and up to 2,500 ft (760 m) of section contains overpressured producing zones in the fairway wells.

Reservoir performance is significantly enhanced by vertical fractures and initial fluid-pressure gradients, some of which exceed 0.8 psi/ft. The crude has a high paraffin content resulting in pour points above 100°F (37.78°C), gravities of 30–50° API, and an average GOR of 1,000 cu ft/bbl. This unique combination of geologic and hydrocarbon conditions makes it difficult to evaluate the ultimate recovery of the field, which could be more than 250 million bbl.

INTRODUCTION

The Altamont-Bluebell field, in the Uinta basin of northeast Utah, produces crude oil with high pour points from highly overpressured accumulations. Production is from multiple thin Tertiary sandstones, mainly at depths between 8,000 and 17,000 ft (2,440–5,200 m), within an area 45 by 15 mi (72 by 24 km). The field is being developed on 1-mi (1.6 km) spacing. As of June 1974, 210 wells had been completed, and production had increased to an average of 60,000 BOPD. Further significant increases in production will depend on the solution of transportation problems related to the movement of large volumes of extremely waxy crude oil with high pour points from this relatively remote location.

SIGNIFICANCE

The Altamont-Bluebell trend contains a major accumulation which is unique because of the character of reservoir and fluid properties and the highly disseminated occurrence of producible oil. These factors have resulted in unusual engineering and evaluation problems. The most important of these factors are: (1) very low matrix porosities enhanced by postlithification fractures; (2) multiple thin productive zones with abnormally high fluid pressures; (3) undersaturated waxy crude with pour points of over 100°F (37.78°C); (4) production derived from intervals up to 2,500 ft (760 m) thick in the central part of the field; and (5) difficulty in defining field limits laterally and vertically because the trap is purely stratigraphic and there are no simple downdip water levels or sharp facies boundaries to the producing intervals. The presence of sandstone matrix porosity, fractures, high fluid pressures, and multiple producing zones is the key to commercial production. The factors limiting production are (1) facies changes from sandstone to nonporous redbeds, lacustrine shale, and dense carbonate rocks, or (2) the occurrence of capillary water in the sandstones with lowest porosity. A thick series of organic-rich shales and dense carbonate mudstones provides the cap seal.

The field is significant not only as a recent major hydrocarbon discovery, but also as an example of oil accumulation near the center of a deep basin. The pertinence of the latter observation is that Altamont-Bluebell may be an example of a group of deep-basin, organic-shale-related, overpressured accumulations where hydrocarbons are (at least initially) the dominant movable fluid within a large volume of rock. Evidence of significant hydrocarbon potential in this type of setting exists in many deep-basin wildcats; their exploitation awaits only the recognition of sufficient reservoir volume.

EXPLORATION HISTORY

The Uinta basin long has been recognized as one of the most petroliferous basins in the United States, because of the exposures around the basin

[1]Manuscript received, January 17, 1975. Published with permission of Shell Oil Company.

[2]Shell Oil Company, Houston, Texas 77001.

We recognize with appreciation the numerous members of the Shell Oil Company staff, in both the Exploration and Production Departments, who contributed significantly to our knowledge during the formative stages of the exploration effort and the field development. We also thank the Shell management for permission to publish.

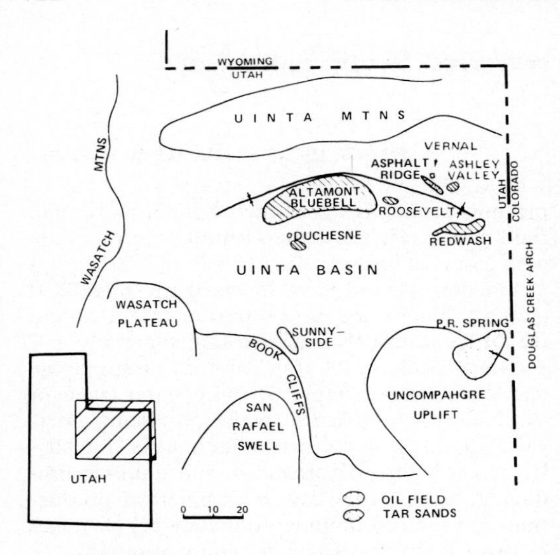

FIG. 1—Index map, Uinta basin, Utah, showing structural elements and location of major hydrocarbon occurrences.

margins of thick "oil shales," gilsonite dikes, and "tar sands" with billions of barrels of oil in place (Fig. 1). The first truly commercial oil discovery in Utah and the Uinta basin was in 1948 at Ashley Valley near Vernal. This field produces from a structural trap in Paleozoic reservoirs on the northeast margin of the basin. Several later discoveries were made in the Tertiary section, such as Roosevelt in 1948 and Red Wash and Duchesne in 1951. Exploratory activity waned and then increased again in the early 1960s, but no significant reserves were found except for the western extension of Red Wash to the Wonsits area in 1964. During 1967, shallow productive zones at Bluebell were discovered on a low-relief structural nose. Development in this field could not be supported at the original well spacing, and spacing was subsequently increased to 320 acres.

Prior to the discovery of Altamont in 1970, Red Wash was the only major field in the basin. Red Wash field also produces waxy crude from a stratigraphic trap in Tertiary rocks; however, the productive sandstones occur at much shallower depths than those at Altamont and have normal fluid and reservoir conditions (Koesoemadinata, 1970). The generally poor development-success ratios in the basin were due to rapid productivity declines from thin producing zones and to severely reduced porosities at the greater depths. Abundant noncommercial shows were found. A few wells in the central part of the basin indicated elevated formation-fluid pressures but did not prove to have further producing potential. Most published views of the Tertiary stratigraphy based on the earlier drilling activity suggested that a rather limited Tertiary stratigraphic potential remained to be tested. No doubt the reason for these views was the layer-cake stratigraphic concept whereby basin-margin redbed clastic units were correlated across the basin. From such an understanding of the stratigraphy, the top of the Wasatch facies became the general base of drilling for several years.

In October 1969, Mountain Fuel Supply initially completed their Cedar Rim No. 2, Sec. 20, T3S, R6W, which discovered what now appears to be the western edge of the Altamont trend. Shell Oil Company, in December 1969, spudded the Miles No. 1, Sec. 35, T1S, R4W, which was completed in May 1970 for 1,150 BOPD from 79 ft (24 m) of perforations between 12,341 and 12,942 ft (3,761–3,944 m) with an estimated 26 ft (7.9 m) of net "pay."

The Miles No. 1 established production at abnormally high pressure (over 0.7 psi/ft) from multiple zones, and also had an extensive interval of shows. Of greater importance is the fact that the well indicated the presence of only thin Wasatch redbeds with oil-bearing lacustrine strata below. The fact that oil shows and pressures were still increasing at total depth led to subsequent deeper drilling, which firmly established the presence of continuous lacustrine facies in the Green River Formation (Eocene) and the underlying Flagstaff (Eocene-Paleocene) in the central basin area (Fig. 2). This expanded section with oil-bearing potential disproved the belief that there was no opportunity for Tertiary production below the Wasatch, and it became clear that the Colton, Wasatch, and North Horn Formations were basin-margin facies and that the limited area of Flagstaff lacustrine facies seen in outcrops at the western margin of the basin expanded in the subsurface to the east.

In April 1971, Gas Producing Enterprises deepened their Powell No. 3 in the Bluebell field from shallow producing zones in the Green River Formation to a new total depth of 12,530 ft (3,819 m). This deepening extended the Altamont producing zones eastward 13 mi (21 km), into the Bluebell field. Current wildcat and development drilling continues along the field margins in an effort to determine economic limits of these producing zones. Greatest potential for extension is now toward the east.

UINTA BASIN GEOLOGIC HISTORY

Concurrent with the Laramide orogeny along the Wasatch Mountains on the west, the generally north-south-trending Late Cretaceous shoreline receded eastward (Young, 1966). Within the evolving Late Cretaceous coastal plain, increasing subsidence in northeastern Utah, southern Wyoming, and adjacent Colorado created a major area of internal drainage. In the Uinta basin, the lacustrine environment initially attracted inflow of clastic material primarily from the south. A thick wedge of northward-thinning coarse clastic sediment (North Horn, Colton, Wasatch facies) was deposited, which interfingered with organic-rich lacustrine clays and carbonate muds (Green River, Flagstaff facies). Later, clastic inflow from the Uinta Mountains to the north became dominant (Fig. 2). The total thickness of the sequence deposited from initiation of internal drainage to the first major Eocene expansion of lacustrine facies (Tgr 3 marker; Fig. 2) exceeds 8,000 ft (2,440 m; Fig. 3). Widespread carbonate and shale beds which resulted from this lacustrine expansion provide a good series of markers which aid in reconstruction of the more complicated relations of underlying facies.

During late Paleocene and early Eocene time, northerly derived clastics were deposited as a wedge of southward-thinning redbeds (Fig. 4). Because of the general textural similarity, color, and associated marginal-lacustrine sediments, it has been the practice to assign these rocks in the subsurface to the Wasatch Formation. They are only partly equivalent in age to the smaller Colton tongue of clastic redbeds on the south flank

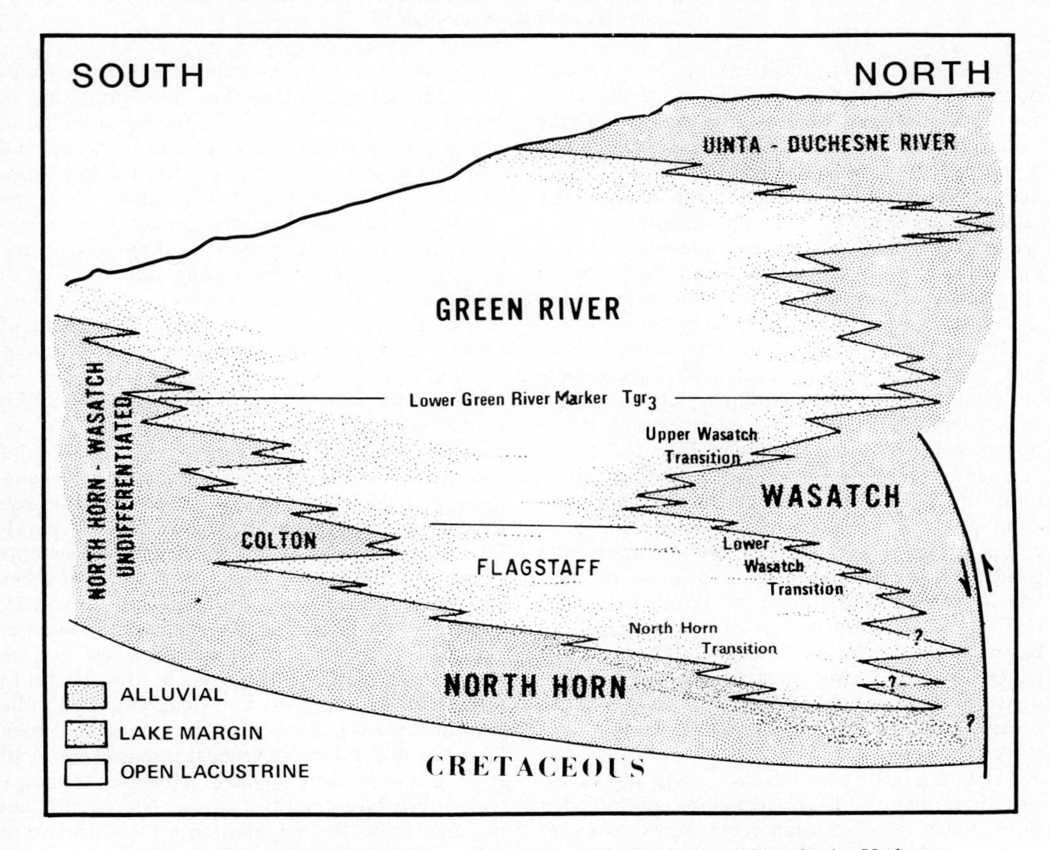

FIG. 2—Schematic stratigraphic section of lower Tertiary across Uinta basin, Utah.

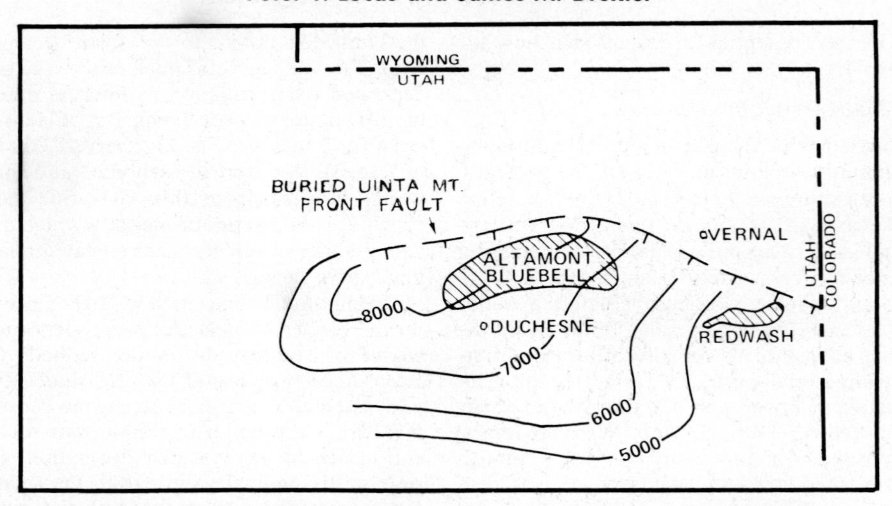

FIG. 3—Isopach map of lower Tertiary from base of Tertiary to lower Green River marker, northern Uinta basin, Utah.

of the basin and, in contrast, are underlain by an open-lacustrine, organic shale facies of the Flagstaff. The Altamont-Bluebell trap exists where the northerly derived Wasatch sandstones pinch out within the lacustrine depocenter. Where north- and south-derived sediments have crossed the lake and intermingled, as at the eastern and western ends of the basin, oil entrapment is erratic and volumes are small; the accumulations depend on geometry of individual sandstone units rather than on a more regional facies relationship.

During middle Eocene time, subsidence continued but clastic influx waned and organic-rich, "oil-shale" carbonate sediment spread more widely over the basin. This intermontane continental sedimentation continued into the Oligocene and then virtually ceased until Quaternary uplift of the Colorado Plateau. Additional differential uplift of the basin margins resulted in truncation of the entire Tertiary section across the south flank of the basin and deposition of Quaternary gravels across the north flank. This latest uplift produced a shift of the structural axis of the basin from the more central east-west trend which existed during most of the depositional phase to an axial deep directly adjacent to the presently buried Uinta Mountains frontal fault (Fig. 3).

The Uinta basin area has had a long and varied structural history. Early tectonism preserved a thick belt of Precambrian rocks across parts of the Uinta Mountains. To the south during middle Paleozoic time, the Uncompahgre became a major basement uplift. However, most important in framing the Uinta basin and influencing fault and fracture trends was the Laramide orogeny and later rejuvenation of the Uinta Mountains. Compression in the west caused extensive thrusting which provided a source of sediment; later, relaxation resulted in extension faulting, fracturing, and localized igneous intrusions. Recurrent uplift in the Uinta Mountains area from early Tertiary to Pleistocene time was expressed by pulses of clastic sedimentation and by burial of major mountain-front faults.

Within the central part of the basin, no significant faulting is documented at the surface, although orthogonal fracture sets are found throughout. However, more interesting for evaluation of subsurface conditions are the solid-hydrocarbon-filled fractures which generally parallel the trends of the major structural elements (Crawford and Pruitt, 1963). These veins or dikes are nearly vertical and are filled with ozocerite, gilsonite, and wurtzilite originating from organic-rich layers generally shallower (and less mature chemically) than those related to the Altamont trend. The dikes suggest that significant hydraulic pressures were generated in the rich source-rock layers so that extensive vertical fractures up to 10 ft (3 m) wide were opened. However, analogy with subsurface fractures of the Altamont-Bluebell field must be tempered with the observation that most of the Altamont-Bluebell fractures have

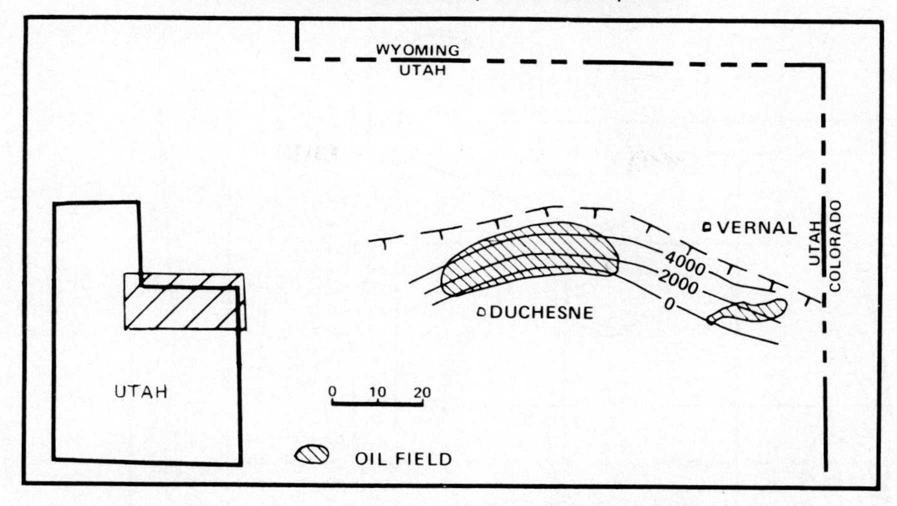

FIG. 4—Isopach map of north-derived Wasatch redbeds below lower Green River marker.

significant mineralization lining the fracture walls. This is not the general case in the dikes filled with solid hydrocarbons.

FIELD CHARACTERISTICS

Structure

Structural closure plays no part in entrapment of hydrocarbons at Altamont-Bluebell; however, the regional dip provides the setting for updip porosity pinchouts. The productive limits of the field extend more than 5,000 ft (1,525 m) downdip to essentially the deepest part of the basin (Fig. 5). To the north, a seismically mapped, buried fault is present adjacent to the basin deep and parallel with the Uinta Mountains front. Seismic records show over 6,000 ft (1,830 m) of displacement at the base of the Tertiary section, but little, if any, displacement is recognizable in the stratigraphic column above the level of the top of abnormal formation pressures. The fault appears to be a steep reverse fault along which major displacement was more or less synchronous with initiation of deposition of the Wasatch redbed wedge. Some growth continued during deposition of the lower Green River beds. Both the fault displacement and the impermeable redbeds probably contribute to retention of the abnormal fluid pressures of the field in the mountainward direction.

South of the area of major mountain-front faulting, neither seismic nor well data indicate faulting which might control the field outline. A few minor faults indicated by seismic data may influence local reservoir continuity, but this has not yet been established. In addition, faulting is not sufficiently common or of such magnitude as to have a meaningful causal relationship to the fracturing which occurs throughout the field.

Fractures

Fractures in the reservoirs of Altamont-Bluebell are essential for commercial flow rates. Core studies indicate a significant frequency of nearly vertical, mineral-lined open fractures throughout the field. Most nonmineralized breaks in recovered cores have been interpreted as having been induced by the coring process. In a collection of cores from 3 to 4.5 in. (7.6–10 cm) in diameter, more than 90 percent of the fractures are within 10° of vertical and 5 percent fall within 40° of horizontal. The pattern of occurrence is about 50 percent single fractures within any core diameter (Fig. 6) and 13 percent as crossing sets (Fig. 7). The remainder occur as multiple, subparallel fractures (Fig. 8). No significant results were obtained from limited attempts to determine fracture directions. The dominant directions of reservoir communication probably will not be determined until more data are available on asymmetry in reservoir drainage and pressure communication between wells. In the older, shallow Bluebell accumulation (northeast corner of the Altamont trend), the communication proved

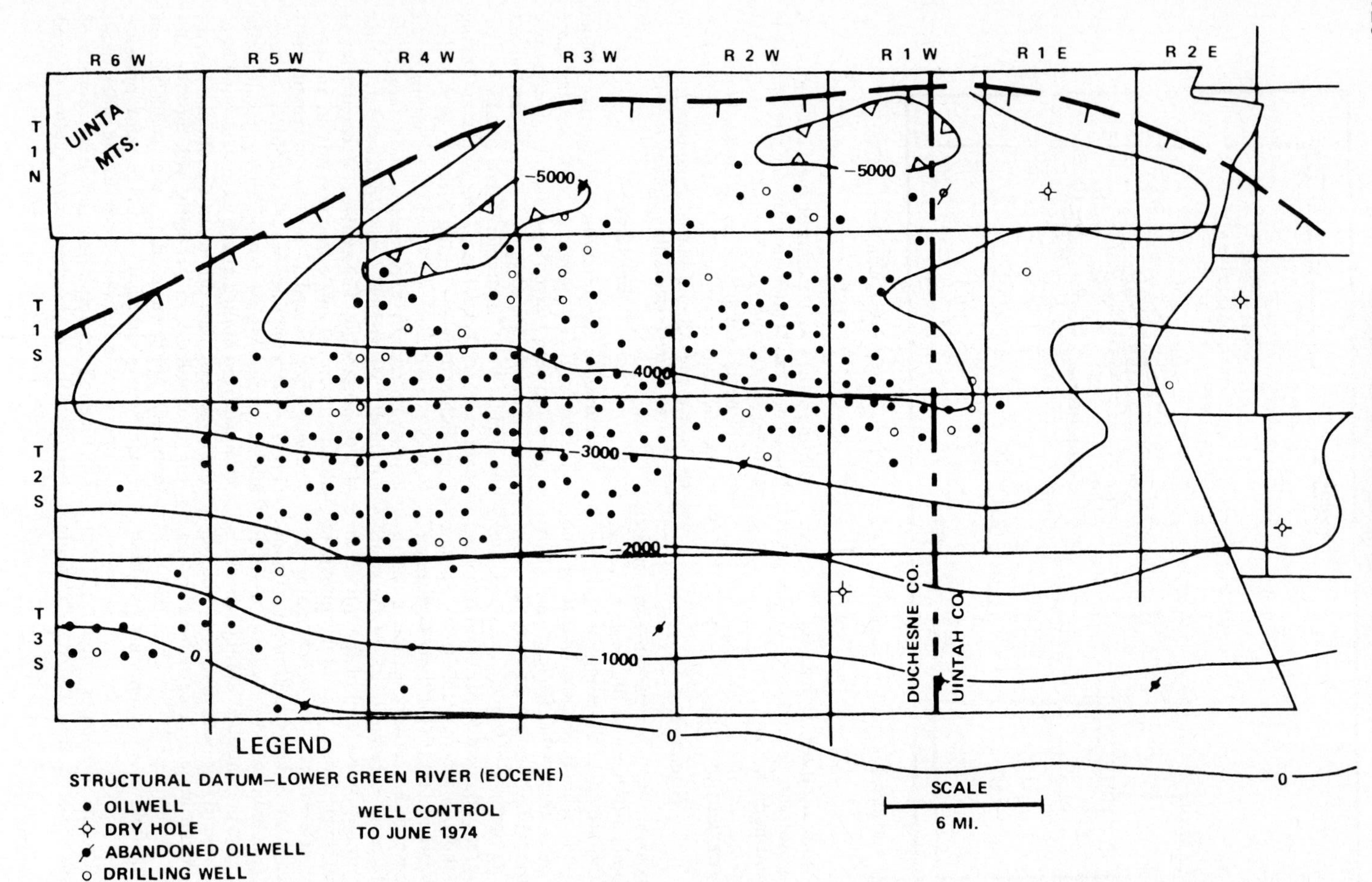

FIG. 5—Structure map of top of lower Green River marker, Altamont-Bluebell field, Utah. Structural nose across east end of field is site of original shallow Bluebell production.

to be east-west, approximately parallel with the axis of the structural nose.

Fracture frequency is controlled significantly by lithology. The dense carbonate mudstones are most fractured, sandstones less so, and shales are least fractured. Because vertical fractures are dominant, nearly 90 percent of fractures examined terminate within the recovered cores, either abruptly at stylolites or depositional lithologic boundaries (Figs. 9, 10), or gradually at more subtle lithologic transitions from sandstone or carbonate rock to shale. From 25 to 30 percent of the original fractures observed are open and fluid filled. These open fractures have average wall separations of about 0.5 mm and are approximately 50 percent infilled with coarse crystalline minerals. The mineral fill is predominantly calcite but, where fractures cut through sandstones, significant amounts of quartz were deposited prior to calcite deposition.

The maximum frequency of fractures appears to be in the zone of interbedded thin sandstones, carbonate rocks, and lacustrine shales. The redbeds on the north and the more massive shale and carbonate mudstone sections on the south are relatively unfractured. However, the optimum producing trend is offset northward from the area of thin sandstone interbeds toward the redbed wedge, where thicker bedded and coarser sandstones provide essential matrix porosity for economically adequate reservoir storage capacity. Production from the fractured mudstones in the central lacustrine facies declines rapidly in spite of high initial reservoir pressures.

As is common in all attempts to analyze subsurface fracture distributions, the interpretation of results can be significantly biased and limited by the nature of material available for study. Because small-diameter cores rarely recover rock with large fractures preserved in subsurface position, determination of actual spacing and orientation of open fractures in the horizontal plane becomes an imprecise statistical exercise. It is concluded from the best data available that *open* fractures in the central Uinta basin are probably oriented in a more or less orthogonal network, spaced a few feet apart. A more subtle observation is that porous sandstones appear to have

FIG. 6—Photograph of plug (1 in. or 2.5 cm in diameter) from core taken at 10,742 ft (3,275 m), showing vertical fracture in sandstone. Oil-stained fracture is lined with euhedral quartz overgrowths overlain by patches of calcite druse. Fracture permeability is 600 md through estimated average opening of 0.3 mm. Sandstone matrix porosity is 5.5 percent and permeability is less than 0.1 md. Core from Shell 1-13B4 Myrin Ranch, Sec. 13, T2S, R4W, Duchesne County, Utah.

FIG. 7—Vertical view of core (2.5 in. or 6.3 cm diameter), showing nearly vertical intersecting fractures in nonreservoir sandstone with matrix porosity of 2.8 percent. Walls of partially open fractures are lined with quartz and calcite druse. Core from Shell 1-36A3 Ute, Sec. 36, T1S, R3W, Duchesne County, Utah.

higher frequencies of both total fractures and open fractures than nonporous beds of equal thickness. If this observation is true, it might suggest that fracture generation is related to pore-fluid pressure as well as to rock strength.

No unique mode of fracture generation has been proved. Fractures clearly postdate rock burial, compaction, and lithification, but they predate hydrocarbon emplacement. This timing is determined on the basis of the similarities of fracture-filling mineralization and reservoir-rock cementation, and on the basis of textural fabrics, which indicate no oil inclusions or interference phenomena. The unusually systematic change in oil composition and gravity in producing zones at increasing burial depths suggests "primary" migration into the fractured reservoirs from the adjacent sequence of organic lacustrine shales. No additional reservoirs or migration phenomena are known which could have provided opportunity for secondary migration of oil into the presently producing zones. As a result, fracture propagation, fracture mineralization, and entry of the oil charge probably occurred during a relatively short time span during the later phases of basin subsidence and sediment loading. The preferred interpretation for creation of the deep subsurface fractures, in order to satisfy both the timing and the observed physical character and distribution, is that they formed during subsidence through a process combining preferential fracturing of less ductile, porous, generally thin beds in an area of post-thrusting crustal relaxation and high pore-fluid pressures.

Rock-mechanics theory indicates that high pore pressures contribute to yield. Pore pressure would be provided most readily by hydrocarbon generation from the load-bearing kerogen layers in the highly organic-rich rock sequence. However, this mechanism alone does not satisfy the interpreted sequence of (1) rock compaction and lithification, (2) fracturing, (3) fracture mineralization, and (4) oil migration. As a consequence, the concept that pore-fluid pressures are an essential stimulus to rock yield within the constraints of the Altamont-Bluebell setting requires the initiation of abnormal fluid pressures during expulsion of connate waters in the late phases of compaction but before significant oil generation has taken place. By this time, permeability must have been low enough to cause fluid pressure buildup. The mechanism implies that the stratigraphically deepest fractures were formed first as sediment overburden increased and the overpressured envelope expanded upward with increasing maturation of the source rocks. Continued burial maintained the oil-generation process and increased the magnitude of abnormal pressures.

Reservoir Porosity

Sufficient open fracture space has not been observed to account for the volumes produced from the better wells in the field (19 wells had produced more than 500,000 bbl of oil each as of January 1, 1975). Core and mechanical-log data indicate that reservoir storage capacity is provided primarily by multiple low-porosity sandstones averaging 5 percent porosity (range, 3–10 percent). The highest matrix porosities are in sandstones with low clay content and little or no calcite cement. These are either fine-grained, well-sorted (Fig. 11), or poorly sorted sandstones with pebbles. In contrast, the uniformly lower porosities are present in the more lacustrine calcite-cemented sandstones (Fig. 12). All sands were highly compacted prior to quartz cementation.

Figure 13 shows the electric-log characteristics of a typical productive interval in the field. Producing zones are dispersed over long intervals in

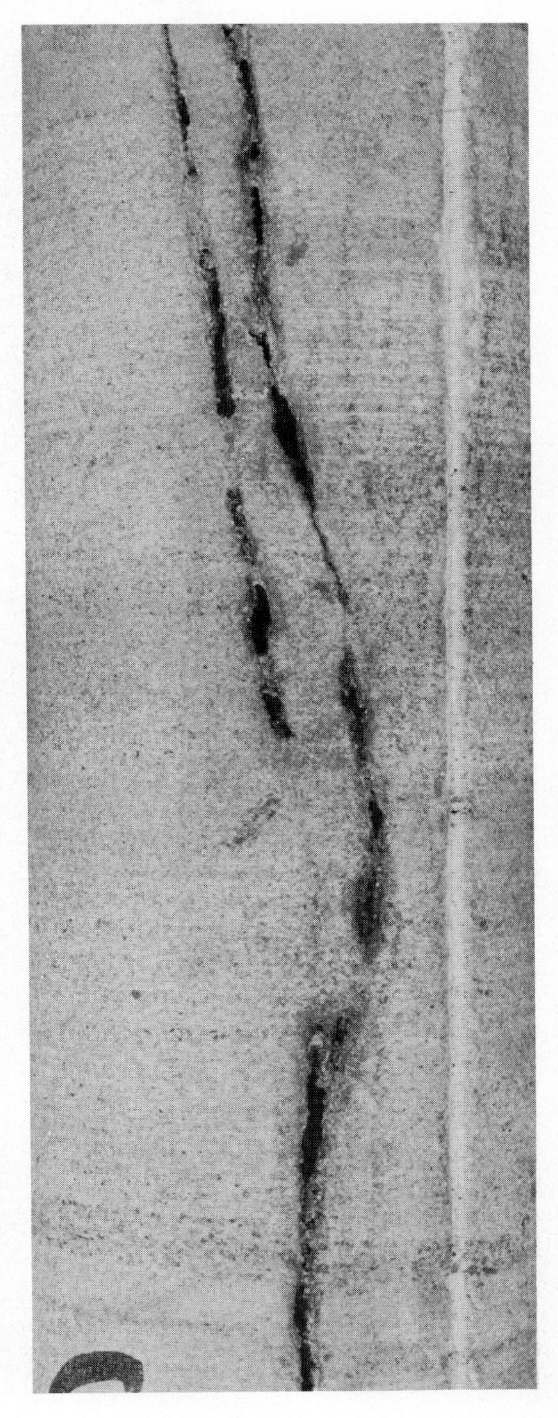

→

FIG. 8—Sandstone core with partially open, oil-stained fracture. Bifurcating, irregular fracture occurs within a 21-ft (64 m), very fine- to medium-grained sandstone interval. Sandstone matrix porosity is 4 percent, permeability is less than 0.1 md, and residual oil saturation is 20 percent. Fracture recovery was 5 ft (1.5 m) before it reached core margin at top and base. Original fracture opening is about 30 percent infilled and healed with quartz and calcite druse. Note slight vertical displacement across fracture. From 11,526 ft (3,513 m), Shell 1-21B4 Hunt, Sec. 21, T2S, R4W, Duchesne County, Utah.

FIG. 9—Polished core slab of lower Green River dolomitic limestone with fracture partially filled with calcite. Subtle downward compositional and textural changes from nonorganic to organic carbonate rock control fracture terminations at lower stylolite. Multiple hairline fractures are completely healed. From 8,376 ft (2,553 m), Shell 1-23B4 Brotherson, Sec. 23, T2S, R3W, Duchesne County, Utah.

the central part of the field or are limited to a single zone of several thin productive intervals at the field margins. Logs of potential producing zones are characterized by a subdued SP curve and by the high resistivity normally associated with hydrocarbon saturation. Commercial production appears to depend on the presence of fractures, which are not reflected on the logs. Because normal logging techniques have not made it possible to recognize fractures, indication of total producing potential requires completion and production testing. Full-bore spinner surveys taken several times over an extended period in the same borehole commonly indicate varying flow rates and different contributing zones from within the gross productive interval.

FIG. 10—Core slab of interbedded black, organic-rich shale and carbonate mudstone showing discontinuous, open vertical fracture with only hairline indication of fracture interconnection. These short open-fracture segments are calcite-druse lined, but are open horizontally through entire core width (4 in. or 10 cm). From 10,643 ft (3,244 m), Shell No. 1 Murdock, Sec. 26, T2S, R5W, Duchesne County, Utah.

Reservoir Pressures

Fluid pressures are essentially hydrostatic from a depth below near-surface topographic effects to below the lower Green River marker (about 8,000 –10,000 ft [2,400–3,000 m]). Below this, pressure gradients increase to a maximum of over 0.8 psi/ft (equivalent to 16 lb/gal mud weight) in the central part of the field. In the transitional interval, the rate of increase of pressure is as high as 3 psi/ft. As a result, even closely spaced reservoirs in this interval have considerable pressure differential. The areal distribution of maximum fluid pressures encountered throughout the field is shown in Figure 14. The indicated values encompass all stratigraphic levels and are based on a combination of borehole pressure measurements, drill-stem-test pressures, and drilling-mud weights. The base of the Tertiary stratigraphic column is generally nonpermeable; therefore, few reliable data are available as to the distribution of pressures below the producing zones. Indications are that reduced fluid pressures again approaching hydrostatic occur in the lowermost Tertiary. Additional information on the problems of drilling and completion at Altamont-Bluebell caused by the combination of multiple-fractured producing zones with differential pressures is given by Baker and Lucas (1972), Findley (1972), and Bleakley (1973).

FIG. 11—Photomicrograph of highly compacted, fine- to medium-grained quartz sandstone with relict oolitic dolomite rock fragment. Secondary quartz overgrowths have reduced porosity to 8 percent. Sandstone was oil stained. From 12,545 ft (3,824 m), Shell No. 1 Miles, Sec. 35, T1S, R4W, Duchesne County, Utah. Magnification 8×.

FIG. 12—Photomicrograph of highly compacted, fine-grained quartz sandstone with quartz overgrowths and pore-filling calcite cement. Nonreservoir rock has porosity of about 3 percent. From 11,571 ft (3,527 m), Shell No. 1 Miles, Sec. 35, T1S, R4W, Duchesne County, Utah. Magnification 20×.

Production

Initial well productivities are at flow rates up to 5,000 bbl/day with gas/oil ratios ranging from 1,500 cu ft/bbl (4,250 m^3/bbl) in the updip part of the field to 500 cu ft/bbl (1,415 m^3/bbl) downdip. Wells are currently flowing at an average of about 600 bbl of fluid per day. Reservoir drive mechanism is liquid expansion–solution gas. Gas-saturation pressures range from about 5,000 psi in the oldest reservoirs to 2,600 psi in the shallow,

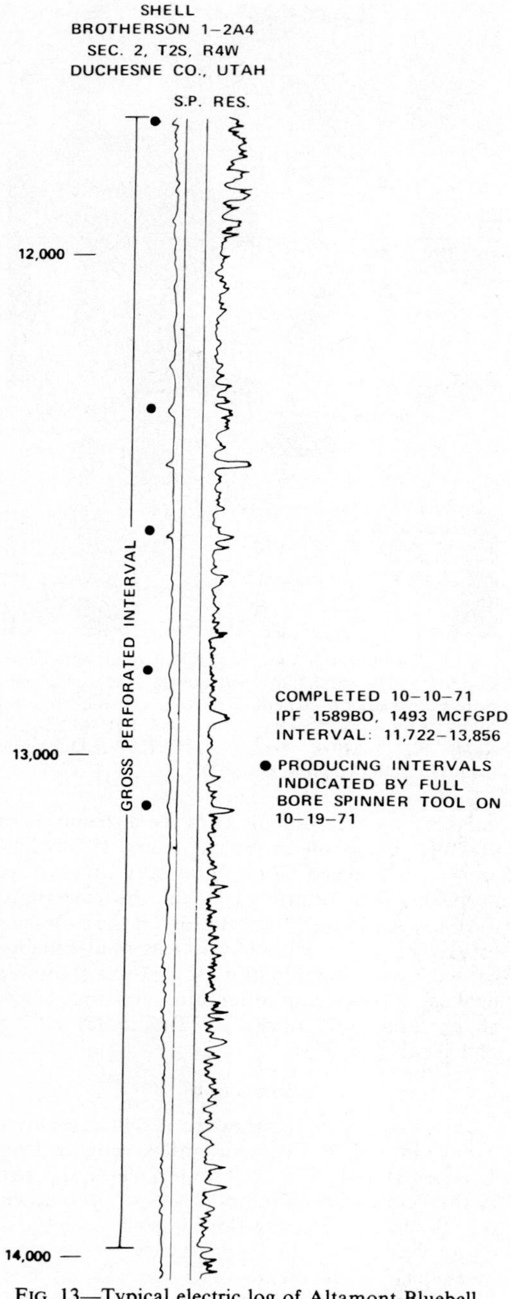

FIG. 13—Typical electric log of Altamont-Bluebell producing well.

more normally pressured reservoirs. With minor exceptions, the initial fluids produced under elevated reservoir pressures are hydrocarbons essentially free of water. As the pressure is drawn down, water production commences and gradually increases. The lower limit of matrix porosity for production of oil is about 3 percent. However, residual oil saturation is not evident in all sandstones with porosity above this value, although fractures passing through these sandstones produce oil. Because identifiable free-water levels are absent in the field, water production is assumed to result from drainage of the sediment blocks between fractures as the pressure is reduced. Liquid expansion provides the drive mechanism. Water is probably contributed from both the unstained sandstones and from the partial water saturation of oil-bearing sandstones. At present, water is about 30 percent of the total fluid produced, and the percentage is increasing.

Oil gravities range from 30 to 50° API and pour points range from 95 to 115°F (35–46°C). In general, oil gravity, pour-point temperature, and gas-saturation pressure increase with depth. The increase in gravity is coincident with a change from a naphthenic "immature" crude oil to a highly paraffinic crude oil. The more naphthenic crude oils are black, whereas the paraffinic oils become lighter in color grading to olive brown and then to honey yellow as the wax content increases. The thermal gradient for the field is approximately 1.5°F/100 ft (0.87°C/30 m).

FIELD MODEL

The trap and reservoir concepts are shown diagrammatically in Figures 15 and 16. Lacustrine shales and carbonate mudstones provide the source rock, updip seal, and cap seal for the accumulation. Transitional facies between continental redbeds and the sealing rocks are characterized by thin beds (generally 10 ft [3 m] or less) of very fine- to medium-grained, white sandstones, ostracodal limestones, and light gray to pale green waxy shales. The pressure gradient of the field reaches a maximum of slightly over 0.8 psi/ft in the central area. Away from this area, maximum pressures decline to hydrostatic updip where the facies change to nonreservoir lacustrine shales and "tight" carbonate rocks and, laterally, where the clastic beds from northerly and southerly sources intertongue.

The unusual reservoir conditions (vertically separated, fractured, low-porosity producing zones

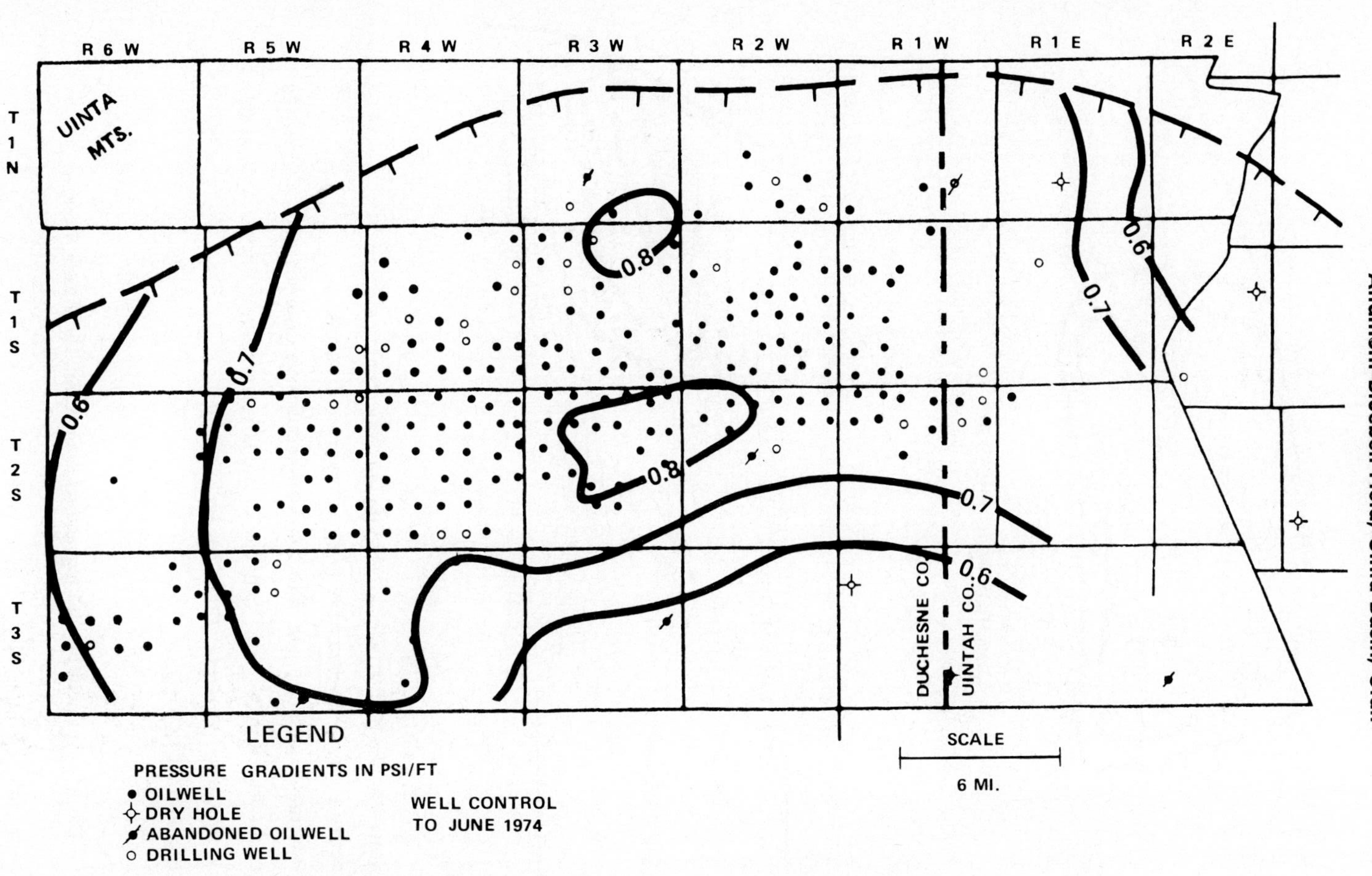

FIG. 14—Maximum observed pressure gradients, Altamont-Bluebell field, Utah. Data based on mud weights, hydrocarbon shows, lost circulation, and measured pressures. Maximum values occur at different stratigraphic levels within gross measured interval.

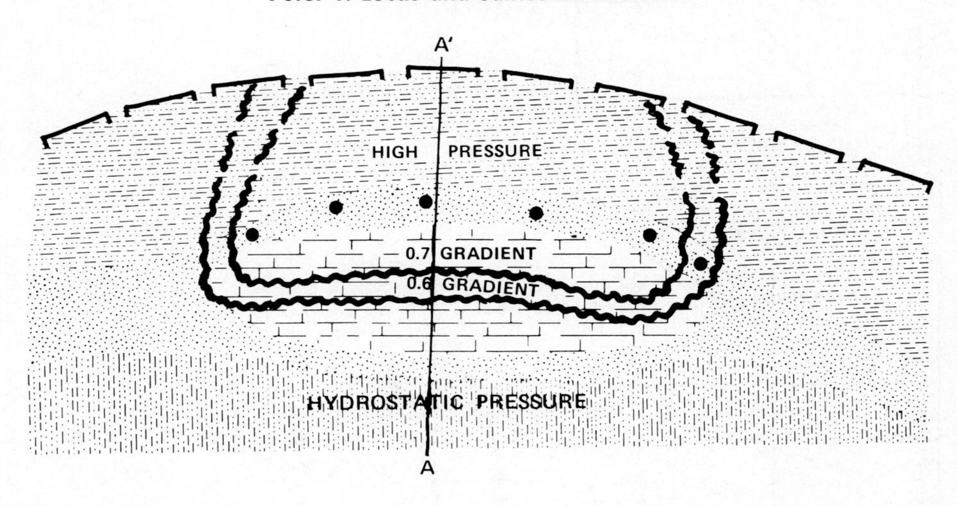

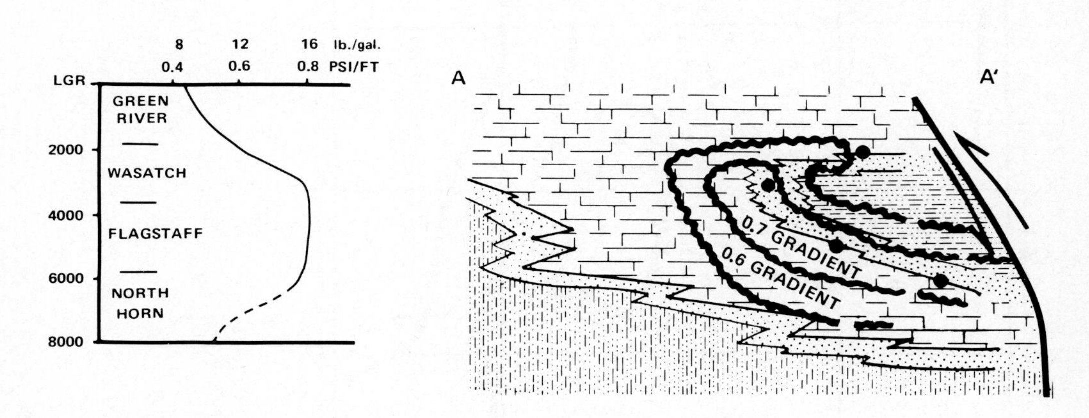

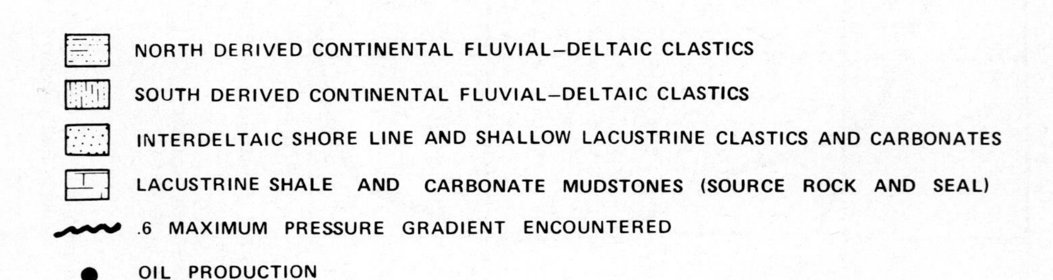

FIG. 15—Altamont-Bluebell trap and reservoir model, Uinta basin, Utah, showing pressure distribution relative to lithology.

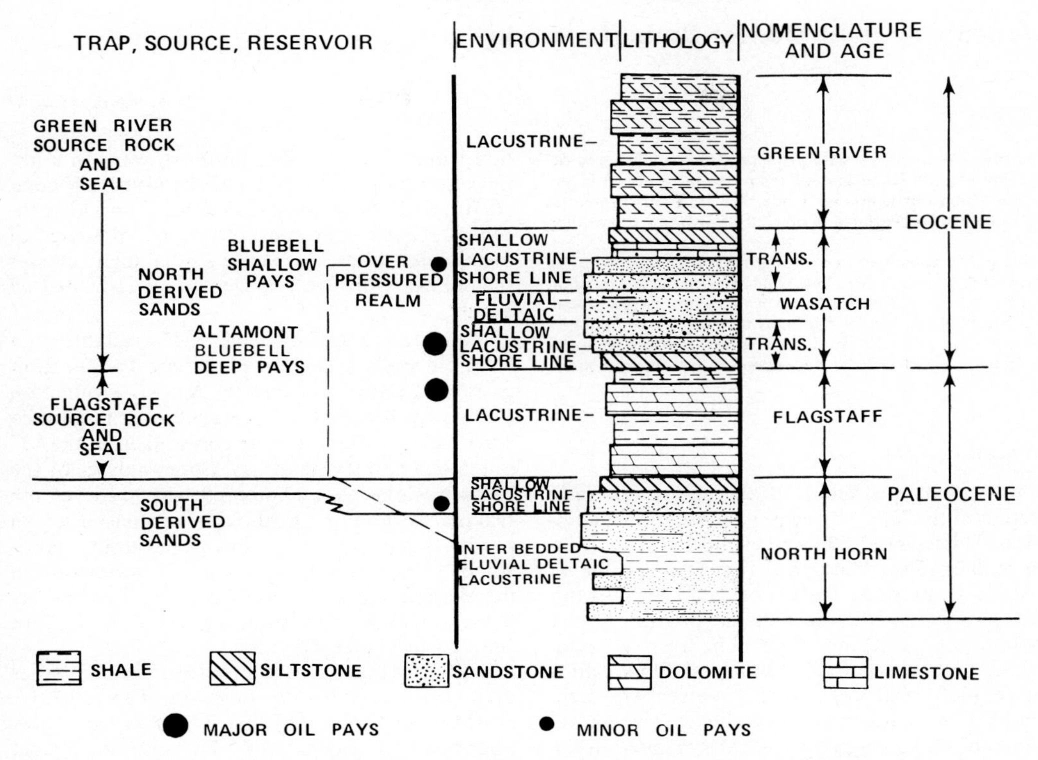

FIG. 16—Stratigraphy, nomenclature, and environmental relations of northern Uinta basin.

of variable lithology) and fluid properties (undersaturated, high-pour-point oil) do not lend themselves to normal reservoir-engineering techniques. Ultimate recoveries for individual wells are predictable only after a considerable production history. Even then, the ultimate oil/water ratio is uncertain. Production history to date indicates that the larger well recoveries (over 500,000 bbl) will occur mostly within the 0.7-psi/ft pressure-gradient envelope and updip from the maximum thickness of the wedge of Wasatch redbeds.

The field encompasses over 350 sq mi (907 km^2) and may have an ultimate production of more than 250 million bbl (Lucas, 1973). Because many facets of reservoir character and performance are not fully understood, the final economic value and ultimate drilling density are still uncertain. It is clear, however, that the Altamont-Bluebell field represents one of the most widely dispersed major oil accumulations to be exploited.

REFERENCES CITED

Baker, D. A., and P. T. Lucas, 1972, Major discovery in Utah: Strat trap production may cover over 280 square miles: World Oil, v. 174, no. 5, p. 65-68.

Bleakley, W. B., 1973, How Shell solves Uinta basin problems: Oil and Gas Jour., v. 71, no. 6, p. 45-50.

Crawford, A. L., and R. G. Pruitt, 1963, Gilsonite and other bituminous resources of central Uintah County, Utah: Utah Geol. and Mineralog. Survey Bull. 54, p. 215-224.

Findley, L. C., 1972, Why Uinta basin drilling is costly, difficult: World Oil, v. 174, no. 5, p. 77-81.

Koesoemadinata, R. P., 1970, Stratigraphy and petroleum occurrence, Green River Formation, Red Wash field, Part A: Colorado School Mines Quart., v. 65, no. 1, 77 p.

Lucas, P. T., 1973, Altamont—a major fractured and overpressured stratigraphic trap (abs.): AAPG Bull., v. 47, no. 4, p. 791.

Young, R. G., 1966, Stratigraphy of coal-bearing rocks of Book Cliffs, Utah-Colorado: Utah Geol. and Mineralog. Survey Bull. 80, p. 7-21.

Wattenberg Field, Denver Basin, Colorado[1]

R. A. MATUSZCZAK[2]

Abstract Wattenberg field produces from a large gas accumulation (1.1 Tcf estimated recoverable reserves) in the "J" sandstone of Cretaceous age. The trap was formed in the delta-front environment of a northwesterly prograding delta.

The gas is contained in a stratigraphic trap straddling the axis of the Denver basin. Reservoir characteristics are poor, and fracturing by artificial means is necessary to test a well.

Large extensions to the field may be made in the future by following the trend of the delta-front environment. Industry exploration efforts will be dependent on gas-price economics.

Introduction

Wattenberg gas field is located in the Colorado portion of the Denver basin. Its present area comprises 978 sq mi (2,530 km^2) lying between Denver and Greeley, Colorado (Fig. 1). The Denver basin is a Laramide feature oriented north-south and paralleling the east flank of the Rocky Mountain Front Range uplift. The basin is asymmetric, with a steep (10° dip) west flank and a gentle (½° dip) east flank. Wattenberg field straddles the basin axis. It produces from an accumulation of gas stratigraphically trapped in the "J," or Muddy, sandstone of Cretaceous age (Fig. 2). Estimated recoverable reserves from the "J" sandstone of Wattenberg field are estimated to be 1.1 Tcf of gas. In addition, the shallow Upper Cretaceous Hygiene sandstone is oil- and gas-productive in local areas within the field.

The purpose of this paper is to present the history of Wattenberg field from exploration conception to date, tracing both evolution of geologic thought and working procedures used to develop the field.

Exploration Concept, "J" Sandstone

A regional and isolith map (Fig. 3) shows that a large lobe of the Cretaceous "J" sandstone was deposited in the Denver-Greeley area by a westerly to northwesterly prograding delta complex. A smaller lobe of sand was deposited in the area south of Denver by a northeasterly prograding delta. Study of outcrops on the west flank of the basin provides further evidence of these two lobes, inasmuch as the "J" sandstone is thin and exhibits characteristics of shallow-marine deposition from the city of Boulder northward to a few miles south of Fort Collins. Haun (1963), Weimer (1970), and MacKenzie (1971) have provided excellent studies that substantiate the presence of these delta lobes. A marine embayment crosses the outcrop northwest of Denver, as indicated on Figure 3.

Examination and evaluation of available data from old wells drilled on the Union Pacific Railroad land (later optioned by Amoco Production Company) revealed the presence of a large area where all drill-stem tests or cores taken in the "J" sandstone had shows of gas. Core analyses of the "J" sandstone were compared with those of the Dakota Formation of the San Juan basin, which produces gas from very low-permeability sandstones of Cretaceous age. The "J" sandstone in the Wattenberg area is similar to the Lower Cretaceous Dakota Sandstone reservoir of the San Juan basin (Table 1).

Supporting data were provided by an earlier discovery at Roundup field, in T2N, R60W, northeast of Denver (Fig. 3). This small gas field, discovered in August 1967, has several unusual qualities. It produces from a stratigraphically lower part of the "J" sandstone than is productive in the D-J basin "trend" area. The prospect also was located on the basis of minor shows of gas recovered from drill-stem tests made in subsequently abandoned wildcat wells. The major risk involved was the uncertainty of achieving a commercial completion by fracturing techniques. Reservoir characteristics of the "J" sandstone are similar to those of the Dakota Sandstone in the San Juan basin. The discovery well in Roundup field was completed for 2,700 Mcf of gas per day after a fracture treatment.

In the Wattenberg field area, gas shows suggest the presence of a sandstone of marine origin on the northeast edge of the marine reentrant. The

[1]Manuscript received by editor, April 17, 1974. This paper was printed originally, in modified form, in *The Mountain Geologist*, v. 10, no. 3 (July 1973), p. 99-107.

[2]Amoco Production Company, Denver, Colorado 80219.

This paper is printed with permission of Amoco Production Company and *The Mountain Geologist*.

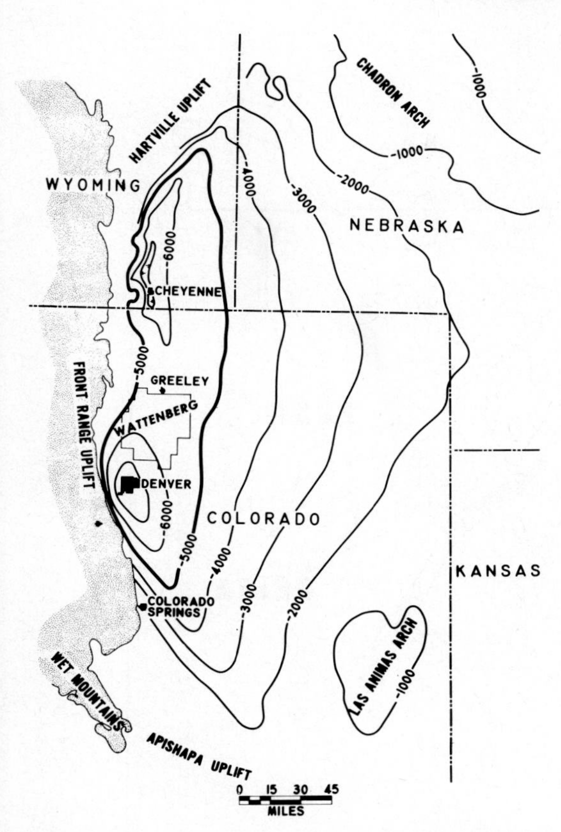

FIG. 1—Structural contour map, Denver basin. Datum is top of Precambrian. C.I. = 1,000 ft.

transition from marine shale at the base of the "J" to sandstone at the top, as seen on electrical logs and in cores, implies the presence of a delta-front sandstone of a regressive marine sequence. The electrical logs run on several old wells in the area are remarkably alike, indicating a similar environment of deposition. As a result of this subsurface evaluation, it was concluded that a delta-front sandstone was present in this area. Because delta-front sandstones tend to be sheet sandstones of possibly large areal extent, a preliminary prospect map was prepared (Fig. 4).

FIRST STAGE OF DEVELOPMENT

The initial step was to outline the possible maximum areal extent of the "J" sandstone reservoir in the Wattenberg area. Of equal importance was

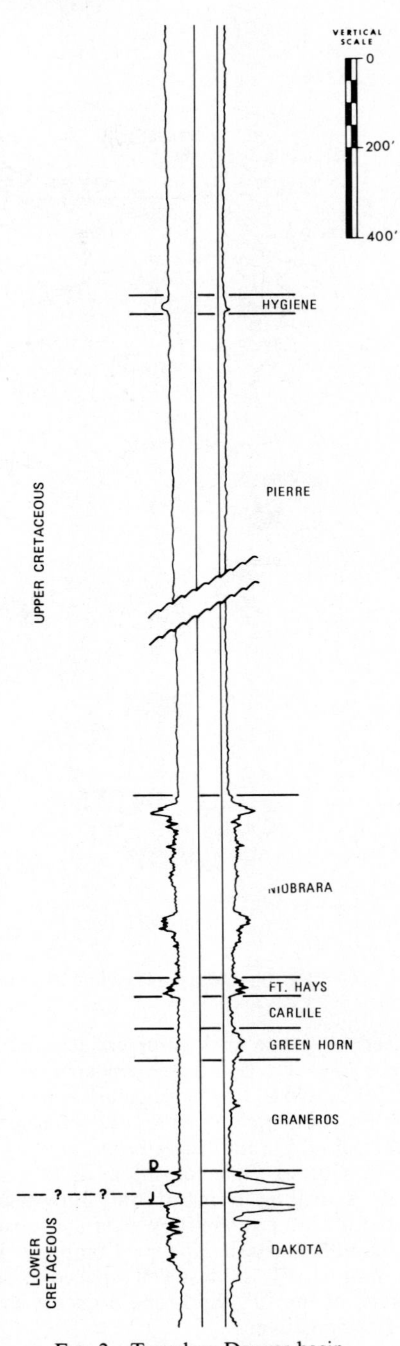

FIG. 2—Type log, Denver basin.

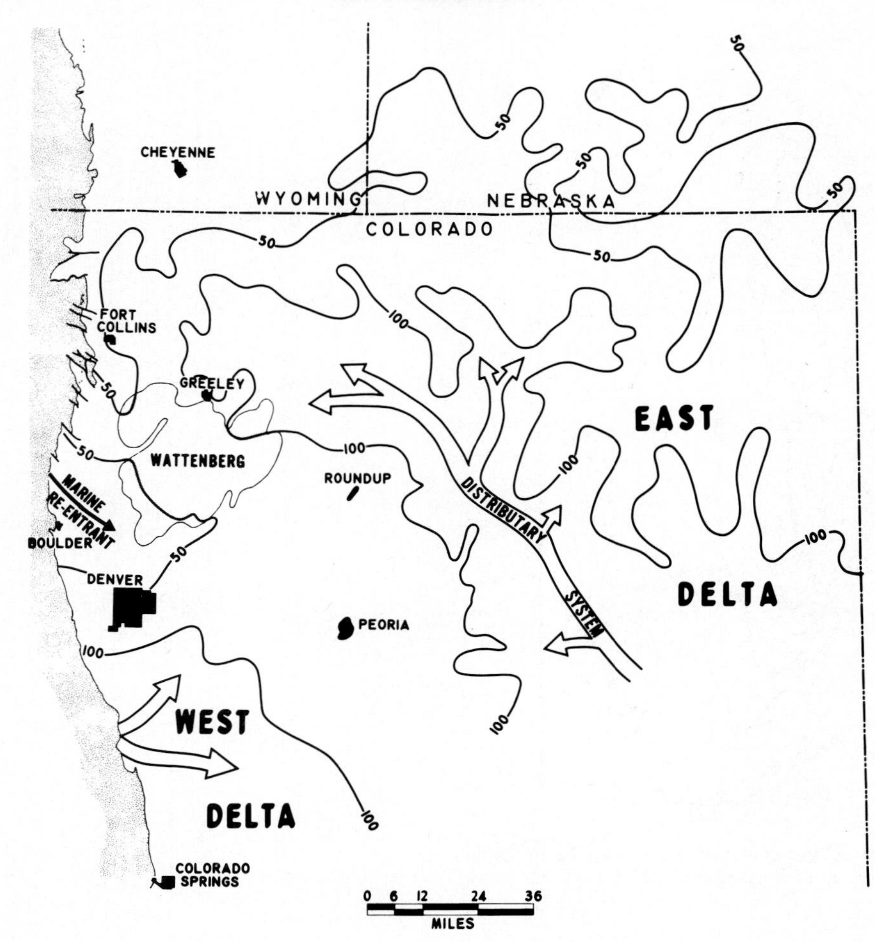

FIG. 3—Delta-sand isolith, "J" sandstone, Denver basin (modified from Haun, 1963).

the need to prove that commercial completion was possible. To test these concepts, six wells were drilled close to the original control wells. Five were completed as small gas wells, and acquisition of additional leases began.

Subsequent to the acquisition of acreage by Amoco, a drilling program was undertaken in partnership with Panhandle Western, a subsidiary of Panhandle Eastern Pipeline Company. Initial plans were to drill 100 test wells spaced to outline the limits of the "J" sandstone prospect as envisioned at that time.

The evaluation of prospective shallow zones also was considered desirable, inasmuch as minor production and shows in these intervals were present in nearby wells. However, to evaluate the Upper Cretaceous Hygiene sandstone (Parkman, Sussex, Shannon sandstone equivalents?) properly by means of samples, cores, and drill-stem tests would have been very costly and would have considerably slowed down the program for the deeper "J" sandstone objective. Denver basin wells normally are drilled with native mud and water to at least the Niobrara Formation. This technique facilitates rapid drilling; penetration rates of 2–3 ft/min (0.6–0.9 m/min) are not uncommon through the shallow formations. Under such conditions, sample quality is poor, and proper core

Table 1. Comparison of Productive Sandstones in San Juan and Denver-Julesburg Basins

Well Name and Location	Status	Interval Analyzed[a] (ft)	Formation	Range of Permeability (md)	Water Sat. (%)	Porosity Range (%)	DST Data
San Juan Basin, New Mexico							
Pan American Duff Gas Unit C-1 Sec. 27, T30N, R12W San Juan County	IP 2,502 Mcf gas/day from Dakota	6,152-6,162	Dakota	0.01-0.07	ND	2.8-9.6	
		6,228-6,248	Dakota	<0.01-3.7[b]	ND	4.7-8.9	
Pan American Gallegos Can. Unit No. 195 Sec. 33, T29N, R12W San Juan County	IP 9,516 Mcf gas/day from Dakota	5,850-5,864	Dakota	<0.1-1.3	23.2-90[c]	4.1-8.8	
		5,932-5,939	Dakota	<0.1-0.2	17.7-31.9	6.4-8.0	
		5,952-5,962	Dakota	<0.1-0.1	38.9-74[d]	6.2-9.6	
Denver-Julesburg Basin, Colorado							
Round Rouse Monaghan No. 1 Sec. 8, T3S, R65W Adams County	"D" sandstone oil production[e]	8,363	"J"	0.99	52.7	5.5	DST 8,308-8,453 ft. Open 1½ hr; SI 30 min. GTS 15 min. No gas. Rec. 550 ft oil and 60 ft oil-cut mud. SIP 3,500 lb.
		8,367-8,399	"J"	0.94-7.1	11.7-43[f]	5.7-12.3	
California No. 1 Hayes Sec. 20, T5N, R66W Weld County	Abandoned[e]	7,856-7,905	"J"	0.0-0.07[g]	7.4-69[h]	5.5-11.8	

[a]Water saturation, porosity, and permeability from core analysis.
[b]1 ft. Mostly 0.02 md or less.
[c]Mostly 30.0± percent.
[d]Mostly 45-50 percent.
[e]"J" sandstone probably productive.
[f]Mostly 15-25 percent.
[g]Accuracy of lower permeability determinations questionable.
[h]Mostly 30-45 percent.

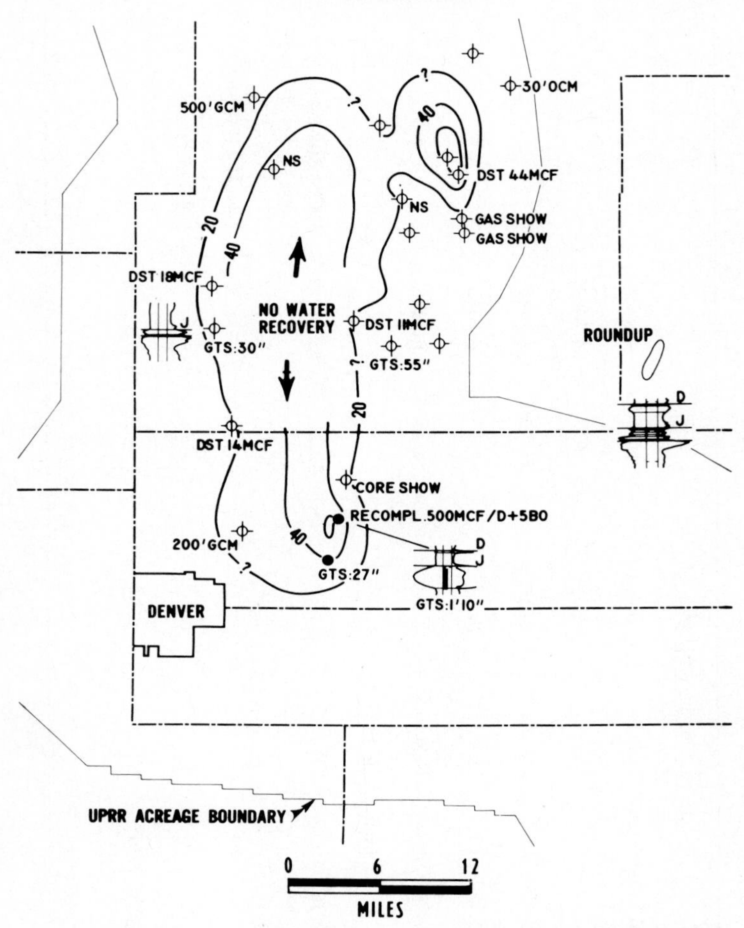

FIG. 4—Isopach map of lower "J" sandstone. Preliminary prospect map, Wattenberg field. C.I. = 20 ft.

points and drill-stem-test intervals are difficult to estimate. Therefore, it was decided to use mud-gas detecting units as an exploration method for the shallow formations above the "J" sandstone. It was believed that gas-detection equipment, combined with a suitable mechanical logging program, would provide sufficient data to make possible the subsequent location of a series of shallow wells to test the most prospective areas. Discoveries to December 1972 are shown on Figure 5.

Cumulative production to the end of 1973 was 10 Bcf of gas and 260,000 bbl of oil. Most of the wells are shut in. Gas plants, gathering system, and other facilities are under construction. Development drilling is continuing so that production will be ready when equipment is installed.

PRESENT STATUS OF "J" SANDSTONE EXPLORATION

The discovery well for the field was the Tom Vessels No. 1 Grenemeyer, SW 1/4, SW 1/4, Sec. 26, T1N, R67W, Adams County, Colorado. Initial potential, flowing, was 485 Mcf of gas plus 10 bbl of oil per day through a 20/64-in. choke. The well was completed in July 1970 at a total depth of 8,315 ft (2,537 m).

The drilling of 100 "J" sandstone wells by Amoco Production Company and Panhandle Western resulted in the completion of 72 wells capable of gas production from the "J" sandstone. Other companies and individuals have drilled and completed over 30 additional wells in the field. The present configuration is shown on Figure 6, which is a net-"pay" isopach map based on mechanical log analyses.

During the drilling program, 26 cores of the "J" sandstone were obtained from the Wattenberg field area, and several were obtained outside the area. Interpretation of the cores by Tohill (1972) verified the original concept of the delta-front sheet sand.

Trapping Mechanism in Wattenberg Field

Data from cores and mechanical logs provided insight into the trap responsible for the gas accumulation in Wattenberg field. The trap on the west and south sides is formed by the pinchout of the reservoir sandstone into a thin, "tight" siltstone and silty sandstone. The trap on the northeast side is formed by a loss of permeability caused by an increase in siliceous cement.

A total-sand isolith map shows that the "J" sandstone in Wattenberg field thickens in a northeast direction (Fig. 7). It is presumed that the main source of sand was in that direction; however, this portion of the sandstone body is

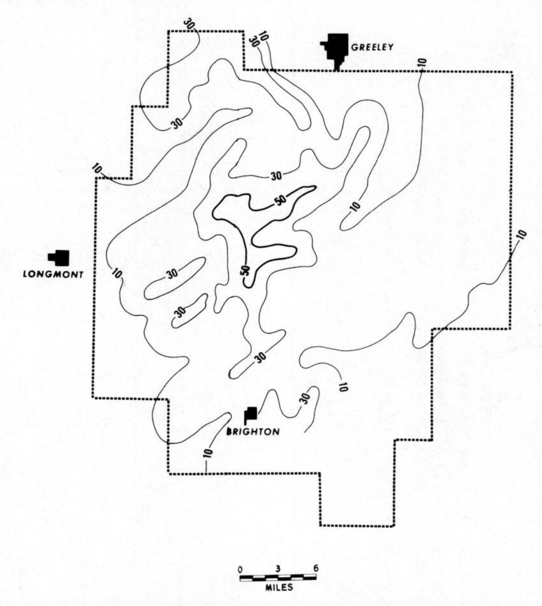

Fig. 6—Wattenberg field, net-"pay" isopach map, based on log calculations; 8 percent or more porosity, 60 percent or less water saturation. Contours are in feet.

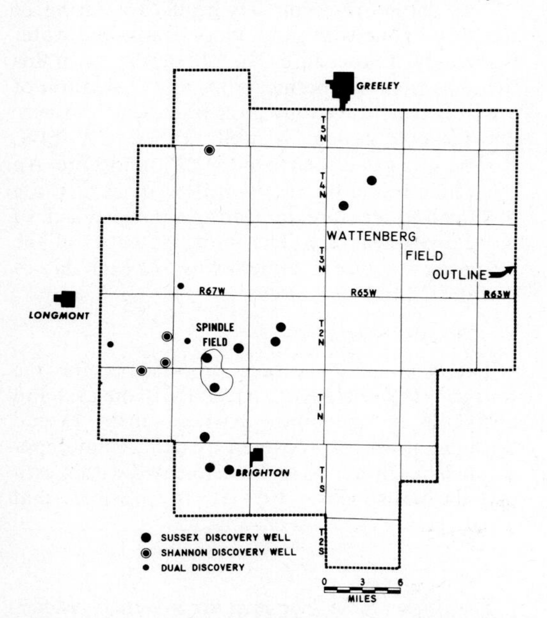

Fig. 5—Sussex-Shannon (Hygiene) discovery wells to December 1972, Wattenberg field.

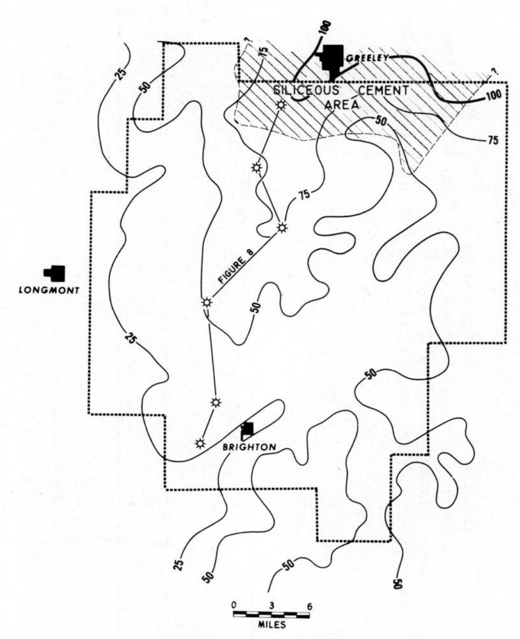

Fig. 7—"J" sandstone isolith, Wattenberg field. C. I. = 25 ft. Location of cross section (Fig. 8) is shown.

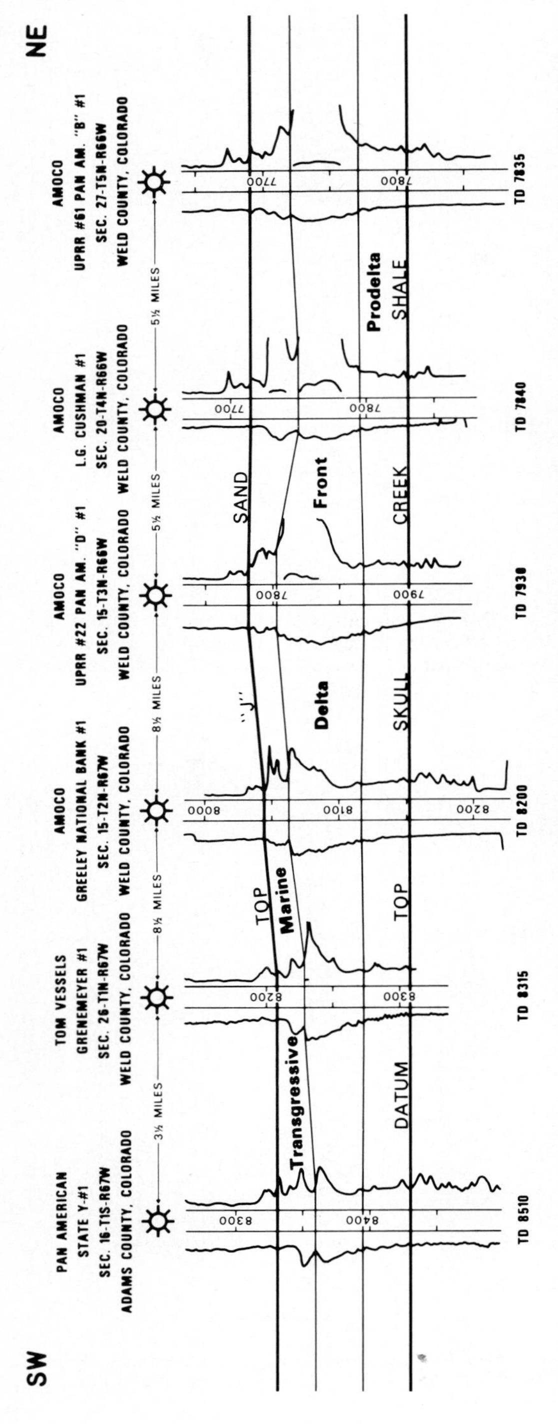

now tightly cemented in what originally was the thickest and coarsest grained reservoir rock. A possible explanation is that secondary silica cementation was caused by a pH contrast resulting from expulsion of alkaline marine waters into acidic fresh distributary waters during postdepositional compaction (Fisher et al, 1969, Fig. 118). The trap on the east side of Wattenberg field is formed by loss in permeability caused by the presence of both siliceous cement and clay.

Mechanical logs indicate that the "J" sandstone can be divided into two distinct members over most of the field area. Locally, these two parts merge into a single unit. Figure 8 displays this two-part division. Examination of cores indicates that this division is caused by an increase in dispersed-clay content, presence of clay balls, and increase in number of scattered thin shale laminae found in the "J" sandstone. The division is made at the contact between delta-front and transgressive-marine sediments. The contrast on mechanical logs is more apparent than real, because the low-resistivity, low-SP zone, which resembles shale in log character, is usually a sandstone. Figure 9 is a display of electric-log and core data. It also includes the environmental interpretation.

Table 2 presents some statistical data for the "J" sandstone reservoir. Hydraulic fracturing of the "pay" zone with sand, glass beads, and water is a standard procedure. The "J" sandstone in this field has such low permeability that evaluation of a well is conclusive only after fracturing and testing through casing. A well in Sec. 33, T1N, R67W, was gas-drilled into the "J" sandstone. An open-hole test of the reservoir flowed gas at a rate too small to measure, indicating a decided lack of natural permeability. The initial potential of the well after fracture treatment was 573 Mcf/day of gas and 22 bbl/day of condensate.

Future Possibilities

The geologic conditions responsible for the trap in Wattenberg field, particularly the east and northeast permeability barrier, may extend throughout the area covered by delta-front deposition. Thus, a trend can be delineated within estimated limits (Fig. 10). It is possible that

←

FIG. 8—Southwest-northeast cross section, Wattenberg field, showing division of "J" sandstone into two members. Location is shown on Figure 7.

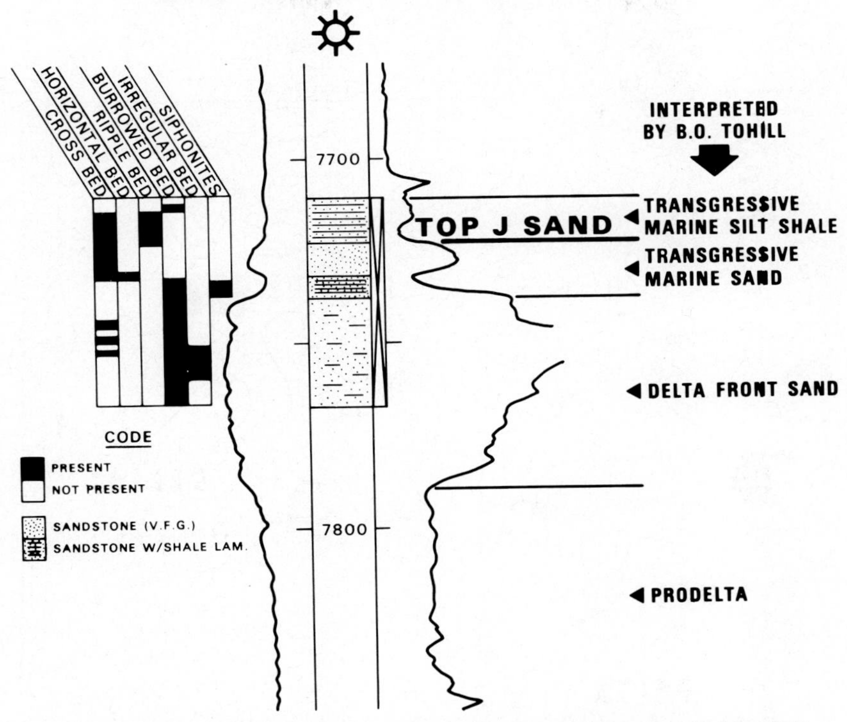

FIG. 9—Environmental interpretation of electric log and core data (Tohill, 1972).

Table 2. Basic Reservoir Data, "J" Sandstone

Parameter	
Average porosity	9.5 percent
Average permeability	0.1–0.2 md
Connate water saturation	44 percent (?)
Recoverable reserves	1.1 Tcf
Field size	283,000 acres
Average pay thickness	25+ ft
Original reservoir pressure	2,750 psi
Depth range	7,350–8,500 ft
Initial potentials	100–3,575 Mcf/day

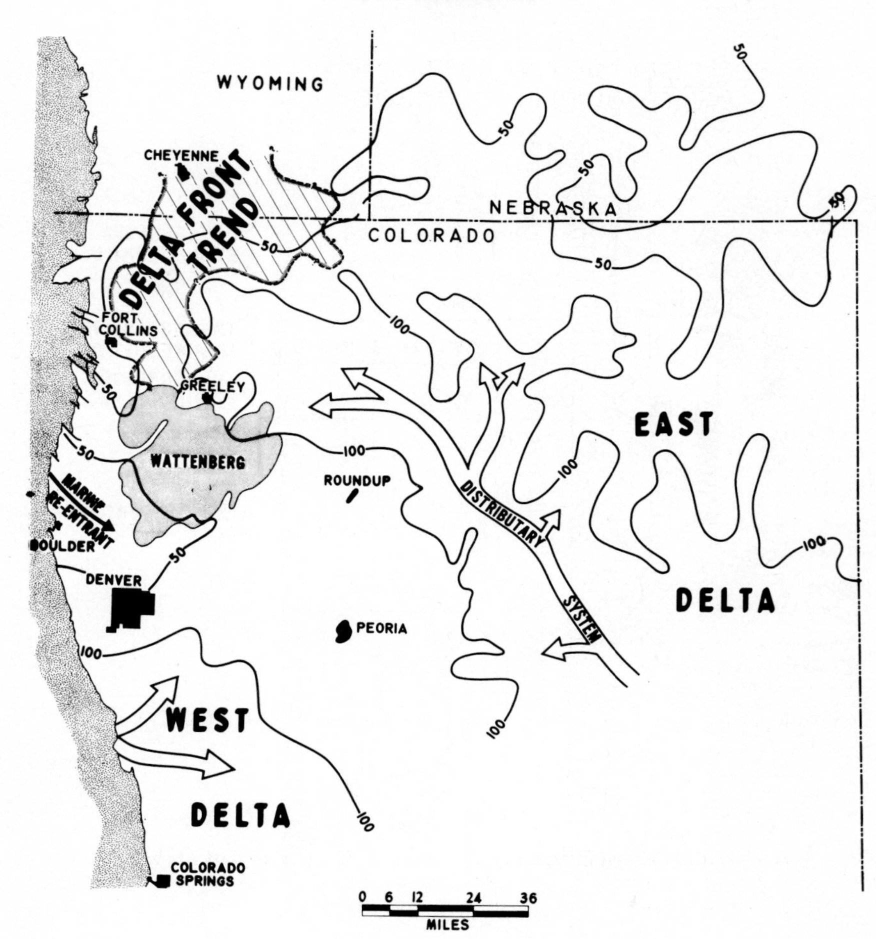

FIG. 10—Denver basin, showing area of delta-front deposition.

considerable extensions to the gas-productive area will be made.

Extension of commercial production is dependent only in part on sufficient geologic evaluation. It must be assumed that future discoveries will be similar to existing wells and, therefore, will be of low to moderate deliverability. Gas prices necessarily will affect the timing and extent of further industry exploratory efforts.

References Cited

Fisher, W. L., et al, 1969, Delta systems in the exploration for oil and gas: Texas Univ. Bur. Econ. Geology, 78 p.

Haun, J. D., 1963, Stratigraphy of Dakota Group and relationship to petroleum occurrence, northern Denver basin: Rocky Mtn. Assoc. Geologists Guidebook, 14th Ann. Field Conf., p. 119-134.

MacKenzie, D. B., 1971, Post-Lytle Dakota Group on west flank of Denver basin, Colorado: Mtn. Geologist, v. 8, no. 2, p. 91-131.

Tohill, B. O., 1972, Depositional environments and hydrocarbon traps in "J" sandstone (Lower Cretaceous), Denver basin, Colorado (abs.): Denver, Colorado, AAPG-SEPM Ann. Mtg. Prog., p. 658-659; 1972, AAPG Bull., v. 56, no. 3, p. 658-659.

Weimer, R. J., 1970, Dakota Group stratigraphy, southern Front Range, South and Middle Parks, Colorado: Mtn. Geologist, v. 4, no. 3, p. 157-184.

Big Wells Field, Dimmit and Zavala Counties, Texas[1]

R. L. LAYDEN[2]

Abstract The Big Wells (San Miguel) and Big Wells (Lo East) fields, 75 mi (120 km) southwest of San Antonio, Texas, are productive from stratigraphic traps in sandstones of the San Miguel Formation (Upper Cretaceous). Ranging in producing depths from 5,200 to 5,800 ft (1,585–1,768 m), Big Wells (San Miguel) field covers approximately 30,500 surface acres (25,400 oil-producing and 5,100 gas-producing acres) in an area 14 mi (23.5 km) long and 3–4 mi (4.8–6.4 km) wide. Estimated original oil in place is 200 million bbl. Since discovery of the field in January 1969, 210 oil wells (on 80-acre spacing) and 9 gas-cap wells have been completed. Maximum daily production of 21,000 bbl was reached in March 1972. Daily production in mid-1973 was 14,000 bbl. In December 1973, daily production was 11,665 bbl. The cumulative production through 1973 was 17,476,275 bbl of oil plus 19,851 MMcf of gas. Ultimate recovery is estimated to be 43 million bbl of liquids plus 60 Bcf of gas.

The Big Wells (Lo East) field covers approximately 14,000 surface acres in an area 8 mi (12.9 km) long and 2 mi (3.2 km) wide. Producing depths range from 5,600 to 5,900 ft (1,707–1,798 m), 200 ft (61 m) lower in the section than the producing zone in Big Wells (San Miguel). Fifty-three wells had been completed in this reservoir by the end of 1973 and were producing 1,745 bbl/day. Through 1973, cumulative production was 2,327,485 bbl of oil plus 1,569 MMcf of gas. Estimated original oil in place is estimated as 35 million bbl. Ultimate recovery is expected to be 7 million bbl of liquids plus 4 Bcf of gas.

INCEPTION AND DISCOVERY

Following leads in evaluating drilling of the Olmos Formation (Upper Cretaceous) and the Wilcox Group (Eocene) in western Webb County, Texas, Sun Oil Company discovered the Northeast Big Wells field in southeastern Zavala County in January 1969. In 1967 and 1968, exploration for shallow gas in the Olmos and Wilcox sandstones was under way in western Webb County (Figs. 1, 2). To evaluate the possible extent of the gas-bearing strata, regional mapping of the structure at the top of the Olmos was carried out for Webb, Dimmit, Zavala, LaSalle, and Frio Counties. A large area of structural flattening was mapped in northeastern Dimmit County and southeastern Zavala County. The flattening is on trend with the large Pearsall anticline in central Frio County on which oil is produced from the Austin Chalk and from Olmos and Navarro sandstones. Within the area of flattening, Sun found most of the acreage available for leasing for less than $10 per acre. A 14-section block of acreage was assembled. In northeastern Dimmit County, on the southeast edge of the area of structural flattening, the Gulf No. 1 Racine (1)[3] had been drilled in 1956. After fracture treatment it tested 12 BOPD from a San Miguel sandstone, and was abandoned as noncommercial.

Although the first well, the Sun No. 1 Erin Bain Jones (2), was planned primarily as a 5,000-ft (1,524 m) Olmos test, Sun decided to drill an additional 500 ft (152 m) to test the San Miguel section (Fig. 2). No shows were found in the Olmos. The San Miguel sandstone (at 5,333–5,362 ft or 1,625–1,634 m) was 198 ft (60 m) higher than in the Gulf No. 1 Racine (1), 3.5 mi (5.6 km) southeast. Sidewall-core analysis indicated possible oil production, but only gas was recovered on openhole tests. The casing was perforated at the bottom of the sandstone from 5,354 to 5,360 ft (1,532–1,534 m), and on January 8, 1969, the well was tested for an initial potential of 1,388 Mcf of gas per day. Because the well was 30 mi (48 km) from the nearest pipeline, the daily production of 1,388 Mcf did not arouse much interest. After fracture treatment, the well was tested for an open-flow potential of 19,248 Mcf of gas per day plus 24.6 bbl of condensate. No pressure decline was indicated after 10 days of testing at rates of 2 MMcf/day. Thus the well apparently was not producing from a typical, very small San Miguel accumulation.

Downdip wells drilled 1 mi (1.6 km) southeast of the discovery gas well established oil production in the San Miguel sandstone. Diamond-core analysis indicated "tight" sandstone. Coring in the Sun No. 1-A Jones (3) indicated a 33-ft (10 m) interval of sandstone, 18 ft (5.5 m) of which had measurable permeability; 11 ft (3.3 m) had permeability less than 0.1 md. Maximum permeability of 5 md was measured. Five oil wells were drilled and completed for 58–90 BOPD using

1Manuscript received, August 16, 1974; revised, December 2, 1974. Publication approved by Sun Oil Company.

2Sun Oil Company, Houston, Texas 77001.

3Numbers in parentheses throughout the text refer to numbered wells on location map (Fig. 3).

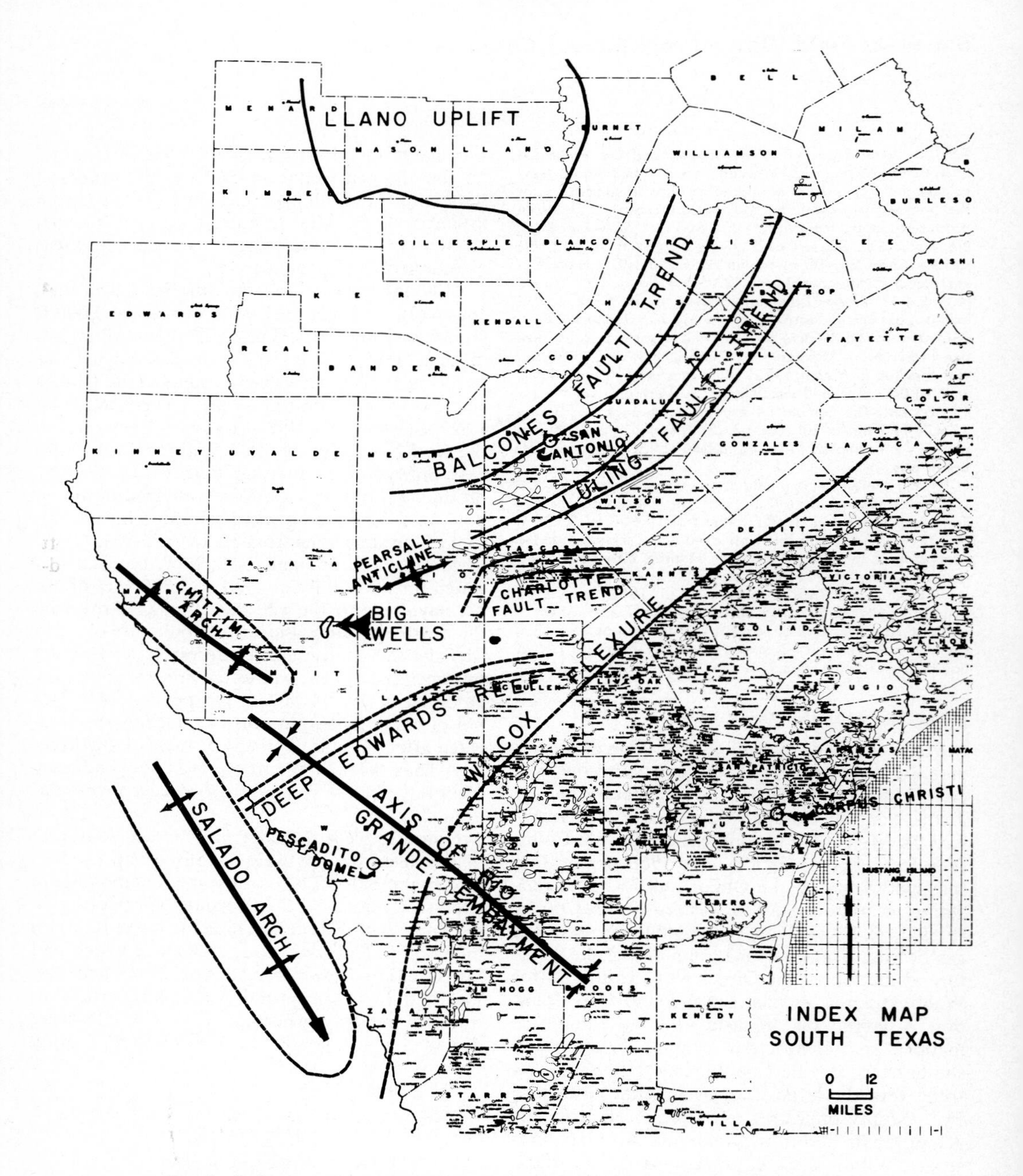

FIG. 1—Index map showing location of Big Wells field in relation to regional geologic features of South Texas.

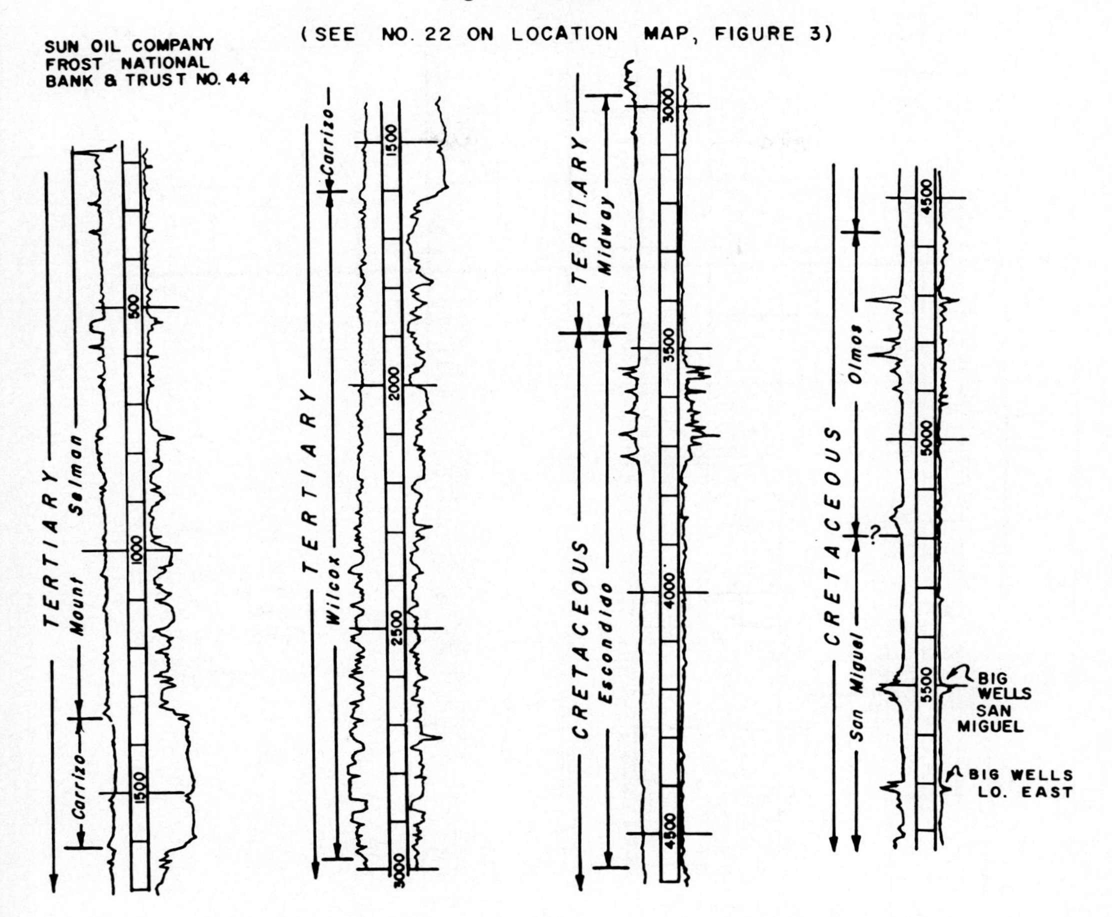

FIG. 2—Big Wells field, type log, showing productive San Miguel sandstones.

conventional fracture-treatment procedures. The field was named "Northeast Big Wells."

After establishment of oil production, a possible stratigraphic trap was mapped as extending from the Nueces River in northeast Dimmit County 30 mi (48 km) northeastward through central Frio County. By the early part of 1970, drilling had defined the northeast extent of sandstone development in southeastern Zavala County. Five oil wells were drilled in southeastern Zavala and northeastern Dimmit Counties in the first stage of development. Production from these wells rapidly declined to 15–27 BOPD, causing concern as to the economics of further development.

Beginning of Large Fracture Treatments

Of the five oil wells completed in the original stage of development, the Sun No. 1 Jeff Baggett (4), the best producing well, had declined in early 1970 to 27 BOPD. This well was given a fracture treatment called "Super Frac" using hot oil and treated water capable of carrying large volumes of "frac" sand. Fracture treatment used 50,000 gal of dispersant, 4,000 lb of 20–40 mesh sand, and 225,000 lb of 10–20 mesh sand. The potential after fracturing was 142 BOPD. The other four wells were given similar treatments with comparable results. Production declined, but sustained

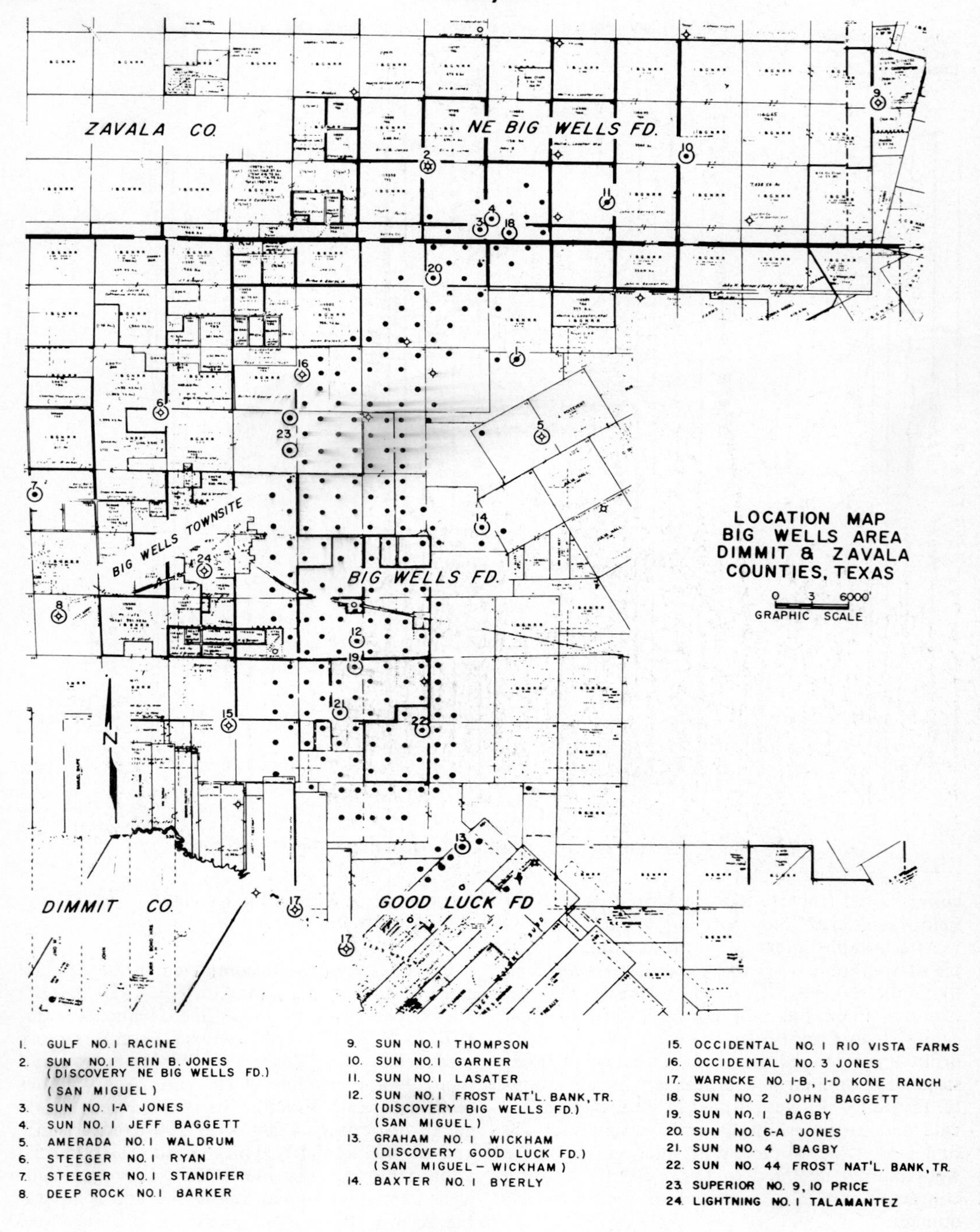

FIG 3—Map showing locations of wells referred to by number in text.

rates were at much higher levels than those prior to the Super Frac treatment.

Extension Drilling—Big Wells (San Miguel) Discovery

On March 16, 1970, the Sun No. 1 Frost National Bank Trustee (Marrs McLean Estate; 12) was spudded. It was 7 mi (11.3 km) south of production in the Northeast Big Wells field. Subsurface control from Olmos drilling indicated the No. 1 Frost National Bank would be on strike with the San Miguel oil production to the north. From mud-log depths 5,374–5,470 ft (or 1,638–1,667 m), a 96-ft (29.3 m) San Miguel section was described as fine-grained, glauconitic, slightly calcareous sandstone with oil saturation. Electric-log analysis indicated that the San Miguel extended from 5,379 to 5,471 ft (1,639.5–1,667.6 m) and had 32 ft (9.8 m) of net productive sandstone. Although the San Miguel was approximately 60 ft (18 m) lower than anticipated in the No. 1 Frost National Bank, it was on strike with the old Gulf No. 1 Racine (1), which had produced 12 BOPD from the San Miguel. Also, sandstone porosity and permeability were much better in the No. 1 Frost National Bank. Potential before fracture treatment through perforations from 5,401 to 5,426 ft (1,647–1,657 m) was 53 BOPD through a 3/8-in. choke; no water was produced, the gas-oil ratio was 486/1, and tubing pressure was 0/30 lb. This was the first well in the area to flow naturally from the San Miguel. After a large hot-oil fracture treatment using 71,000 gal of dispersant and 338,000 lb of sand, the well flowed 125 BOPD through a 9/64-in. choke with tubing pressure of 500 lb. On test 30 days after completion, the flow rate was 136 BOPD, with no water, through a 9/64-in. choke with tubing pressure of 500 lb; the gas-oil ratio was 275/1. Following completion of the No. 1 Frost National Bank, Sun began an active development program of the Big Wells (San Miguel) field, using first two and, later, three rigs.

Acreage Situation

In addition to a Sun block of 30,000 acres, other operators held considerable acreage in the area (Fig. 4). Superior held three sections in the central part of the area, two sections on the southwest end, and 8,000 acres on the east and downdip. Del Ray and Champlin held six sections between Northeast Big Wells and Big Wells fields. The Frost National Bank, trustee for McLean-Bowman, held 4,600 unleased acres in the east-central and southern parts of the area. Double U of San Antonio acquired this acreage.

Good Luck Field Discovery

In September, 4 months after the Sun No. 1 Frost National Bank was drilled, the Graham No. 1 Wickham (13) was drilled 3 mi (4.8 km) to the southeast and completed from a sandstone 200 ft (61 m) below the productive San Miguel section at Big Wells field (Fig. 5). Twenty feet (9 m) of sandstone was productive, and the well was completed for 130 BOPD after fracturing. Graham named the field the "Good Luck field." Additional drilling extended this production to the north and east, and the name was changed to "Big Wells (Lo East) field." In the Graham well, the San Miguel sandstone between 5,628 and 5,647 ft (1,715–1,721 m) was oil saturated and extended the oil column downdip to −5,178 ft (−1,578 m), which is the base of the sandstone in this hole. This was 203 ft (62 m) lower than the lowest well on the northwest, establishing an oil column of more than 400 ft (122 m). No water column is associated with the Big Wells (San Miguel) reservoir.

Development

By April 1, 1971, 2 years and 2 months after completion of the Sun No. 1 Jones gas discovery, 146 oil wells had been drilled. Sun Oil had completed 83 wells producing 10,000 BOPD. Total field production was 12,000 BOPD. The Big Wells and Northeast Big Wells fields were combined and called the "Big Wells (San Miguel) field."

By May 1972, a total of 210 wells, drilled on 80-acre spacing, had been completed in the Big Wells (San Miguel) field. Maximum daily production of 21,000 bbl of oil and 24 MMcf of casinghead gas was reached in March 1972.

By the end of 1973, 195 wells were producing 11,665 BOPD and 24 MMcf of casinghead gas. Nine gas wells were completed in the gas cap. Cumulative production was 17,476,275 bbl of oil and 19,851 MMcf of gas. The original bottomhole pressure was 2,550 lb. It had declined to 1,500 lb by the end of 1973. In September 1973, injection of casinghead gas plus gas-cap gas was begun.

A refrigeration-type gas plant was built by Sun Oil Company in the early stages of field development. This plant processed 24 MMcf/day, much higher than its designed capacity, and produced 1,200 bbl/day of liquids.

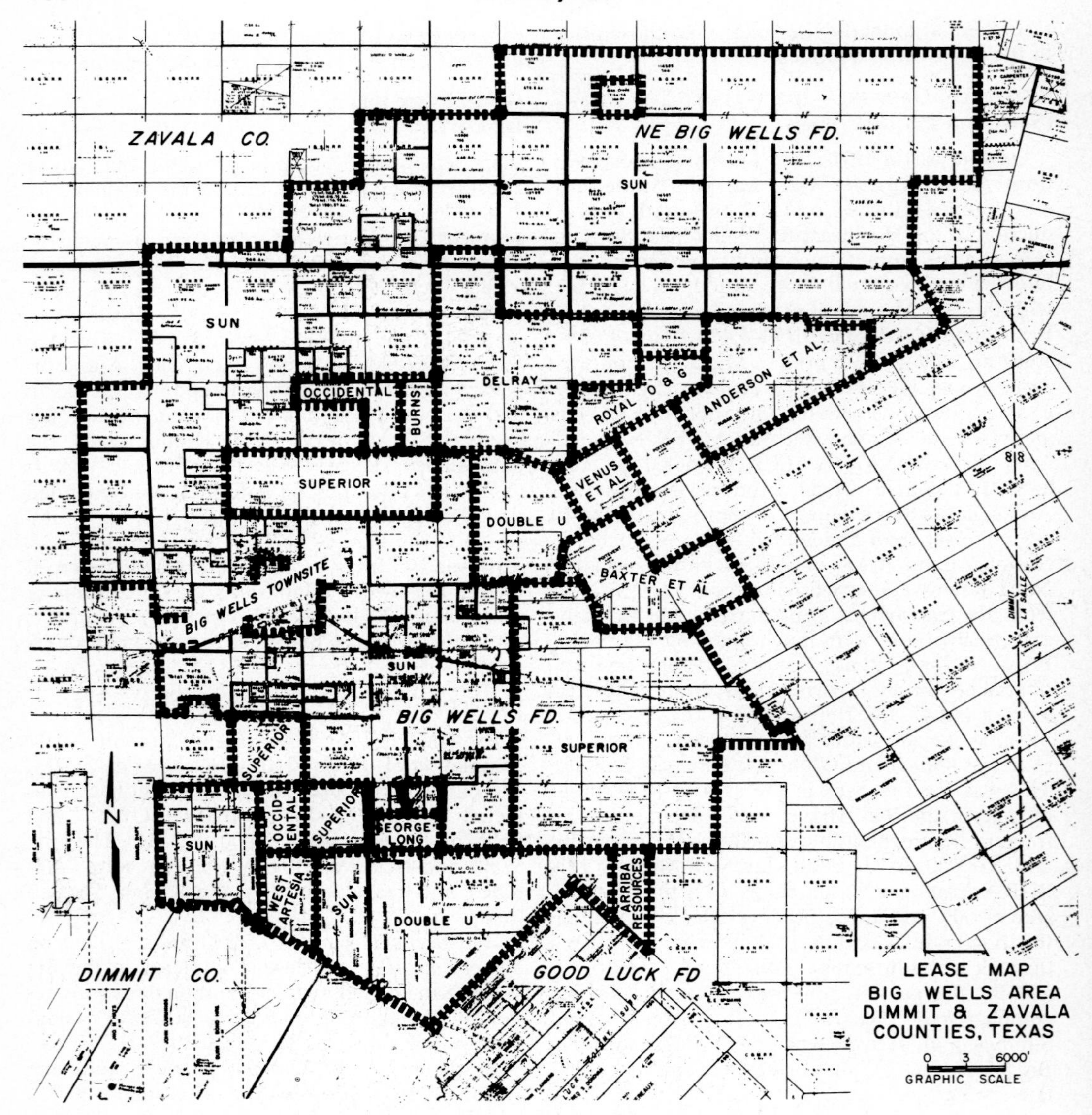

FIG. 4—Map showing major lease holdings, Big Wells area.

A second plant, a refrigeration-absorption plant, was built later. In May 1974, the plant processed 27 MMcf of casinghead gas and 6.5 MMcf of gas-cap gas per day. It produced 2,400 bbl/day of liquids, recovering 98–99 percent of the liquids in the gas. Since September 1973 the gas has been injected into the reservoir.

STRATIGRAPHY

The producing sandstone in Big Wells (San Miguel) field is interpreted to be an offshore bar, pinching out to the east and west from a maximum development of 32 ft (9.8 m) of net sandstone. Lithology of the sandstone interval is

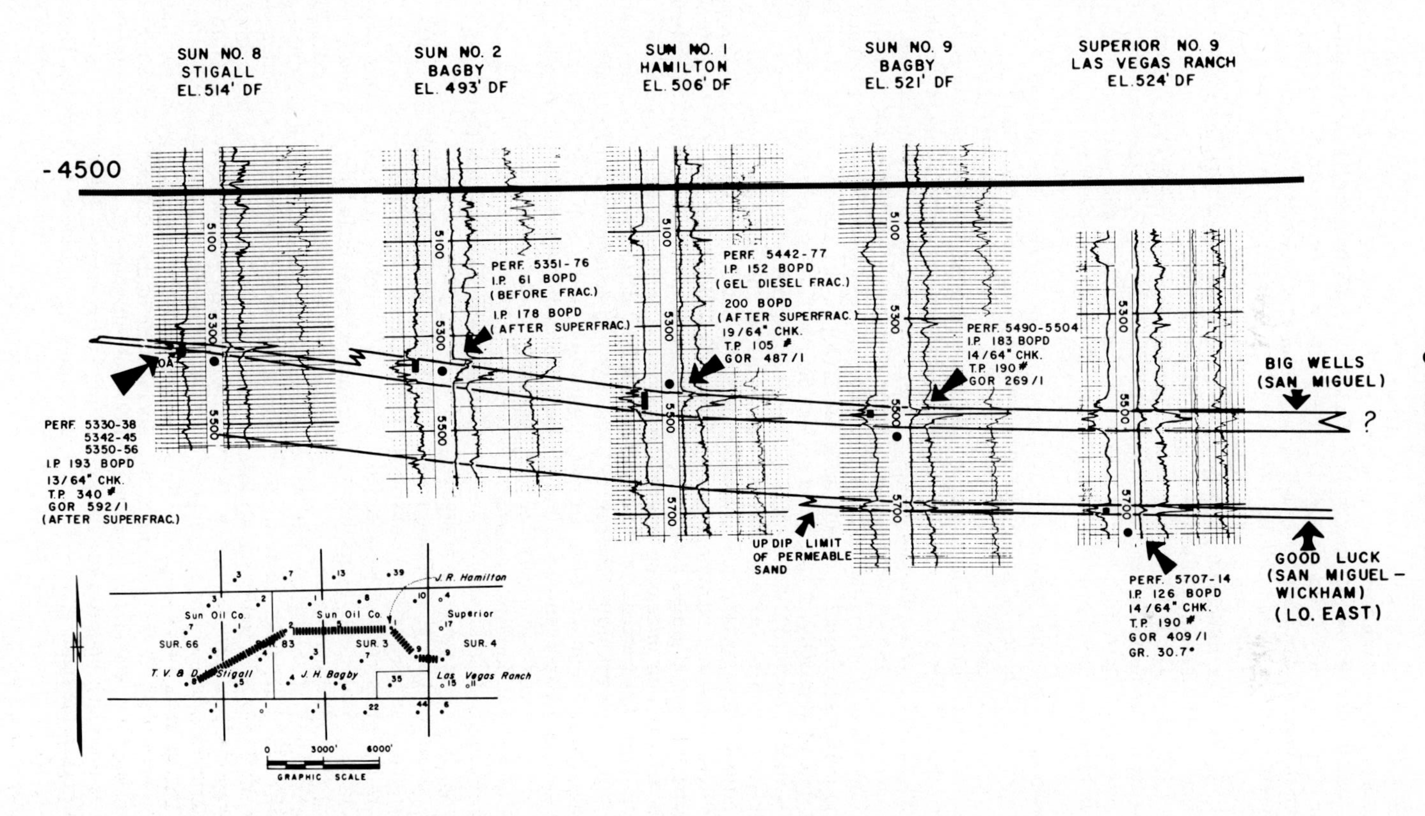

FIG. 5—Cross section of southern part of Big Wells field showing relation of Big Wells (San Miguel) sandstone to Good Luck (Wickham) sandstones.

shown on Figure 6. Figure 7 is a structure map of the top of the San Miguel sandstone; it also shows the sandstone limits. The overall interval in which sandstone is present ranges from 30 to 96 ft (9.1–29.3 m). The thickness axis of the bar roughly follows the present strike. An oil column extending from −4,740 ft (−1,445 m) to at least −5,178 ft (−1,578 m) was indicated midway through development. The gas/oil contact now is placed at −4,760 to −4,780 ft (−1,451 to −1,457 m) in wells on the northwest end of the field; it varies within separate fault blocks (Fig. 7). No water has been encountered in the reservoir.

Figure 6 depicts a typical San Miguel sandstone section. Beginning at the top of the San Miguel sandstone there is a zone of oil-saturated, very fine-grained siltstone with low porosity and almost no permeability. This section ranges in thickness up to 30–40 ft (9.1–12.2 m) in the southern part of the field, but is totally absent in the northern part. It commonly is characterized by fair SP deflection on the IES log and porosity on the density log, but its resistivity is low. Where the San Miguel section is fractured, the top of the fracture extends to the top of the oil-saturated siltstone. Very little, if any, production comes from this upper unit. Below the siltstone, a very fine-grained sandstone interval grades downward into a sandstone section with increasing quantities of shale inclusions. The percentage of shale inclusions increases downward until shale constitutes 80 percent of the section in a matrix of very fine-grained, porous, permeable, and oil-saturated sandstone. This lower unit commonly has SP deflection and indication of porosity on the density log, but the resistivity decreases as the shale inclusions increase. This decrease in resistivity resembles a water contact, and the lower part of this section is defined from log analysis as having high water saturation. This section contributes substantial production. Shales above and below the sandstone are of shallow-marine origin.

This sequence is typical of an offshore bar. Although no detailed study has been made to determine whether the San Miguel sandstone at Big Wells actually represents a bar deposit as described in the literature, core descriptions suggest such a deposit (Davies et al, 1971.)

Porosity and permeability vary from north to south and within the field. Generally, porosities and permeabilities are much better in the southern and western part of the field.

In the Sun No. 1-A Jones (3) in the north end, the San Miguel has two zones of permeable sandstone. The core-analysis data are shown in Table 1. The well was cored in a water-base mud. The Sun No. 2 John Baggett (18), east of the Sun No. 1-A Jones, was cored in an oil-base mud and had very similar core-analysis figures, but the water saturation was reduced to 40 percent. As a result, the estimate of oil in place per acre-foot was 500 bbl in the water-base mud core and 800 bbl in the oil-base mud core.

The best reservoir conditions are present in the Sun No. 1 Bagby (19) in the southern end of the field. Core-analysis data are shown in Table 1. The highest measured permeability was 97.6 md, with 5 ft (1.5 m) of sandstone having average permeability of 66 md. The storage capacity per acre-foot was measured at 936 bbl.

There are areas in the east-central part of the field in which the reservoir characteristics of the sandstone are as poor as those on the north. Along the west, updip side of the field, the sandstone is thin but has better porosity and permeability; the result is better wells. The Sun No. 6-A Jones (20) in the northwestern part of the field has only 8 ft (2.4 m) of net sandstone compared to 19–20 ft (5.8–6.1 m) of productive section in wells just to the east. The wells with the greater sandstone development are all low-volume wells (below 50 BOPD). The No. 6-A Jones flowed 292 BOPD on completion. Several other wells along the west side of the area have similar high rates of production. Better permeability and nearness to the gas cap (energy source) explain this better production. The updip edge of the bar may have been exposed above water with resultant cleaner, less shaly sandstone and better permeability.

Drilling and Completion Practices

Wells in the area were drilled to 5,600–5,800 ft (1,707–1,768 m) in approximately 1 week. In general, 4½-in. casing was set to total depth. The logging program included an induction electrical log, a compensated formation-density log with gamma ray log, and a caliper log. At first, circulation of cement to the surface was attempted, but it commonly failed because of breakdown of the Carrizo-Wilcox sandstones from 1,400 to 3,000 ft (427–914 m). The Carrizo sandstone, at 1,400–1,700 ft (427–518 m), is the main freshwater producing zone in the area and must be protected. A DV multiple-stage cementer was placed in the casing string around 3,000 ft (914 m) to assure cementing from there to the surface. Through experience, a cementing technique was developed which resulted in almost 100 percent success in

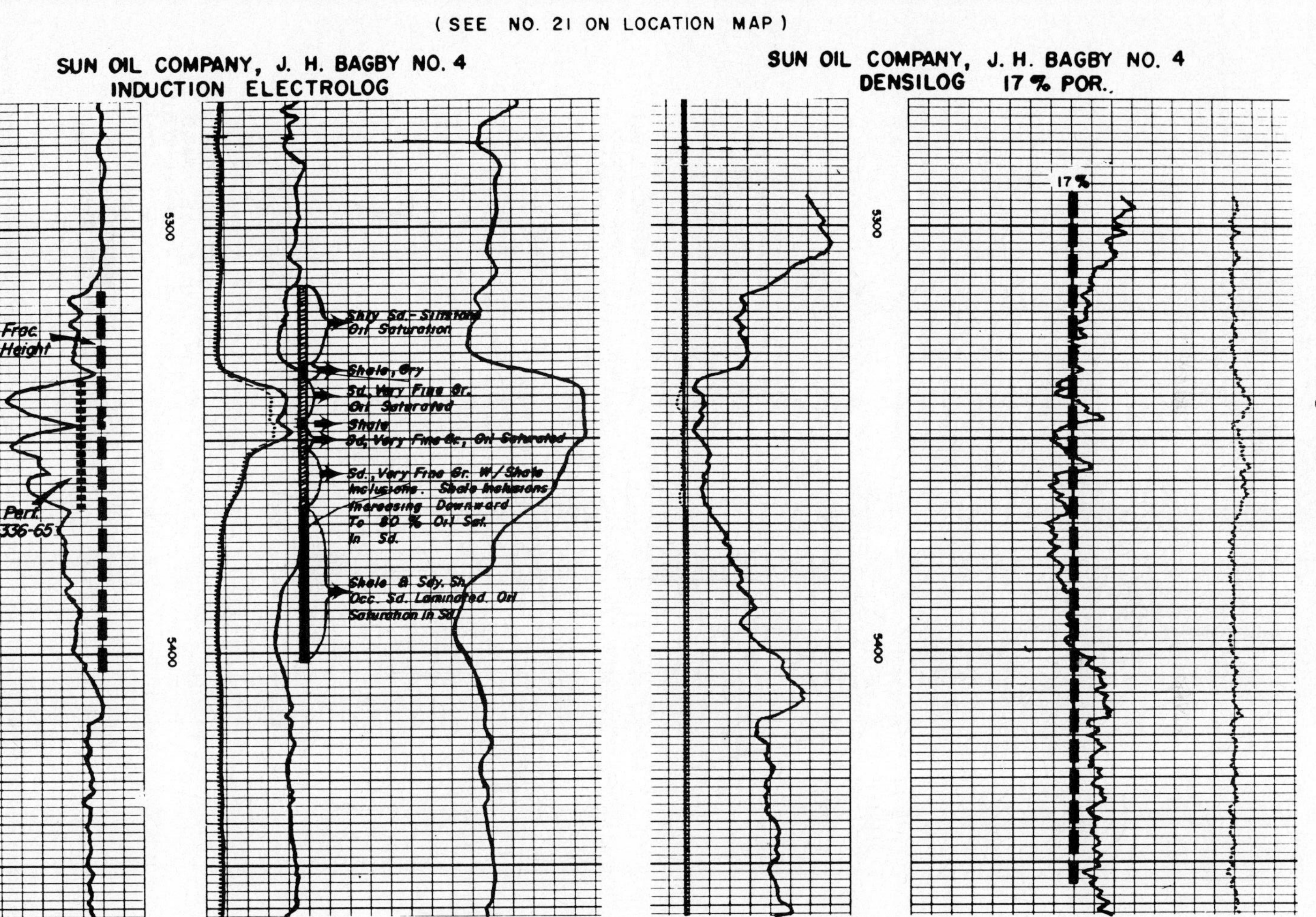

FIG. 6—Typical induction electrical log and Densilog of Big Wells (San Miguel) sandstone interval, usual perforated interval, and fractured interval. Seventeen-percent porosity line is shown on Densilog.

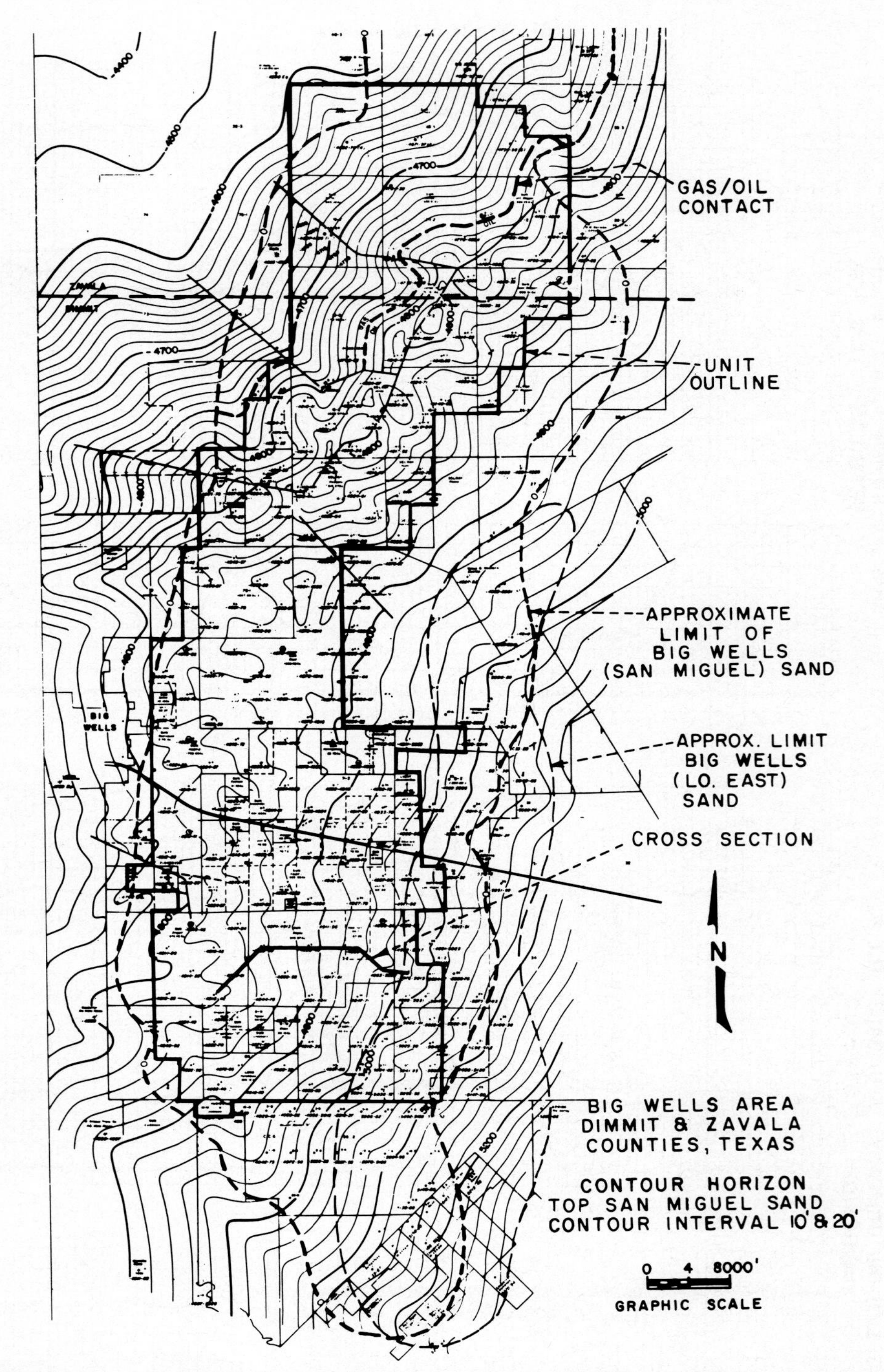

FIG. 7—Structure map of top of Big Wells (San Miguel) sandstone showing limits of sandstone after full development of field. Outline of Big Wells Unit is shown. Structure map prepared by Alan Rutherford, Houston.

Table 1. Core Analyses, Big Wells Field

Well and Depth (ft)	*Av. Perm. (md)*	*Av. Por. (%)*	*Oil Sat. (% of Por.)*	*Water Sat. (% of Por.)*
Sun No. 1–A Jones				
5,429–5,435	1.68	20	23	46
5,441–5,458	2.67	18	24	49
Sun No. 1 Bagby				
5,368–5,394	26.10	18	29	33
5,399–5,403	6.63	20	24	39

circulating cement from total depth to surface, eliminating cost and bother of a DV tool.

A typical well in the field cost $60,000 to drill and complete. The perforated interval usually included all of the sandstone section with evident SP deflection and resistivity. An average fracture treatment used 500 gal of mud acid, 25,000 gal of dispersant (hot oil), 4,000 lb of 20–40-mesh sand, and 55,000 lb of 10–20-mesh sand. On large fracture treatments, horizontal fractures were calculated as extending up to 900 ft (244 m); for a smaller, more common treatment, fractures extended 350–400 ft (107–122 m). A typical completion was for 185 BOPD through a 3/64-in. choke with tubing pressure of 225 lb, gas-oil ratio of 492/1, and gravity of 35°. The allowable, based on 80-acre spacing, was 171 bbl per calendar day in 1973, but it varied according to the market-demand factor. When 100 percent production allowable was granted by the Texas Railroad Commission, the Big Wells field allowable was penalized to 80 percent because of excess gas production until all the gas was utilized.

To obtain the greatest rate of production and sustained full allowable rates, all wells require the hot-oil, large-sand-volume fracture treatment. Several wells were originally treated with a gel diesel fracture. In each of these wells, the production declined to less than 100 BOPD. At first, decline was thought to result from poor sandstone properties; however, after Super Frac (hot-oil) treatment, wells originally treated with the gel diesel fracturing became full-allowable wells. One well producing 80 BOPD after the gel diesel fracture was treated again and flowed 204 BOPD after Super Frac treatment.

Unitization

On July 1, 1974, the Big Wells Unit was placed in operation (Fig. 7). This unit involves 18,518 surface acres—14,762 oil and 3,756 gas. The unit covers 61 percent of the oil in place (original oil in place, 145,268,200 STB), but contributes 95 percent of the production. As of September 1973, 22 gas-injection wells were scattered throughout the field. In addition, water was being injected in wells near the gas/oil contact to isolate the gas cap from the oil column. Gas injection was designed primarily as a pressure-maintenance project. From the unit area, primary ultimate recovery was estimated to be 28,976,000 bbl of oil, 59,378 MMcf of gas, and 3,403,000 bbl of plant liquids. Under the gas/water injection program, ultimate recoverable reserves are estimated to be 35,644,000 bbl of oil, 19,216 MMcf of gas (prior to blowdown), and 7,466,000 bbl of plant liquids.

Reservoir properties are: average porosity, 18.9 percent; average permeability, 7.46 md; average water saturation, 54.4 percent; bottomhole pressure as of June 1, 1973, 1,305 psi; average net thickness of oil sand, 17.1 ft (Snyder, 1975).

Summary

The Big Wells (San Miguel) field extends 14 mi (23.5 km) north-south and 3.4 mi (4.8–6.4 km) east-west (Fig. 7). The oil column totals 420 ft (128 m) and the gas column, 180 ft (55 m), giving a total producing column of 600 ft (183 m). A total of 210 oil wells on 80-acre spacing and 9 gas wells on 640-acre spacing was completed. Surface acres of the field are 30,500—oil productive,

25,400, and gas productive, 5,100. Original oil in place is estimated as 200 million bbl.

The Big Wells (Lo East) field covers 14,000 acres in an area 8 mi (12.9 km) long and 2 mi (3.2 km) wide (Fig. 7). Fifty-three wells drilled on 80-acre spacing were completed. Original oil in place is estimated as 35 million bbl. Ultimate recovery from this reservoir is estimated to be 7 million bbl of oil plus 4 Bcf of gas.

Big Wells is one of the largest fields found in Texas in the last 10 years. Ultimate liquid recovery from the two reservoirs is estimated to be 50 million bbl. Parts of the field could be considered economically favorable to produce under old fracturing and producing techniques, but modern "frac" treatments have resulted in a much better rate of production in good areas and economically attractive rates of production in areas not producible under older techniques.

Selected References

Davies, D. K., F. G. Ethridge, and R. R. Berg, 1971, Recognition of barrier environments: AAPG Bull., v. 55, no. 4, p. 550-565.

Layden, R. L., 1971, The story of Big Wells: Gulf Coast Assoc. Geol. Socs. Trans., v. 21, p. 245-256.

Snyder, R. E., 1975, Big Wells—Sun's example of modern field development: World Oil, v. 180, no. 1, p. 75-78.

Geology of Fairway Field, East Texas[1]

ROBERT T. TERRIERE[2]

Abstract The Fairway field, Anderson and Henderson Counties, Texas, is a major oil field in a reef and reef-associated facies of the Lower Cretaceous James Limestone Member of the Pearsall Formation. The present productive limits are controlled largely by structure. The location of the reef also was influenced by a contemporaneously growing structure, so the reservoir can be considered to have a combination structural-stratigraphic trap, both physically and genetically.

The James Limestone Member consists of several limestone types, differentiated on the basis of texture and fossil content in cores. Maps of the distribution of these rock types during successive stages of reef growth show that the main core of the reef, dominated by frame-building organisms, was in the northwest part of the field. The frame-builders were a closely associated suite of corals, stromatoporoids, algae, and rudistids. By about the middle of the time of development of the James reef, the center of growth was at its maximum, and smaller satellite reefs appeared in the southeast and southwest. A facies dominated by large bivalves occupied much of the area between centers of growth of the main frame-builders. The south-central part of the field was an area of persistent accumulation of carbonate sands and gravels. Carbonate muds and muddy sands were the dominant facies elsewhere.

Porosity and permeability are present in all of the limestone types but are higher on the average in the associated limestones than in the reef proper. The porosity is largely secondary, although it is in part the result of enlargement of primary pores.

INTRODUCTION

The Fairway field is located in Anderson and Henderson Counties, Texas, approximately on the axis of the East Texas basin (Fig. 1). The principal oil and gas production is from the James Limestone Member of the Pearsall Formation, although smaller amounts have been produced from the Ferry Lake Anhydrite and the Rodessa and Sligo Formations. All of these units are of Early Cretaceous age (Fig. 2). Only the James is included in the present study.

The reservoir rock in the James Limestone Member is an unusually good example of a subsurface reef complex. Despite widespread recognition of the importance of reefs in petroleum exploration, relatively few limestone bodies are sufficiently cored or exposed—or have original textures well enough preserved—to demonstrate convincingly a reef origin. Material for the present study included cores from about 35 wells in various parts of the Fairway field. Electric logs from about 200 wells were used to determine the overall shape of the limestone body and the structure and thickness of various units.

The field was discovered in 1960. By the end of 1974 it had produced 123,703,000 bbl of oil and had estimated reserves of 76,234,000 bbl (Oil and Gas Jour., 1975, p. 118). The oil is 48° API gravity. Other reservoir properties and production practices are cited by Perkins (1964) and by Calhoun and Hurford (1970).

In the immediate area of the Fairway field are two much smaller fields. For this report, the Frankston field, just east of Fairway, is considered a part of the Fairway field; however, the reservoirs probably are separated by a fault. The Isaac Lindsey field, just southwest of Fairway, is an unimportant gas field producing from the James limestone.

The trap in the Fairway field is a combination structural and stratigraphic trap. The location of the reef complex in the James was at least partly determined by a growing structure. Later vertical movements modified the shape and extent of the trap.

The major stratigraphic units in the vicinity of the field are shown in Figure 2. The Hosston Formation, which overlies the generally accepted Jurassic-Cretaceous boundary, is a terrigenous unit that contains redbeds and conglomerate, especially in the lower part. Above the Hosston are three limestone units—the Sligo, James, and Rodessa—separated by intervals of dark gray calcareous shale. The limestone units range from very dark and argillaceous to lighter colored and relatively pure. The purer facies of the Sligo and Rodessa are largely grainstones, some of them oolitic and others composed of fossil fragments, oncolites, and Foraminifera. The James is regionally extensive as a thin argillaceous limestone. Where the

[1]Manuscript received, February 27, 1975. Published with permission of Cities Service Oil Company.

[2]Cities Service Oil Company, Tulsa, Oklahoma 74110.

The writer thanks the many colleagues who assisted or supported various phases of the study, especially T. L. Broin, M. K. Horn, N. P. Leiker, W. K. Pooser, B. A. Silver, and K. F. Wantland, Jr.

reef facies is well developed in the Fairway field, the James is as thick as 250 ft (75 m) and is relatively pure and porous.

Overlying the Rodessa Formation is the Ferry Lake Anhydrite, which consists of anhydrite with interbeds of shale and limestone. The limestone is largely argillaceous, but isolated thin beds are oolitic and contain intraclasts, pellets, small mollusks, and Foraminifera. These limestone beds apparently are the reservoirs for the small amount of oil that has been produced from the Ferry Lake, although cementation and partial replacement by anhydrite have reduced their porosity and permeability.

James Limestone Member

Structure and Thickness

Structure of the top of the James limestone reef is shown in Figure 3. The basic configuration is that of a southeast-plunging nose, truncated on the northwest by a group of faults and somewhat modified by other faults on the east and northeast.

The thickness of the James limestone, including argillaceous limestone at the base, is shown on Figure 4. The thickness generally parallels the structure. The overall parallelism of isopach lines and contours of the top of the reef appears to be the result of localization of reef growth by the developing structure. The thickness differences are much less than the elevation differences, so the shape of the top of the limestone is not simply depositional. The conclusion that the structure was growing during the time of deposition is also supported by isopach maps of thin arbitrary intervals above and below the reef. These maps show patterns similar to those of the reef thickness and structure, especially with regard to considerable thickening northwest of the Fairway field. The shales and shaly limestones in these intervals surely were deposited in horizontal layers, so the thickness variations must reflect active structural movement during deposition.

Salt domes are present in the general area of the field (see Fig. 1), and it is probable that the persistent vertical movements were caused by

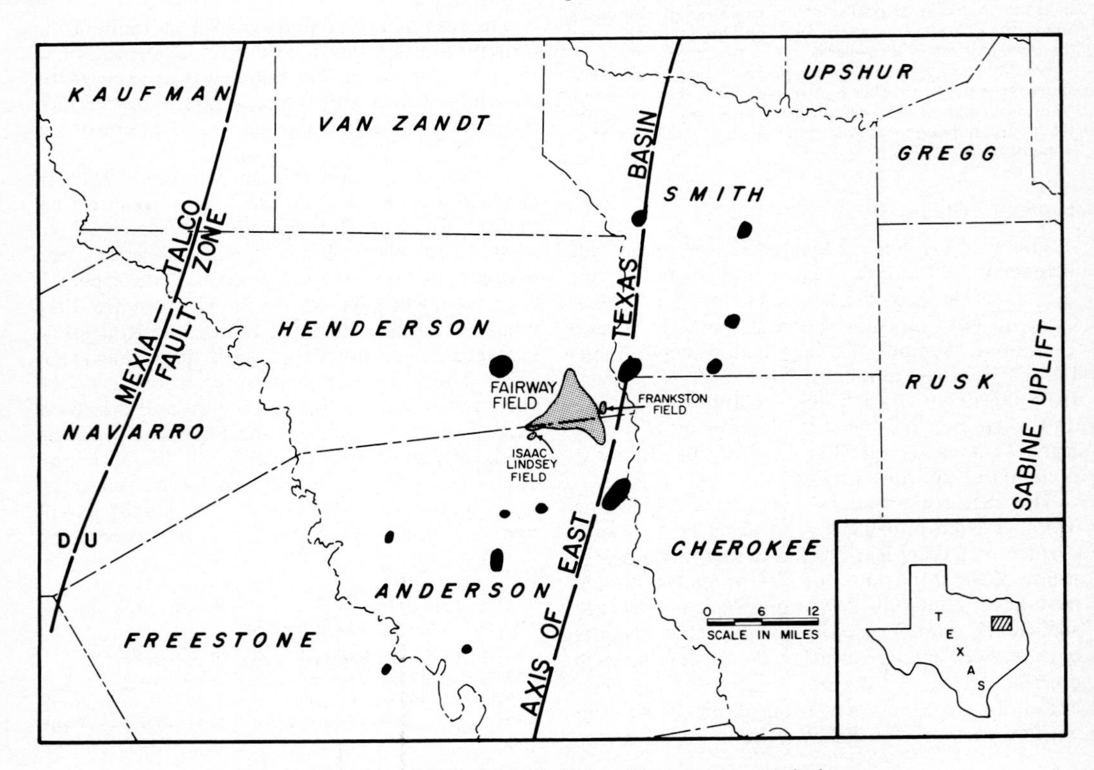

FIG. 1—Location of Fairway field, Texas. Black spots are salt domes.

flowage of the underlying salt into the domes. It has been suggested by geologists working in the area that the Fairway field is on an "interdomal high" still underlain by the original thick salt after flowage of salt from adjacent areas into the domes. Others believe that the area is a "turtle-shaped structure" (Trusheim, 1960, p. 1533) underlain by relatively thick post-salt deposits that had filled a topographic low during early salt flowage.

Lithologic Varieties

For purposes of showing the major rock types in the field, a somewhat informal limestone classification was established based on megascopic description of slabbed cores. During description of the first few cores, it was seen that certain varieties of limestone kept recurring. Each major variety was defined as a limestone type and identified by a number. Description of additional cores resulted in some modification and additions to the system. The present scheme seems to work well in expressing the major limestone types in the Fairway field and in showing vertical and lateral changes.

The rock types recognized are as follows:

Type I. Calcarenite and calcirudite
- a. Conglomerate and well-sorted carbonate sandstone (grainstone of Dunham, 1962)
- b. Muddy carbonate sandstone (packstone)

Type II. Predominantly micritic limestone, mostly with intermixed carbonate sandstone and shell fragments (wackestone and mudstone)

Type III. Limestone containing large bivalves
- a. Many unbroken shells, commonly upright in the rock
- b. Broken shells, some algal-coated

Type IV. Limestone characterized by colonial organisms such as corals, stromatoporoids, rudistids, and algae

Type V. Dark argillaceous limestone

Type I

The limestones included in Type I are the carbonate sandstones and conglomerates—the definitely clastic limestones (see Figs. 5, 6). The coarsest grains are mostly intraclasts (terminology of Folk, 1962). They apparently were not transported far, although many are well rounded. Nearly all seem to have been lithified when transported; none is squashed as would be the case if reworked while soft. Among the pebble-sized fossils, clam fragments appear to be the most abundant, but other types of fossils seen in the reef are also present, as are many unidentifiable fragments. Sand-size grains are more difficult to identify. They include intraclasts, pellets, and fossils; the relative abundance of each is uncertain because much of their character has been obliterated by micritization. Many are the intermediate-type grains, neither intraclasts nor pellets, that have been called "lumps" or "pelletoids" (called "peloids" by McKee and Gutschick, 1969, p. 101, 555). The interstitial material includes both micrite and spar cement. Many limestones have microspar cement, which appears to be mostly primary rather than recrystallized micrite.

Type I limestones can be further subdivided into Types Ia and Ib. The calcirudites and rela-

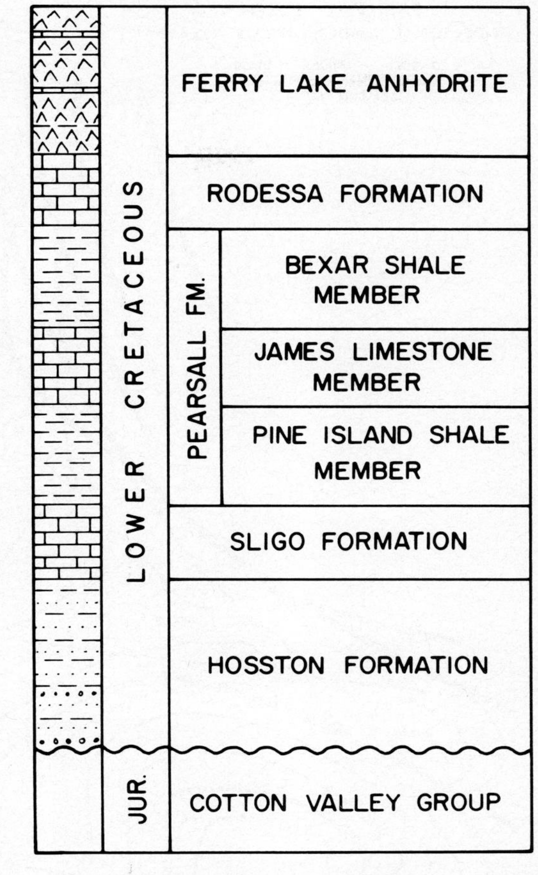

FIG. 2—Partial columnar section, Fairway field area.

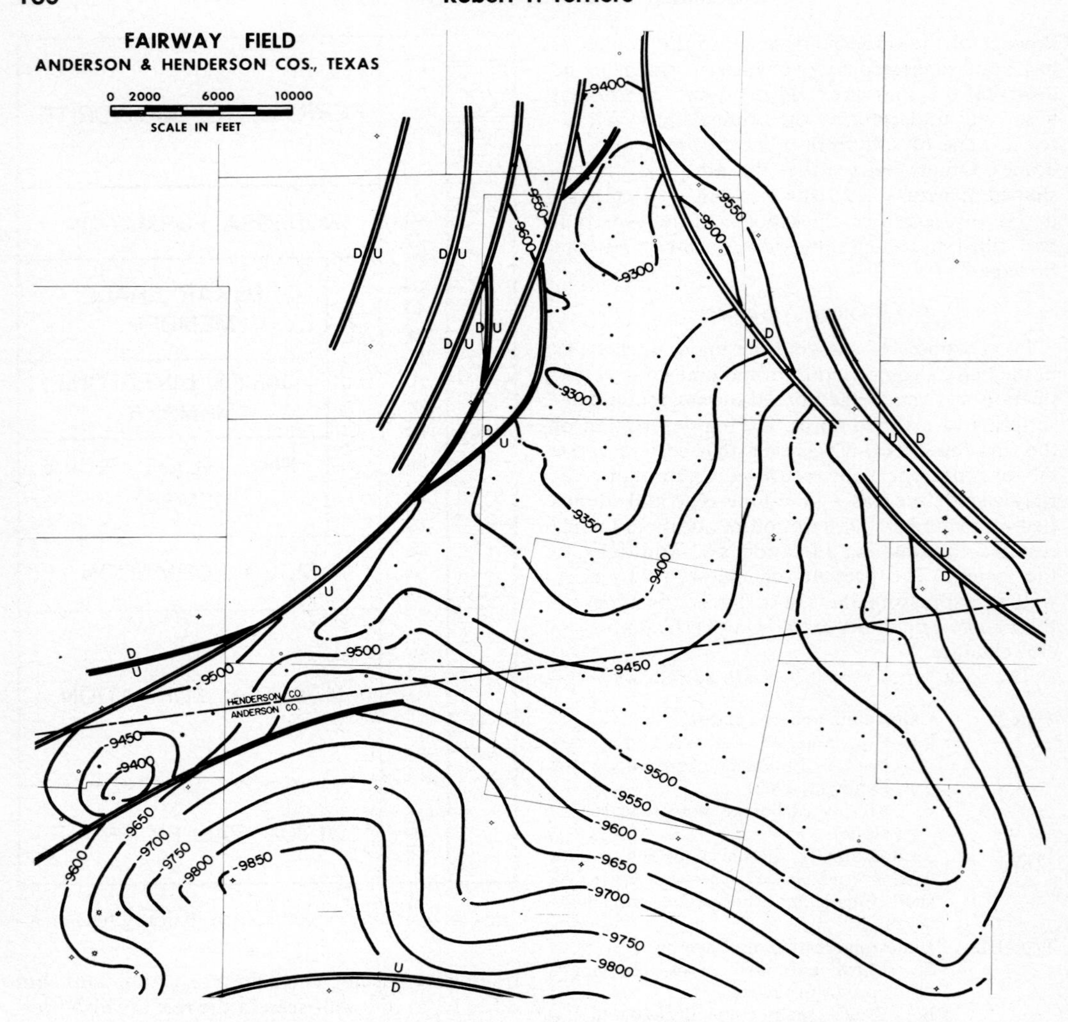

FIG. 3—Structure of top of James Limestone Member. Depths are in feet below sea level.

tively coarse calcarenites are characterized by spar or microspar cement and rounded grains. They constitute the limestone Type Ia and were deposited under conditions of considerable wave action, probably as marine bars.

The limestones classed as Type Ib are transitional into the calcareous mudstones of Type II. They have an appreciable proportion of micrite matrix and are generally poorly sorted and not well rounded. Environmentally, they also are intermediate between Types Ia and II. For convenience in description, rare intervals of alternating Types Ia and II were also classed as Type Ib.

Type II

Type II limestones are composed predominantly of micrite (Figs. 7, 8). They include the wackestone and mudstone of Dunham's (1962) terminology and the calcilutite of older usage. Some rocks in Types III, IV, and V are also more than 50 percent micrite, so Type II is a "wastebasket" group for micritic limestones that do not fit into one of the other categories. Moreover, some rocks that megascopically appear to be micrite proved to contain many pellets and other grains when studied in thin section.

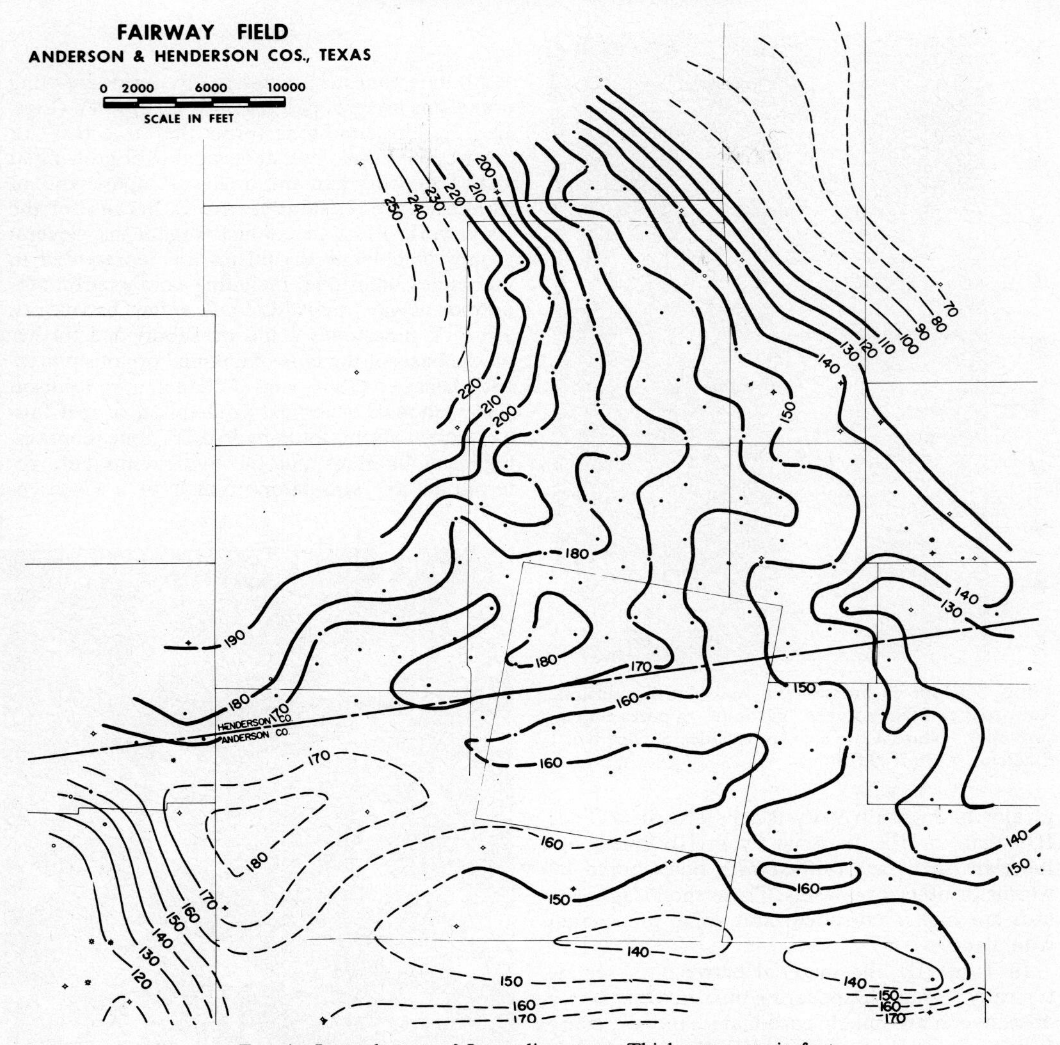

FIG. 4—Isopach map of James limestone. Thicknesses are in feet.

Environmentally, most Type II limestone represents quiet-water deposition. It may have been deposited partly in sheltered intra-reef locations and partly in deeper water.

Type III

Some limestones in the Fairway area are distinctive because of their content of very large bivalves (Figs. 9-11). These constitute limestone Type III. Some of the clams appear to be about 6 in. (15 cm) across, although the limited size of the core prevents viewing of whole specimens. They have not been formally identified, but many resemble the genus *Chondrodonta.* Perhaps a variety of biologic types is grouped together, but smaller shells are specifically excluded. Some of the large clams are in upright, presumably growth position in the rock (Fig. 9). In other intervals, shells that are considerably broken seem to be mostly from the same type of very large clam (Figs. 10, 11).

Type III limestones are subdivided into Types IIIa and IIIb. Rocks of Type IIIa contain many relatively unbroken shells, many of which are upright. Type IIIb designates rocks containing broken shells. Many fragments are impossible to identify even as to the size of the original shell, so

FIG. 5—Type Ia limestone, a well-sorted grainstone (calcarenite). Black sootlike "gilsonite" in pores emphasizes rock texture. Core slab from Cities Service No. 1 Emerson, 9,876 ft (3,010 m).

greater biologic diversity is represented in Type IIIb than in IIIa. In some Type IIIb limestones the clam shells are riddled with holes bored by worms or other organisms. These shell fragments also are highly corroded, and some are coated with algae.

In Type IIIa the material between the clams typically is smooth-appearing micrite. Only rarely does it contain much carbonate sand or many fossils. The matrix of Type IIIb limestones contains sand-size carbonate grains, but carbonate mud predominates. Although Type IIIb is transitional into Type I limestone, borderline cases are surprisingly uncommon.

Some Type III limestones are transitional into Type IV. They contain some colonial corals but very few other frame-building organisms such as stromatoporoids. Type III limestone also borders on Type IV limestone in an environmental sense. The large clams that characterize it lived around the edges of the reef core and spread over the surrounding sea bottom, possibly at very shallow depths.

Type IV

All limestones characterized by frame-building organisms are grouped togehter in Type IV (Figs. 12, 13). This limestone forms the "true reef" or "reef-core" lithology that is capable of growing at angles steeper than the angle of repose and of forming wave-resistant structures because of the framework effect of colonial organisms. Several groups of colonial organisms are represented in the James limestone, including corals, stromatoporoids, algae, rudistids, and a few bryozoans. Type IV limestones could be subdivided further on the basis of the type of colonial organism present. Achauer (1967) and Achauer and Johnson (1969) showed a vertical progression of reef faunas from a stromatolite-hydrozoan zone (containing algal banding and the hydrozoans here referred to as "stromatoporoids") to a *Chondro-*

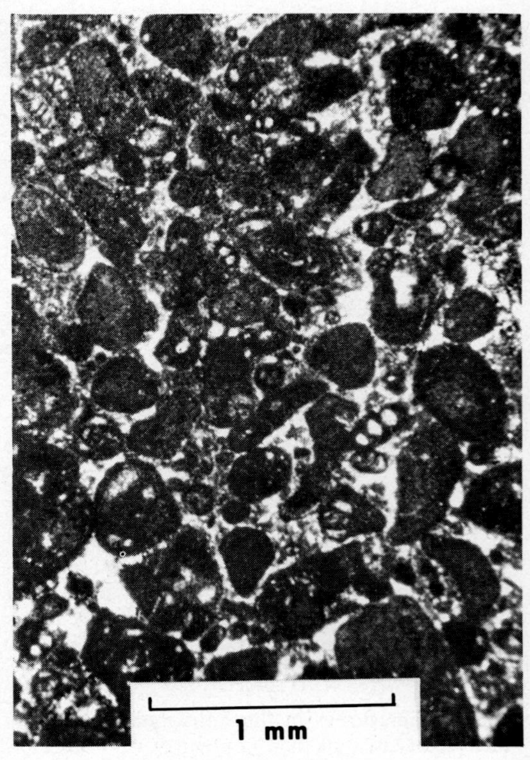

FIG. 6—Type Ia limestone in thin section. Grains are pellets, intraclasts, and Foraminifera. Unidentifiable micritized fossil fragments are common in this lithology. From Hunt No. 1 West Poynor, 9,925 ft (3,025 m).

donta zone, and then to a rudistid zone. For the present study, however, the limestones containing frame-building organisms have all been grouped under Type IV.

Type V

Type V includes the limestones that are dark colored and impure (Figs. 14, 15). It is gradational into other varieties of limestone and into shale, and dividing lines are hard to place. For cores in which a differentiation of Type V limestone from calcareous shale was difficult, the decision was based on fissility. Rocks with a conchoidal or irregular fracture across the bedding were classified as Type V, and those that split along bedding planes were classified as shale.

The separation of Type V limestone from other limestone types was based largely on color, which also corresponds well with the amount of insolu-

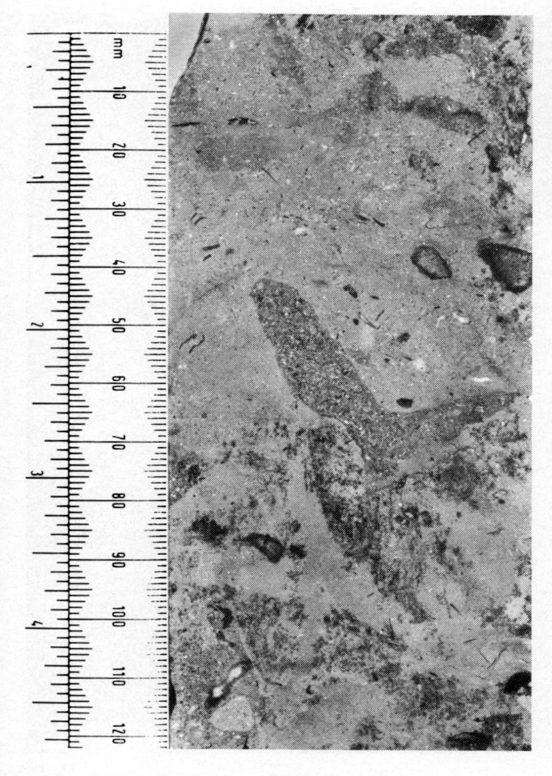

FIG. 7—Type II carbonate mudstone containing burrows filled with muddy carbonate sandstone. Core slab from Atlantic No. 1 Larue, 9,999 ft (3,048 m).

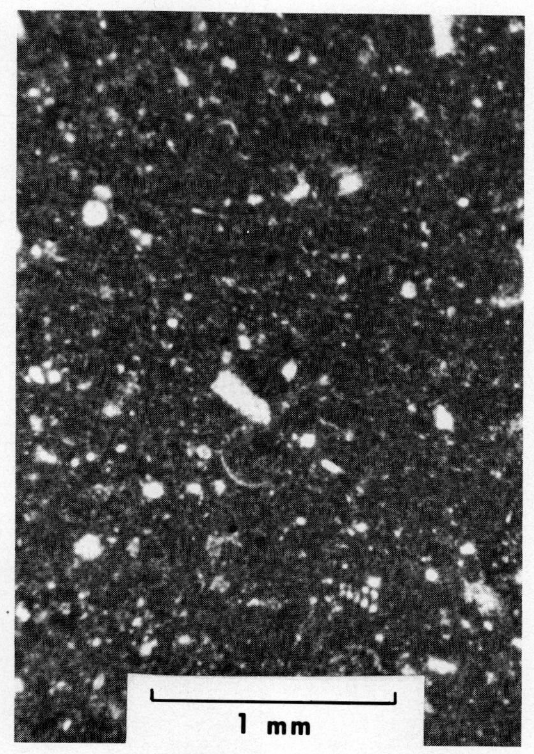

FIG. 8—Thin section of Type II carbonate mudstone containing scattered small fossils. Many such thin sections contain considerable sand-size material and suggestions of pellet texture. From Cities Service No. B-1 Miller, 9,851 ft (3,003 m).

ble material. Limestones of Type V are medium dark gray or darker, based on comparison of a freshly broken surface with the rock-color chart. Texturally, Type V limestones most commonly grade into the carbonate mudstones of Type II, but samples intermediate between Type V and Type Ib (muddy sandstones) are also common. Thin sections show that sand-size carbonate grains are more abundant than would be suspected from viewing hand specimens. Some Type IV limestones are fairly dark colored, but little difficulty was encountered in separating them from Type V.

Core-slab surfaces of Type V limestone that have been etched with hydrochloric acid have a dirty brownish-gray color, apparently caused by a surface residue of clay and by slight oxidation. Many of them show irregular sedimentary struc-

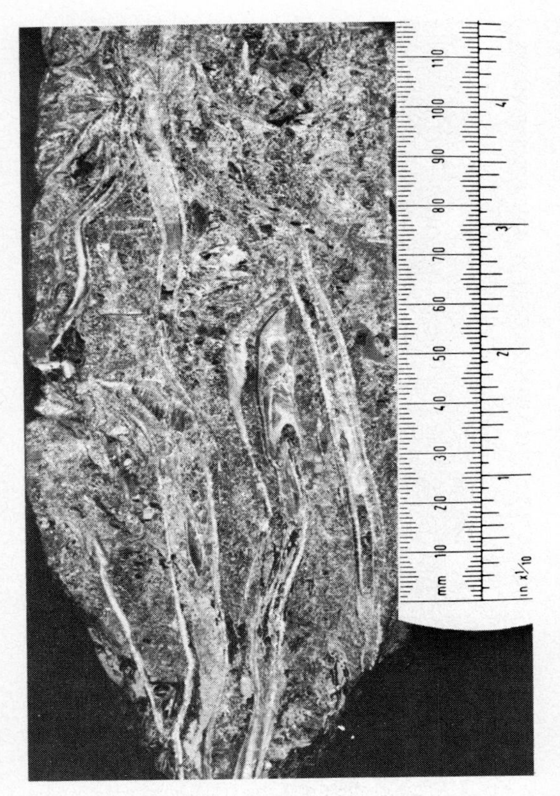

FIG. 9—Type IIIa limestone containing large bivalve shells upright in rock. Core slab from Cities Service No. C-1 Miller, 9,901 ft (3,018 m).

tures, probably from slight slumping of the sediment while soft, from compaction, and from burrowing organisms. Others have spheroidal to irregular concretionlike masses that are slightly more calcareous than the rest of the rock. Some of the Type V limestone contains a few megafossils, mostly small clams. Pyrite is common, and quartz silt is more abundant than in the cleaner limestone types. Some very dolomitic Type V samples from above and below the reef resemble the other Type V rocks.

Type V includes essentially all of the offreef James limestone. This type has not been divided into subvarieties, but recognition of meaningful variations in these rocks can be important in reef exploration. For example, a core of dark argillaceous limestone from a dry hole less than 1 mi (1.6 km) northeast of Fairway field contains sand-size, relatively pure intraclasts in Type V limestone at the stratigraphic level of the reef (Figs. 16, 17). The well was drilled before the reef was discovered, and in retrospect the intraclasts can be recognized as evidence of a nearby buildup, although none are identifiable as definite reef fragments.

Other Limestone Varieties

Some relatively minor limestone types are distinctive enough to merit special mention, though they are not differentiated as major rock types. At one time a limestone type was established for biomicrite containing many megafossils, separate from the types containing large clams (Type III) or frame-builders (Type IV). These fossiliferous limestones contain moderate to small-sized bivalves, gastropods, echinoderm fragments, and unidentifiable shell fragments, all of which are widespread in the James. Experience has shown

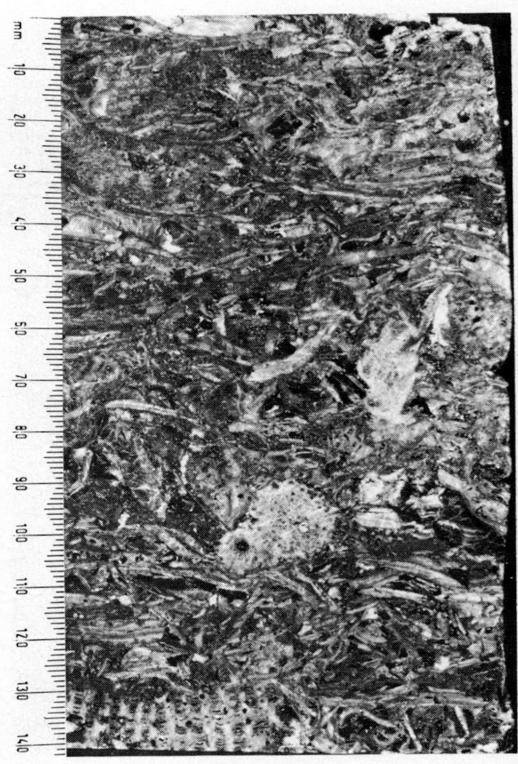

FIG. 10—Type IIIb limestone containing many broken bivalve shells and a few small corals. Core slab from Cities Service No. C-1 Miller, 9,891 ft (3,015 m).

that they are abundant enough to form a distinctive rock type in only a very few places, and the subdivision was dropped from the classification.

Some of the cores contain many oncolites, mostly pebble-sized pisolites composed of wavy algal bands enclosing rounded shell fragments or intraclasts (Fig. 18). Oncolites are numerous enough in some rocks to form a distinct limestone variety. The oncolite-rich lithology tends to overlap Type IIIb limestone by virtue of the presence of progressively thicker algal coatings on broken clam fragments. The volume of oncolite limestone is small, so oncolites were noted as a subsidiary constituent in rocks classified as one of the other types, usually IIIb. Environmentally, the oncolites seem to have formed in areas of relatively slow deposition, as indicated by their association with bored and corroded clam fragments; it is supposed that algae were better able to coat grains that were not buried too rapidly.

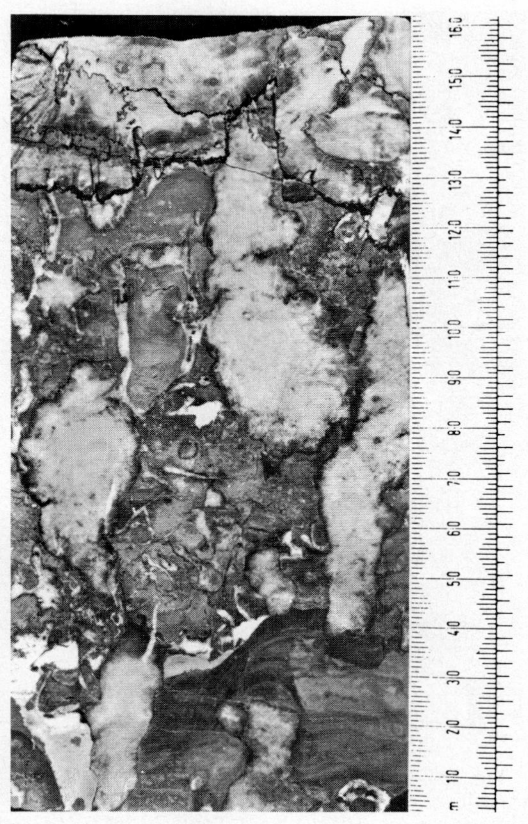

FIG. 12—Fingerlike or plumelike stromatoporoid colonies in Type IV limestone, "core" facies of reef. Dark bands of probable algal origin are locally visible in mud between colonies. Core slab from Cities Service No. C-1 Ellis, 9,936 ft (3,028 m).

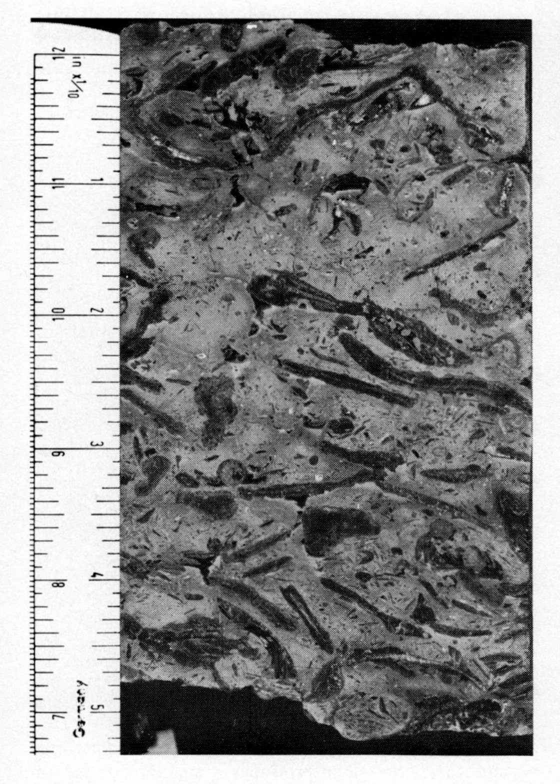

FIG. 11—Type IIIb limestone in which fragments of large mollusks are corroded and bored and heavily coated with algae. Core slab from Cities Service No. C-1 Miller, 9,886 ft (3,013 m).

Limestones rich in Foraminifera have not been classified separately. Foraminifera are sufficiently abundant in some cores to form distinctive rock types, but these are conveniently treated as varieties of Type I and Type II limestones, and the abundance of Foraminifera is noted separately. There is a remarkable relative scarcity of foraminifers in rocks of Type III and Type IV. Miliolids are the most abundant group of foraminifers in and near the reef, but other types are present also, especially in offreef areas. The Foraminifera are concentrated in the upper part of the James. It is largely speculative whether the environment in which they flourished was specifically provided

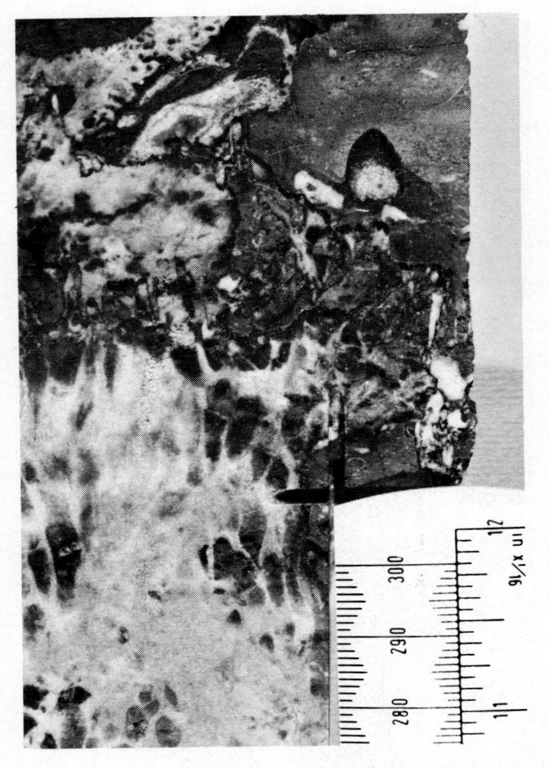

FIG. 13—Type IV limestone containing a largely recrystallized coral colony. Algae and a small rudistid(?) also are present. Atlantic No. 1 Truitt, 10,264 ft (3,128 m).

by the reef itself, or whether some essentially independent ecologic factor such as regional water depth or temperature was favorable only during the latter part of reef growth.

DISTRIBUTION OF LIMESTONE TYPES

The three-dimensional distribution of the rock types is most conveniently shown by a series of maps, each indicating the lithologic distribution during a different vertical interval of reef growth. Ideally, the separation of intervals should be by time planes correlated from well to well. Unfortunately, no internal marker horizons can be recognized, and artificial "time" intervals had to be used. The engineering committee concerned with oil recovery in the field used sonic logs to correlate porosity zones between wells (Fairway Technical Comm., 1969) and mentioned geologic correlations by Achauer. Such correlations can be made locally but become questionable at best over wider areas and are doubtful markers of contemporaneity.

Above and below the reef, many thin electric-log units are easily recognized across the field. The number and remarkable continuity of these units, which are composed of calcareous shale and argillaceous limestone, are convincing evidence that the lithologies above and below the reef parallel time planes. Subdivision of the reef limestone itself was done by dividing the interval between the highest marker below the reef and the top of the reef into six equal units (Fig. 19). These were treated as approximately contemporaneous units in the various wells, a working hypothesis that helps in visualizing the succession of facies patterns. The lowest one sixth of the interval between the markers is below the reef proper and has not been cored; it was labeled "unit 0" and was not used further. The other divisions, from the base up, were called "units 1 to 5." Their thickness varies from well to well in proportion to variation in the thickness of the entire James interval.

FIG. 14—Type V argillaceous limestone. Acid etching of smoothed core slabs produces a surface coating of clay much lighter colored than natural rock and brings out structures not otherwise easily seen. This core slab shows considerable burrowing; from Cities Service No. C-1 Miller, 9,825 ft (2,995 m).

The lithology in each "time" unit was determined from cores, and the dominant lithology was used where several limestone types were interbedded. The lateral boundaries shown between facies on the map were influenced by the abundance of subsidiary lithologies in mixed intervals. Electric logs were used to aid in refining lateral changes between shale and limestone in unit 1. The resulting maps are shown as Figures 20–24.

Unit 1

The distribution of the limestone types in the first of these units, representing initial stages of reef development, is shown on Figure 20. The apparent center for growth of frame-building organisms was in the northwest part of the present Fairway field, where a small area of Type IV limestone was deposited. The shape and extent of this reef framework are uncertain because of limited core control, but basic patterns of limestone-shale distribution are apparent from electric logs. Subordinate amounts of limestone with frame-builders in wells outside the shaded area also help in fixing approximate limits.

The same reasoning, plus somewhat better core control, places centers of calcarenite distribution

FIG. 15—Core slab, more typical than that in Figure 14, of dark argillaceous nonreef facies of James limestone (Type V). From Cities Service No. B-1 Miller, 10,047 ft (3,062 m).

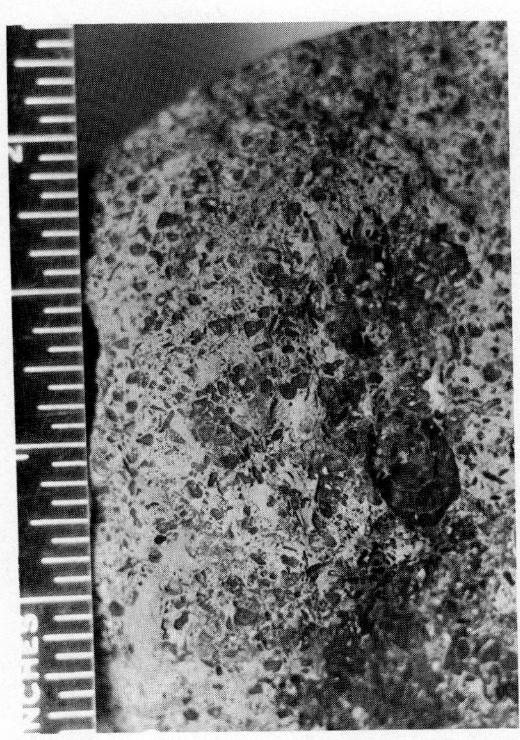

FIG. 16—Broken core surface that has been etched with dilute hydrochloric acid. Rock is Type V limestone, dark colored and very clayey, and etching has left a light gray insoluble clay residue standing out on surface. Intraclasts of purer limestone have dissolved somewhat, leaving smooth surfaces with no clay residue. Intraclasts could be seen on natural rock surface only with difficulty. From Cities Service No. 1 Xenia Miller, 10,138 ft (3,090 m).

in the south-central and southern parts of the field. The northern reef-core facies and southern calcareous sandstone facies are separated by an area of carbonate and terrigenous mudstone.

Unit 2

The second of the subdivisions of the James limestone reef is characterized by a greatly expanded reef core in the northwest and north-central part of the area and by the presence of relatively pure limestone over areas where argillaceous micrite had been deposited previously (Fig. 21). Calcarenites of limestone Types Ia and Ib are notably widespread, and the principal area of Type Ia deposition lies along an east-west barlike feature.

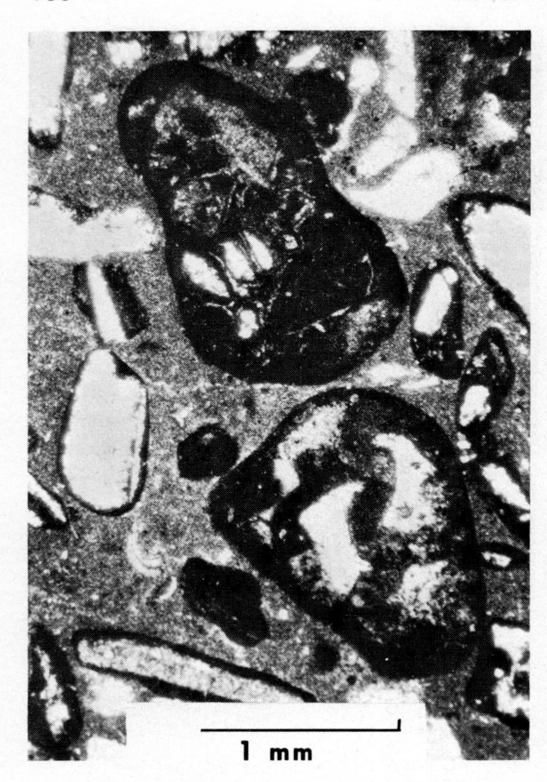

FIG. 17—Thin section of Type V limestone (see Fig. 16). Limestone intraclasts are in matrix of micrite and clay. Plane-polarized light. From Cities Service No. 1 Xenia Miller, 10,137 ft (3,089 m).

Unit 3

Limestones characterized by large fossils reached their maximum extent during Unit 3 deposition (Fig. 22). The main reef core of Type IV limestone in the northwest actually diminished slightly in size, but new centers of reef-core facies appeared in the southwest and southeast parts of the field. In addition, Type III limestone extended over a broad region, and large bivalves dominated the fauna between the centers of growth of corals, stromatoporoids, etc. Cross sections through the field show that this Type III limestone formed a definite unit that is traceable over several square miles. The ecologic significance of this extensive area of growth of large clams is uncertain, but its lateral extent would suggest that the seafloor was fairly flat. Perhaps the reef had grown nearly to sea level, which caused a pause in upward growth and a spreading out over a larger area. The common occurrence of borings and algal coatings in Type IIIb limestones may indicate relatively slow deposition. Flattening of the reef top and perhaps restriction in the interior would result in shift of growth of frame-builders to the margins of the reef. Extensive Type Ia limestone along the southern edge of the field indicates persistent wave action in that area.

Unit 4

During unit 4 deposition, growth of large organisms was somewhat more restricted and was largely confined to a band along the northeast side. The main reef core in the northern part of the field had shifted slightly to the northeast from earlier positions. Comparison of Figures 20 and 23 will show that areas that were offreef clayey limestones to the northeast of the reef in unit 1 were within the center of growth by the time of unit 4 deposition. Satellite centers of growth persisted, especially one on the east side of the field. On the southwest tip of the present field, an area of satellite reef growth in unit 3 was supplanted by Type III limestone in unit 4.

Another striking change during unit 4 deposition was the sudden influx of large numbers of Foraminifera, principally miliolids. The presence of numerous Foraminifera is indicated by an "F" on the maps. The foraminifers did not invade the centers of growth of the frame-builders but were abundant in the sheltered, and perhaps slightly restricted, area in the reef-complex interior. They are very numerous also in the sandstones in the southern part of the field, where they may have been concentrated partly by waves and currents.

Oncolites, the pisolite-like algal grains, are also relatively abundant in this interval. Their distribution is not shown on Figure 23, but it generally parallels that of the abundant Foraminifera.

Unit 5

The final stage of the reef is marked by further restriction of the reef core and the continued presence of abundant foraminifers and oncolites (Fig. 24). The northern center of growth of the frame-builders had almost disappeared, and its remaining remnant was farther northeast than before, overlying older offreef sediments. The satellite reefs on the east and southwest persisted in small areas. As indicated by Achauer (1967), rudistids largely supplanted stromatoporoids as frame-builders in this upper part of the reef complex. In the center of the field, micrites were deposited in greater abundance than before, and calcarenites were less common.

Porosity and Permeability

The porosity in the Fairway field is largely secondary, although primary porosity and permeability have influenced distribution of the secondary pores. In some samples, porosity is due chiefly to dissolution of grains, whereas in others matrix porosity is dominant. In still others, porosity is nonselective regarding fabric, occurring in both grains and matrix.

To compare the reservoir qualities of the various types of limestone, a tabulation was made of rock type versus porosity and permeability for 13 cores from which both lithologic descriptions and core analyses were readily available. The averages from this tabulation are shown in Table 1.

It is emphasized that there are many sources of error in this procedure, and the values should be considered approximate. They include values both from good reservoir rock and from denser zones within the reef. Rocks of Type V, the dark, argillaceous nonreef limestones, were not included because cores of obviously nonporous rocks are rarely analyzed in the core laboratories. This same bias in sampling also tends to increase the apparent porosity and permeability values of other relatively nonporous limestone types, because values that would be very low were not obtained. Permeability values have very great sample-to-sample variation, so the averages of permeability are less reliable than those for porosity. Other uncertainties arise from the fact that the analyses were run by three different laboratories using somewhat different techniques and methods of reporting.

Despite these problems, the figures can be used as a rough measure of the relative reservoir properties of the different types of limestone. Each of the rock types contributes to the reservoir. This is worth emphasis because of the widespread impression that reefs are far better reservoirs than are other carbonate rocks, and because of the tendency to equate the word "reef" with rocks containing frame-building organisms.

The highest average porosity and permeability values are for Type Ia and Type IIIb. Type Ia, the well-sorted calcarenite and calcirudite, has had considerable reduction of original intergranular porosity by cementation, but enough has remained to permit relatively easy development of secondary porosity. Type IIIb contains many fragments of mollusks, which probably were partly aragonite. The fragmental nature of the rocks and the mineralogic instability of the shells would both favor development of secondary porosity.

FIG. 18—Limestone composed of pisolites that apparently are oncolites formed by algal coating of shells. This lithology is transitional into algal-coated mollusk facies of Figure 11. From Cities Service No. 1 Miller, 9,863 ft (3,006 m).

Conversely, Type II limestones have relatively low porosity and permeability, probably because of low primary permeability. Type Ib limestone is intermediate between Type Ia and Type II, both in lithology and in porosity and permeability. In Type IIIa, large bivalves are relatively unbroken and commonly are in a micrite matrix; thus, it is intermediate in lithology between Types IIIb and II, as is indicated also by porosity.

Type IV limestones, those with frame-building organisms, are relatively low in porosity. The organisms are in a matrix of micrite, and many are surrounded and partly coated by dark bands of algal origin (Achauer and Johnson, 1969). The frame-building organisms themselves, especially the stromatoporoids, do not seem to have been easily leached, and their occurrence in relatively thick beds may also have minimized leaching.

As mentioned, the nonreef limestones of Type V have rarely been analyzed, but there is no

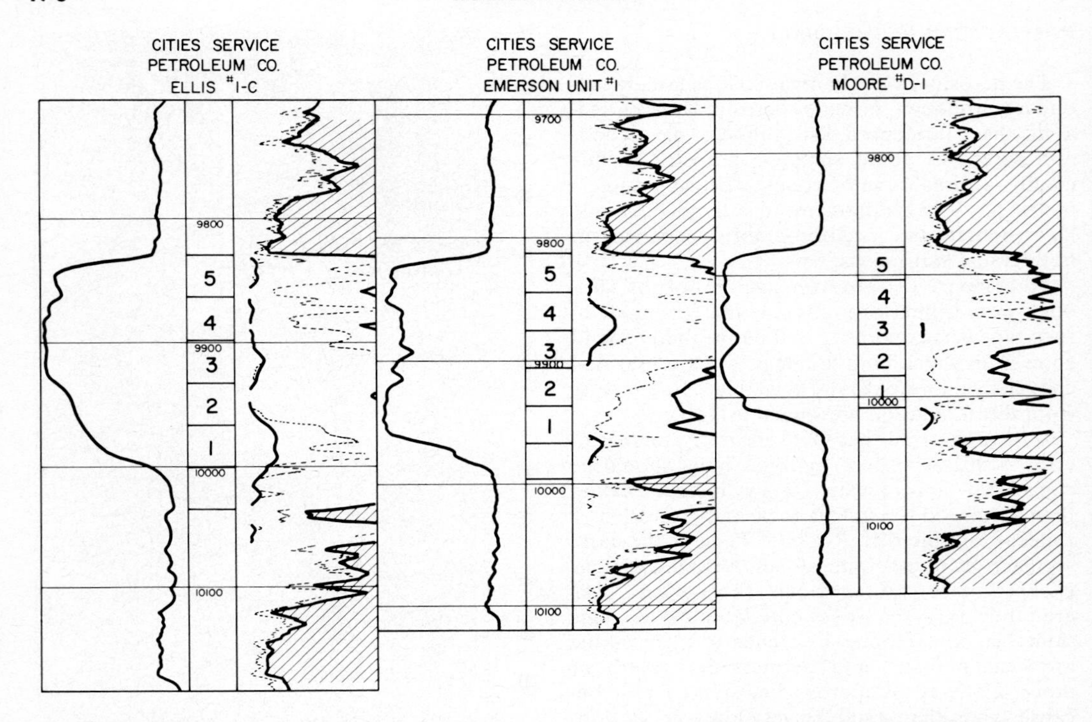

FIG. 19—Electric logs of James limestone interval from three representative Fairway field wells in northwest, central, and southeast parts of field. An arbitrary space under resistivity curve has been shaded to emphasize detail with which pre-James and post-James sections can be correlated. Within James limestone, subdivisions used for Figures 20-24 are numbered 1 through 5.

doubt that their porosity and permeability are very low.

Additional Features of Reef Limestone

Accessory minerals are a relatively minor component of the James limestone. The mineral dolomite is relatively scarce in the reef, and none of the rock is dolomite. The mineral occurs in two forms—as coarse white dolomite within the reef complex itself and as a very finely crystalline variety that is largely restricted to the argillaceous nonreef rocks. The coarse dolomite is mostly a replacement and cavity fill in megafossils. Petrographic evidence indicates an early postdepositional origin. Some of the limestones that contain this coarse dolomite also have small amounts of anhydrite.

The finely crystalline dolomite is most abundant in argillaceous pre-reef and post-reef beds and rarely exceeds a few percent of the rock volume. The impermeability of the rocks would suggest an early origin for this fine dolomite, but no direct evidence is available.

Chert is virtually absent from the James Limestone Member. Trace amounts of authigenic silica occur as tiny quartz crystals, most of them with nuclei of detrital quartz silt. The detrital quartz itself is largely confined to the argillaceous nonreef rocks; it is present in the reef only locally and in minor amounts.

Many parts of the James contain a black organic material, much of it a powdery "soot." Elsewhere, the material is less dispersed and is present in shiny masses with conchoidal fracture. Apparently both types are forms of the same material; they have been called "gilsonite" in some core and sample descriptions. They occur within secondary pores and hence have entered the rock after its original deposition. It seems probable that the original hydrocarbon to enter the pores has further matured into the present high-gravity oil and the sootlike black residue.

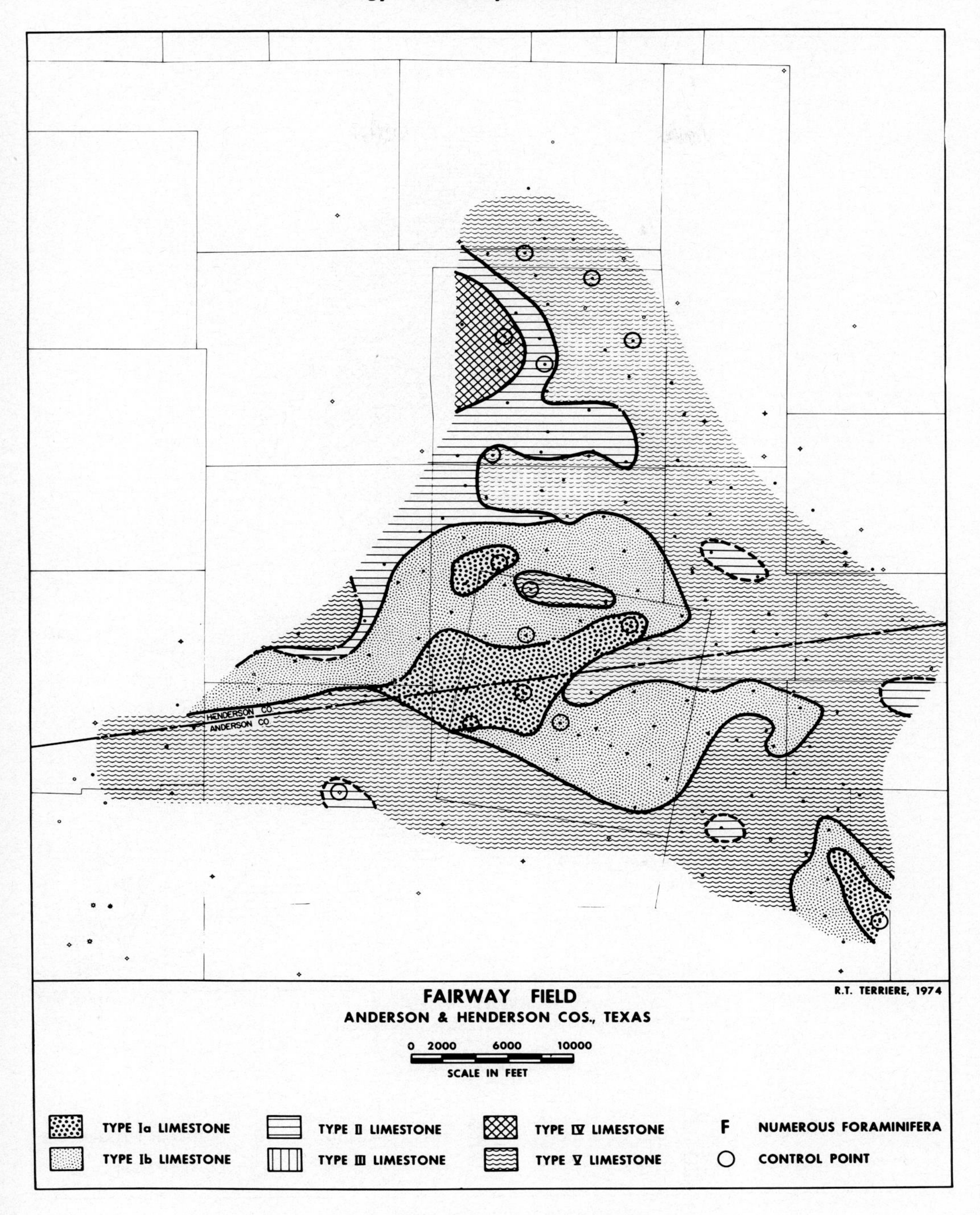

FIG. 20—Distribution of limestone types in unit 1.

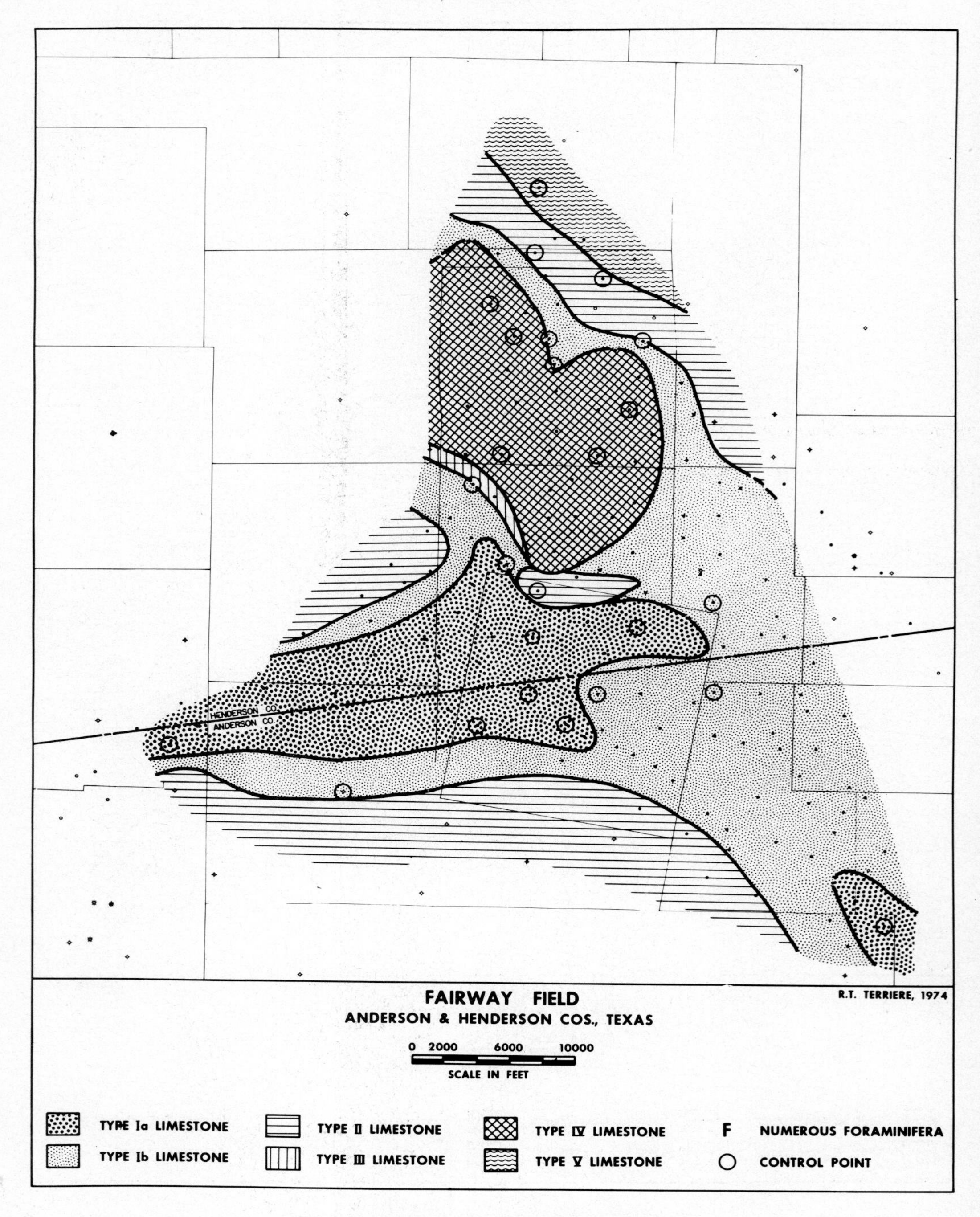

FIG. 21—Distribution of limestone types in unit 2.

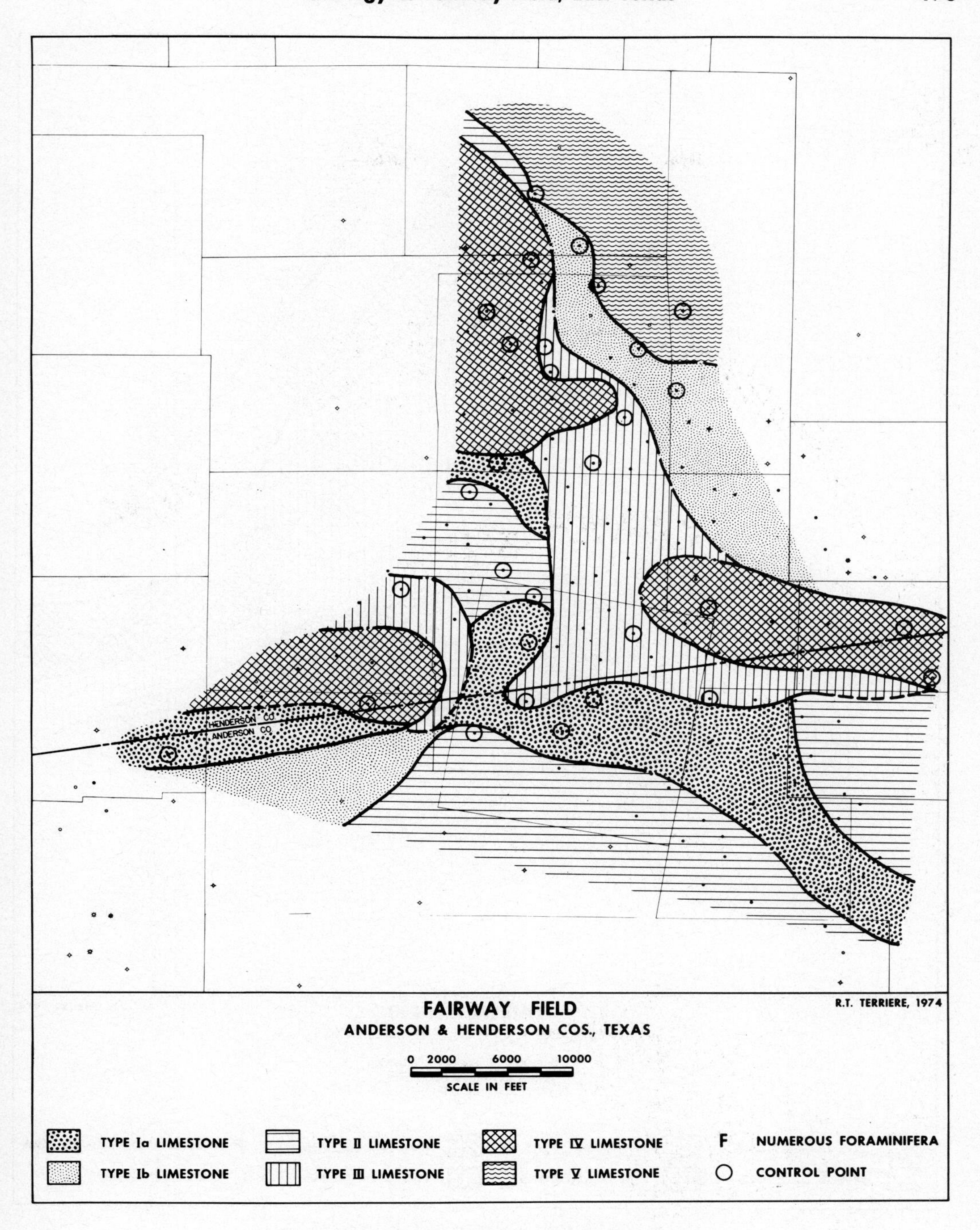

FIG. 22—Distribution of limestone types in unit 3.

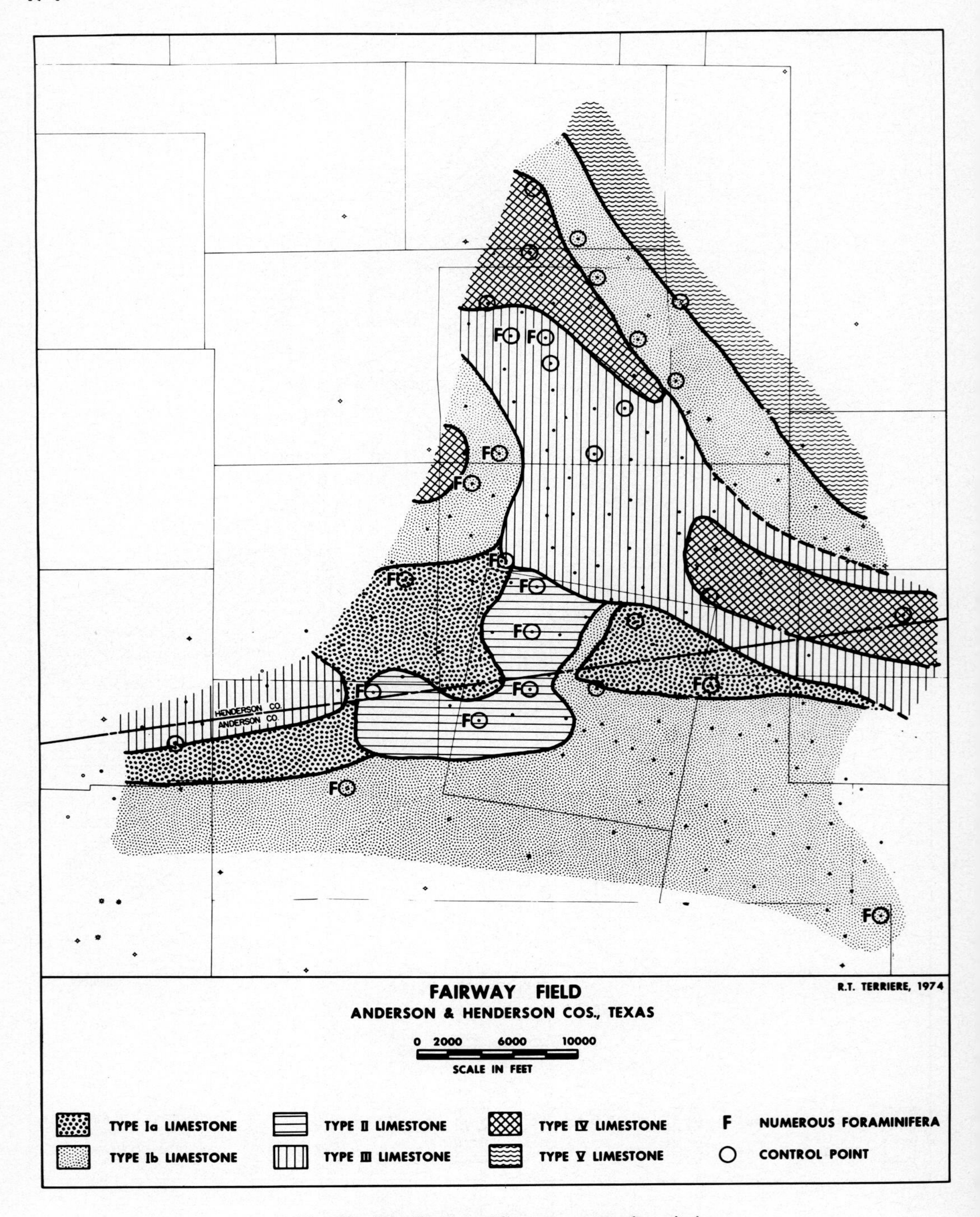

FIG. 23—Distribution of limestone types in unit 4.

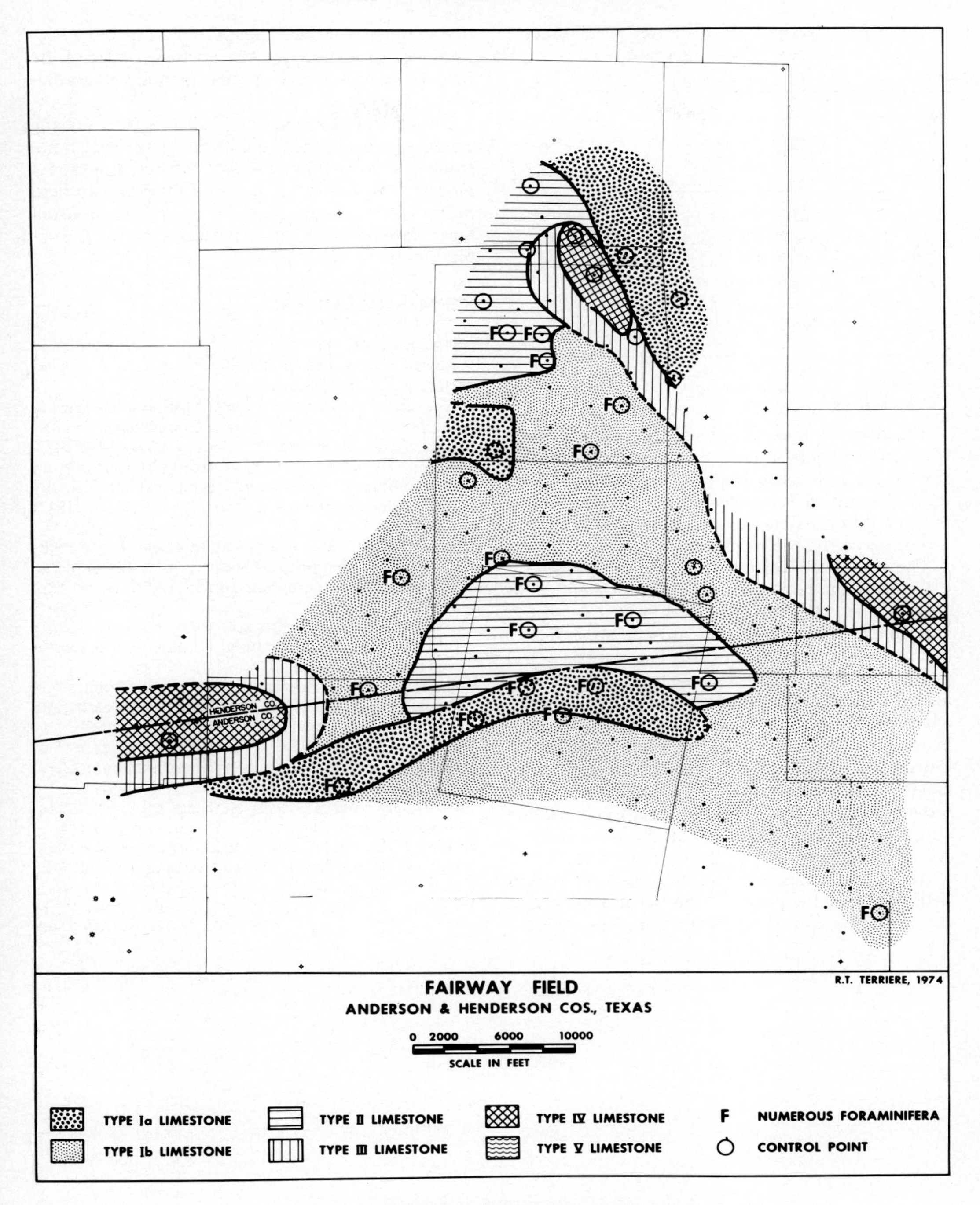

FIG. 24—Distribution of limestone types in unit 5.

Table 1. Average Porosity and Permeability Values, Fairway Field

Limestone Type	*No. of Values*	*Average Porosity (%)*	*Average Permeability (md)*
Ia	276	10.8	37.2
Ib	322	9.4	14.0
II	185	7.2	8.0
IIIa	87	9.2	28.6
IIIb	84	10.8	43.7
IV	106	8.2	12.4

CONCLUSIONS

The reef complex that forms the reservoir rock for the Fairway field grew on a slight topographic rise over a continuously positive structure. Later, more pronounced vertical movement resulted in faulting that restricts the present field area to only a portion of the reef complex.

The earliest deposition of the reef proper was at the northwest corner of the present field. This center of growth expanded and moved eastward, and smaller centers of reef growth appeared to the southeast and southwest. The south-central part of the field area was the site of persistent accumulation of carbonate sands and pebbles. Muddy carbonate sands and muds extended over broad areas around and between the other facies. During maximum development of the reef proper, the large interior area between centers of growth was inhabited by large bivalves. Near the end of reef growth, miliolid Foraminifera and oncolites were abundant in this interior area.

All of the limestone facies in and associated with the reef have porosity and permeability, and all contribute to the oil production. The productive limits are defined mostly by faults and structural elevation. Only on the northeast edge of the field is there a facies change to nonreef argillaceous limestone.

Because of the sparsity of drilling outside the productive area, the areal limits of the reef limestone are not well known. The thickest James reef limestone lies west and north of the Fairway field proper. Perhaps additional oil is present in structural traps in other parts of this same reef complex.

REFERENCES CITED

Achauer, C. W., 1967, Petrography of a reef complex in Lower Cretaceous James limestone (abs.): AAPG Bull., v. 51, p. 452.

——— and J. H. Johnson, 1969, Algal stromatolites in the James reef complex (Lower Cretaceous), Fairway field, Texas: Jour. Sed. Petrology, v. 29, p. 1446-1472.

Calhoun, T. G., and G. T. Hurford, 1970, Case history of radioactive tracers and techniques in Fairway field: Jour. Petroleum Technology, v. 22, p. 1217-1224.

Dunham, R. J., 1962, Classification of carbonate rocks according to depositional texture, *in* W. E. Ham, ed., Classification of carbonate rocks: AAPG Mem. 1, p. 108-121.

Fairway Technical Committee, 1969, Fairway (James lime) unit stratification study: Unpub. rept. to operators in Fairway field.

Folk, R. L., 1962, Spectral subdivisions of limestone types, *in* W. E. Ham, ed., Classification of carbonate rocks: AAPG Mem. 1, p. 62-84.

McKee, E. D., and R. C. Gutschick, 1969, History of the Redwall Limestone of northern Arizona: Geol. Soc. America Mem. 114, 726 p.

Oil and Gas Journal, 1975, Here are the big U.S. reserves: Oil and Gas Jour., v. 73, no. 4, p. 116-118.

Perkins, S. L., 1964, Fairway field, *in* Occurrence of oil and gas in northeast Texas: East Texas Geol. Soc. Pub. 5, v. 1, p. 13-25.

Trusheim, F., 1960, Mechanism of salt migration in northern Germany: AAPG Bull., v. 44, p. 1519-1540.

Upper Smackover Reservoirs, Walker Creek Field Area, Lafayette and Columbia Counties, Arkansas[1]

CALVIN A. CHIMENE[2]

Abstract This study has made possible the zonation of two major upper Smackover reservoirs productive at Walker Creek field, Lafayette and Columbia Counties, Arkansas, and a reconstruction of the depositional environments that were present contemporaneously throughout this area.

Petrographic examination of etched core plugs, petrographic-microscope studies of thin sections, core descriptions, and consultation with authors of published papers permitted a grouping of upper Smackover carbonate rocks into facies more or less environmentally controlled and similar to those already established in the literature. Minor changes in petrographic groupings were necessitated by the additional data and the numerous core descriptions incorporated in this paper; otherwise, the facies groupings of W. F. Bishop generally have been followed.

Distribution of nonskeletal particles in the sea which existed during deposition of the upper Smackover in this area resembled in many aspects that at the present northeastern tip of Yucatán.

The presence of discrete sandstone bodies and their carbonate equivalents permitted lithologic markers to be correlated throughout this area of predominantly carbonate deposition, resulting in a unique interpretation of depositional environments. Major zonations of oolite bars have been confirmed by pressure data, and dense dark limestones useful as a diagnostic facies appear to have environmental significance. Certain diagnostic phenomena disclosed in thin-section analysis indicate possible eolianite lithification in the vadose or phreatic zone.

The Walker Creek field produces from a structurally controlled stratigraphic trap containing nearly 100 million bbl of oil in place plus 100 Bcf of recoverable gas.

INTRODUCTION

The area studied includes Walker Creek and Lake Erling fields in Lafayette and Columbia Counties, Arkansas (Fig. 1). The area is on the north side of a graben and related fault zone which trend east-west near the Arkansas-Louisiana border. This structural feature was discussed fully in a paper by Bishop (1973). In Late Jurassic time, during deposition of the upper Smackover, the site of the present Walker Creek field was on the gently dipping shelf slope between a narrow open-marine basin on the south, in the area of the graben, and a broad coastal shelf on the north. The stratigraphy and other aspects of several fields productive from the Smackover have been discussed previously (Bishop, 1968, 1969, 1971a, b, c; Crow, 1958; Dickinson, 1968, 1969; Vestal, 1950).

A detailed petrographic study was made of limestone cores from the upper 200 ft (61 m) of the Smackover. This interval includes the two main reservoirs producing at Walker Creek, Dooley Creek, and Lake Erling fields. Etched core plugs and selected thin sections from 1,512 ft (460.9 m) of core were examined (Table 1).

With the use of additional data from lithologic logs, electric logs, and core descriptions, the section was subdivided petrographically into facies which were more or less environmentally controlled. The published facies descriptions by Bishop (1968, 1971b) were followed in general; some slight departures which were made in the mapped facies will be explained in the section on petrography.

Because most of the data included in this paper come from the upper Smackover, which is productive at Walker Creek—and because the stratigraphy and structure of the region have been well described in previously published papers—this report is limited to a discussion of the Walker Creek environs.

A siliciclastic zone separating the two main producing reservoirs in the western part of the field (Fig. 2) has been the subject of much discussion and conflict by the Geologic Subcommittee on Walker Creek Field Unitization. The data in this paper are the results of an attempt to establish the regressive nature of these two upper Smackover reservoirs and to determine their separation in the western part of the field.

The Geologic Subcommittee finally divided the Walker Creek reservoirs into five subzones (two

[1]Manuscript received, March 12, 1974; revised, May 24, 1974.

[2]Total Leonard, Inc., Houston, Texas 77002

The writer acknowledges the patient cooperation of Max Carman, University of Houston, for his assistance in the use of their optical instruments; W. F. Bishop, for his help in coordinating facies interpretations; Karl Schneidau, for constructive criticism and valued suggestions; and friends at Austral Oil Company without whose assistance the paper could never have been completed. Thanks also are expressed to the Austral Oil Company, which allowed me the time and expenses to complete this project.

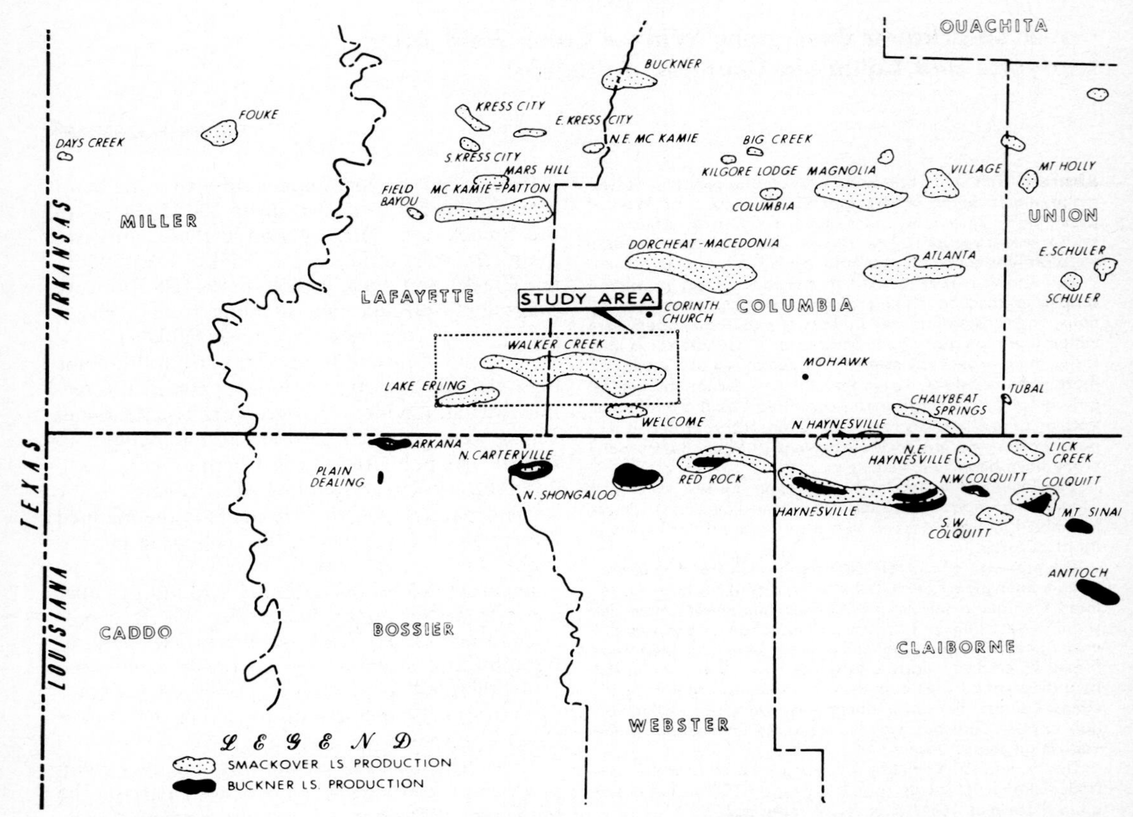

FIG. 1—Index map, North Louisiana and South Arkansas, showing relation of area studied to other areas of important Jurassic calcarenite reservoirs (modified from Bishop, 1971b).

in stratigraphic Zone 1 of the upper Smackover and three in Zone 2) on the basis of electric-log correlations and pressure differentials recorded in production histories. Pressure data confirmed the major separation between Zones 1 and 2 in the western part of the field. The Geologic Subcommittee's reservoirs 4 and 5 are in Zone 1 and reservoirs 1, 2, and 3 are in Zone 2.

Smackover allochems are almost entirely nonskeletal and are distributed similarly to those presently forming on the northeastern coast of Yucatán (see Figs. 24, 25). The Walker Creek area is sufficiently large for overall paralic depositional fluctuations to be noted within the framework of two complete cycles of detrital-carbonate-detrital deposition.

After deposition of the argillaceous, pelletal carbonate mudstones of the basal upper Smackover, an influx of siliciclastic material began, but this influx ceased abruptly. Salinity and turbulence became favorable for intensive oolite accretion, and the reservoir facies of Zone 2 was deposited. There was then regression from the area which now forms the western part of the field, and a repetition of the siliciclastic influx occurred. After this influx, additional regression occurred and the environment returned to one favorable for oolite accretion, and the reservoir facies of Zone 1 then was deposited seaward from that of Zone 2 (Fig. 3). Continued regression and increasing amounts of fine siliciclastic material caused a transition in deposition to evaporite-flat sediments (Buckner mudstone).

Hydrocarbons are produced at Walker Creek field from two nonskeletal calcarenite bars at the top of the Smackover Formation. Production is

Table 1. Wells in Walker Creek Field Area, Arkansas

WELL NO.	OPERATOR	FEE	LOCATION	INTERVAL EXAMINED (FEET)
1	Austral	#3 IPCO	33/19-24	11,101-11,465 Core Description
2	Austral	#1 IPCO	34/19-24	10,905-11,363 Core Description
3	Austral	#2 IPCO	35/19-24	10,971-11,122 Core Description
4	Austral	#4 IPCO	36/19-24	11,037-11,100 Core Description
5	Austral	#1 Hayes	36/19-24	11,155-11,218 Core Description 11,289-11,354 Core Description
6	Austral	#A-3 IPCO	27/19-24	11,043-11,103 Examined
7	Austral	#A-1 IPCO	22/19-24	10,826-11,018 Examined
8	Austral	#A-2 IPCO	22/19-24	10,975-11,066 Examined
9	Mosbacher	#1 IPCO	23/19-24	10,953-10,985 Core Description
10	Austral	#A-4 IPCO	23/19-24	10,973-11,030 Core Description
11	Lyons et al	#1 IPCO	23/19-24	Not Cored
12	Hunt	#1 IPCO	14/19-24	10,845-10,942 Core Description
13	Austral	#B-1 IPCO	11/19-24	10,679-10,785 Examined
14	Bodcaw	#11 Fee	13/19-24	Not Cored
15	KWB	#1 IPCO	24/19-24	10,952-10,973 Core Description
16	KWB	#2 IPCO	24/19-24	10,916-10,978 Core Description
17	Bodcaw	#9 Fee	24/19-24	10,914-10,967 Core Description
18	Bodcaw	#9-A Fee	24/19-24	10,854-10,986 Core Description
19	Bodcaw	#10 Fee	18/19-23	11,264-11,389 Core Description
20	Bodcaw	#1 Boyette	17/19-23	10,806-10,934 Core Description
21	Arkla Gas	#1 Burton Heirs	16/19-23	10,814-10,828 Examined
22	Lyons	#1 Herndon	16/19-23	10,743-10,865 Examined
22a	Tenneco	#1 Kosek Unit	15/19-23	10,711-10,733 Core Description
23	Broyles	#1 Burton	15/19-23	10,710-10,810 Core Description
24	Bodcaw-Broyles	#1 Sorrels	14/19-23	10,762-10,864 Core Description
25	Partee et al	#1 Brigham	18/19-22	10,795-10,857 Core Description
25a	Partee et al	#1 Kosek	18/19-22	10,772-10,828 Core Description
26	Woods	#1 Teutch	17/19-22	10,737-10,778 Core Description
27	Arco	#1 Cabe	21/19-22	Not Cored
28	Arco	#1 McFaddin	22/19-22	10,954-11,013 Core Description
29	Partee	#1 Disotell	27/19-22	10,911-10,952 Core Description
30	Arco	#1 Bodcaw Unit	21/19-22	10,843-10,960 Core Description
31	Marathon	#1 Henderson	20/19-22	10,754-10,833 Core Description
32	Woods	#1 Knight	19/19-22	10,751-10,867 Core Description
33	Woods	#1 Stuart	24/19-22	10,757-10,902 Core Description
34	Woods	#3 Stuart	13/19-23	10,691-10,810 Core Description
35	Woods	#2 Stuart	14/19-23	10,680-10,810 Core Description
36	Woods	#2 Blackwell	23/19-23	10,851-10,972 Core Description
37	Bodcaw	#3 Fee	22/19-23	10,867-10,944 Examined
38	Chapman	#1 Hughes	21/19-23	10,810-10,910 Examined
39	Oakland	#1 Umphries	20/19-23	10,819-10,917 Examined
40	Bodcaw	#5 Fee	19/19-23	11,111-11,205 Core Description
41	Bodcaw	#4 Fee	20/19-23	Not Cored
42	Chapman	#1 Burton	20/19-23	10,864-10,983 Examined
43	Chapman	#1 Whitehead	21/19-23	10,840-11,000 Examined
44	Bodcaw	#2 Fee	22/19-23	10,858-10,962 Examined 10,904-11,009 Examined
45	Woods	#1 Blackwell	23/19-23	10,847-11,032 Core Description
46	Woods	#1 McDole	23/19-23	10,864-10,970 Core Description
47	Woods	#1 Jack	23/19-23	10,910-10,996 Core Description
48	Woods	#2 Jack	26/19-23	10,911-10,971 Core Description
49	Woods	#1 Powell	24/19-23	10,846-10,950 Core Description
50	Bodcaw	#1 Pickler	19/19-22	10,818-10,947 Core Description
51	Bodcaw	#1 Kosek	20/19-22	10,777-10,905 Core Description
52	Pennzoil	#A-1 Bodcaw	28/19-22	10,870-11,002 Core Description
53	Marathon	#1 Boucher	28/19-22	10,910-10,975 Core Description
54	Pennzoil	#1 Kosek	29/19-22	10,873-11,004 Core Description
55	Pennzoil	#1 Nations	29/19-22	10,814-10,981 Core Description
56	Pennzoil	#1 Pickler	30/19-22	10,834-10,976 Core Description
57	Pennzoil	#2 Pickler	30/19-22	10,912-11,014 Core Description
58	Woods	#4 Stuart	25/19-23	10,893-10,958 Core Description
59	Hunt	#1 Ham	25/19-23	10,995-11,020 Core Description
60	Clegg	#1 Stewart Ham	26/19-23	Not Cored
61	Bodcaw	#A-1 Fee	27/19-23	Not Cored
62	Bodcaw	#1 Fee	27/19-23	11,021-11,055 Core Description
63	Chapman	#1 Kosek	28/19-23	10,841-10,971 Examined
64	Bodcaw	#12 Fee	28/19-23	10,873-11,022 Core Description
65	Bodcaw	#13 Fee	33/19-23	11,065-11,122 Core Description
66	Chapman	#1 Helms	29/19-23	10,873-10,995 Core Description
67	Austral	#1 Shirey	29/19-23	10,912-11,012 Examined
68	Stanolind	#1 Bodcaw	29/19-23	10,988-11,073 Core Description 11,176-11,225 Core Description
69	Pennzoil	#1 Woodard	35/19-23	Not Cored
70	Midroc	#1 Slappy	32/19-22	Not Cored
71	Forgey	#2 Fraizer	33/19-22	Not Cored

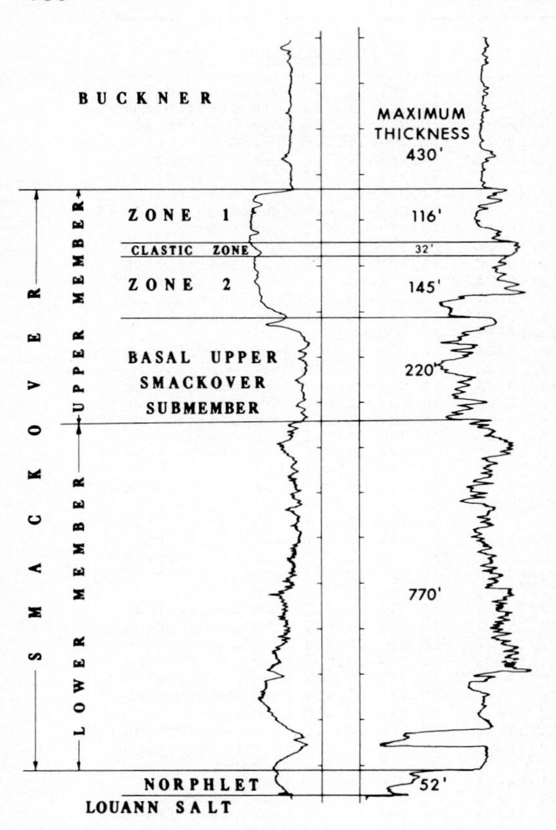

FIG. 2—Composite electric log showing Jurassic stratigraphic nomenclature used and relation of two reservoir zones productive of hydrocarbons.

limited updip by the absence of permeability and downdip by the oil-water contact.

On the Smackover shelf slope, calcarenite deposition was concentrated on beaches and on turbulent shoals localized by prior structural movement. Isopach maps of facies intervals in the upper Smackover at Walker Creek indicate a distinct pattern of depositional environments, probably controlled by bathymetric irregularities.

The reservoir facies have primary intergranular porosity. They have been subdivided into two zones separated by siliciclastic rocks and admixed siliciclastic rock and transitional limestone. Below the lower zone there is a thick body of argillaceous, pelletal carbonate mudstone; above the upper zone the Smackover grades into the overlying Buckner mudstone.

STRUCTURE

The structure of the Walker Creek field as mapped at the top of the Smackover (Fig. 4) can be described best as a roughly arcuate series of three en echelon, east–west-trending, low-relief ridges separated by narrow synclinal areas and broken in the southern portion of the field by broader troughs. The ridges are distinct on the west but appear to coalesce on the east.

Two major oolitic reservoir beds roughly parallel these ridges and pinch out updip to form the traps for the hydrocarbon accumulation in this field. Where the ridges coalesce on the east, so do the reservoir oolites. Neither the eastern nor western limit of the field is known at present.

Structure maps of the top of the Smackover Formation (Fig. 4) and the top of the basal upper Smackover (Fig. 11) show the same basic structure, with relief diminishing upward in the section.

Structural crests and synclinal axes remained constant despite rapid calcarenite buildups on ancient highs. Faulting prior to calcarenite deposition may be the cause of the generally east-west trend of structural configuration in the area. Structural growth is believed to have diminished after Buckner deposition, and the region gradually was tilted southward and, later, westward.

The most striking structural feature on the map is the steep north turnover with dips approaching 5 ½° in Secs. 13, 23, and 24, T19S, R24W, in Lafayette County, an area of *maximum control.* This dip was used as a guide in the structural mapping of the entire field and was found to conform well to facies isopachs. A narrow synclinal area, trending roughly east-west through wells 11, 14, and 19, separates well 12, in which the reservoir-equivalent section has no porosity, from productive wells 16 and 18.

There has been conflicting information on the exact depth of well 41. It has been mapped here according to the original bottomhole survey, but the operator contends that the top of the Smackover is 49 ft (15 m) lower, and his designated top was used by the Walker Creek Subcommittee.

PETROGRAPHY

In an attempt to construct facies-distribution maps, which reflect depositional environments, it was necessary to synthesize a facies nomenclature adaptable to core descriptions by others, published material, examinations of core plugs, and petrographic slide data. Published nomenclature

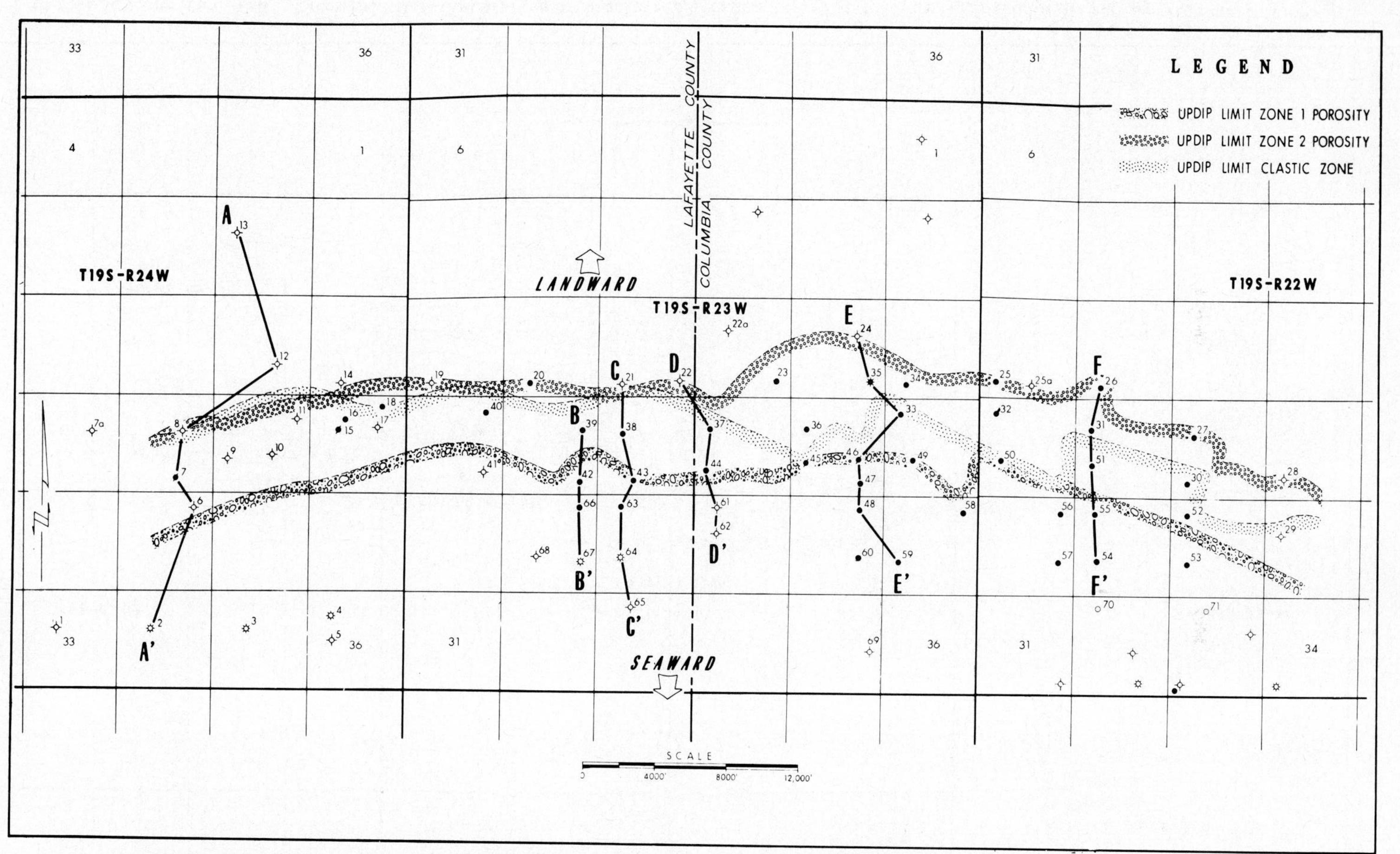

FIG. 3—Cross-section location map of Walker Creek field area, Lafayette and Columbia Counties, Arkansas. Six north-south stratigraphic cross sections are oriented at nearly right angles to depositional strike. Map shows relation of updip limit of porosity to updip limit of clastic zone.

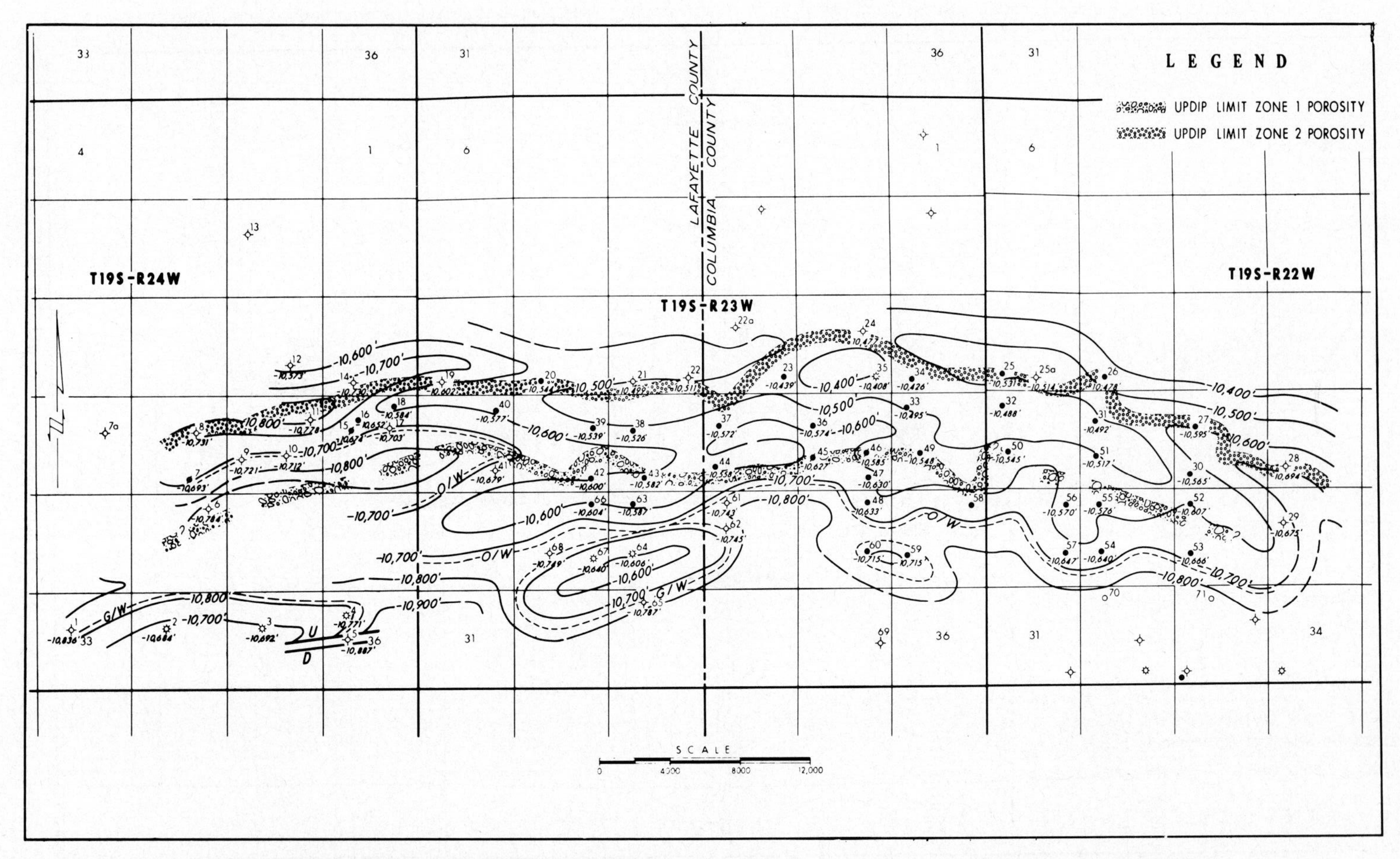

FIG. 4—Structure map, Walker Creek field area. Datum is top of Smackover Formation. C.I. = 100 ft (30 m). Original oil-water and gas-water contacts are shown by dashed lines. Updip porosity limits are shown for both main producing zones. (See Table 1 for well names.)

by W. F. Bishop was satisfactory in most cases, because the facies he described also reflect depositional environments in other areas.

In preparation of this paper the following procedures were used.

1. Numerous cores were described at the well sites.
2. Core plugs were etched, described, and logged, and thin sections were cut from key plugs in an attempt to confirm certain correlations.
3. Facies analyses—based on color, size of grains, shape of grains, sorting, porosity, lithology, cement type, and interpreted energy requirements—were made to standards compatible with published descriptions and core descriptions.
4. Where necessary, empirical projections of facies types were used in wells for which data based on examination of petrographic slides and core descriptions were not available.
5. Core-analysis permeabilities were tabulated.

Etched core plugs taken at 1-ft (30.5 cm) intervals from wells listed in Table 1 were examined. A total of 1,512 ft (460.9 m) of core was examined in the laboratory, and 91 thin sections of representative parts of the core plugs were examined under petrographic microscopes.

The upper Smackover limestones are divided into four major facies (Bishop, 1968), as follows.

1. *Transition facies*—a limestone gradational into siliciclastic material; includes a shaly subfacies composed largely of pellets and recrystallized oolites in a matrix of shale- and/or silt-size anhydrite or dolomite, and a sandy subfacies composed of pellets with abundant quartz sand grains and sparry calcite cement.

2. *Reservoir facies*—composed of oolites and superficially coated pellets with spar cement and primary intergranular porosity (high-energy environment).

3. *Mixed facies*—composed of subequal quantities of all grain types with spar cement and mud matrix (medium-energy environment).

4. *Pellet-mudstone facies*—composed of pellets in a mud matrix which commonly is dolomitized. This facies also contains minor amounts of argillaceous dolomite and "disturbed mud" (low-energy environment).

In the numerous areas where facies data come solely from core descriptions by others, an attempt was made to assign the data to previously published facies classifications. For the reservoir facies and for the siliciclastic beds of the transition facies, this was quite simple; however, the various classifications of transition, mixed, and pellet-mudstone facies require some adjustment.

Additions and changes to previously published facies are as follows:

The *reservoir facies* is modified from that described by Bishop (1968) to include only the zones where porosity and permeability are present. Where no permeability is present, the rest of Bishop's reservoir facies was placed in the transition, mixed, or an added dark limestone facies.

The *mixed facies* is considered to include the light-colored, nonpermeable part of Bishop's reservoir facies that is completely cemented with spar.

A *dark limestone facies* is added to include rocks described by others from cores. It consists mainly of dark gray to black pellet mudstone with a few inclusions of dark-colored mixed and transition beds. A "disturbed-mud" subfacies is included for mapping purposes. The dark limestone facies is believed to have been deposited in a low-energy restricted environments.

The *transition facies* is modified to include a rock type consisting of large, gray micritic lumps lacking any internal structure. Many of these lumps have sutured contacts and are loosely to tightly cemented with spar. In places these diagenetically altered limestones exhibit minor amounts of permeability but have no resemblance to the reservoir facies.

Facies Based on Combined Core Descriptions by Others

A facies is not necessarily a single lithologic unit. It may be a combination of such units whose nature is regarded as a criterion of conditions which controlled their formation. Using other people's core descriptions of lithologic units, and attempting to place these units within facies groupings, is at times subjective, especially in the cases of transition and mixed facies.

The following descriptions are of the various lithologic types placed within the facies groupings based on core descriptions by others.

Transition facies—limestone, medium gray, dense, hard, "tight," with some scattered stylolites and shale inclusions; well-cemented, with low porosity and no permeability; in places, finely oolitic with calcite-filled matrix, locally sandy and finely granular; contains lignitic plant fragments and fossils.

Reservoir facies—limestone, generally medium gray, well-sorted; oolitic with some pisolites; slight to excellent intergranular porosity and permeability; spar cement; few scattered fossils;

locally light to dark gray, finely to coarsely oolitic, well- to poorly sorted.

Mixed facies—limestone, medium gray, slightly oolitic and pisolitic, hard, "tight," dense; scattered stylolites; porous but not permeable; a few pebbles; tightly cemented with spar; generally poorly sorted.

Pellet-mudstone facies—mudstone, gray, carbonate or finely oolitic, not porous or permeable. Generally, additional data from logs are needed to pick this facies with certainty. From core descriptions alone, it may be confused with the transition facies in many places. Dolomite and very finely crystalline dolomitic limestone are included in this facies.

Dark limestone facies—limestone, dark gray to black, pelletal, with a few algal pisolites; locally fine to medium crystalline; rarely containing fine pebbles of lithographic limestone in calcite matrix; scattered fossils; poorly sorted; in places argillaceous and rarely sandy; very low porosity and no permeability. Also included in this facies group is the "disturbed-mud" subfacies, a mottled, dark gray, dolomitic limestone which appears to have been burrowed.

Local Stratigraphy

LOWER SMACKOVER, NORPHLET, LOUANN (FIG. 2)

Well 7 is the only well which completely penetrated the lower Smackover member and the Norphlet Formation. The Louann was reached at a depth of 12,122 ft (3,694.7 m), but was not fully penetrated. The Norphlet consists of 52 ft (15.8 m) of red shale and pink, medium- to fine-grained, partly shaly, nonporous sandstone with some frosted grains. The lower Smackover consists of 770 ft (235 m) of brown to dark gray, crystalline limestone with a few traces of oolitic or pelletal structure.

UPPER SMACKOVER MEMBER

The upper Smackover member (Fig. 2) lies directly below the Buckner Formation. Previous data indicate that Smackover facies were deposited downdip contemporaneously with Buckner facies updip (Dickinson, 1968, 1969; Bishop, 1968, 1969, 1971b, 1971c); in this area the contact is gradational. This member has been differentiated into four units on the basis of lithologic and electric-log characteristics. The upper part of the member, including Zones 1 and 2 and the clastic zone, locally exhibits detrital-carbonate-detrital cycles of deposition (Figs. 5, 6, 8). A limestone present in most wells at the Buckner-Smackover contact may be a transitional deposit, but it was more convenient to map it within the zones with which it correlated (Figs. 5–9). Changes in the treatment of this limestone would not materially affect the mapping.

Basal Upper Smackover Submember (Fig. 2)

The "basal upper Smackover" (submember of the upper Smackover member) is a local term used to describe an interval between the lower Smackover and the lowermost producing zone. It consists of very fine-grained, dark, argillaceous, pelletal to crystalline limestone containing anhydrite nodules and numerous pelecypod molds. Locally, sandstone and shale are present in this unit (Figs. 5, 9). Oolitic limestone is absent. There is generally a distinct lithologic change at the top which is easily correlated on electric logs (Figs. 5–10). This submember ranges in thickness from 200 to 300 ft (61–91 m) and has been fully penetrated by only five wells in the area (7, 25A, 57, 59, and 68). The top of the submember is used as a datum for the cross sections (Figs. 5–10) and the paleobathymetric chart (Fig. 12).

A structure map was constructed for the top of the basal upper Smackover (Fig. 11). This map shows an average dip of approximately 100 ft/mi (19 m/km) across the field. By use of this structural base and a simple grid based on 100 ft/mi south dip, and another grid based on 30 ft/mi (5.7 m/km) west dip (Sligo structure) for the western part of the field, a schematic paleobathymetric chart of the pre–Zone 2 seafloor was constructed (Fig. 12). Based on the data from wells 61 and 62, it is apparent that the saddle which formed in Secs. 26 and 27, T19S, R23W, continued to subside during deposition of the lower Buckner, as indicated by the presence of approximately 150 ft (46 m) of additional Buckner section in these wells.

Several low-relief prominences were present on the ancient shelf (Fig. 12). These may have been related to early Smackover faulting or structural uplift via salt movement, but subsurface control is lacking for a positive determination of local tectonism. A prominent bathymetric ridge trends generally from west to east through Sec. 23, T19S, R24W, and the north half of Secs. 19, 20, 21, and 24, T19S, R23W, thence striking southeasterly through Secs. 19, 20, and 28, T19S, R22W. A ridge of less relief, broken by minor troughs, also extends west to east through the northern parts of Secs. 28–30 and the southern parts of Secs. 22–24, T19S, R23W. Two minor bathymetric features

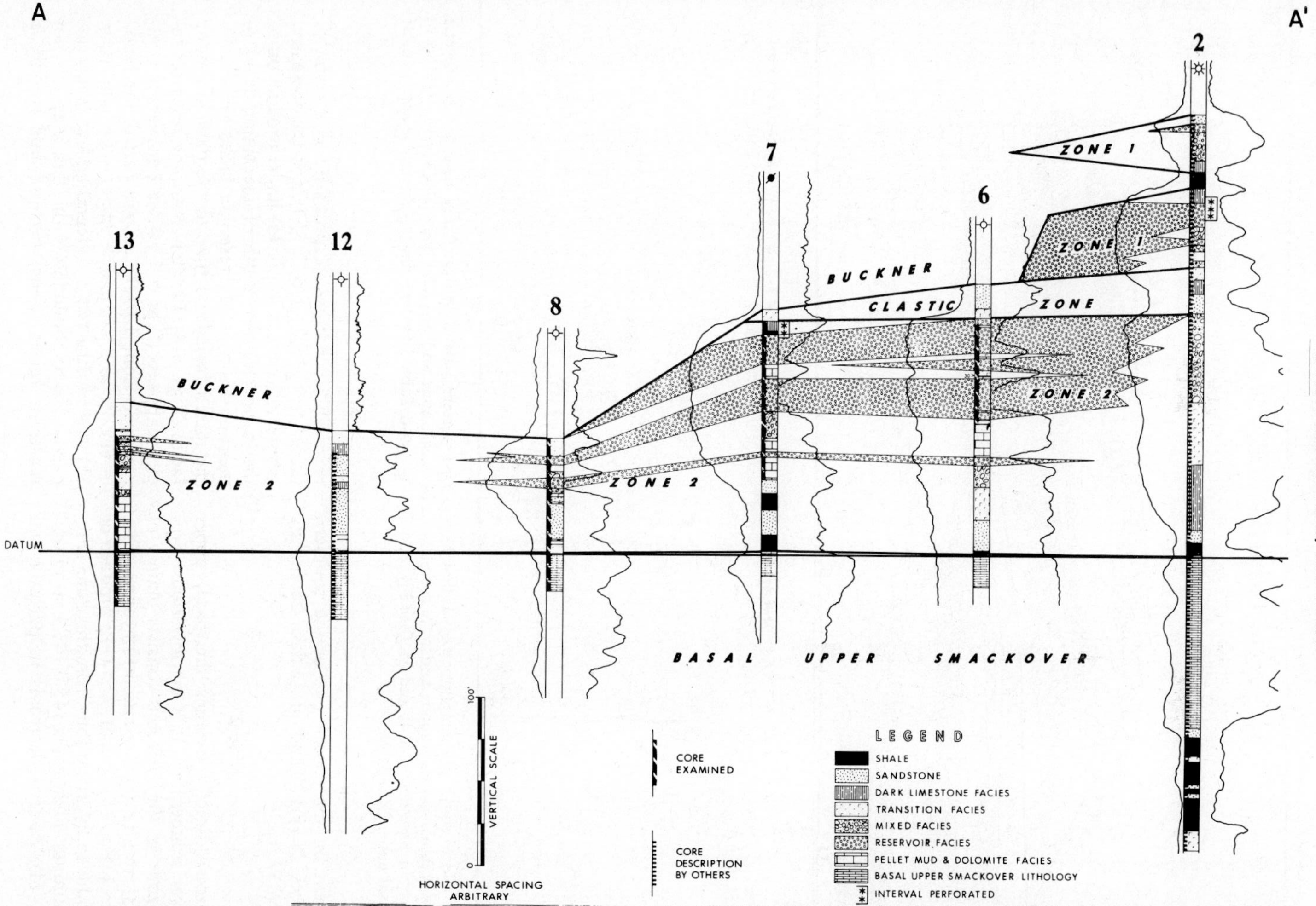

FIG. 5—North-south stratigraphic cross-section *A–A'* across westernmost part of Walker Creek field area, Lafayette County, Arkansas. Section shows reservoir-facies pinchout and regressive offlap of Zone 1. Datum is approximate top of basal upper Smackover lithology. (See Fig. 3 for location and Table 1 for well names.)

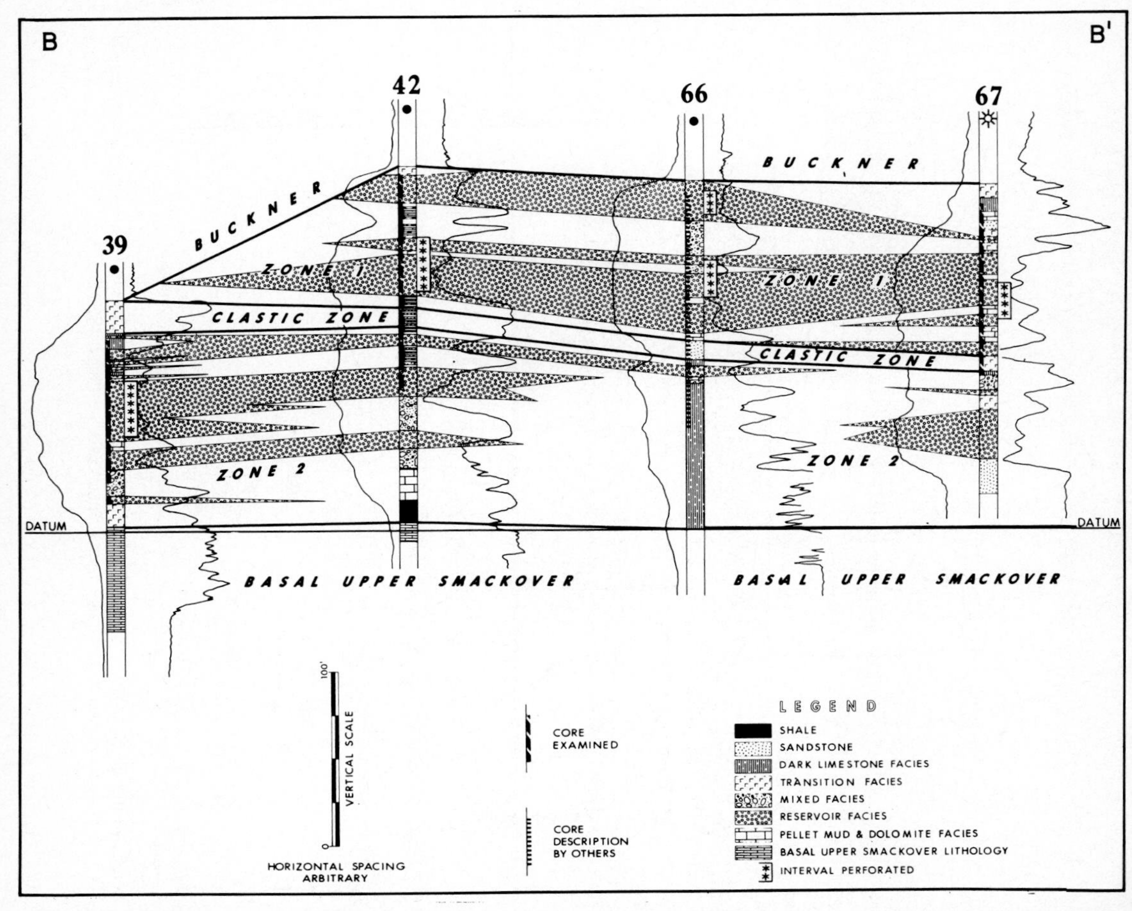

FIG. 6—North-south stratigraphic cross-section *B–B'* across western part of Walker Creek field area, Lafayette County, Arkansas, showing good sandstone correlations in clastic zone and downdip change of reservoir facies to dark limestone between wells 42 and 66. (See Fig. 3 for location and Table 1 for well names.) Datum is approximate top of basal upper Smackover.

are present on the south—one in the southern half of Secs. 28 and 29 and one in Secs. 25 and 26, T19S, R23W.

Zone 2

This zone directly overlies the basal upper Smackover submember. It is the lowermost productive zone in the area and consists mainly of oolitic calcarenite grading into a full range of related facies (Fig. 13). Twenty-six wells produce or have produced hydrocarbons from this zone (Fig. 13). Maximum thickness is 145 ft (43.9 m) in well 52 (Fig. 14). Zone 2 decreases in porosity and permeability northward, primarily as a result of facies changes (Figs. 5, 7, 8, 9), and the decreased permeability (zero in places) limits production to the north (Fig. 4). South of the major development of this zone, the reservoir facies grades into the mixed facies and then to the dark limestone facies (Figs. 6–8). Farther south, the reservoir facies reappears (Figs. 6, 7). Zone 2 appears to coalesce with Zone 1 in the eastern part of the field.

The Zone 2 composite facies schematic (Fig. 13) was constructed by integrating five facies isopachs. Facies boundaries on this map do not represent facies limits, because this is only a

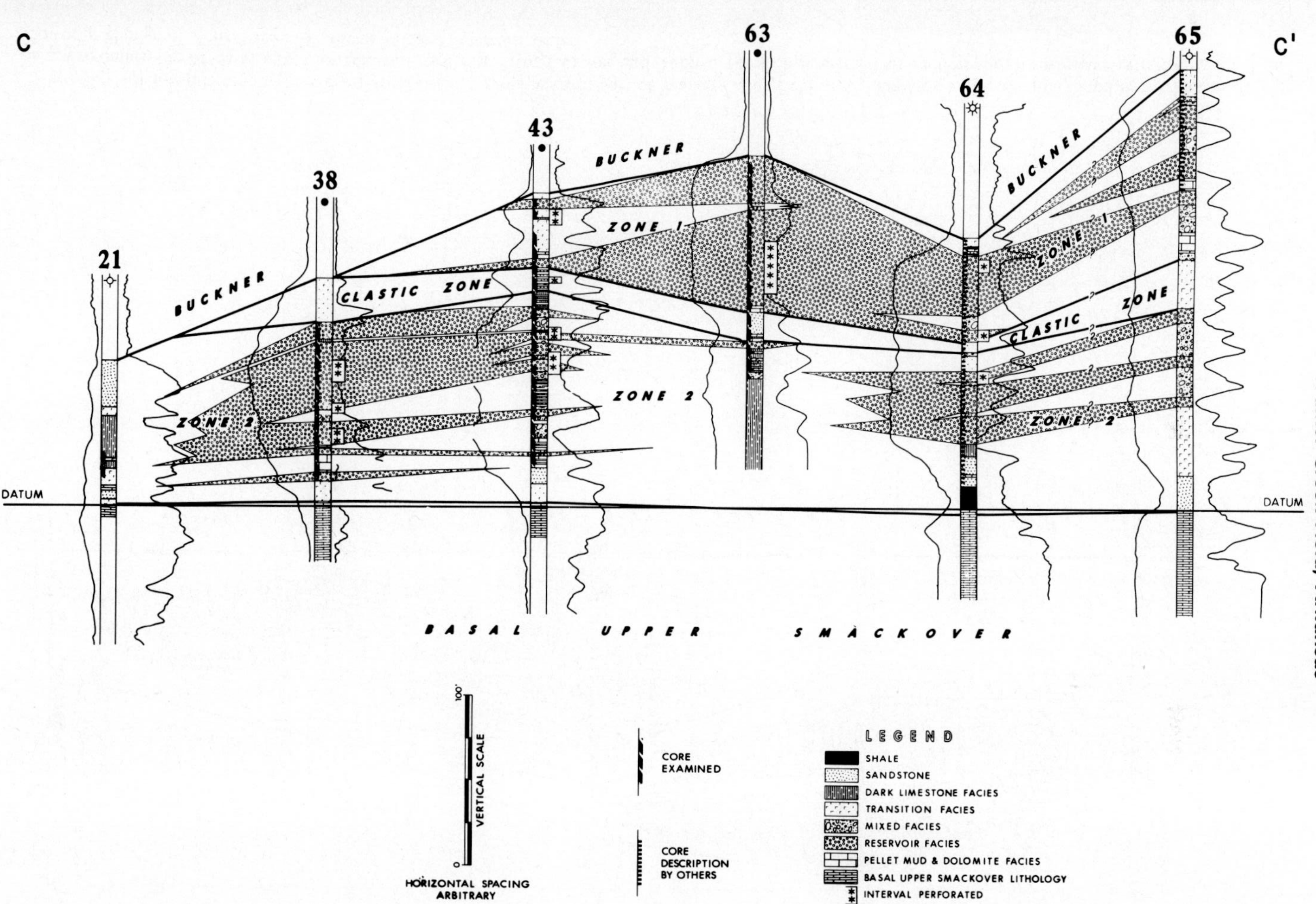

FIG. 7—North-south stratigraphic cross-section *C–C′* across western part of Walker Creek field area, Lafayette County, Arkansas. Datum is approximate top of basal upper Smackover. (See Fig. 3 for location and Table 1 for well names.) This section shows critical area of correlations of clastic zone between wells 43 and 63. Correlations to well 64 are questionable.

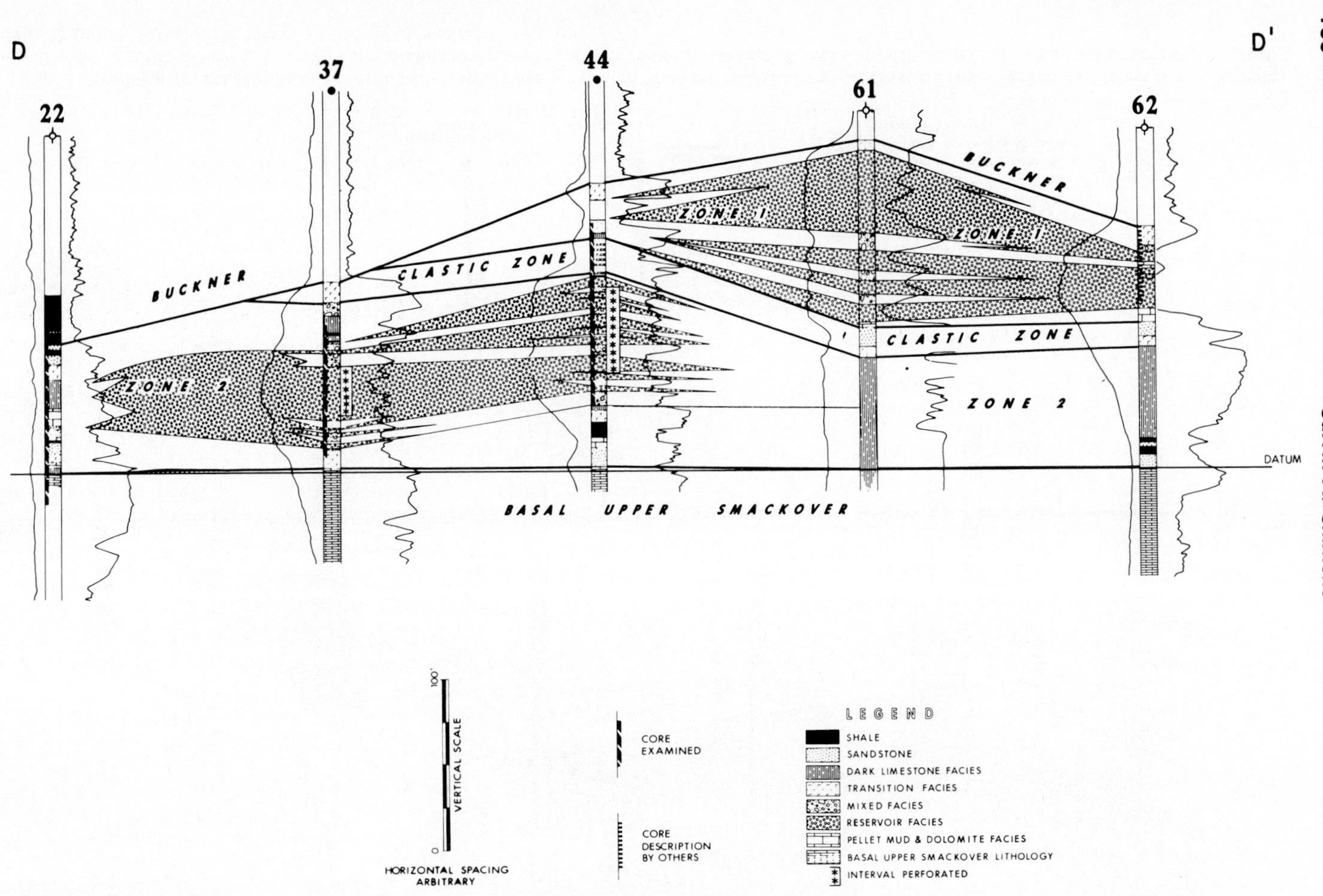

FIG. 8—North-south stratigraphic cross-section *D–D'* aross western part of Walker Creek field area, Lafayette and Columbia Counties, Arkansas. Datum is approximate top of basal upper Smackover. (See Fig. 3 for location and Table 1 for well names.) This section shows regressive offlap of Zone 1 and downdip change of Zone 2 reservoir facies to dark limestone facies.

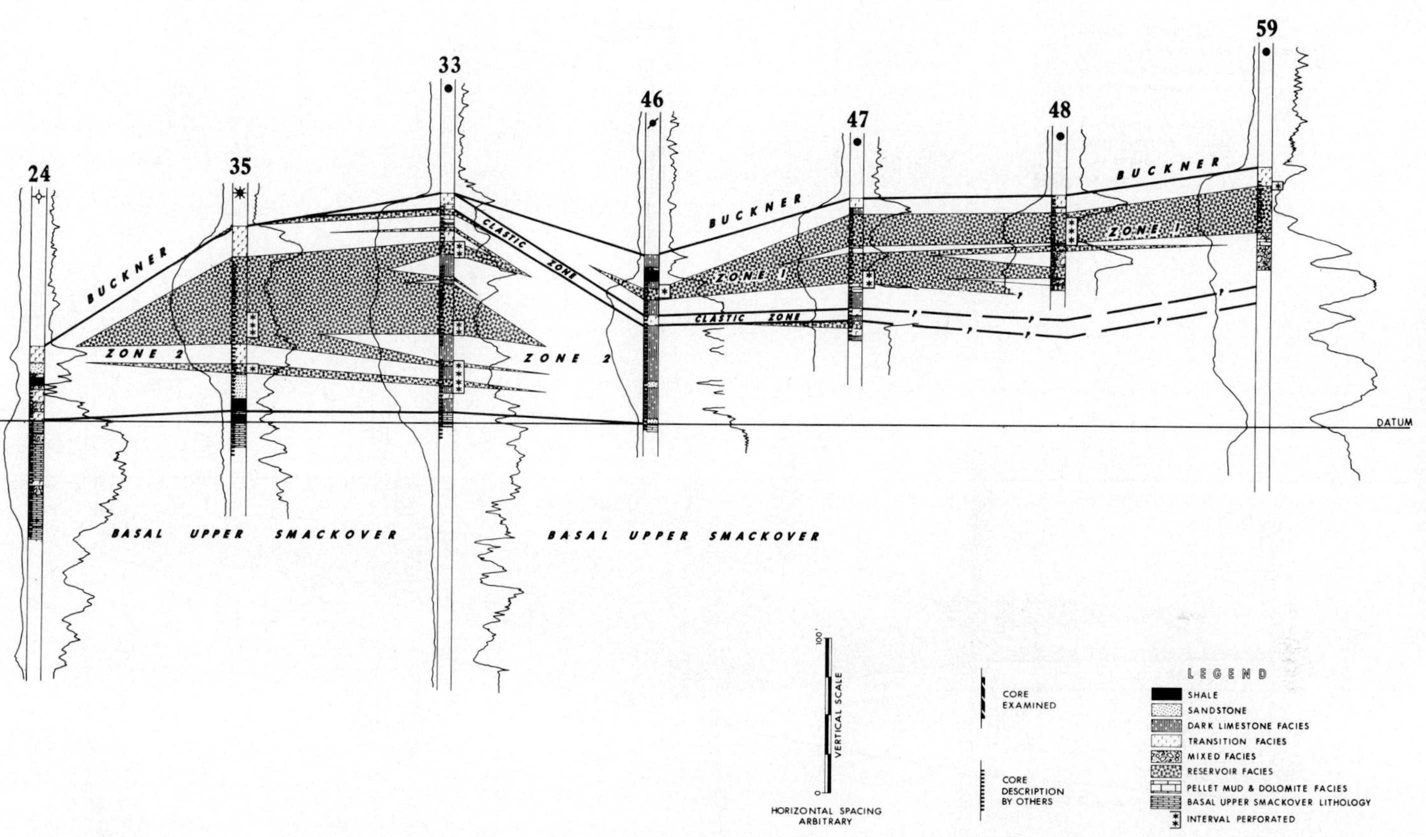

FIG. 9—North-south stratigraphic cross-section *E–E'* across Walker Creek field, Columbia County, Arkansas, showing abrupt updip pinchout of reservoir beds of Zone 2 and abrupt downdip change to dark limestone. (See Fig. 3 for location and Table 1 for well names.) Datum is approximate top of basal upper Smackover.

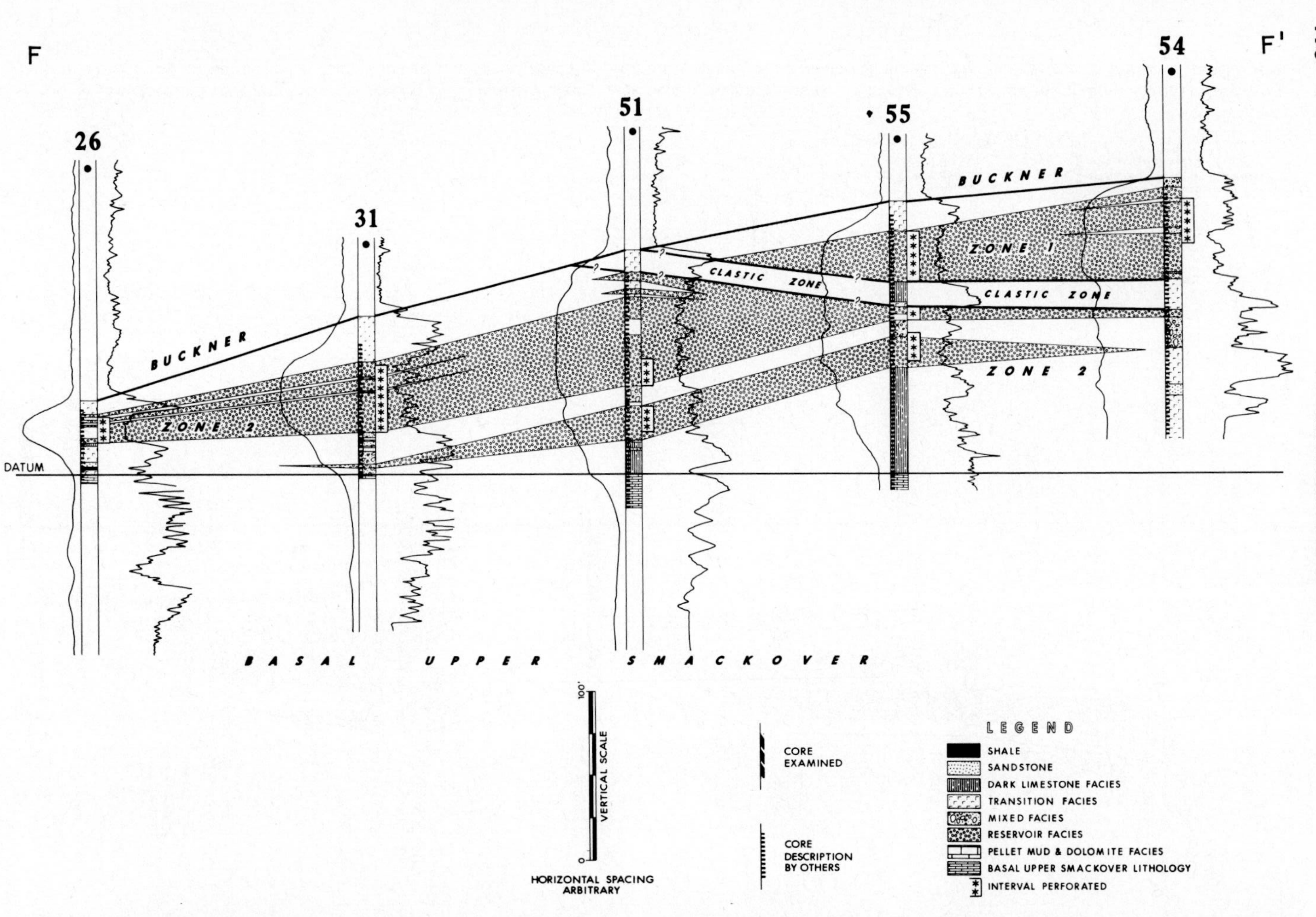

FIG. 10—North-south stratigraphic cross-section *F–F'* across eastern part of Walker Creek field area, Columbia County, Arkansas. (See Fig. 3 for location and Table 1 for well names.) Datum is approximate top of basal upper Smackover. Section shows lack of definition of clastic zone and probable merging of Zones 1 and 2 in this part of field.

generalized depositional synthesis of Zone 2. A comparison of Figures 12 and 13 readily shows the relations between bottom topography and facies development.

In the lower part of Zone 2, lithologic changes across the area indicate a fluctuating depositional environment, possibly on an eroded surface. Dark limestone, pellet-mudstone, reservoir, and mixed facies all directly overlie the basal upper Smackover in places in the updip areas. A very fine-grained, dark-colored pelletal limestone is present in the lower areas between the prominences in Secs. 23, 24, 27, and 29, T19S, R23W, and at Lake Erling field (Fig. 13).

A calcarenite bar trending east-west through Sec. 23, T19S, R24W, on the west, and Sec. 28, T19S, R22W, on the east, was developed on the northern ridge; the dark limestone facies was deposited contemporaneously north of the bar in a narrow restricted basin. At the same time, a mixed facies was being deposited on the updip and downdip edges of this oolite bar. The mixed facies has a maximum thickness in Secs. 19–23, T19S, R23W, directly south of the bar, and grades into dark limestone in the low areas on the south. There was an additional area of mixed facies deposition over the southern feature in Secs. 28 and 29, T19S, R23W, and another at Lake Erling field (Fig. 13).

Minor amounts of light-colored pellet mudstone, dolomitized carbonate mud, and dolomite (Fig. 15) occur throughout Zone 2 but are omitted in the mapping. They are associated with fine-grained limestones and suggest an evaporative, or high-salinity, restricted environment.

Eolian dunes probably were present near the crest of the oolitic bar, as evidenced by examples of both differential solution and corrosion of oolites and replacement by anhydrite and silica (Figs. 16–21). Siliceous replacement is associated with meteoric water, which was probably present in these rocks owing to their proximity to a vadose environment. Examples of meniscus cement, also indicative of eolianite lithification, could be found in the reservoir facies, but the petrographic data in themselves are insufficient for positive identification of a vadose environment. Considering the regressive relation of Zone 1 to Zone 2 and the continued regression during Buckner deposition, both zones probably were subjected to a vadose and/or phreatic environment during or near the end of their deposition.

Some of the mixed facies and reservoir facies may be variations of the same general type of depositional environment (high-energy offshore bar), differing only in lithification processes which occurred in vadose or phreatic areas of depositional-postdepositional environments.

Clastic Zone

The clastic zone, which separates Zones 1 and 2, was found to be a mappable unit over most of the area, although it was difficult to correlate east of well 58 and may be absent over the eastern part of the field (Fig. 10).

The clastic zone averages 15–20 ft (4.5–6 m) in thickness. It pinches out toward the northern part of the field and reaches a maximum thickness of 32 ft (9.8 m) in the southern part. This zone consists mainly of a fine-grained, calcareous to noncalcareous sandstone that grades into sandy limestone of the transitional facies. Locally, dark shales and, in a few places, a dark pellet-mudstone facies are present.

Wells in the western part of the field had a maximum development of sandstone (Figs. 7, 8). Limestone associated with the sandstone was, with four exceptions, a shaly and sandy transitional facies. In wells 2 and 9, cores described by others indicate thin intervals of dark limestone in the clastic zone (Fig. 5); this may be due to minor correlation or core-depth errors. Relatively thick intervals of dark limestone were reported from wells 55 and 56 (Fig. 10) in the eastern part of the field, where clastic-zone correlations are difficult. The thicker dark limestone in this zone may indicate thinning or disappearance of the clastic zone in this area.

In wells 43, 44, and 63 (Fig. 22), sandstone from the clastic zone was cored and thin sections were made. Thin-section studies showed the sandstone from all three wells to have identical petrographic characteristics. These wells are located at a critical juncture of the development of reservoir facies in Zones 1 and 2 (see Figs. 7 and 8).

The updip limit of detrital material is south of the depositional crest of Zone 2 (Figs. 14, 23) and roughly coincident with the downdip limit of the Zone 2 reservoir facies (Fig. 13); thus these reservoir rocks may have been emergent during a part of clastic-zone deposition.

The clastic zone is proposed as a time marker because of its stratigraphic position, thickness, and lithologic similarities. Core data were lacking to determine if the transitional limestone, mapped within the upper part of the clastic zone, is gradational into the Buckner as previously shown (Bishop 1968, 1969, 1971a, b, c). Its inclusion in

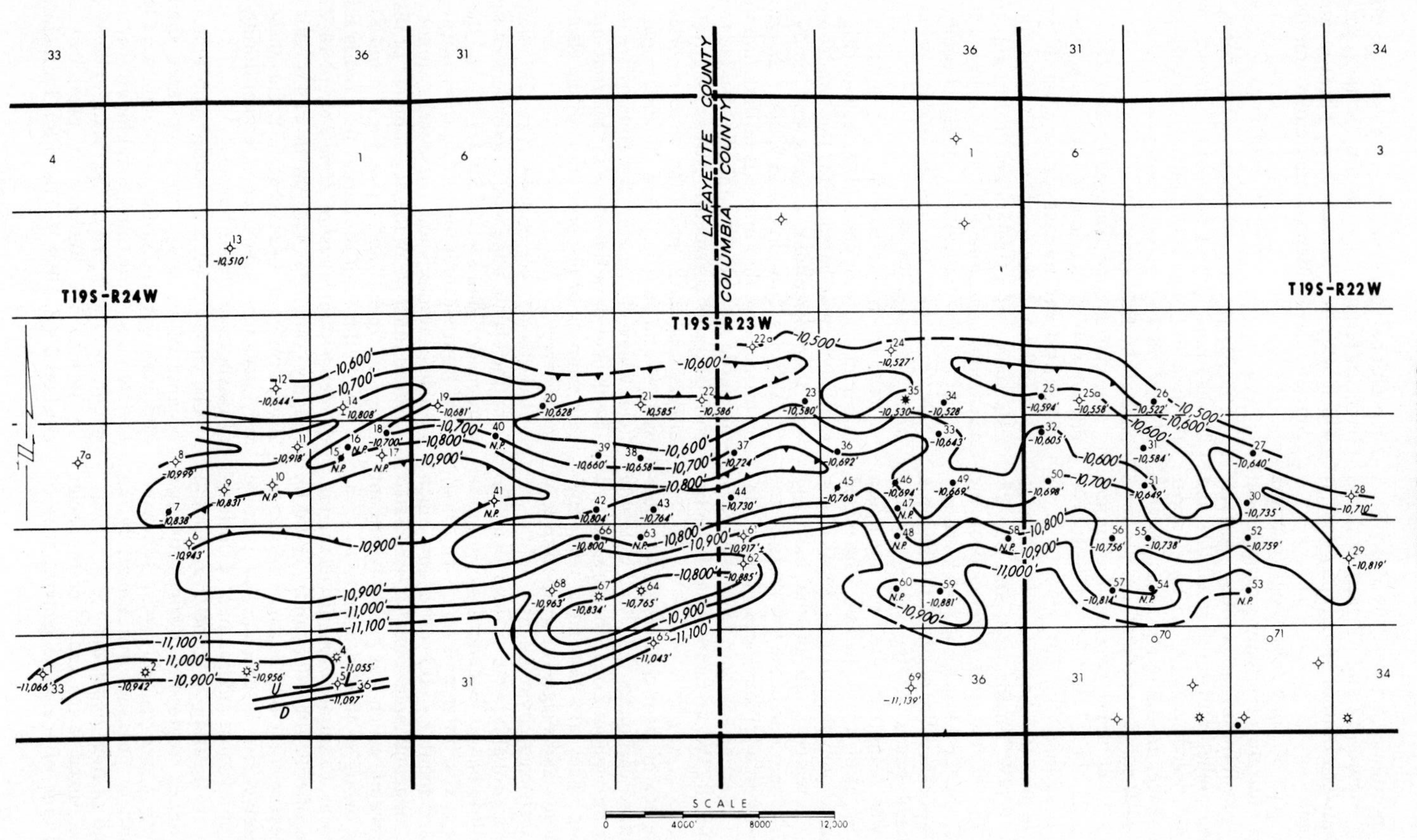

FIG. 11—Structure map, Walker Creek field area. Datum is near top of basal upper Smackover. C.I. = 100 ft (30 m).

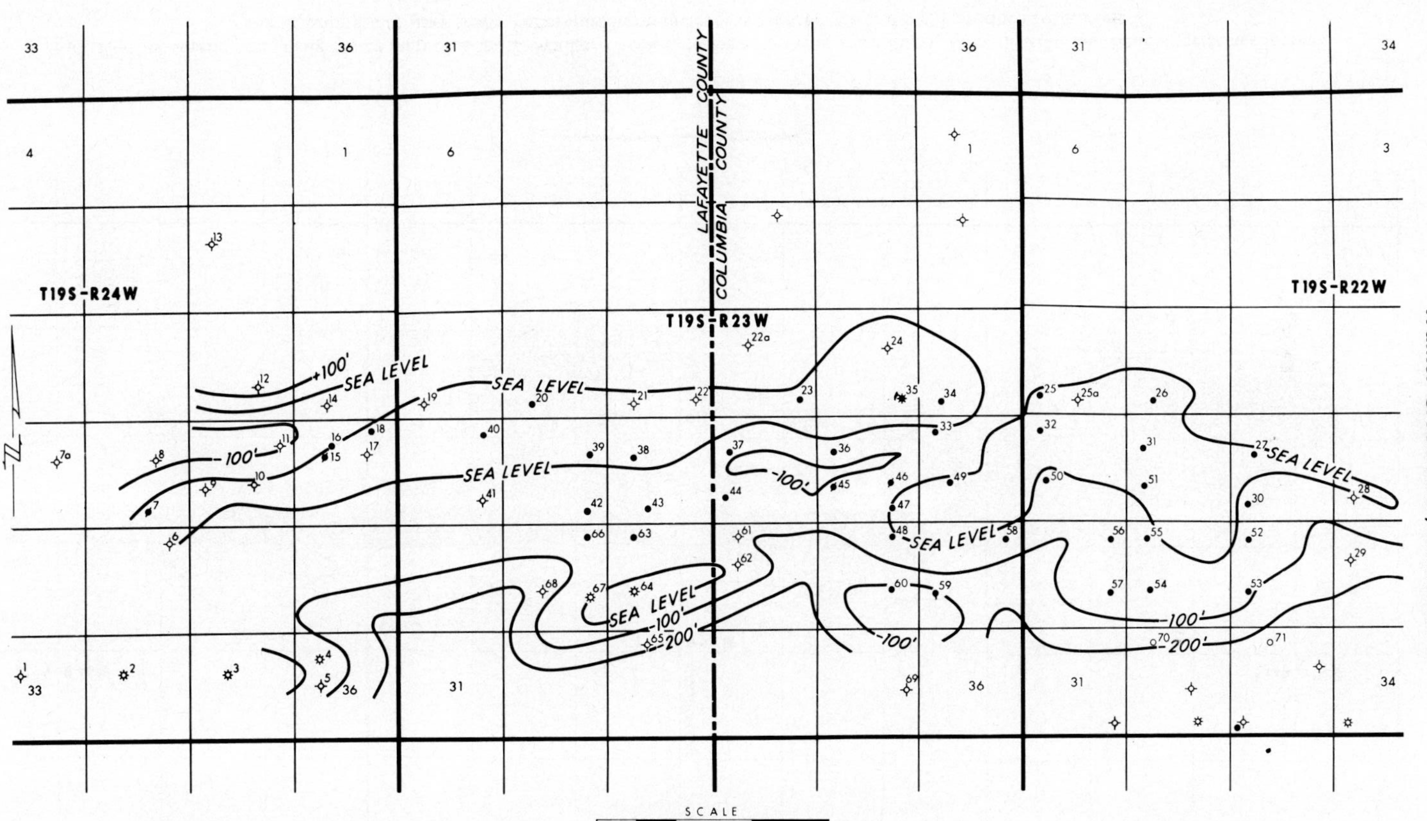

FIG. 12—Schematic paleobathymetric chart, Walker Creek field area, indicating three ridges on west coalescing toward east. This chart represents accentuated approximation of pre-Zone 2 seafloor configuration. C.I. = 100 ft (30 m).

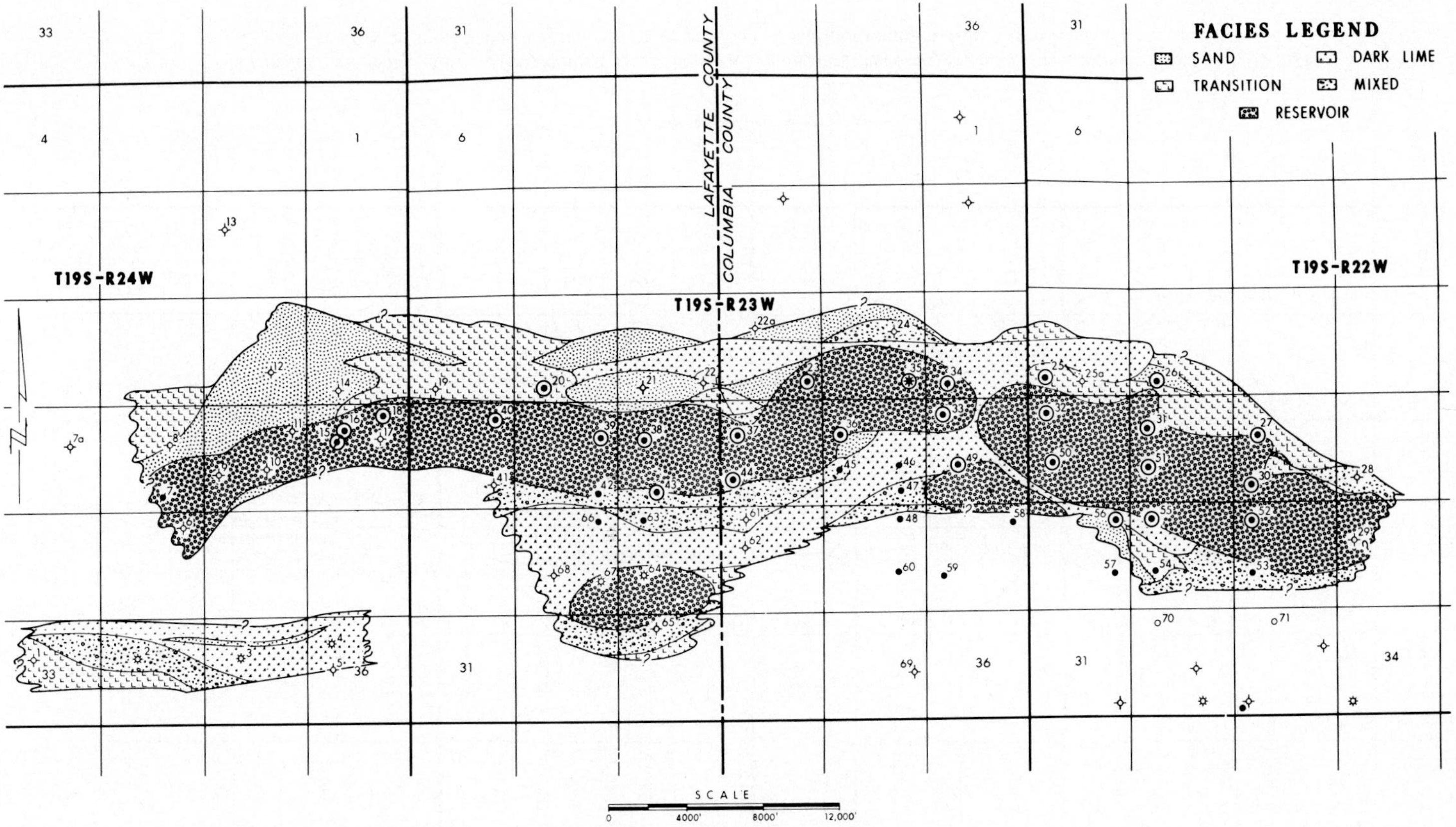

FIG. 13—Schematic composite facies map of Zone 2, Walker Creek field area, showing generalized facies distribution. Sandstone facies includes minor amounts of gray shale. Pellet-mudstone facies was omitted. Circled wells produce from Zone 2.

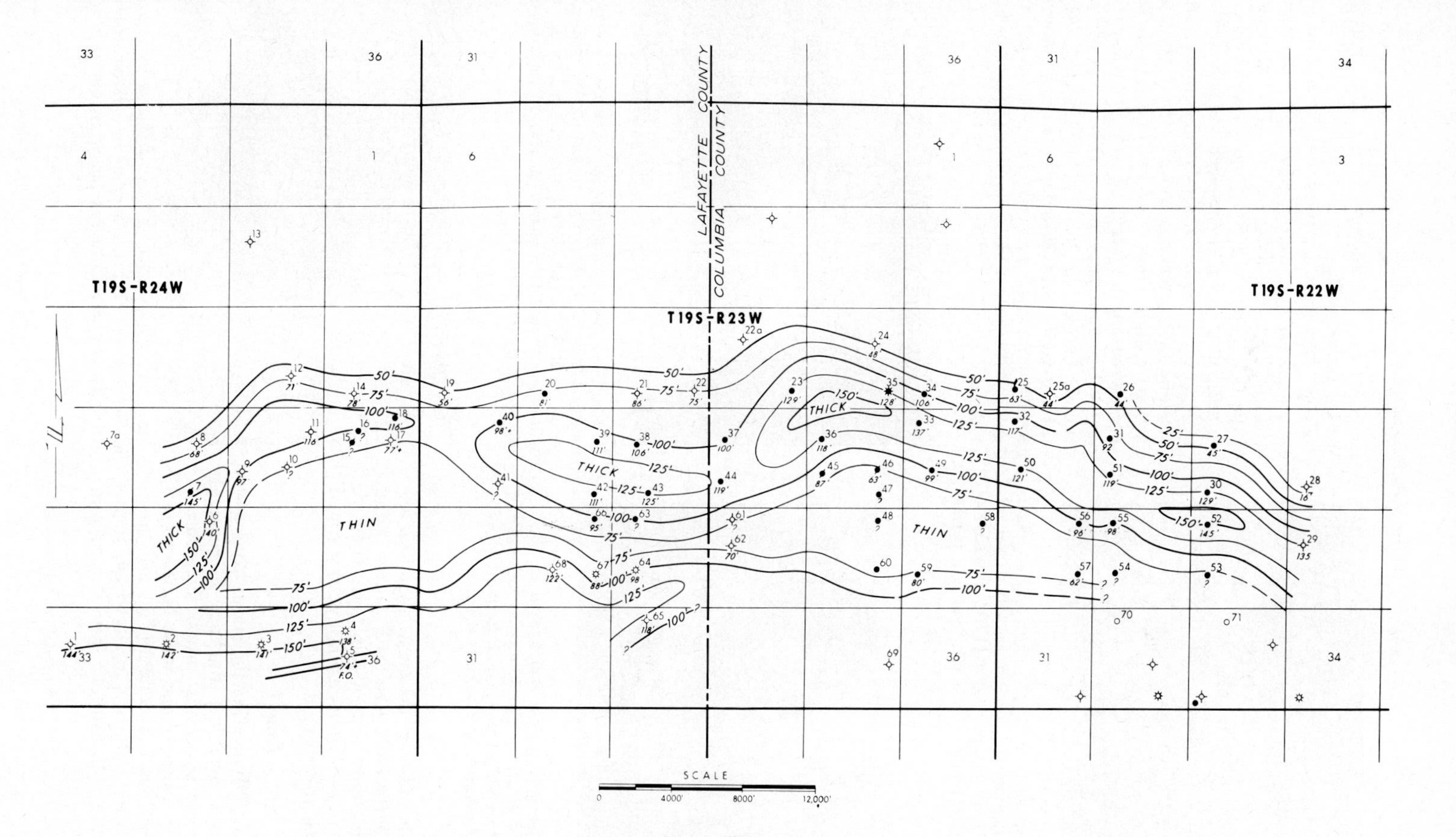

FIG. 14—Isopach map of Zone 2, Walker Creek field area, showing crest of depositional buildup. C.I. = 25 ft (7.6 m).

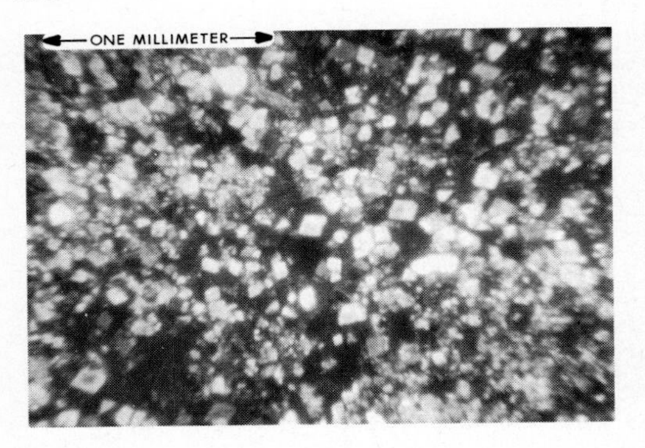

FIG. 15—Thin section of very fine-grained, light-colored dolomite rhombs in association with darker colored micrite from well 6, 11,084-11,085 ft (3,378.4-3,378.7 m), crossed nicols. Porosity 1.2 percent; permeability 0.0 md.

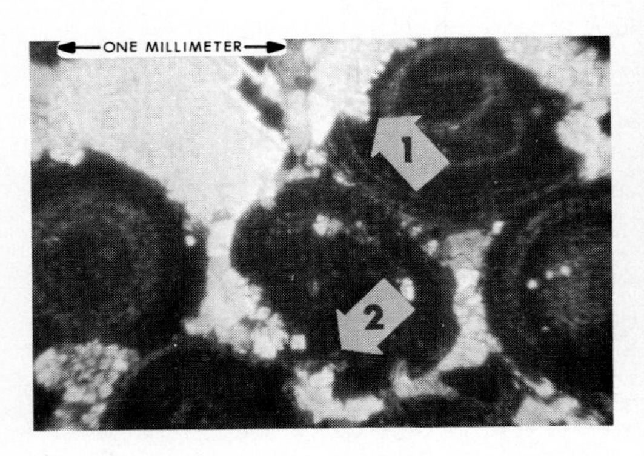

FIG. 16—Zone 2 transition facies, well 8, 10,981–10,982 ft (3,347-3,347.3 m), crossed nicols; dark-colored oolites replaced by anhydrite (1) and dolomite rhombs (2). Porosity 5.1 percent; permeability 0.0 md.

the clastic zone indicates the possibility of an unconformable surface between the Smackover and Buckner Formations in the updip areas (Figs. 5, 7, 8), but this unconformity cannot be confirmed on the basis of present data.

Zone 1

This zone directly overlies the clastic zone and grades into the overlying Buckner Formation (Fig. 2). It is the uppermost productive zone in the area and consists of the same facies groups and rock types found in Zone 2 (Fig. 24). Maximum thickness is 116 ft (35.4 m), in well 4. Twenty wells produce hydrocarbons from this zone in Lake Erling field and in the southern part of Walker Creek field.

Zone 1 apparently merges with Zone 2 in the eastern part of the field (Figs. 10, 24). Regressive offlap of Zone 1 is indicated in the western part of the field (Figs. 5, 8, 9). Zone 1 reservoir beds are difficult to correlate between wells 66 and 67 owing to the presence of sandstone and the dark limestone facies near the top of Zone 1 (Fig. 6). Carbonate dunes, like those in Zone 2, probably were present on this beach. Reservoir-facies sediments also were deposited on the Lake Erling structure in Secs. 33–36, T19S, R24W (Fig. 24).

Mixed-facies beds were deposited contemporaneously with reservoir facies in the adjacent areas, probably as a result of minor changes in water depth.

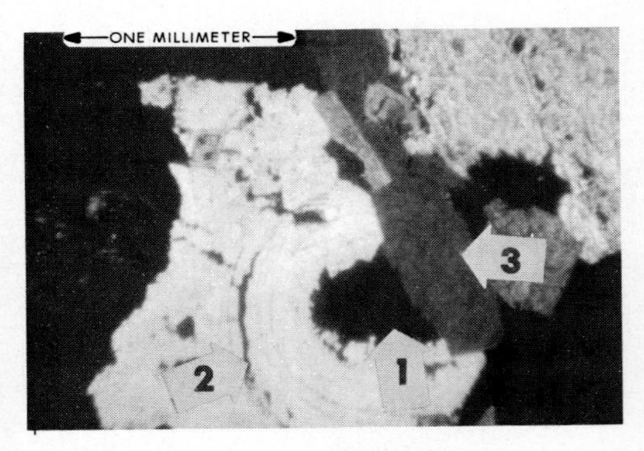

FIG. 17—Zone 2 transition facies, well 8, 10,985-10,986 ft (3,348.2-3,348.5 m), crossed nicols; indistinct center of dark-colored oolite (1) is replaced on left by light-colored secondary quartz (2) and on right by orange-colored anhydrite (3).

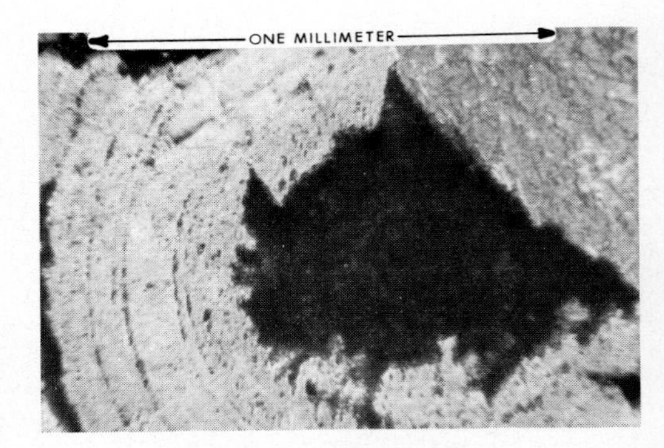

FIG. 18—Enlargement of part of Figure 17, crossed nicols; shows remnants of banding in dark-colored oolite which has been partially replaced by quartz.

Transition beds and siliciclastic sediments were deposited prior to and after oolite deposition on the old bathymetric highs, and darker limestones were deposited contemporaneously in the updip restricted and downdip low-energy environments.

Zone 1 deposition ceased as a result of continued regression, and the influx of fine siliciclastic material of the Buckner Formation increased.

Comparable conditions are apparent at present where dark-colored Holocene sediments are forming in the Nichupte Lagoon landward from the Isla Cancún oolitic beach on the northeast coast of Yucatán (Fig. 25). These lagoonal sediments have a carbon content up to 60 times that of carbonate mud deposited outside the lagoon (Ward et al, 1972). The organic carbon is principally derived from plant material. It is believed that organic content is the major cause of the coloration of the dark limestones in the Zone 1 and Zone 2 beds.

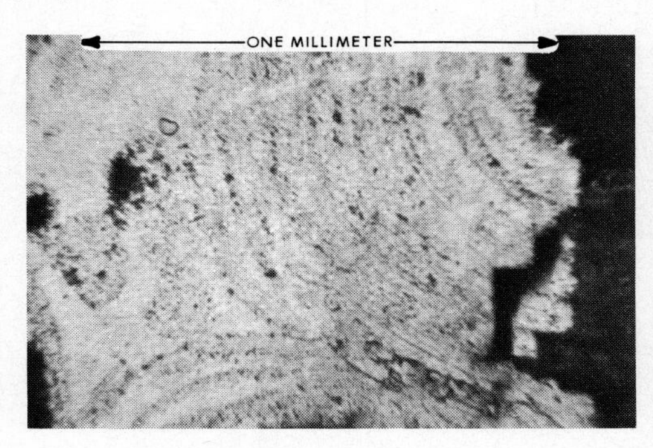

FIG. 19—Enlargement of part of Figure 18, crossed nicols; shows banded remnants of two dark-colored oolites replaced by light-colored quartz.

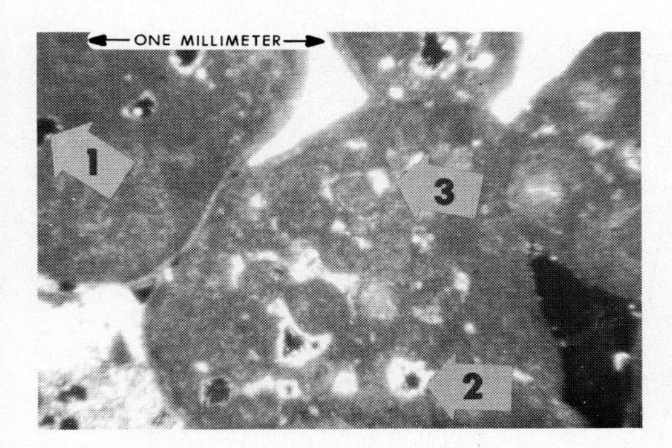

FIG. 20—Reservoir facies from Zone 2, well 7, 10,930-10,931 ft (3,331.4-3,331.7 m), crossed nicols; shows calcareous lumps with individual vugs (1). Some vugs lined by calcite druse partially replacing vugs (2), and some vugs completely filled (3). Porosity 10.7 percent; permeability 64 md.

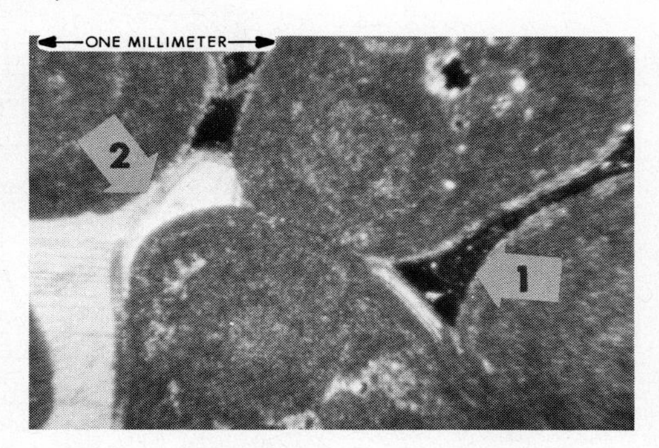

FIG. 21—Reservoir facies from Zone 2, well 7, 10,934-10,935 ft (3,332.6-3,332.9 m), crossed nicols. Dark areas are porosity (1); light areas are blocky spar cement. Pressure-solution phenomena are evident by broken outside lamellae of superficially coated lump (2). Porosity 10.9 percent; permeability 203 md.

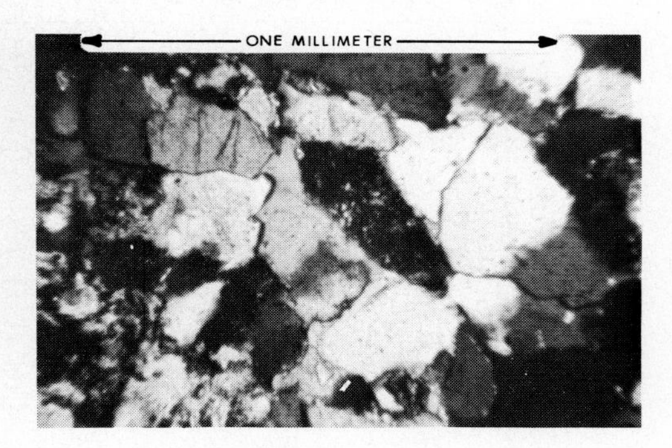

FIG. 22—Sandstone from clastic zone, well 63, crossed nicols; shows interlocking grains caused by pressure solution. Sandstone is clean, very fine grained, well sorted, and about 90 percent quartz; shown as angular white, gray, and black grains with minor amounts of pleochroic calcite cement.

Directly seaward from the lagoon lies the Isla Cancún, approximately 8 mi (2.4 km) long and less than 0.25 mi (0.06 km) wide (Fig. 25). The island is composed largely of fine- to medium-grained, oolitic carbonate sand forming partly lithified dunes up to 55 ft (16.8 m) high. The dunes grade into a white oolite beach. Sea-bottom deposits approximately 0.5 mi (0.15 km) seaward from the beach grade from oolitic facies into a mixed facies of gravelly sand (Ward and Brady, 1973).

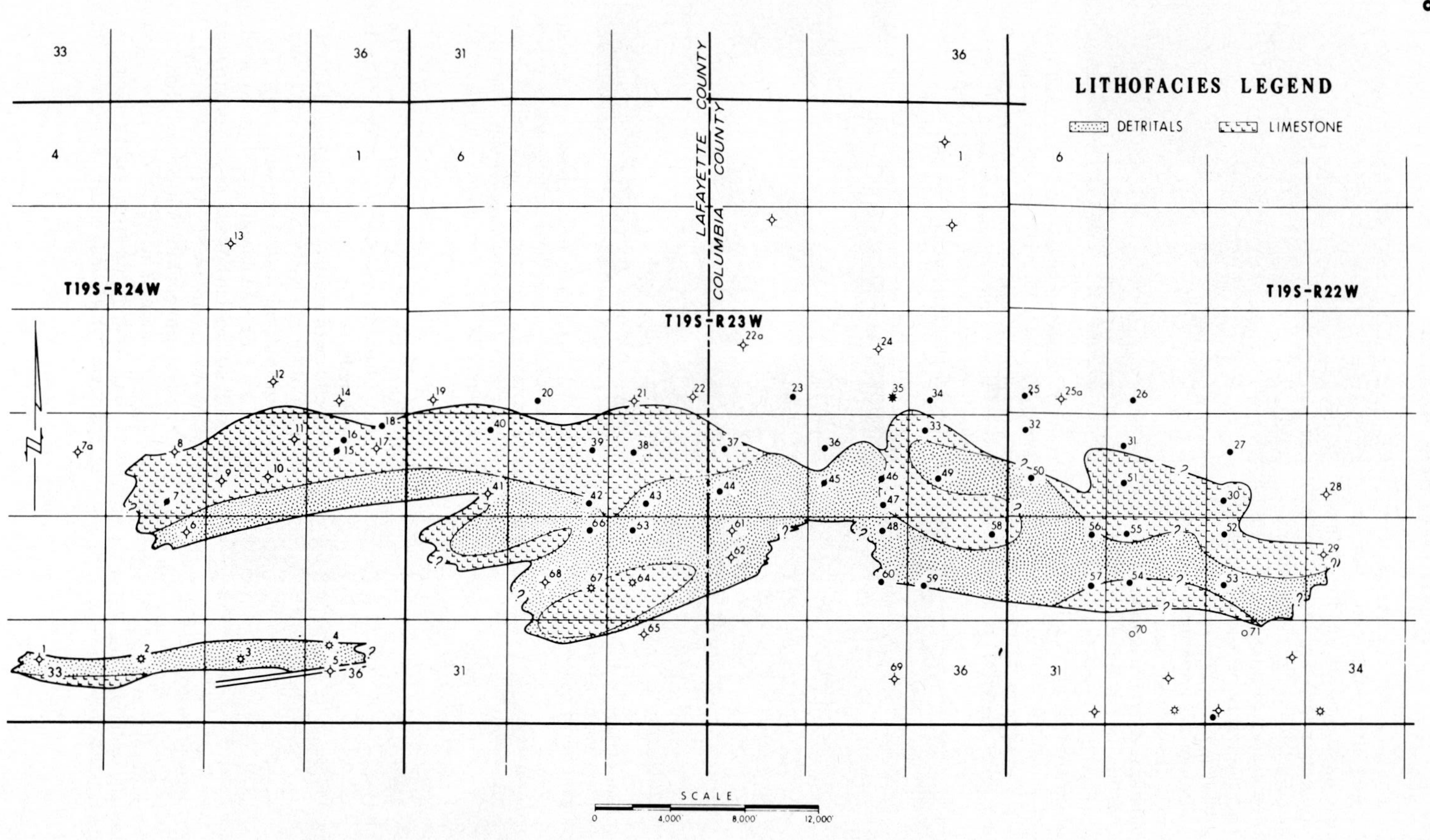

FIG. 23—Schematic composite lithofacies map of clastic zone, Walker Creek field area, Lafayette and Columbia Counties, Arkansas, showing generalized distribution of detrital rocks and limestone.

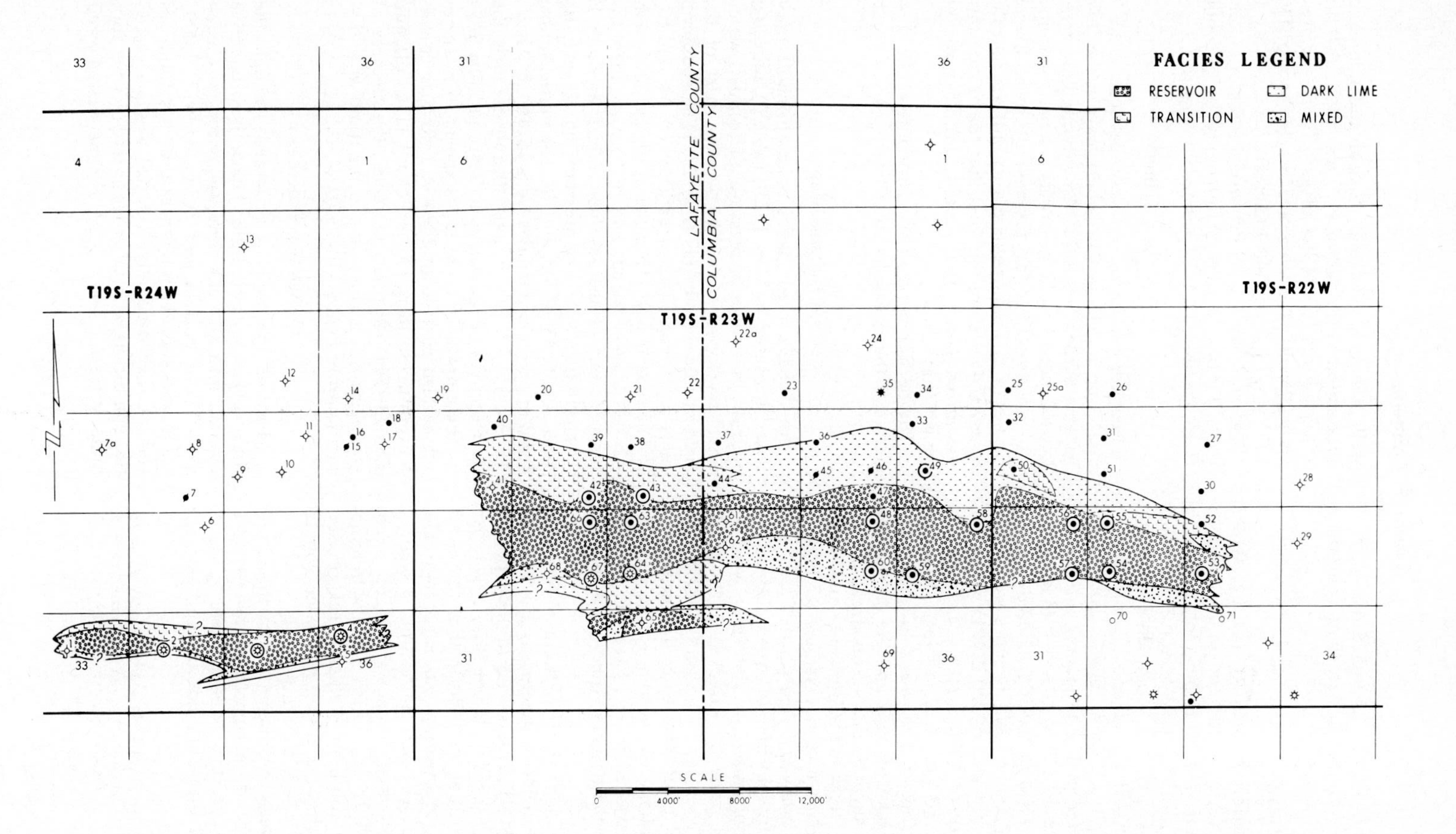

FIG. 24—Schematic composite facies map of Zone 1, Walker Creek field area, showing generalized distribution of facies types. Detrital rocks and pellet-mudstone facies were omitted. Circled wells produce from Zone 1.

Comparison of Figures 24 and 25 shows a striking resemblance in general size, shape, and distribution of facies types.

PRODUCTION DATA[3]

The Smackover Formation in the Walker Creek field was discovered to be productive in the H. A. Chapman No. 1 Helms, located in the NE ¼, Sec. 29, T19S, R23W, Lafayette County, Arkansas. This well was completed in March 1968, flowing 534 bbl of oil and 1,076 Mcf of gas per day through a 12/64-in. choke; gas/oil ratio was 2,015 cu ft/bbl. The producing interval was perforated from 10,870 to 10,885 ft (3,313–3,317 m) and from 10,910 to 10,930 ft (3,325–3,331 m). In July 1968, 320-acre drilling units were established and an allowable of 320 BOPD with a limiting gas/oil ratio of 2,000 cu ft/bbl was set by the Arkansas Oil and Gas Commission. Originally, gas produced from the field was sold to the Louisiana Transit Gas Company and the oil was sold to Falco, Inc. Currently the oil is being sold (1) to Cities Service Pipeline Company and being moved by pipeline to Cities Service Refining in Lake Charles, Louisiana, and (2) to Falco, Inc., being moved by pipeline to Atlas Refining in Shreveport, Louisiana. First oil sales were for $2.95/bbl; this price increased to $5.20/bbl in late 1973. The gas presently sells for $0.26–0.45/Mcf after processing for recovery of natural gas liquids. The Arkansas Oil and Gas Commission classified Walker Creek as a volatile oil reservoir on the basis of visual-cell analysis of the reservoir fluid.

Seismic reflections from the Louann Salt indicated north dip in the southern parts of Secs. 20 and 21, T19S, R23E. A lesser north dip is present at the upper Smackover datum (Fig. 19). It was not until the drilling of well 38 that the stratigraphic nature of the trap was suspected. This was later confirmed with the drilling of well 21. The original oil-water contact at Walker Creek field was $-10{,}720\pm$ ft ($-3{,}267\pm$ m), with minor variations. The southern Walker Creek productive structures have a lower gas-water contact ($-10{,}733\pm$ ft or $-3{,}271\pm$ m).

Depth of wells producing from the upper Smackover ranges from approximately 10,850 ft (3,307 m) on the north side of the field to 11,400 ft (3,474 m) on the south side of the field. The wells averaged $300,000 to drill and complete. Production to the end of 1973 was 10,515,000 bbl of oil and 24,500 MMcf of gas. There are presently 39 wells producing on 320-acre spacing. The field was unitized (14,840 acres) effective May 1, 1974; unit operator is Pennzoil Producing Company. Partial pressure maintenance by injection of the produced gas is planned for the unit.

Pressure Declines

Pressure-decline differentials between Zone 1 and Zone 2 wells in the western part of the field became evident by 1971. An intermediate area, located between the two major zonal buildups, had abnormally rapid pressure declines. Facies studies of this intermediate area indicate that the reservoir facies are thin and that they grade into the mixed facies and a sandy lithology of the transitional facies.

The decline curves (Fig. 26) confirm the dual nature of the reservoirs and the proposed correlations.

Isobaric Map

An isobaric map, based on bottomhole pressures measured in May 1973 (Table 2), indicates four distinct zones of pressure differentials which have developed across the western part of the field in east-west bands (Fig. 27). The southernmost band encompasses wells 59, 60, 64, and 67.

The next band toward the north includes wells 42, 47, 48, 49, 58, 63, and 66 and parallels the development of the reservoir facies in Zone 1. The intermediate low-pressure band includes wells 33, 36, 43, 44, 45, and 46. The northernmost band of pressures includes wells 15, 16, 18, 23, 25, 32, 34, 35, 37, 38, 39, and 40 and parallels the development of the reservoir facies in Zone 2.

These northern three zones appear to merge in the eastern part of the field in a manner consistent with the merging of Zone 1 (Fig. 24) and Zone 2 (Fig. 13) reservoir facies, the merging of structural ridges (Fig. 4), and the trend of bathymetric prominences (Fig. 12).

Completion Practices

Most of the wells were cored through the Smackover zone and drilled below the productive interval into a "tight" limestone section. Casing was set on bottom and cemented with 1,000–1,500 sacks of high-temperature, low-fluid-loss cement. The productive interval was perforated through tubing with two to four jet holes per foot under a treated water load. All the wells flowed

[3]Production data are taken from exhibits presented to the Arkansas Oil and Gas Commission by the Engineering Subcommittee on Walker Creek Field Unitization, April 1974.

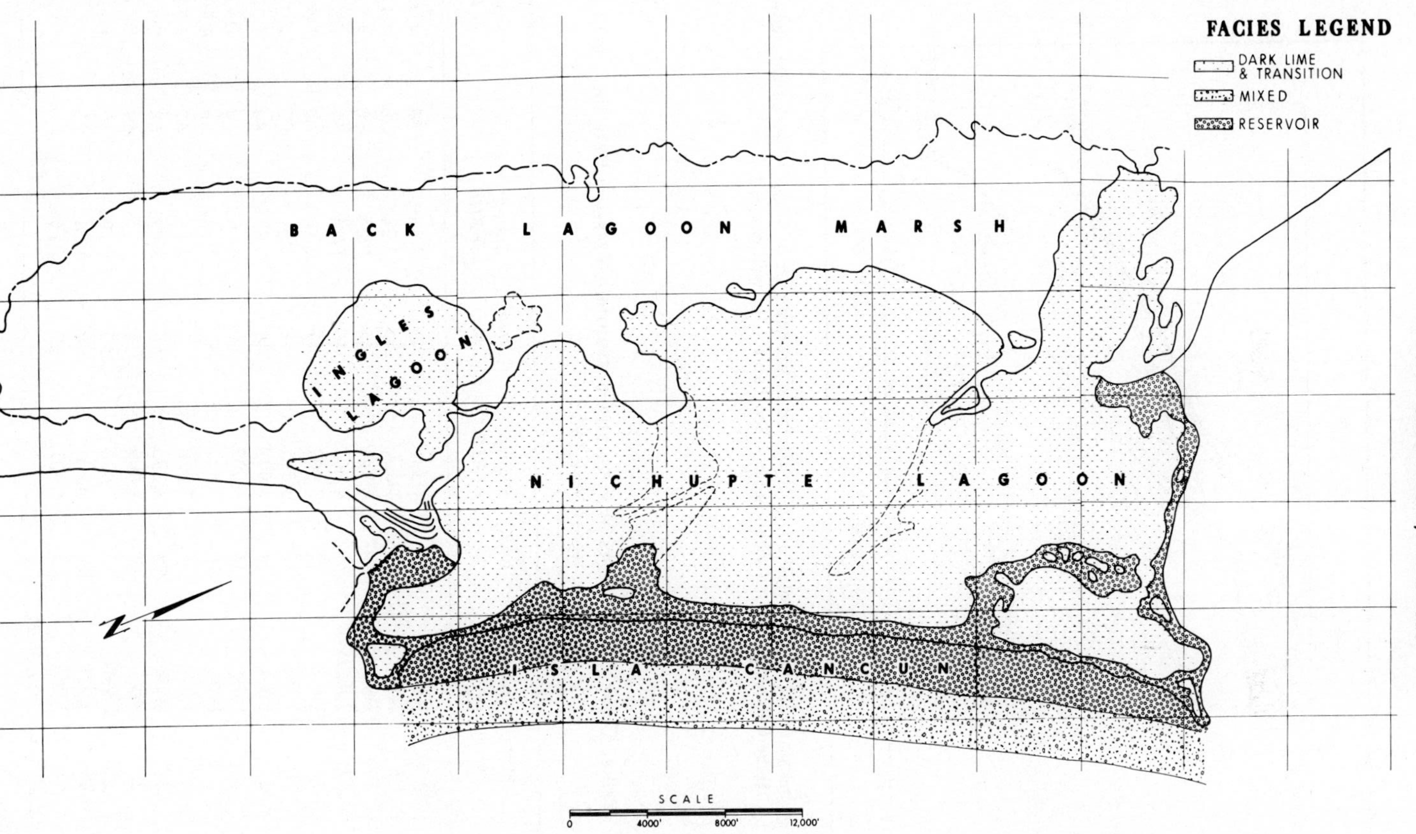

FIG. 25—Map of Isla Cancún and Nichupte Lagoon, northeast Yucatán, showing generalized distribution of Holocene sediments. Scale is same as Figure 24. Modified from Brady (1971) and Ward et al (1973).

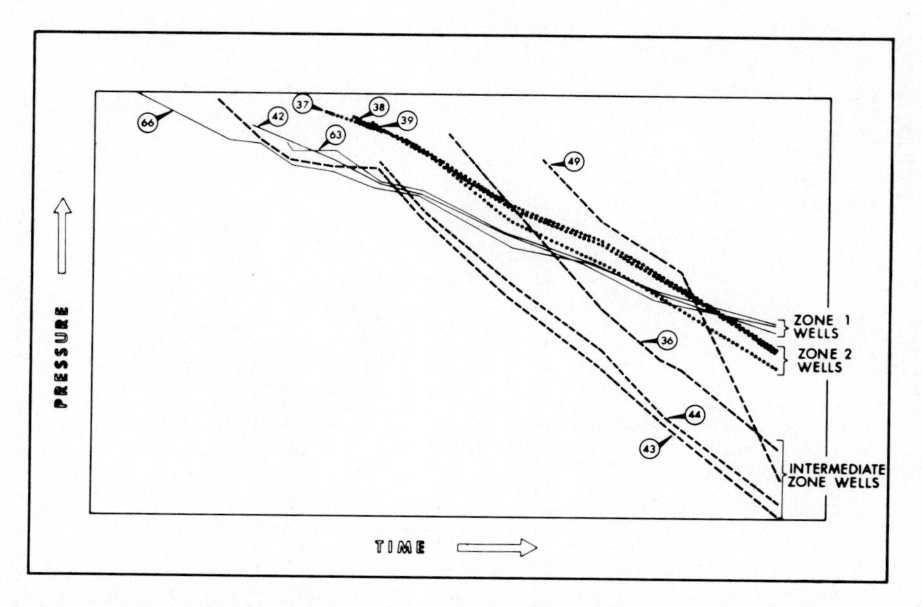

FIG. 26—Bottomhole pressure-decline curves of 10 wells located in western part of Walker Creek field, showing three different pressure-decline rates. (See Table 1 for well names.)

Table 2. Walker Creek Field, Status of Wells and Bottomhole-Pressure Data (in psig), as of May 1973

Well No.	Status* or BHP	Well No.	Status* or BHP	Well No.	Status* or BHP	Well No.	Status* or BHP	Well No.	Status* or BHP
1	D&A	15	P&A	28	D&A	43	3,439	58	4,064
2	ND	16	3,801	29	D&A	44	3,388	59	4,420
3	ND	17	D&A	30	4,141	45	P&A	60	4,530
4	ND	18	3,839	31	4,091	46	P&A	61	D&A
5	D&A	19	D&A	32	3,909	47	4,160	62	D&A
6	D&A	20	1,767	33	3,281	48	4,010	63	4,077
7	P&A	21	D&A	34	3,978	49	3,834	64	4,592
7a	D&A	22	D&A	35	3,995	50	4,109	65	D&A
8	D&A	22a	D&A	36	3,694	51	4,132	66	4,102
9	D&A	23	3,978	37	3,828	52	4,151	67	4,747
10	D&A	24	D&A	38	3,895	53	3,891	68	D&A
11	D&A	25	ND	39	3,899	54	4,141	69	D&A
12	D&A	25a	D&A	40	3,808	55	4,193	70	Loc
13	D&A	26	3,029	41	D&A	56	4,202	71	Loc
14	D&A	27	2,996	42	4,067	57	4,144		

*ND = No data.

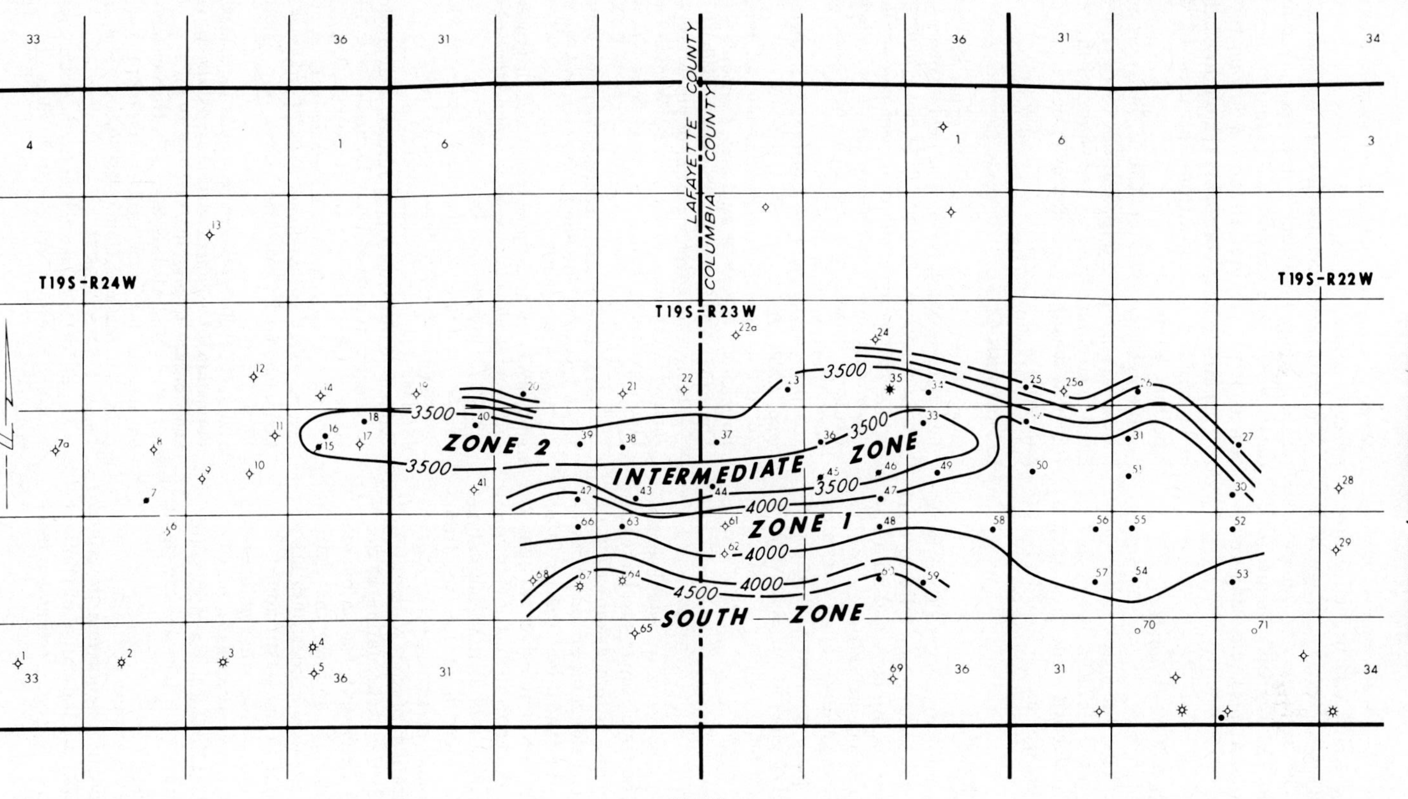

FIG. 27—Isobaric map of Walker Creek field area, showing pressure zones on west side of field merging on east. Based on pressure data, May 1973 (see Table 2). C.I. = 500 psig.

naturally; however, a few of the poorer wells were treated with 15 percent HCl acid to induce flow. The wells are producing through two-stage separator installations. The high-pressure separators are maintained at 400–750 psig, and the low-pressure separators operate at pressures ranging between 40 and 50 psig.

Hydrocarbon-Water Contact

With minor variations, the original oil-water contact in the field was −10,720± ft (−3,267± m). The southern productive structures have a lower gas-water contact at −10,733±ft or −3,271± m.

Reserves

Original oil in place is estimated to have been 96 million STB. Primary oil recovery will approximate 20 percent of the oil in place, and an additional 10 million STB will be recovered under pressure-maintenance operations. Total estimated recovery, excluding plant products, is approximately 30 million bbl of pipeline oil. Recoverable gas is estimated at 100 Bcf.

Conclusions

One purpose of this paper is to establish the clastic zone as the zone which separates the reservoir beds into two bodies. From examination of core plugs in three wells and comparison of petrographic descriptions, the sandstone appears to be identical, but this is not conclusive enough to prove beyond doubt the equivalence of this zone throughout the area. Additional data indicating separation of the two reservoir zones come from (1) the isobaric map and pressure-decline curves which confirm separation of the zones at this stratigraphic level, (2) the cyclic nature of detrital-carbonate-detrital deposition found by mapping and core description, and (3) the detrital component of the transitional limestones associated with the sandstone in the clastic zone.

Establishment of the clastic zone assures the differentiation of the reservoir rocks into at least two zones. The fact that Zone 1 has a more southward updip limit and is overlain by the same Buckner detrital rocks as those overlying Zone 2 indicates a regressive stage of deposition.

Some of the diagenetic phenomena described from petrographic slides are indicative of changes in porosity which may have occurred as a consequence of eolianite lithification in the vadose or phreatic zone.

The presence of certain fine-grained limestones (mostly pellet mudstones) and their distinctive dark color, which possibly is due to increased carbon content, are indicative of a low-energy restricted environment probably related to a calcarenite buildup seaward from the area of dark limestone deposition.

Although the Walker Creek field produces from a stratigraphic trap, seafloor topography related to structure prior to deposition of the reservoir beds was the major factor in development of porosity.

References Cited

Bishop, W. F., 1968, Petrology of upper Smackover limestones in North Haynesville field, Claiborne Parish, Louisiana: AAPG Bull., v. 52, no. 1, p. 92-128.

———1969, Environmental control of porosity in the upper Smackover limestone, North Haynesville field, Claiborne Parish, Louisiana: Gulf Coast Assoc. Geol. Socs. Trans., v. 19, p. 155-169.

———1971a, Geology of a Smackover stratigraphic trap: AAPG Bull., v. 55, no. 1, p. 51-63.

———1971b, Geology of upper member of Buckner Formation, Haynesville field area, Claiborne Parish, Louisiana: AAPG Bull., v. 55, no. 4, p. 566-580.

———1971c, Stratigraphic control of production from Jurassic calcarenites, Red Rock field, Webster Parish, Louisiana: Gulf Coast Assoc. Geol. Socs. Trans., v. 21, p. 125-137.

———1973, Late Jurassic contemporaneous faults in North Louisiana and South Arkansas: AAPG Bull., v. 57, no. 5, p. 858-877.

Brady, M. J., 1971, Sedimentology and diagenesis of carbonate muds in coastal lagoons of northeast Yucatan: Ph.D. dissert., Rice Univ., 288 p.

Crow, N. B., 1958, Mt. Sinai field, Claiborne Parish, Louisiana: Shreveport Geol. Soc. Ref. Rept., v. 4, p. 141-144.

Dickinson, K. A., 1968, Upper Jurassic stratigraphy of some adjacent parts of Texas, Louisiana, and Arkansas: U.S. Geol. Survey Prof. Paper 594E, 25 p.

———1969, Upper Jurassic carbonate rocks in northeastern Texas and adjoining parts of Arkansas and Louisiana: Gulf Coast Assoc. Geol. Socs. Trans., v. 19, p. 175-187.

Vestal, J. H., 1950, Petroleum geology of the Smackover Formation of southern Arkansas: Arkansas Geol. and Conserv. Comm. Inf. Circ. 14, 19 p.

Ward, W. C., and M. J. Brady, 1973, High-energy carbonates on the inner shelf, northeastern Yucatan Peninsula, Mexico: Gulf Coast Assoc. Geol. Socs. Trans., v. 23, p. 226-238.

———et al, 1972, A field trip to northeastern coast of Yucatan: Houston Geol. Soc. Pub., 40 p.

East Cameron Block 270, Offshore Louisiana: A Pleistocene Field[1]

D. S. HOLLAND,[2] CLARKE E. SUTLEY,[2] R. E. BERLITZ,[3] and J. A. GILREATH[3]

Abstract Exploration of the Plio-Pleistocene in the Gulf of Mexico since 1970 has led to the discovery of significant hydrocarbon reserves. One of the better gas fields found to date has been the East Cameron Block 270 field. Utilization of a coordinated exploitation plan with Schlumberger has allowed Pennzoil, as operator, to develop and put the Block 270 field on production in a minimum time.

The structure at Block 270 field is a north-south-trending, faulted nose at 6,000 ft (1,825 m). At the depth of the "G" sandstone (8,700 ft or 2,650 m), the structure is closed; it is elongated north-south and dips in all directions from the Block 270 area. Closure is the result of contemporaneous growth of the east-bounding regional fault.

Structural and stratigraphic interpretations from dipmeters were used to determine the most favorable offset locations. The producing zones consist of various combinations of bar-like, channel-like, and distributary-front sandstones. The sediment source for most of the producing zones was southwest of the area, except for two zones which derived their sediments from the north through a system of channels paralleling the east-bounding fault.

Computed logs were used to convert conventional logging measurements into a more readily usable form for evaluation. The computed results were used for reserve calculations, reservoir-quality determinations, and confirmation of depositional environments as determined from other sources.

Exploration and Production History

Block 270, East Cameron area, is a 2,500-acre (13 km^2) tract approximately 106 mi (170 km) south-southwest of Lafayette, Louisiana, and 65 mi (104 km) from the shoreline on the east side of the East Cameron area (Fig. 1). Average water depth for the block is 170 ft (33 m). The block was acquired in the December 1970 OCS sale by Pennzoil Offshore Gas Operators, Inc., Mobil Oil Corporation, Mesa Petroleum Company, and Texas Production Company for a lease bonus of $32,186,975.

The discovery well, commenced in March 1971, defined gas "pay" in the "A" series (6,470 ft or 1,972 m), "B" series (7,680 ft or 2,340.9 m), "C" series (7,970 ft or 2,429.3 m), and "G" series (8,310–8,710 ft or 2,532.9–2,654.8 m; the "G," "J," "K," "O," and "Q" sandstones). These "pay" zones are shown in Figure 2.

A confirmation well, located approximately 2 mi (3 km) north of the discovery well, was "stubbed" as a platform location in May 1971. Two 8-pile, 18-slot platforms already under construction were assigned to the block. The A platform was set over the No. 2 well "stub" in August 1971. Thirteen wells were drilled from the platform, resulting in nine single and three dual completions. One well was plugged back and redrilled.

The B platform was installed in August 1971. Twelve wells were drilled, resulting in five single and seven dual completions.

In February 1972, a third expendable hole was drilled on the west side of Block 270 in order to evaluate the open acreage to the west.

Production commenced in February 1973 from the A platform and in May 1973 from the B platform.

Average production from the two platforms is 500 MMcf of gas daily. Cumulative production as of February 1, 1975, was 258.6 Bcf of gas and 1,137,885 bbl of liquids. From offset Block 273, cumulative production at the same date was 37.3 Bcf of gas and 109,920 bbl of liquids, and, from Block 255, it was 20.7 Bcf of gas and 89,580 bbl of liquids.

The total cost of acquiring leases, drilling, and equipping the block was $60 million.

Stratigraphic and Structural Setting

Block 270 field is located in the middle Pleistocene (*Angulogerina* B) and lower Pleistocene (*Lenticulina* I) trends. The *Lenticulina* I marker generally occurs several hundred feet above the "G" sandstone. Sandstone percentages in the trend and in surrounding areas are in the 20- to 30-percent range. Paleoenvironment is in Ecological Zones 2, 3, and 4 (lower shelf and upper slope).

[1]Manuscript received, March 17, 1975. Revised from a paper published in *Transactions of Gulf Coast Association of Geological Societies*, v. 24, p. 89–106. Published with permission of GCAGS.

[2]Pennzoil Company, Houston, Texas 77017.

[3]Schlumberger Offshore Services, New Orleans, Louisiana 70112.

Only data from Block 270 wells were used in preparation of this paper. On illustrations, contours shown in surrounding blocks are projections made from Block 270 data.

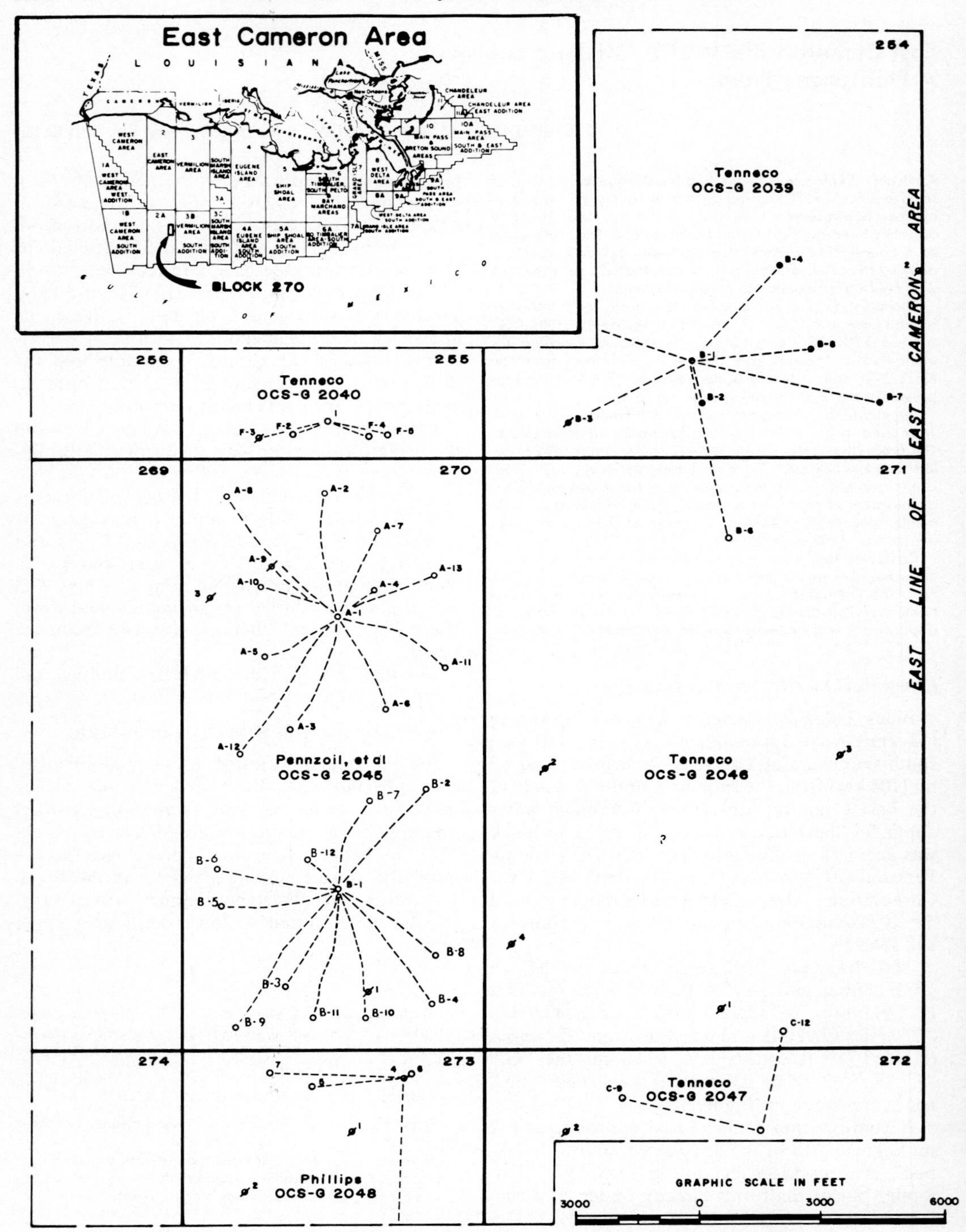

FIG. 1—East Cameron Block 270 field, offshore Louisiana, index map.

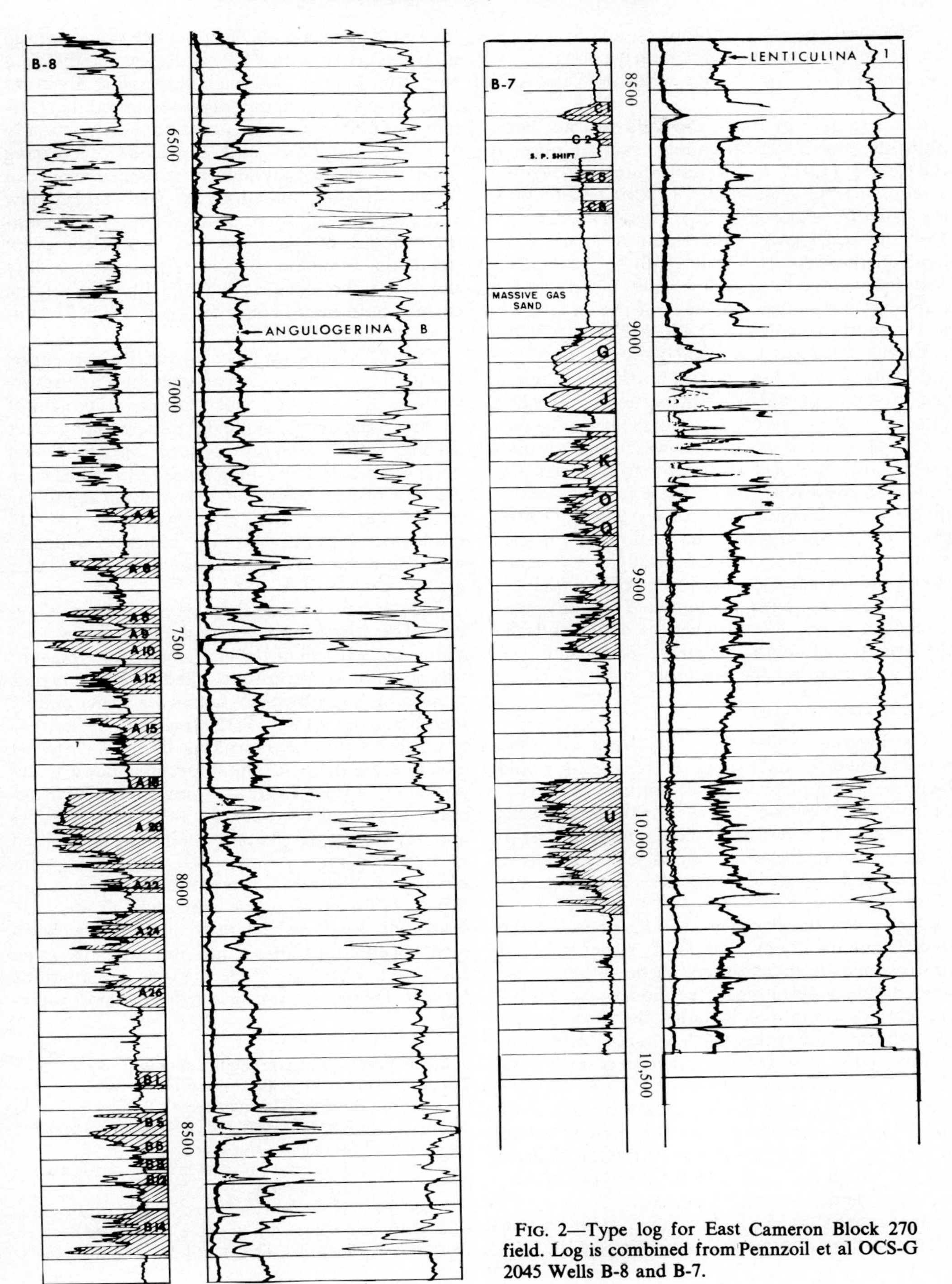

FIG. 2—Type log for East Cameron Block 270 field. Log is combined from Pennzoil et al OCS-G 2045 Wells B-8 and B-7.

Structural growth, faulting, and deposition were contemporaneous, as is clearly shown in the "G" through "Q" sandstone series in Wells A-11 and B-4.

The structure at Block 270 field is a northerly plunging, north-south-trending, faulted nose at the 6,500-ft (1,825 m) "A" sandstone level. At the level of the "G" sandstone (8,700 ft or 2,650 m), the structure is closed by dip (Figs. 3, 4). A major down-to-west growth fault (fault A) bounds the field on the east; this fault, with the associated east dip into the downthrown side, is the primary trapping mechanism. The apex of the structure is in the southeast quarter of Block 270 and at the 8,700-ft ("G" sandstone) level; dip is steeper on the south flank than on the north. Two minor fault systems, parallel with the structural axis and directly west of the crest, are present on the feature. Fault B-1 is a down-to-west fault or fault zone. Fault A-5 is downthrown to the east and appears to be a relief fault. These faults have less than 100 ft (30 m) of throw. They have had little effect on hydrocarbon accumulation, as the gas-water contacts for the "G" through "Q" sandstones are the same on both sides of the faults.

Salt has not been encountered in any well drilled on Block 270, although it could underlie the structure at depth. A shallow, piercement salt dome is present on the southeast in Block 272.

Data Interpretation

The logging program for the Block 270 field was designed so that computer processing would facilitate a complete field evaluation. Resistivity, density, and neutron logs were run on all wells except where prevented by hole conditions. Dipmeter and sonic logs were run on selected wells in the field. All logs were recorded on magnetic tape at the well site.

During the development-drilling phase of the East Cameron Block 270 field, structural and stratigraphic dipmeter interpretations were considered when selecting offset locations. Stratigraphic interpretations included the most probable geometry, the strike, the "shale-out" direction, and the sediment-transport direction of each nonblanket producing sandstone.

Figure 5 compares the SP, dipmeter, and computed logs (SARABAND) for the "G-J-K" zone of Well A-7. It illustrates some of the log characteristics which are used to infer the types of primary sedimentary features present.

The funnel-shaped SP curve, the predominance of current dip patterns, and increasing permeability and decreasing clay percentage from bottom to top imply that the "K" sandstone is a distributary-front deposit. These characteristics are common to progradational depositional units. Current dip patterns (groups of dips with essentially constant directions and downward-decreasing magnitudes) are generated by crossbeds and can be used to infer directions of paleocurrent flow and sediment transport. Since sand-grain orientation should be parallel with the paleocurrent trend, the direction of preferential permeability should be the same as the dip direction of the current patterns in zones such as the "K" sandstone.

The "J" sandstone has a cylindrical SP curve, high permeability, low clay percentage, and several slope dip patterns, which are characteristic of channel sandstones. Slope dip patterns are groups of dips with essentially constant directions and downward-increasing magnitudes. In a channel, dip direction is toward the axis and normal to the strike; therefore, the northeast dip of the "J" sandstone slope dip patterns indicates a northwest-southeast-striking channel with its axis northeast of Well A-7.

The "G" zone represents three depositional cycles. The lower member is a distributary-front sandstone with all of the characteristics typical of progradational deposition. Sediment transport was from southwest to northeast. The middle member, from 9,160 to 9,220 ft (2,791–2,808 m), is a northwest-southeast-striking channel with its axis to the northeast. The upper member was deposited at a slower rate in a low-energy environment. Such sandstones tend to contain more clay and have lower permeability than those deposited in higher energy environments, where more winnowing of the finest sediments occurs.

The sediment-transport patterns of the "G" sandstone zone are illustrated in Figure 6. Sediments were transported into the area from the southwest. There are three possible explanations for this unexpected northeasterly transport direction.

1. The presence of a prodelta southwest of the Block 270 area, positioned so that its northeast quadrant discharged into the area. This northeast quadrant would be analogous to the Main Pass–Pass à Loutre quadrant of the modern Mississippi delta area.
2. Sediment transport into the area by northeasterly flowing longshore currents.
3. Sediment movement deflected from a north-south to a northeast course by a rising salt dome located southwest of Block 270. The growth-fault system associ-

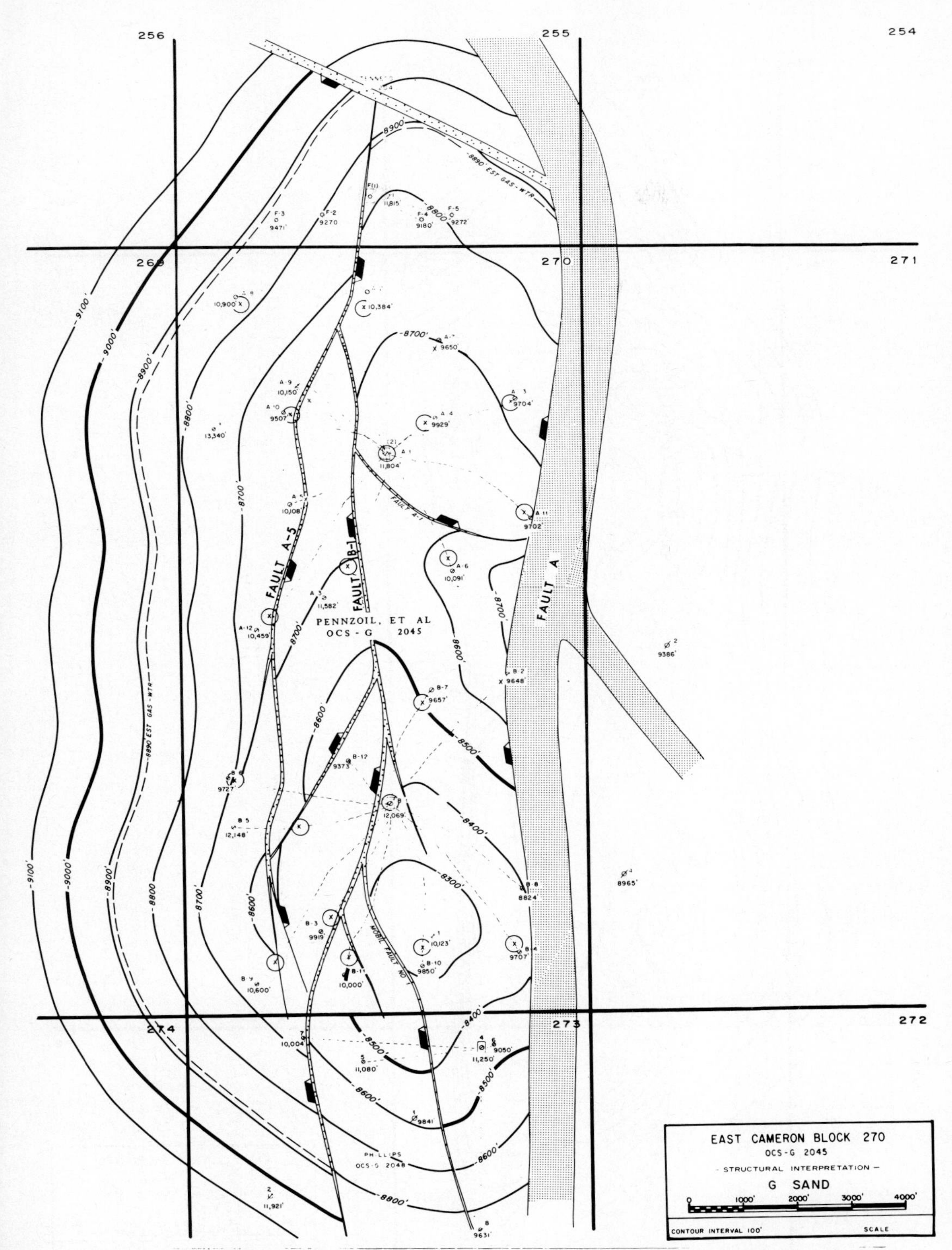

FIG. 3—East Cameron Block 270 structural interpretation at level of "G" sandstone. C.I. = 100 ft.

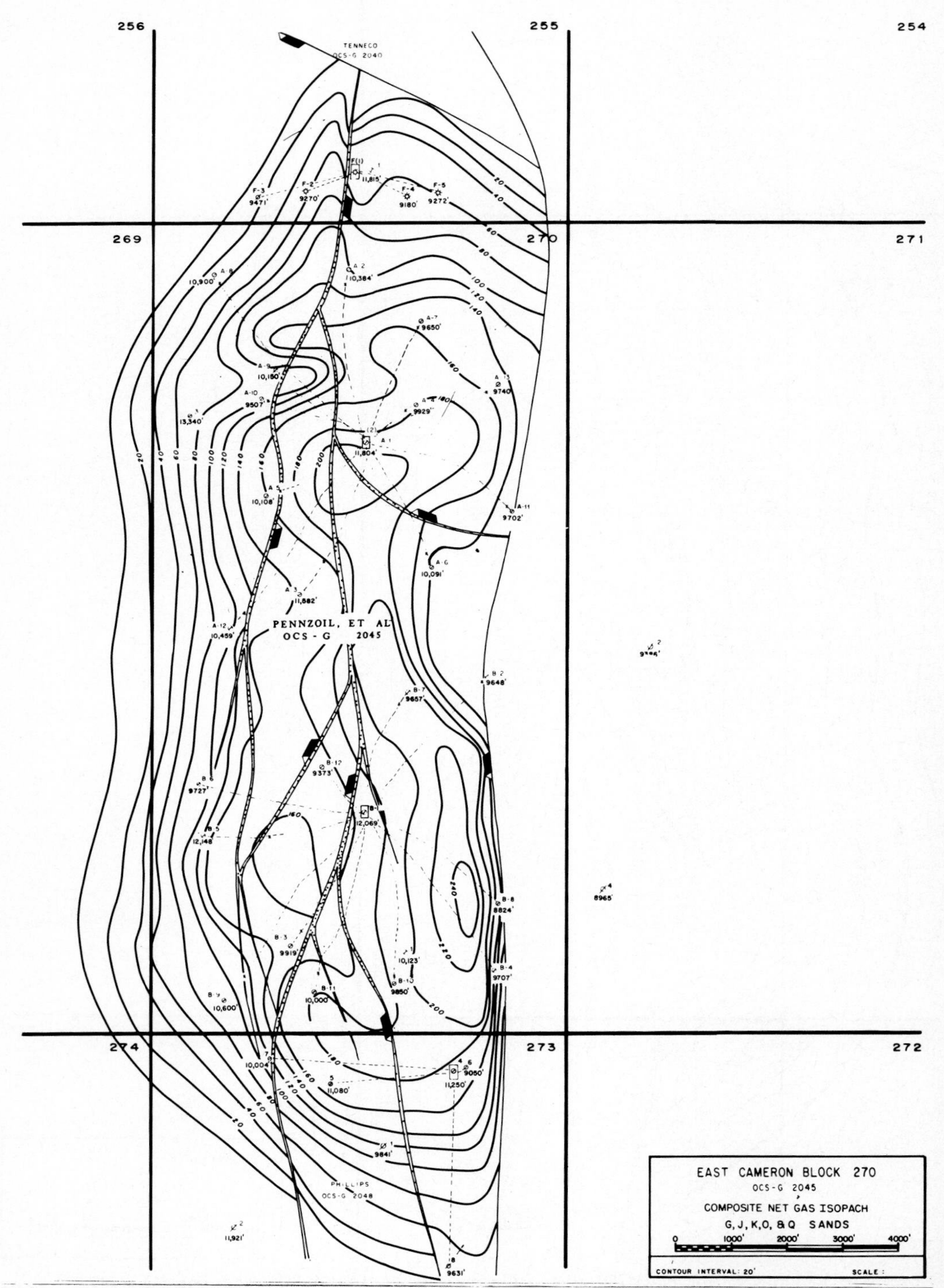

FIG. 4—East Cameron Block 270 field, composite net-gas isopach map of "G," "J," "K," "O," and "Q" sandstones. C.I. = 20 ft.

ated with this dome would have provided additional guidance for the currents bringing sediments into the Block 270 area.

The sands were transported into the area via two channel-like systems. The major system terminated in distributary-front deposits, mostly in the north half of the block. The minor system entered the block parallel with the east-bounding fault, then discharged some of its sand in a northwesterly direction in the south half of the block.

The producing "C" sandstone (Fig. 7) exhibits a bell-shaped SP curve, slope dip patterns in the overlying shales, and decreasing permeability and increasing clay percentage upward. These are some of the characteristics of barlike sandstones. The direction of dip of the slope patterns in the shales overlying a bar is toward the gradation into shale and normal to the strike. Thus the "C" sandstone bar in Well A-7 grades into shale northward and strikes east-west.

The sediment-transport patterns (Fig. 8) indicate continued sediment influx from the southwest at the time the "C" sand was deposited. Some distributary-front sands were deposited, and other sediments were reworked into bars in the northern half of the block. During this postulated glacial period, sand reached the area in reduced amounts, so the bar-building processes incorporated enough clay particles into the bars to cause significantly reduced permeability.

The "B-14" sandstone (Fig. 9) has multiple members in Well A-7. The major portion, from 8,375 to 8,515 ft (2,552–2,595 m), was deposited in a channel with the axis on the northeast and with northwest-southeast strike. This portion is overlain by a sandstone (8,347–8,375 ft or 2,544–2,552 m) deposited in a low-energy environment; it exhibits reduced permeability and increased clay percentage. This body is capped by a distributary-front sandstone. The channel sandstone, in common with many channel sandstones, has high permeability and a low clay percentage.

By the time the "B" zone was deposited, a major change in sediment-transport direction had occurred. Whereas sediments were transported from southwest to northeast during deposition of the "K" through "C" zones, the "B" sands were transported from north to south (Fig. 10).

During deposition of the "A" sands, sediments came from the north via channels parallel with the east-bounding fault in the East Cameron Block 270 field (Fig. 11). The lowermost "A" sands were deposited in the northern half of the block. As the distributary system continued to prograde southward, the middle part of the "A" zone was deposited in the south half of the block. The upper "A" sands were deposited at the southern edge of the block.

The magnetic field tapes containing the other open-hole logs were computer-processed using programs such as "true vertical depth" (TVD) and SARABAND. These computations converted conventional logging measurements into more readily usable forms for evaluation.

True-vertical-depth logs (Fig. 12) were computed using information from a directional survey or a dipmeter. These logs were used to facilitate correlations, mapping, and sand counts in deviated holes.

The computed-log results permitted hydrocarbon detection and provided (1) porosity values corrected for hydrocarbon and shale effects, (2) water-saturation values, and (3) type of hydrocarbons. These outputs, along with the permeability index, silt content, and clay content of sandstones, provide a means of judging sandstone quality. They were used together with dipmeter results to improve stratigraphic interpretations.

The computed logs were converted to true vertical depth (Fig. 13) to facilitate reservoir and mapping studies. The TVD computed-log outputs were used to generate contour maps of porosity, feet of hydrocarbon saturation, permeability, and siltstone volume (Figs. 14, 15, 17, 18).

The "porosity-feet" map for the "G" sandstone (Fig. 14) represents the reservoir volume above the gas-water contact. The contours represent feet of 100-percent porosity and are obtained from the computed log. As expected, the large values occur near the large fault where the sandstones are thickest.

The "hydrocarbon-feet" map (Fig. 15) for the "G" sandstone was derived from the computed-log values of integrated hydrocarbon-feet. The "integrated hydrocarbon-feet" input is the summation of porosity times hydrocarbon-saturation values computed at 6-in. (15 cm) intervals. The map contours represent feet of 100-percent hydrocarbon saturation. Values range from zero at the gas-water contact to a maximum of 18 ft (5.5 m) near the east-bounding fault. The hydrocarbon maximums and minimums are displaced from those on the conventional isopach map (Fig. 16). This shift is the result of porosity changes. Both maps should be considered when choosing drillsites or determining drainage points. No po-

(Text continued on page 226.)

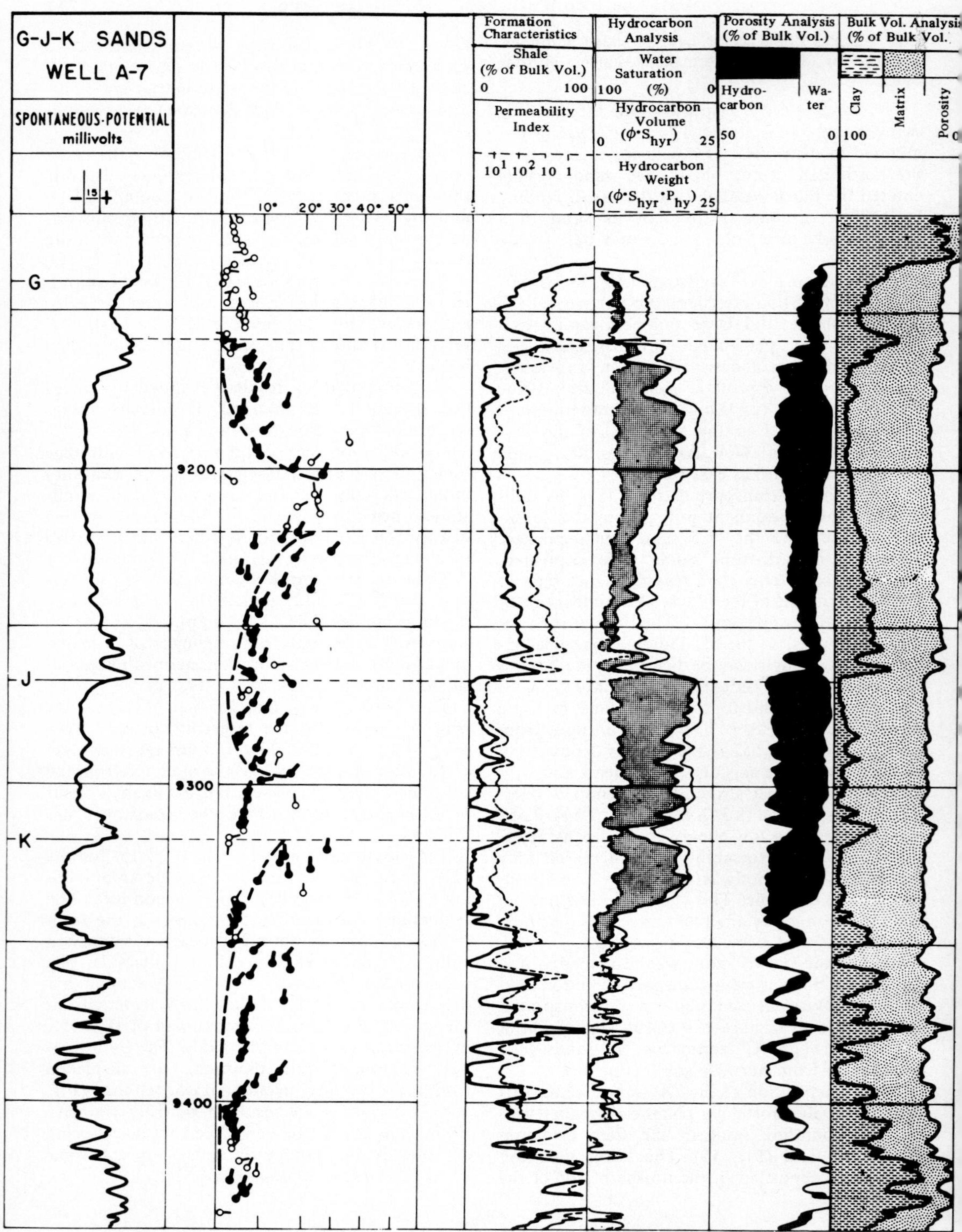

FIG. 5—Comparison of SP, dipmeter, and computer-processed (SARABAND) logs run in "G" through "K" sandstones, Well A-7.

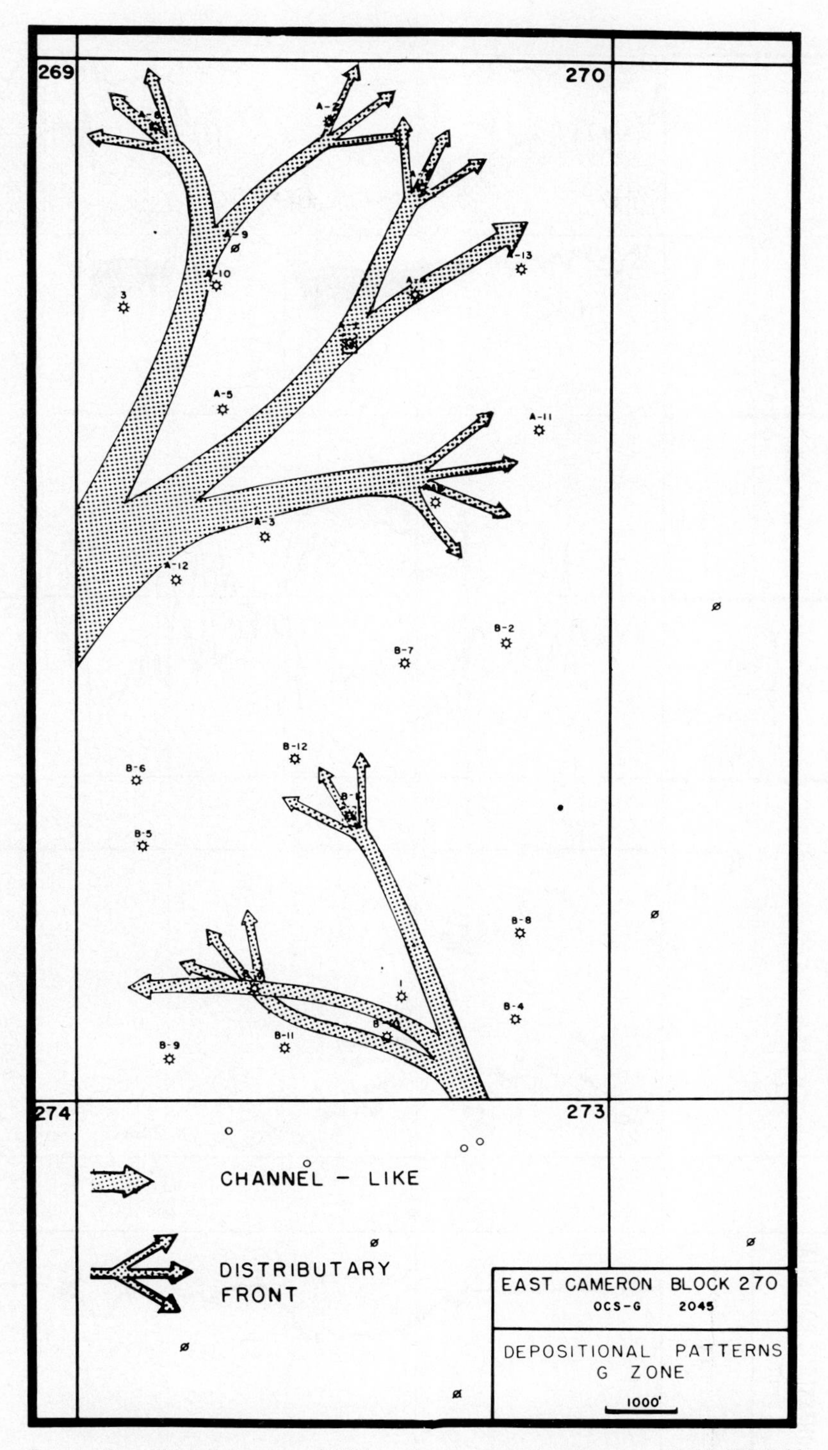

FIG. 6—Depositional patterns of "G" sandstone zone, East Cameron Block 270 field.

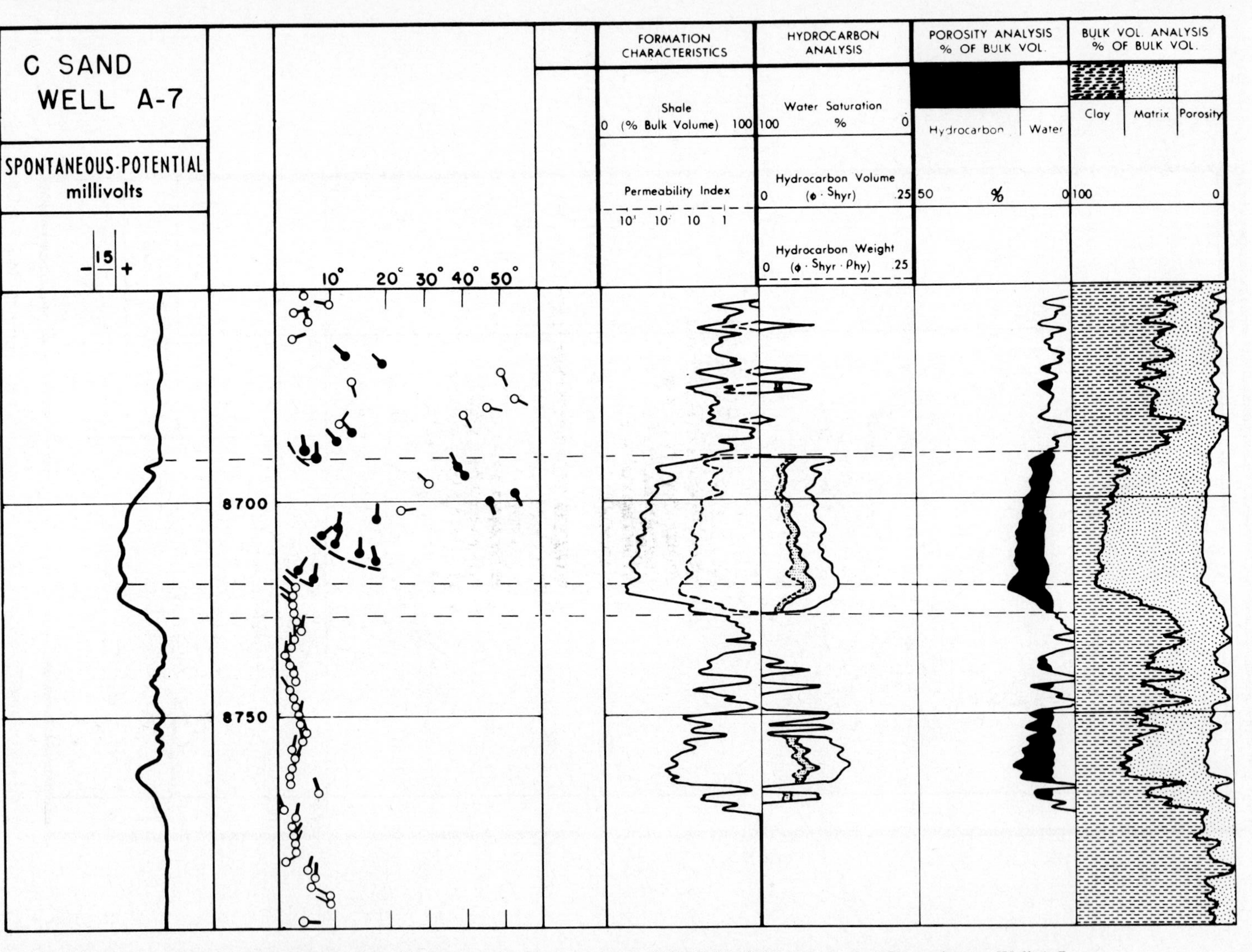

FIG. 7—Comparison of SP, dipmeter, and computer-processed (SARABAND) logs run in "C" sandstone, Well A-7.

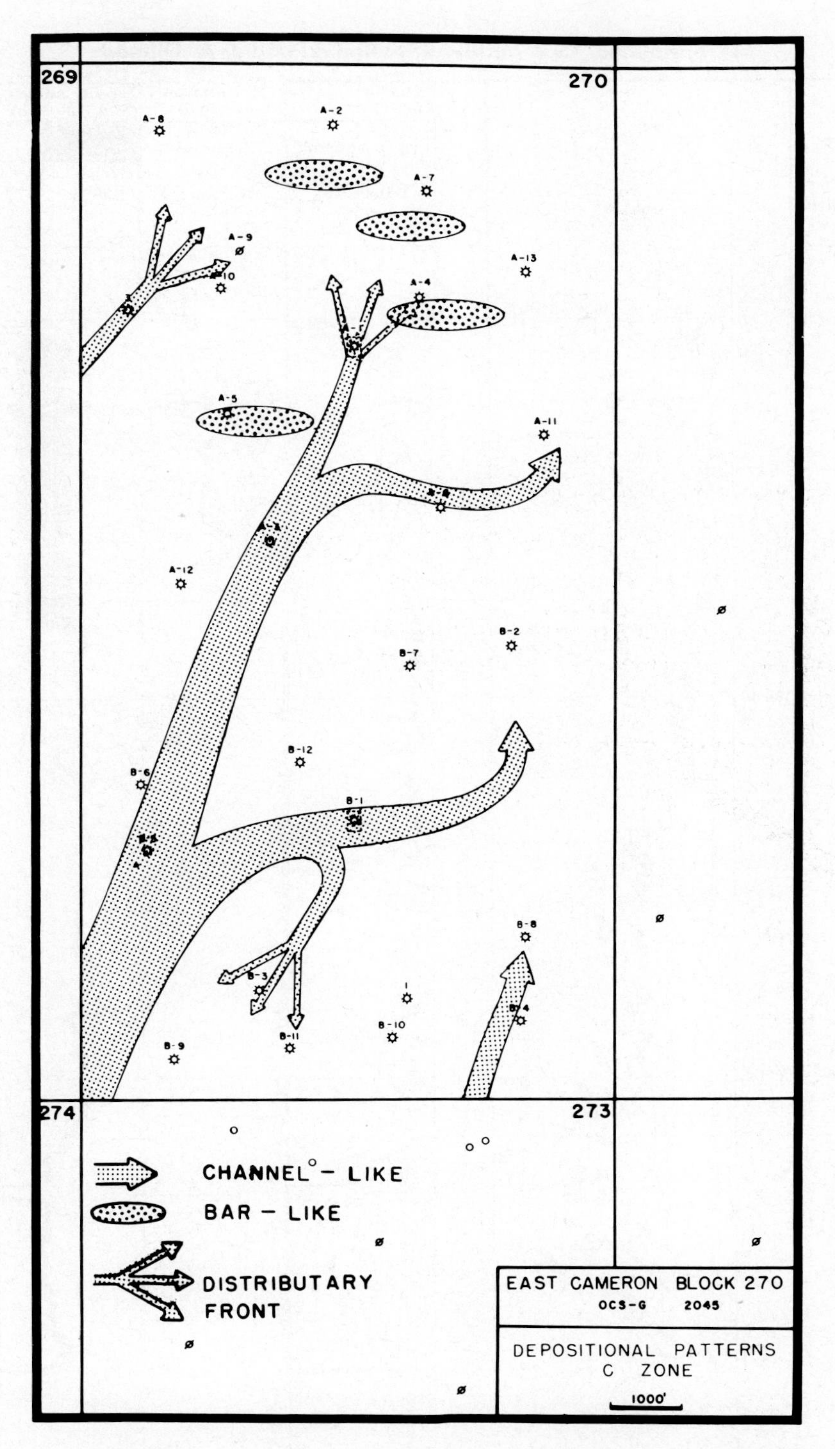

FIG. 8—Depositional patterns of "C" sandstone zone, East Cameron Block 270 field.

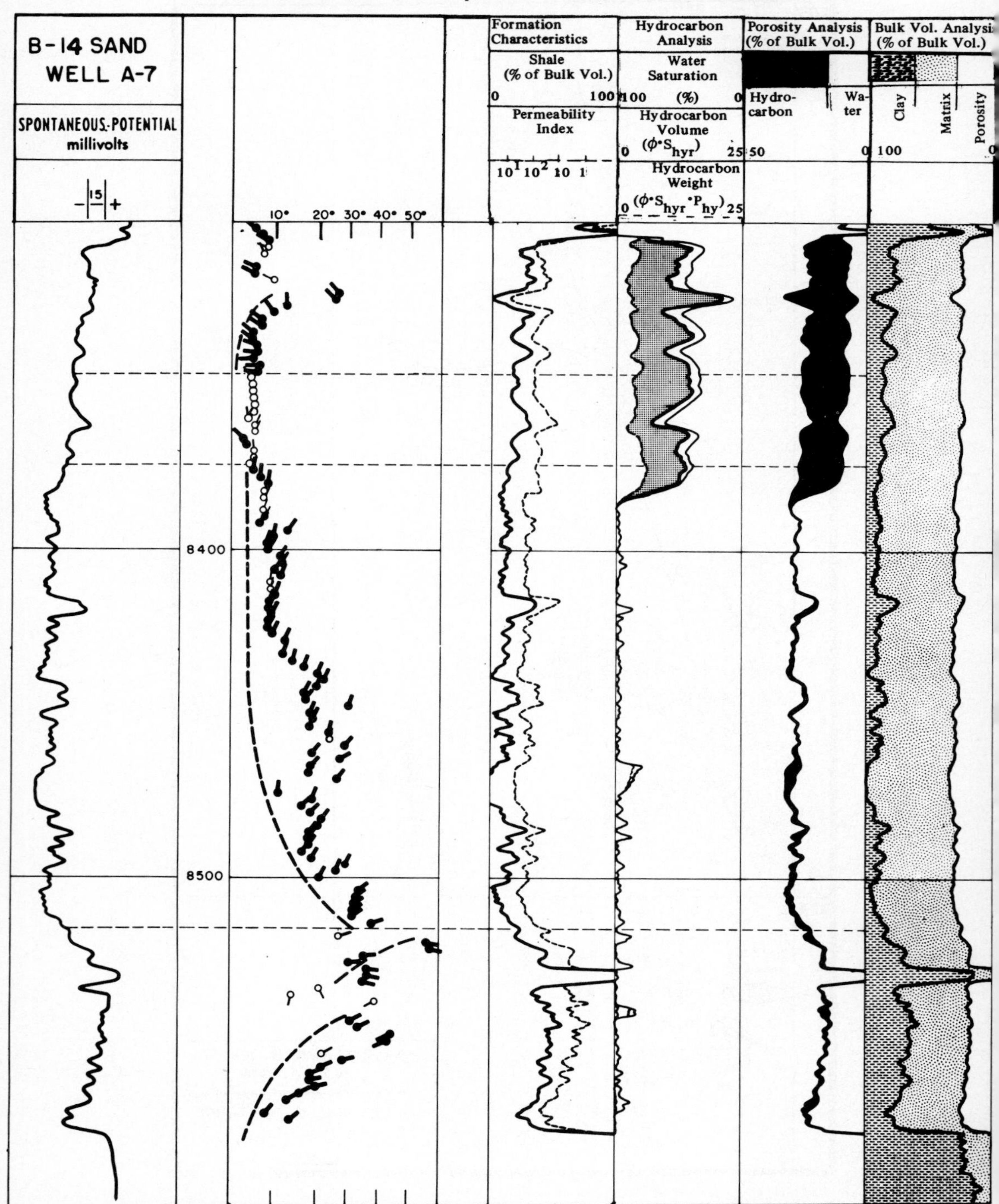

FIG. 9—Comparison of SP, dipmeter, and computer-processed logs run in "B-14" sandstone, Well A-7.

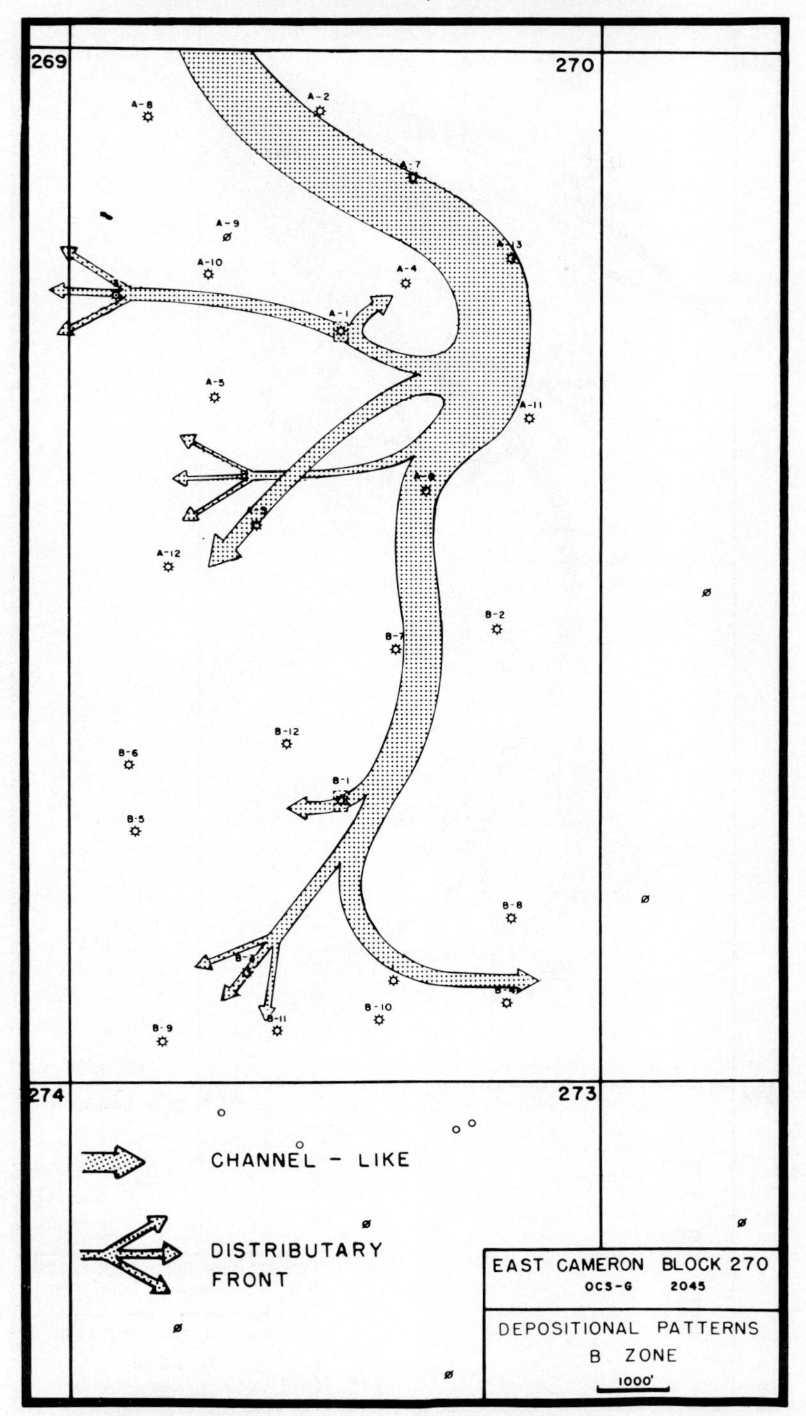

FIG. 10—Depositional patterns of "B" sandstone zone, East Cameron Block 270 field.

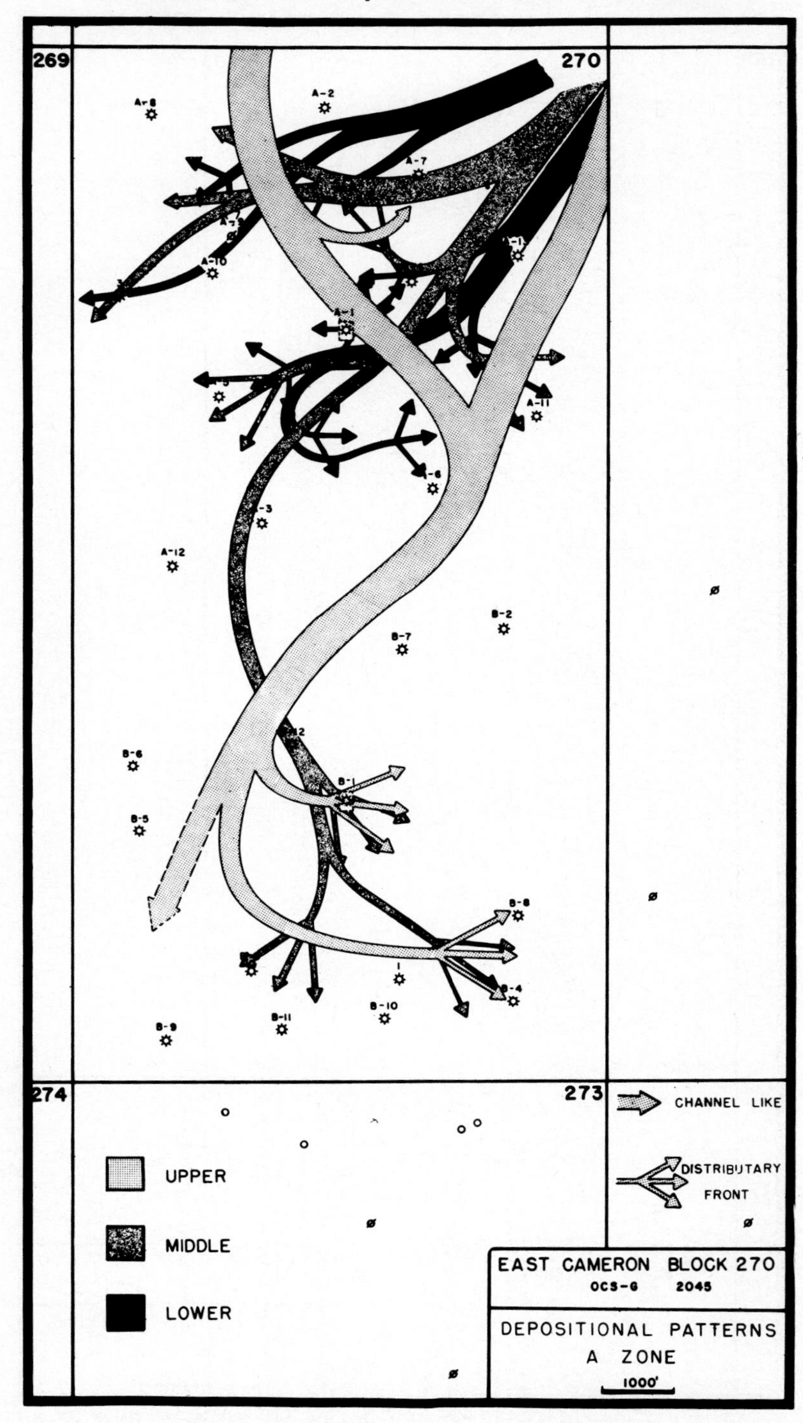

FIG. 11—Depositional patterns of "A" sandstone zone, East Cameron Block 270 field.

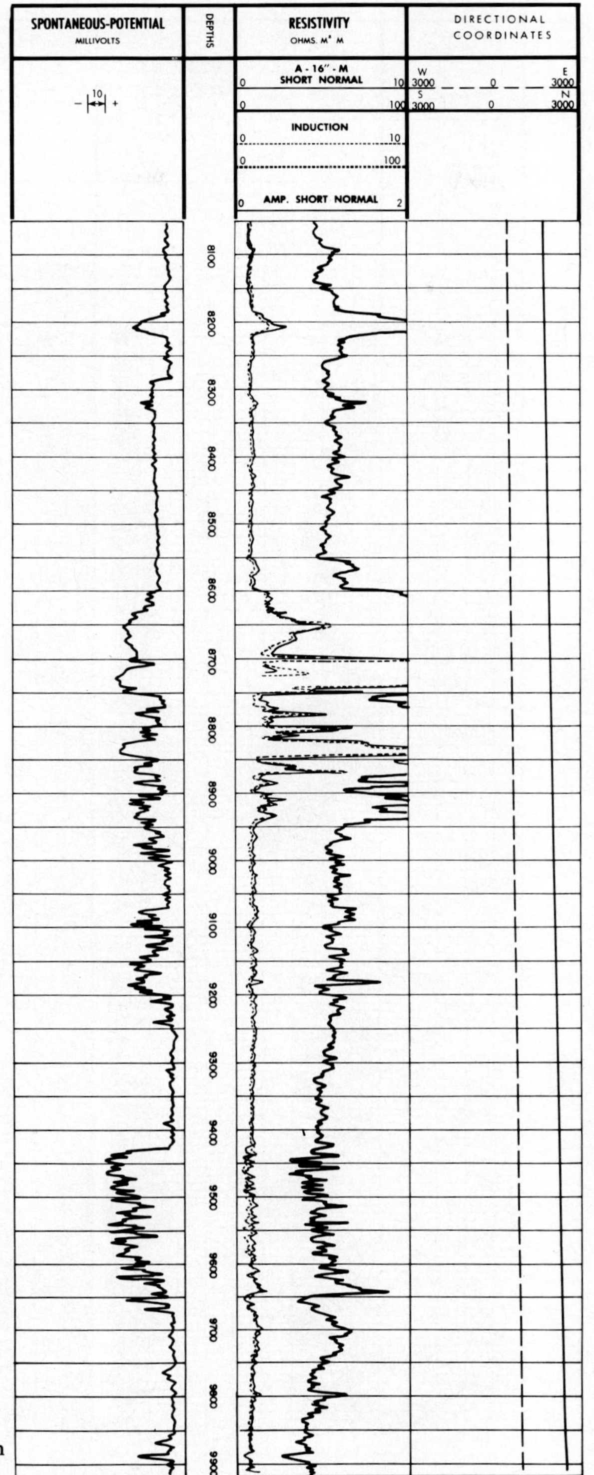

FIG. 12—True-vertical-depth resistivity log, Well B-7.

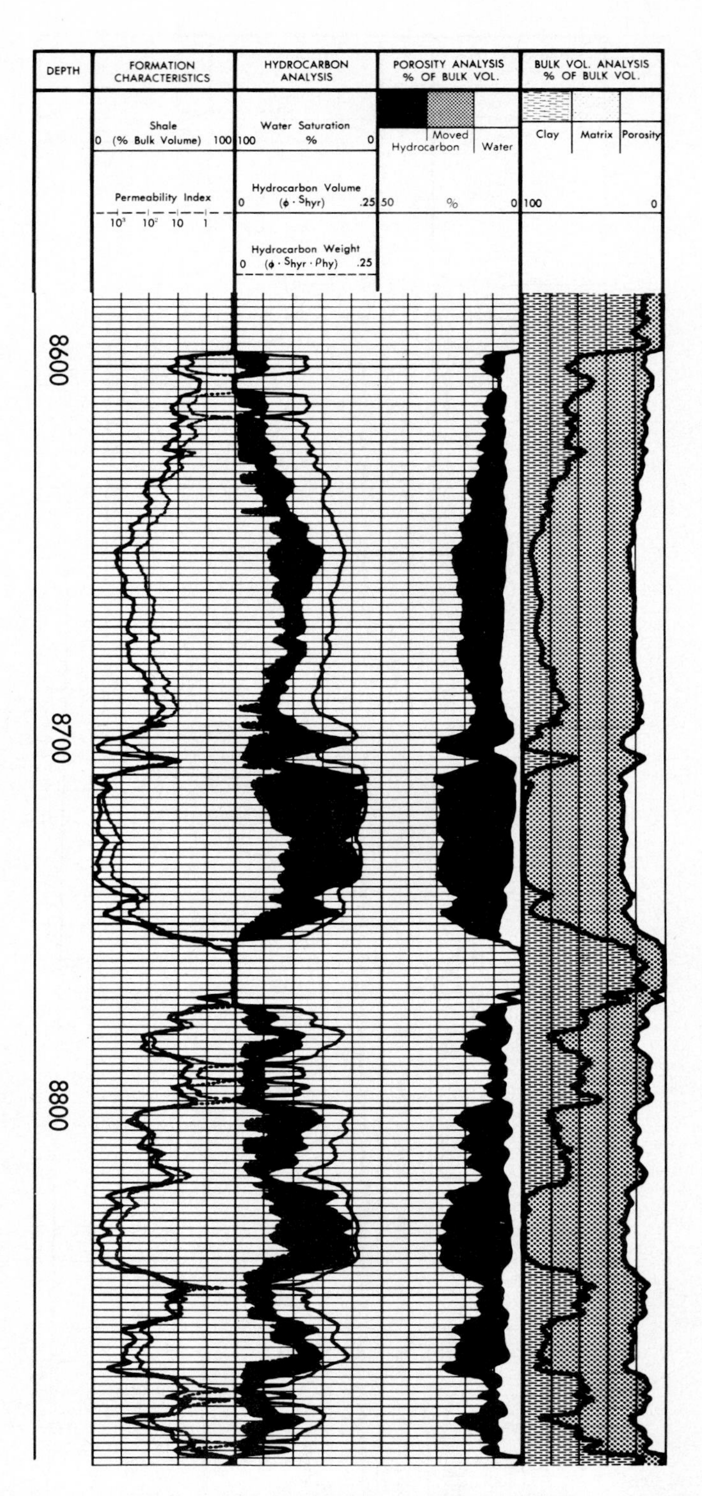

FIG. 13—True-vertical-depth computer-processed (SARABAND) log, Well A-7.

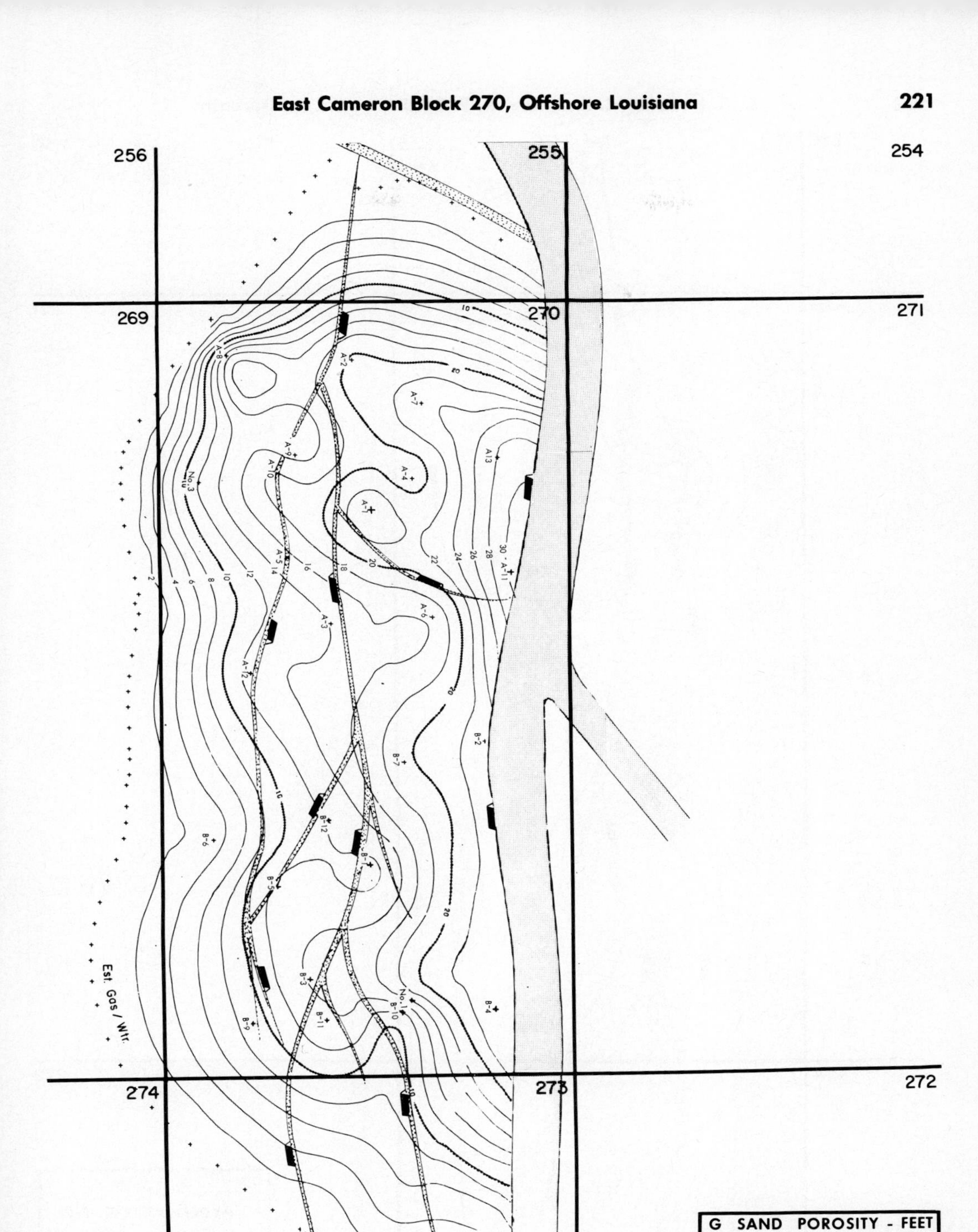

FIG. 14—"Porosity-feet" map of "G" sandstone, East Cameron Block 270 field.

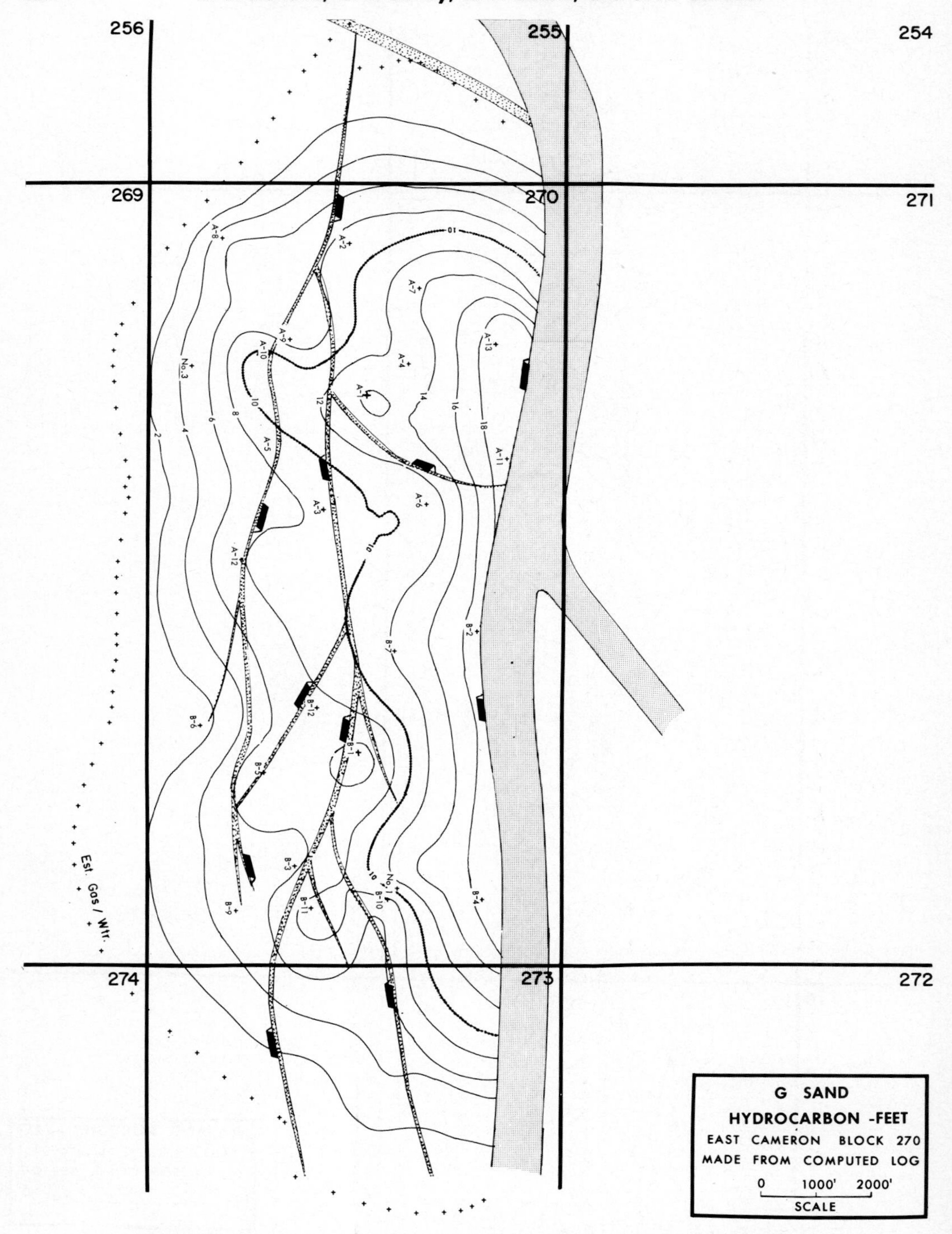

FIG. 15—"Hydrocarbon-feet" map of "G" sandstone, East Cameron Block 270 field.

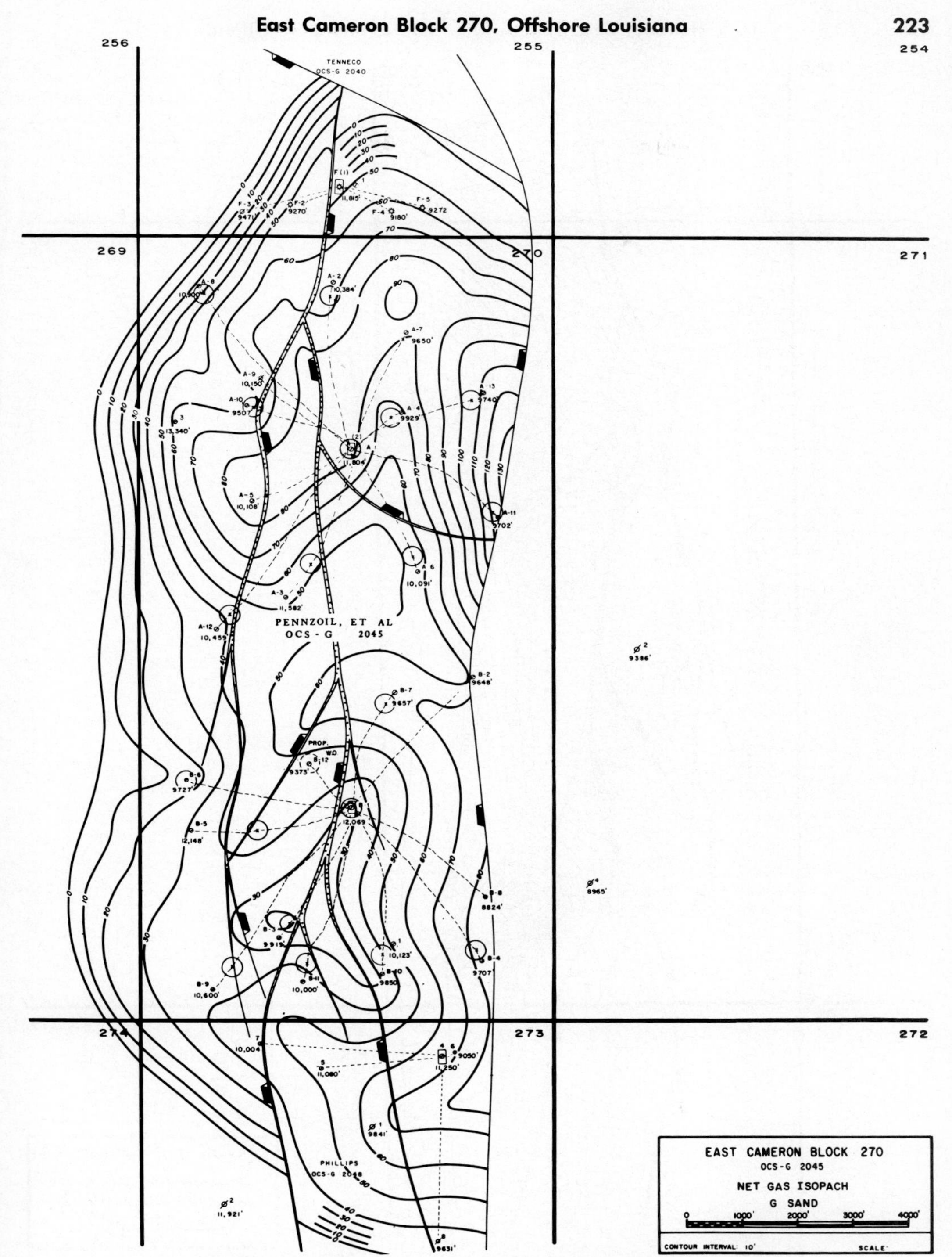

FIG. 16—Net-gas isopach map of "G" sandstone, East Cameron Block 270 field.

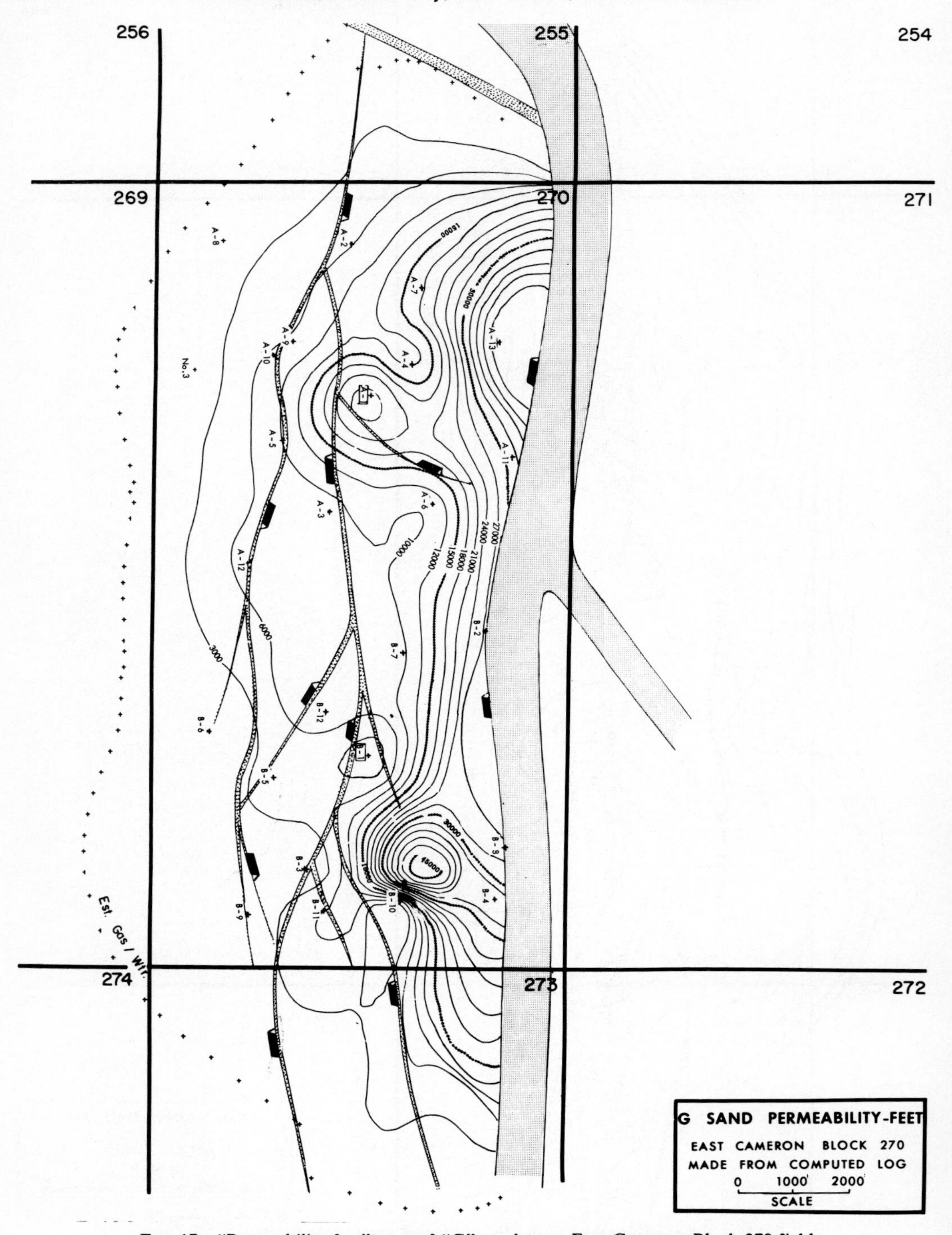

FIG. 17—"Permeability-feet" map of "G" sandstone, East Cameron Block 270 field.

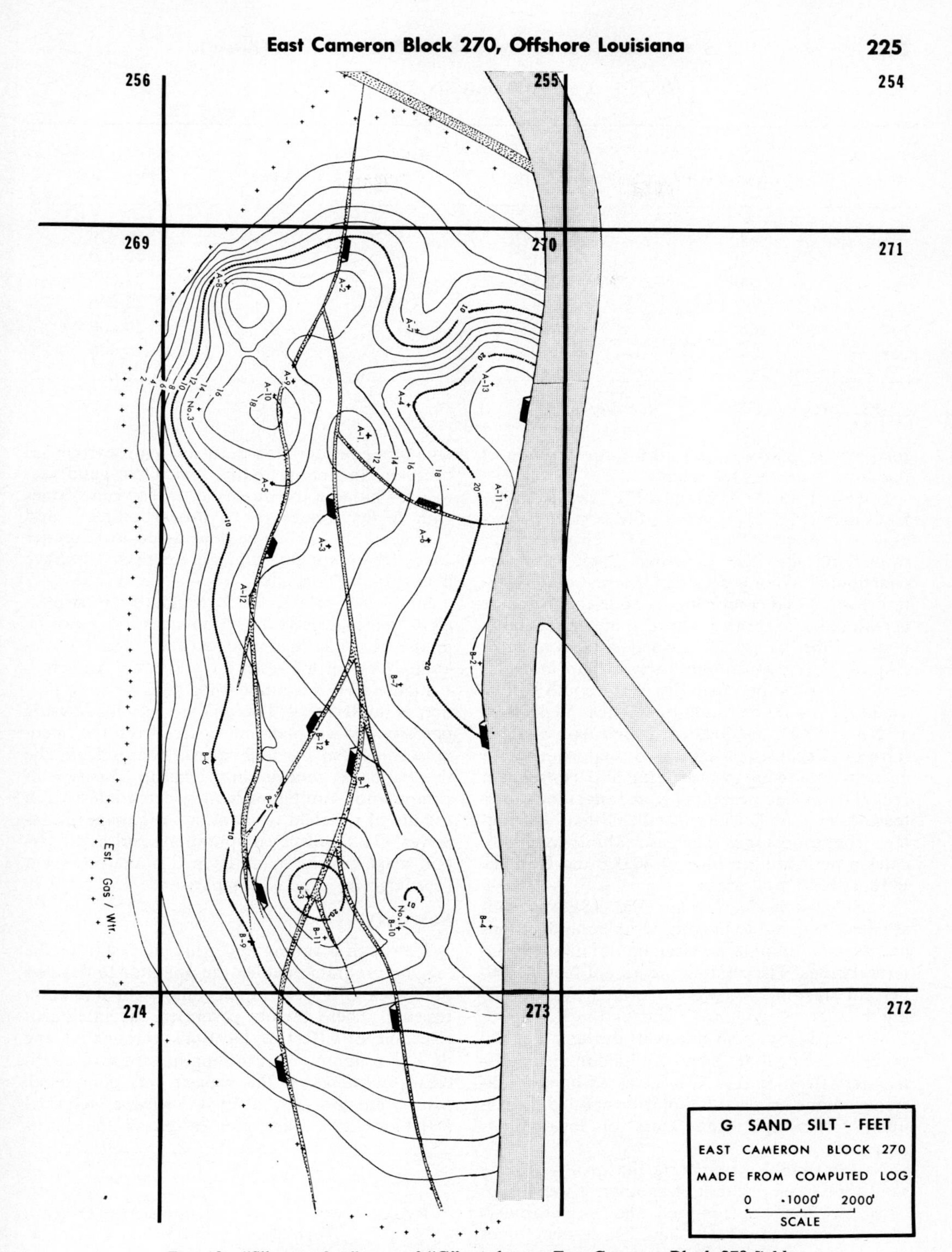

FIG. 18—"Siltstone-feet" map of "G" sandstone, East Cameron Block 270 field.

Table 1. Comparison of Six A-Platform Wells

Well	*Hydrocarbon-Ft*	*Permeability-Ft*	*Type Comp.**	*Zones*	*Prod. Rate (Mcf/Day)*
A-1	21.33	99,714	S	G, J, K	20,000
A-3	20.56	111,514	S-GP	G, J, K	22,000
A-6	24.40	48,866	S	G, J, K	22,000
A-13	25.4	72,442	S-GP	G, J	22,000
A-2	8.8	5,858	S	G	10,000
A-8	4.9	1,520	S	G	8,000

*S = single completion, GP = gravel pack.

rosity-value limit was used in the construction of the "hydrocarbon-feet" map.

Contours on the computed-log "permeability-feet" map (Fig. 17) represent the sum of permeabilities computed at 6-in. (15 cm) intervals throughout the "G" sandstone. This map was constructed as an aid to depositional-pattern interpretation and completion-zone selection. There is reasonable agreement between the permeability map and the "G" sandstone depositional patterns (Fig. 6). The depositional-pattern map indicates that the highest permeability in the south half of the block should be found in and near Wells B-10 and B-1, which are located in the main channel fairway. Because of increased opportunity for current-winnowing of sediment and consequent reduction in clay percentages, channel sandstones tend to have higher permeabilities than distributary-front sandstones. The permeability map indicates a permeability high of 30,000 md-ft in the B-10 and B-1 well areas.

The "siltstone-feet" map (Fig. 18) was constructed as an aid to making completion decisions and as an additional confirmation of dipmeter interpretations. The contours represent feet of 100-percent siltstone and were obtained by integrating the siltstone volume (V siltstone = V shale − V clay) at 6-in. (15 cm) intervals through the "G" sandstone. Siltstone is most abundant near distributary-front fringes. The areas of highest siltstone volume agree with dipmeter-defined distributary fronts and with areas of lowest permeability.

As a means of comparing the quality of one sandstone unit with that of another, it was decided to use "hydrocarbon-feet" and "permeability-feet" from the computed logs. These two values take into consideration porosity, hydrocarbon saturation, and permeability, which are good factors for judging the quality of a reservoir. Zones with higher values of "hydrocarbon-feet" and "permeability-feet" were considered to be better zones for completion and were expected to have better production rates.

Study of Table 1, which compares six A-platform wells, suggests that production rates can be predicted using "hydrocarbon-feet" and "permeability-feet," if the reservoir pressure is known.

Figure 19 is a computed-log analysis for a portion of the B-5 well. This analysis was made using porosity and clay-volume figures from the open-hole computed log and measurements from the thermal-decay tool run in the casing. This type of computation can be made at selected intervals in the life of the field to monitor the changes in reserves. Gas-in-place computations and hydrocarbon maps can be made from this analysis for a "cardiogram" type of comparison.

SUMMARY

Thorough planning and prudent design of the logging program enabled the operator to develop the Block 270 field rapidly and with minimum expense. These factors permitted the intelligent selection of effective locations throughout the drilling program, contributing in turn to the efficiency with which the project was completed. Moreover, the wealth of knowledge acquired from logs and other sources during the devel-

→

FIG. 19—Computed-log analysis (Thermal Decay Tool) of part of B-5 well.

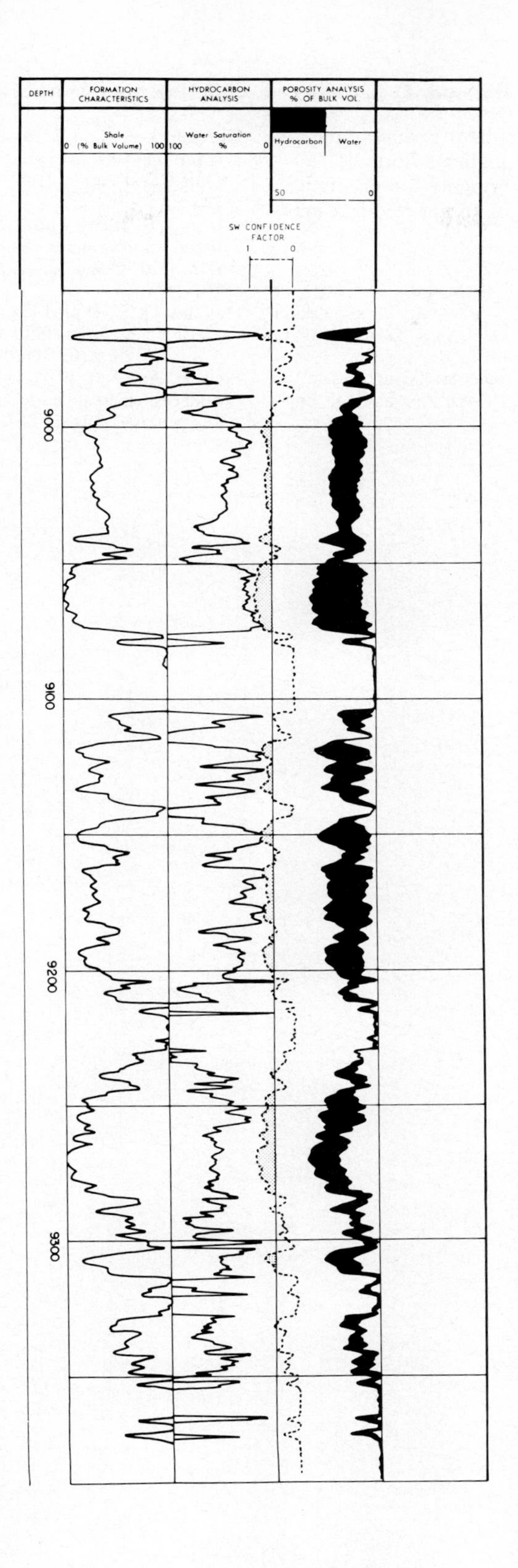
DEPTH
FORMATION CHARACTERISTICS
HYDROCARBON ANALYSIS
POROSITY ANALYSIS % OF BULK VOL
Shale
0 (% Bulk Volume) 100
Water Saturation
100 % 0
Hydrocarbon
Water
50 0
SW CONFIDENCE FACTOR
1 0
9000
9100
9200
9300

opment is available in perpetuity. Such data can contribute greatly to an understanding of reservoir characteristics, thus bringing top production within reach and improving future resource management.

Selected References

Bonham-Carter, G. F., 1967, Diffusion and settling of sediments at river mouths—a computer simulation model: Gulf Coast Assoc. Geol. Socs. Trans., v. 17, p. 326-338.

Briggs, G., 1966, Primary sedimentary structures in the search for petroleum: Gulf Coast Assoc. Geol. Socs. Trans., v. 16, p. 297.

Gilreath, J. A., and R. W. Stephens, 1971, Distributary front deposits interpreted from dipmeter patterns: Gulf Coast Assoc. Geol. Socs. Trans., v. 21, p. 233-243.

——— J. S. Healy, and J. N. Yelverton, 1969, Depositional environments defined by dipmeter interpretation: Gulf Coast Assoc. Geol. Socs. Trans., v. 19, p. 101-111.

Holland, D. S., et al, 1974, East Cameron Block 270, a Pleistocene field: Gulf Coast Assoc. Geol. Socs. Trans., v. 24, p. 89-106.

Poupon, A., et al, 1970, Log analysis of sand-shale sequences—a systematic approach: Jour. Petroleum Technology, July, p. 867-881.

Grand Isle Block 16 Field, Offshore Louisiana[1]

RICHARD J. STEINER[2]

Abstract Grand Isle Block 16 field is a large offshore oil field with an estimated 277 million bbl of ultimately recoverable oil. The field, located in the Gulf of Mexico 7 mi (11 km) offshore from Grand Isle, Louisiana, was discovered in 1948. Oil is trapped in 26 poorly consolidated upper Miocene sandstones located in peripheral fault blocks on the flanks of a shallow piercement salt dome. The Block 16 salt diapir is circular in plan, and has a maximum diameter of 4.75 mi (7.6 km) and a shallow crest at −1,700 ft (518 m). The field is on one of several large, oil-productive piercements located offshore along a deep-seated, regional salt-ridge trend oriented east-west. The salt-ridge trend and a related regional salt-withdrawal syncline to the north cause a reversal from normal south dip to steep north dip across the field. Deposition of the producing section, which thins toward the salt, was contemporaneous with domal growth. The major oil accumulation is on the southeast flank at depths from −6,500 to −11,000 ft (−1,980 to −3,350 m). Wells typically flow 500–800 BOPD. All wells are directionally drilled from a limited number of platforms in 40–60 ft (12–18 m) of water. As a result, not all well-bore penetrations are ideal for 100-percent primary recovery. Therefore, careful planning and secondary-recovery techniques have been utilized for optimum oil recovery and well-bore utility.

INTRODUCTION

The Grand Isle Block 16 oil field is in the Gulf of Mexico approximately 60 mi (96 km) south of New Orleans and 7 mi (11 km) offshore from Grand Isle, Jefferson Parish, Louisiana (Fig. 1). The field lies under Blocks 16, 17, 21, 22, 23, 29, and 30 of the Grand Isle area, in water depths deepening toward the south from 40 to 60 ft (12 to 18 m).

Oil is produced from upper Miocene strata on the flanks of a large, shallow, piercement salt dome. The structure is generally comparable with other piercements found in the Texas-Louisiana Gulf Coast Salt Dome basin. Significant aspects of the Block 16 field lie in its offshore location, reserve size, areal extent, regional geologic setting, and growth history.

DEVELOPMENT HISTORY

Exxon Company, U.S.A. (Humble Oil & Refining Company), acquired leases on 20,000 acres (80.8 km^2) in Grand Isle area Blocks 16, 17, 22, and 23, in September 1946, based on the results of gravity and reflection seismograph surveys. The discovery well, State Lease 799 No. A-1 sidetrack, located in the south-central part of Block 16, was completed in November 1948 and flowed 219 BOPD from a depth of 4,590 ft (1,399 m). Future development revealed a giant oil field with an estimated ultimate recovery of 277 million bbl of oil.

Six years after completion of the discovery well, a total of 16 wells had been drilled and six platforms installed. Two of these platforms were removed later owing to lack of production. Through 1954, drilling had resulted in four shut-in gas wells and two oil wells. The established oil accumulations were limited to the discovery well and a small reservoir in a supra-cap sandstone on the crest of the dome. Minor gas production had been established on the east flank near the C platform and on the northwest flank near the F and G platforms (Fig. 5).

Perseverance was rewarded in July 1955 when the 18th well on the structure found the first major oil reserves on the south flank of the dome. Acquisition of the north half of Block 29 added 2,500 acres (10.1 km^2) to the lease block in February 1957. Subsequent drilling defined multiple oil zones in peripheral fault blocks around the south half of the salt dome. A total of 74 wells had been drilled by the end of 1960. In the 1966 drainage sale, new leases on 3,972 acres (16.8 km^2) were acquired in Blocks 21 and 30 on the southeast flank of the field. Continued development drilling extended production to the southeast onto Blocks 21 and 30. By 1971, the field was developed with a total of 23 production platforms and 364 wells. The very active drilling program from 1960 to 1971, along with installation of associated pro-

[1]Manuscript received, December 6, 1974.

An earlier paper on Grand Isle Block 16 field was published by the New Orleans and Lafayette Geological Societies (Steiner, 1973).

[2]Exxon Company, U.S.A., Production Department Southeastern Division, New Orleans, Louisiana 70161.

The writer thanks Exxon Company, U.S.A., for permission to publish this paper and for providing basic information, facilities, and services to prepare it. J. R. Lively, L. W. Good, M. J. Vaughan, and G. R. Stude are gratefully acknowledged for their constructive criticism during the preparation of this paper. Typing, including several rough drafts, was done by Cynthia Sylvain. Assistance in preparing the illustrations was provided by J. J. LeBlanc and W. D. Currier.

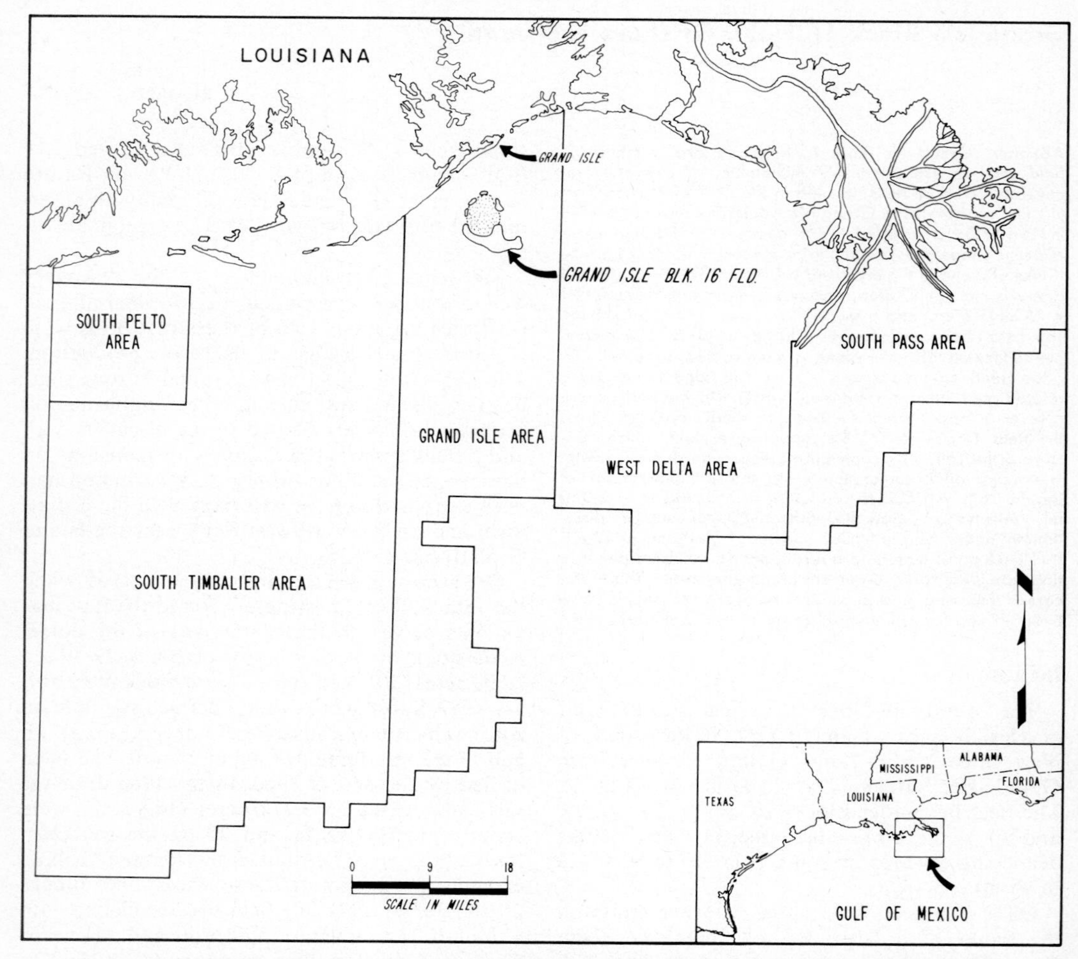

FIG. 1—Index map of a portion of offshore Louisiana, showing location of Grand Isle Block 16 field.

duction platforms and flowlines, provided a steady buildup of annual production. The results of development of the field are shown on the graph of annual and cumulative production in Figure 2.

Regional Setting

The Grand Isle Block 16 salt dome lies on the trend of the Bay Marchand–Timbalier Bay–Caillou Island salt-dome complex. Frey and Grimes (1968) described the complex and concluded (p. 290):

> The Bay Marchand–Timbalier Bay–Caillou Island salt domes are diapiric intrusions or spines on an east-west oriented ancestral salt ridge. The ridge extends approximately 28 mi and encompasses more than 140 mi^2; gravity and subsurface data suggest that the ridge continues eastward to Grand Isle 16 and 18 domes, and westward to the Lake Pelto dome. The salt ridge may be more than 60 mi long.

Figure 3 shows the Bay Marchand salt-ridge trend and, to the north, a related parallel, east-west-oriented, salt-withdrawal syncline called the "Terrebonne trough." The trough trends northeast of the Grand Isle Block 16 and Grand Isle Block 18 domes and then continues easterly to the north flank of West Delta Block 30 salt dome.

A large, buried, down-to-north growth fault forms the south boundary of the Terrebonne syncline in the area between the Grand Isle Block 16

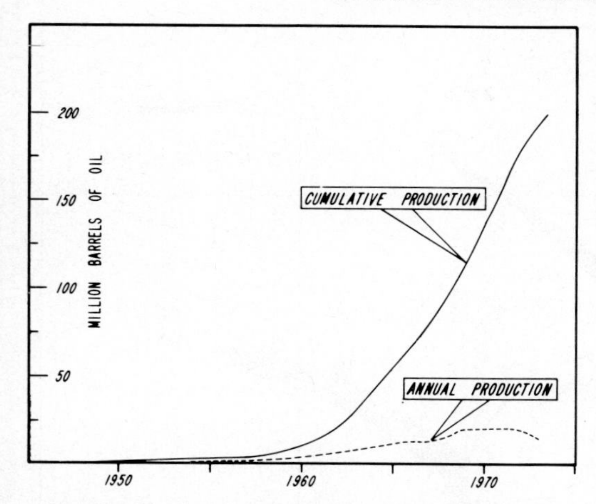

FIG. 2—Graph showing annual and cumulative oil production from Grand Isle Block 16 field.

and Grand Isle Block 18 piercements. This large growth fault influenced sedimentation at the Grand Isle Block 16 field; it effectively bisects the structure into north and south halves. The fault continues east and west as a buried, older regional system on the south limb of the Terrebonne trough and along the salt-ridge trend.

The regional down-to-north growth fault plus subsidence of the Terrebonne trough creates a large reversal from regional dip across the Block 16 field. From north to south, normal south dip reverses to steep north dip into the Terrebonne trough. The structure map of the top of the "Grand Isle Ash" reflects 5,000 ft (1,525 m) of north dip (Fig. 4). The regional dip reversal is also expressed by a northward thickening of the normally pressured sedimentary section across Block 16 field from 11,000 ft (3,350 m) to more than 20,000 ft (7,000 m) in the Terrebonne trough. Deposition contemporaneous with salt movement not only caused thickening toward the trough, but also caused several sedimentary units to become thinner toward the salt.

Domal Structure

At the 10,000-ft (3,050 m) salt contour, the salt stock measures 4.75 mi (7.64 km) north-south and 4 mi (6.4 km) east-west (Fig. 5). The crest of the salt lies at −1,700 ft (−518 m) and is centrally located. In northwest-to-southeast profile, the salt is asymmetric with the steep face toward the south and the more gently sloping face toward the north. Based on seismic refraction surveys, a very large salt overhang is present at 9,000 ft (2,740 m) on the west and southwest flanks. On the north half of the dome, a minor overhang is present between 10,000 and 12,000 ft (3,050 and 3,660 m). Caprock found above −2,000 ft (−610 m) in the central portion of the dome contains sulfur which is mined by Freeport Sulphur Company.

The pierced strata generally dip 20–30° but, near the salt contact, dip is as great as 55°. Large, radially oriented faults divide the field into several peripheral fault segments (see Fig. 4). All of the commercial oil production is obtained from these peripheral fault segments, of which the structu-

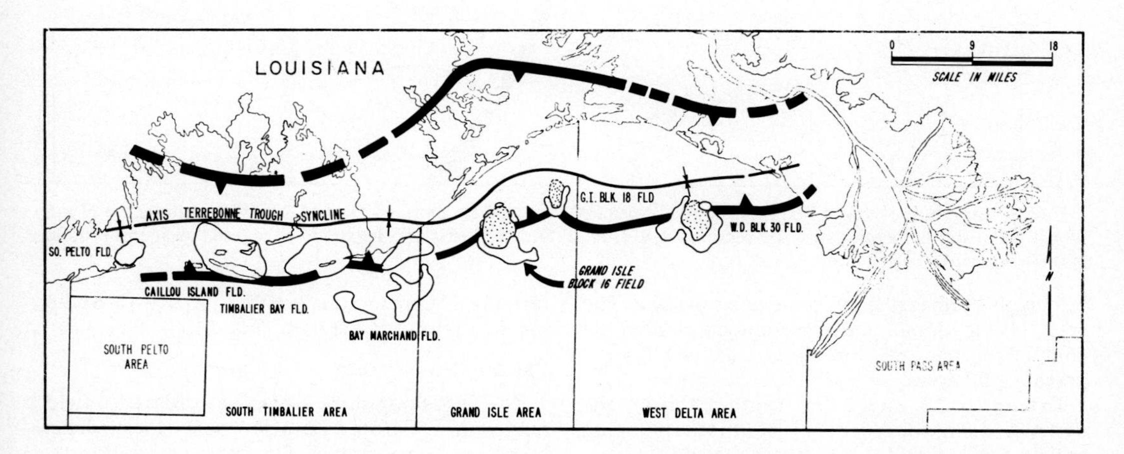

FIG. 3—Regional map showing relation of Grand Isle Block 16 field to Bay Marchand salt-ridge trend and Terrebonne trough regional salt-withdrawal syncline. Heavy lines are faults; ticks indicate downthrown side.

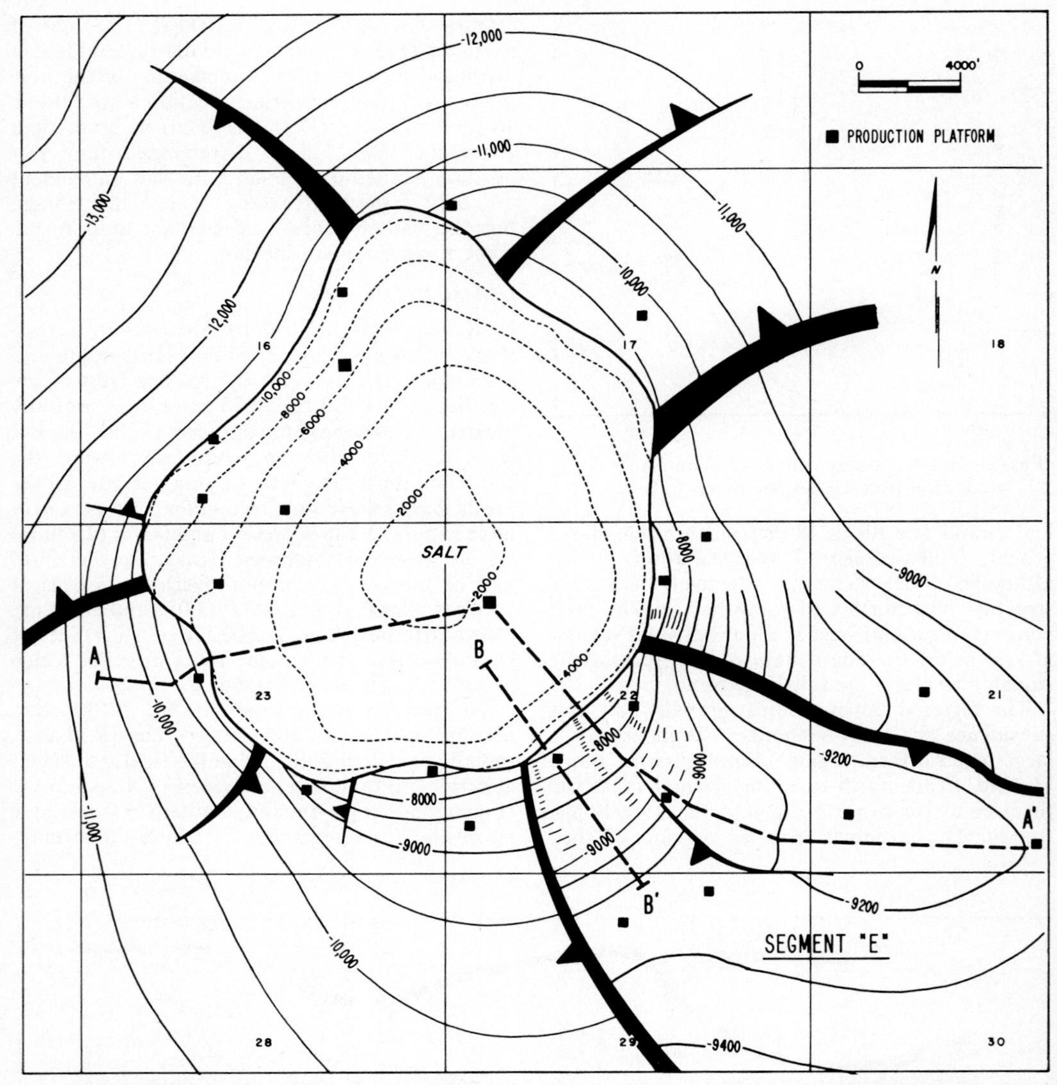

FIG. 4—Structure of top of "Grand Isle Ash" with 500- and 100-ft contours. Dashed 2,000-ft contours are of top of salt. Map shows location of cross-sections *A-A'* (Fig. 6) and *B-B'* (Fig. 8). Segment "E" at southeast end of field is fault segment.

rally high southeast flank is the most prolific. The crest of the dome had one noncommercial oil completion made at −1,700 ft (−518 m) in a supracap sandstone.

Extending 2.5 mi (4 km) southeast from the dome is a broad, low-relief, southeast-plunging anticline. Nearly all the hydrocarbons on this feature are found in stratigraphic traps with updip pinchouts toward the west (Fig. 6).

PRODUCING SECTION

The sandstone-shale sequence in Block 16 field ranges in age from late Miocene to Pleistocene. Ninety-nine percent of the Block 16 hydrocarbons are found in the upper Miocene strata, particularly in the *Bigenerina floridana, Cristellaria* K, and *Bigenerina-Discorbis* biozones (Fig. 7). For reservoir mapping, an alpha-numeric system was utilized to designate all the individual sandstone members in the producing section. Twenty-six separately designated sandstone members are productive in the various fault segments on the flanks of the dome. Nineteen of these sandstone members are productive in the flush "E" fault segment on the southeast flank of the field (Fig. 4).

Depositional environment of the productive section ranged from neritic to upper bathyal. The productive formations on the prolific southeast flank are at depths from −6,500 to −11,000 ft (−1,980 to −3,350 m) and overlie abnormally pressured deeper water units. On the northwest flank, the correlative section has been penetrated to 18,000 ft (5,485 m) and is not abnormally pres-

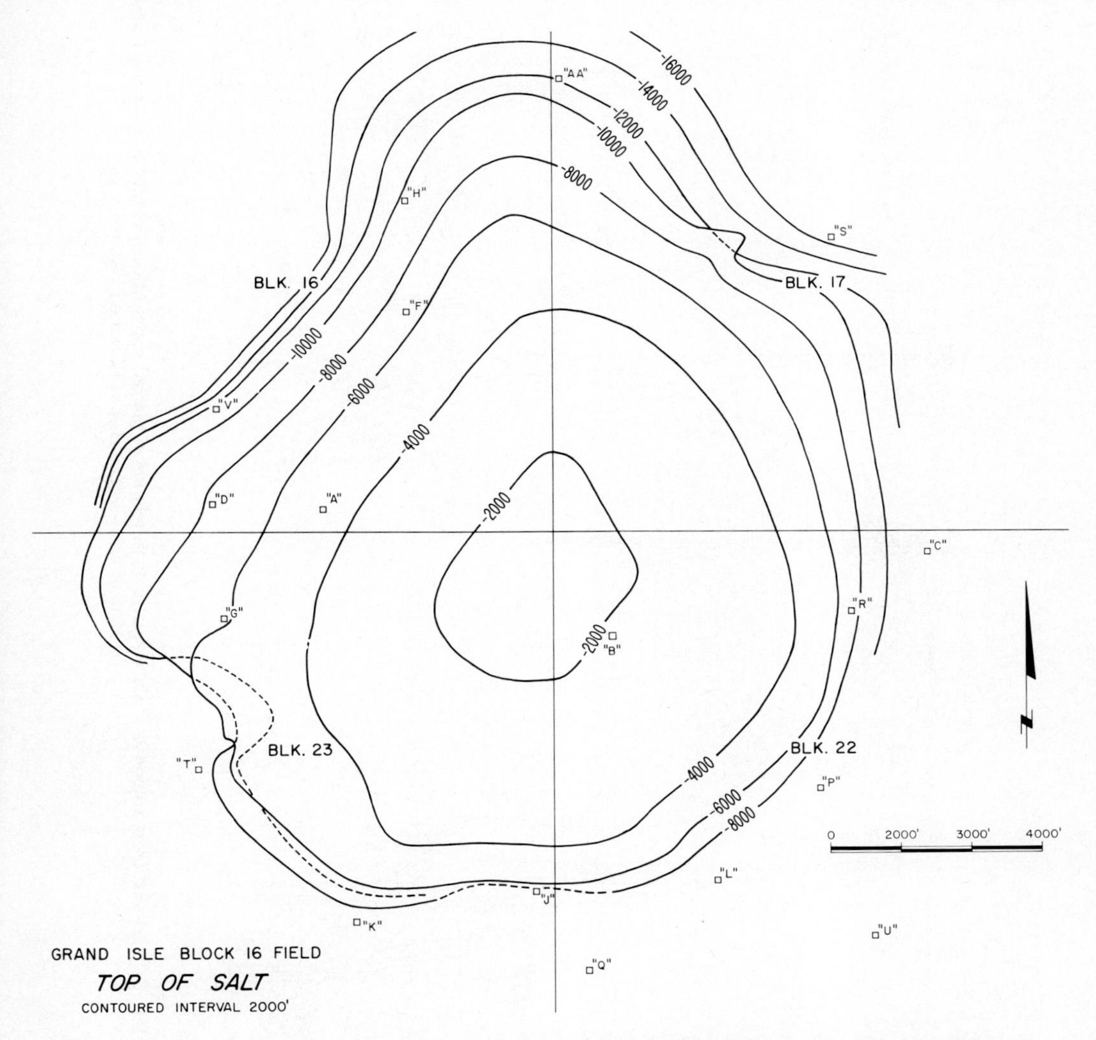

FIG. 5—Structure of top of diapiric salt. Dashed contours on southwest flank indicate salt overhang below 9,000 ft. C.I. = 2,000 ft.

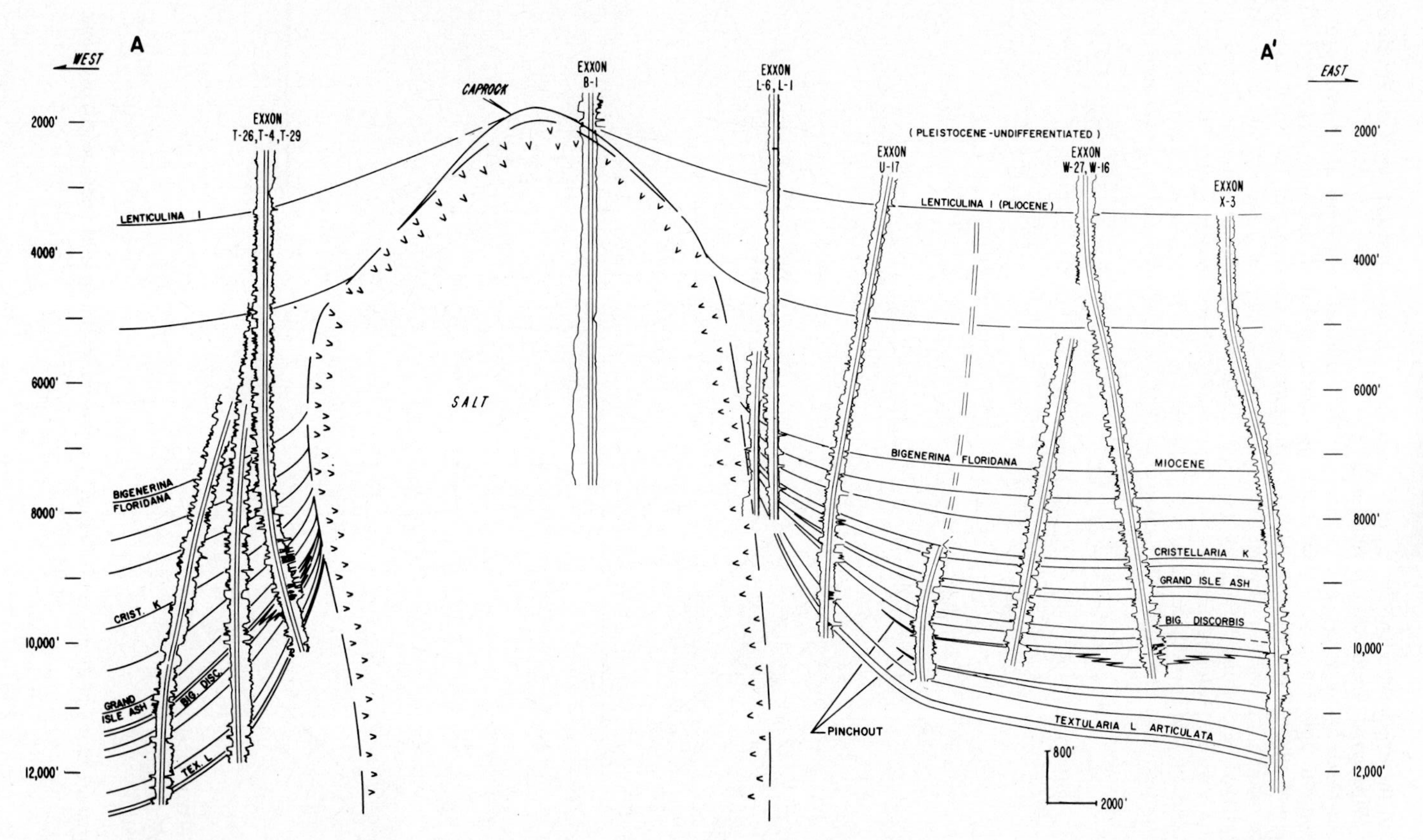

FIG. 6—East-west cross-section *A-A′* showing structural relation of pierced upper Miocene strata and low-relief, southeast-plunging nose. Line of cross-section *A-A′* is shown on Figure 4.

sured. Gas production has been established to − 15,000 ft (− 4,570 m) on the northwest flank.

Below the "Grand Isle Ash," a mass of abyssal shale forms a large concentric band 1,000 ft (305 m) wide around the south flank of the piercement. Similar shale masses, described as diapiric, were reported at Bay Marchand and Timbalier Bay fields by Frey and Grimes (1968). The shale mass may be significant with respect to accumulation of hydrocarbons inasmuch as it is associated with the most productive portion of the Grand Isle Block 16 field; however, the relations are not clear. The shale mass may represent only a gouge zone consisting of older and deeper abyssal shales dragged up by the mobile salt. The "Grand Isle Ash" appears to truncate the abyssal shale mass with an unconformable relation (Fig. 8). Several local unconformities have been identified below the "Grand Isle Ash." Very active contemporaneous growth of the piercement is reflected by the pronounced wedging of the sediments between the "Grand Isle Ash" and the older *Textularia* L unconformity (Fig. 8).

A close association of hydrocarbon accumulation with contemporaneous structural growth is demonstrated by the fact that each sandstone member in this portion of the section is productive in the structurally high "E" fault segment on the southeast flank. The largest single reservoir in the field is segment E of the C-1 sandstone; the reservoir has an estimated 36.8 million bbl of recoverable oil. This major reservoir lies below the "Grand Isle Ash" near the apex of the sedimentary wedge; thus a close association with the maximum contemporaneous structural growth is apparent.

Reservoirs

Table 1 is a generalization of the average reservoir properties of the more widespread blanket sandstones found in the peripheral segments such as fault-block "E," on the south half of the structure. A typical reservoir has 29 percent porosity, 500–800 md permeability, 20 percent water saturation, and a solution gas/oil ratio of 700,000 cu ft/bbl. Reservoir dips increase from 15 to 47° with depth. The decrease in reservoir dip to 20° below the "Grand Isle Ash" is a result of a shift in the radial position of the deeper reservoirs with respect to the salt and abyssal shale mass (Fig. 8).

The upper Miocene sandstones typically are poorly consolidated and friable. As a result, many wells become clogged with sand and cease to flow after a short period of production. Therefore,

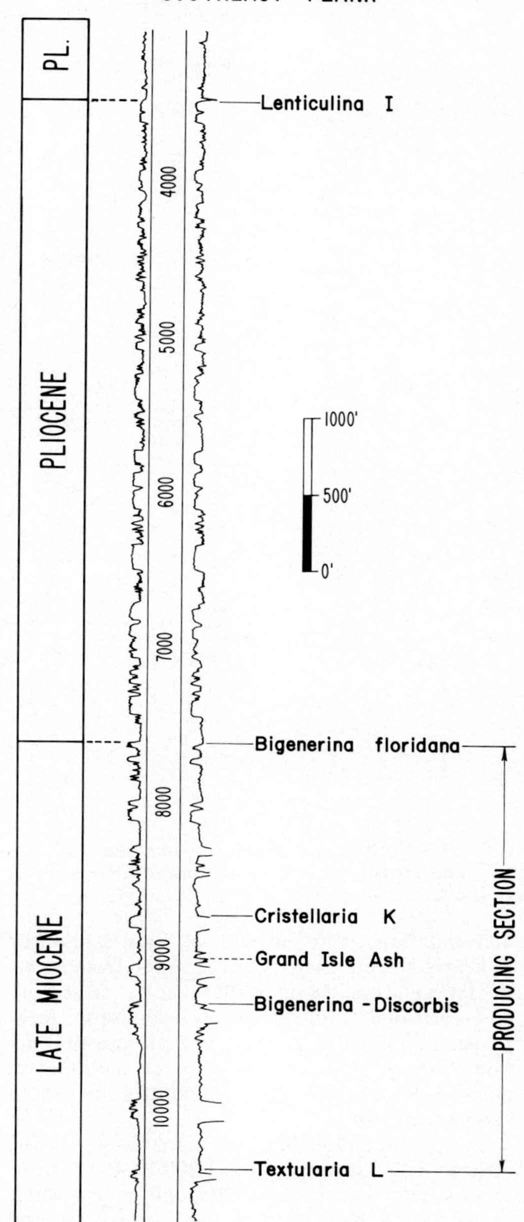

FIG. 7—Type log for southeast flank of Grand Isle Block 16 field.

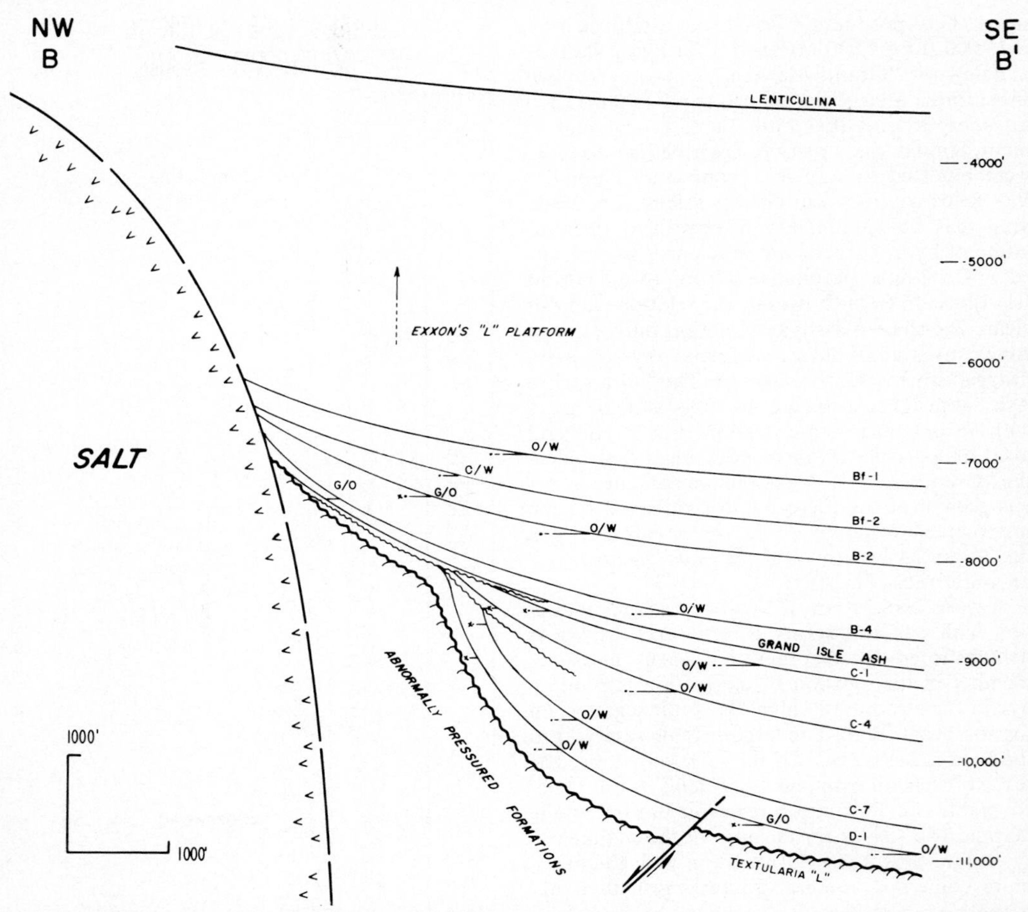

FIG. 8—Northwest-southeast cross-section *B-B'* showing structural relation of major producing sandstones in "E" fault segment on southeast flank of Grand Isle Block 16 salt dome. Line of cross-section *B-B'* is shown on Figure 4.

consolidation treatment with plastics is required in nearly all completions to keep sand from entering the well bore. Many wells must be washed out and retreated with plastic consolidation techniques periodically in order to maintain production. Where possible, gravel-pack completion techniques are also used to control unwanted sand production.

Typical flow rates of consolidated wells are 500–800 BOPD but range up to 1,200 BOPD for reservoirs with high permeability and strong water drives. Wells drilled to a single objective and completed with open-hole gravel-pack techniques have produced up to 4,000 BOPD from reservoirs with strong water drive.

Operations and Secondary Recovery

The field is developed entirely by directionally drilled wells from 23 production platforms installed in water depths ranging from 40 to 60 ft (12–18 m). The logistics of economic field development and drilling of directional wells through multiple reservoirs on the flanks of a piercement salt dome create unique geologic and production problems in Grand Isle Block 16 field.

The number of wells which can be drilled from each platform is affected by the size and location of reservoirs to be reached, in addition to platform size and load limitations. Therefore, careful planning of platform locations and efficient use

Table 1. Reservoir Properties of Major Producing Sandstones, South Half of Grand Isle Block 16 Field

Sandstone	Approx. Oil-Water Contact (ft)	Height of Oil Column (ft)	Av. Reservoir Dip (deg)	Net Sandstone (%)	Porosity (%)	Permeability (md)	Water Saturation (%)	API Oil Gravity	Solution Gas/Oil Ratio (Mcf/bbl)
Bf-1	-6900	350	15	13	29	308	27	37.5	673:1
Bf-1A	-6960	500	16	16	27	195	26	35.3	830:1
Bf-2	-7175	425	20	56	30	696	15	36.1	657:1
B-1	-7575	700	21	30	28	157	17	36.6	615:1
B-2	-7750	550	22	40	29	763	15	35.3	750:1
B-3	-8150	200	25	18	27	194	26	35.3	753:1
B-4	-8525	1150	28	51	30	716	19	34.3	815:1
"Grand Isle Ash" Marker									
C-1	-9080	580	20	36	30	542	19	32.6	841:1
C-4	-9200	600	27	50	28	873	23	31.0	822:1
C-7	-9600	1100	45	28	29	368	19	31.1	805:1
D-1	-9900	1000	47	31	28	434	21	36.8	1109:1

of well bores by making dual or triple completions is imperative. Yet, after the best planning, the number and position of well-bore penetrations in many reservoirs may not be ideal for 100-percent primary recovery of producible oil. Therefore, secondary-recovery techniques are required; such techniques have been applied successfully to a large number of reservoirs in the field.

Gas injection to produce up-structure oil is one of the most common techniques utilized for secondary recovery. In reservoirs where oil is located up-structure from the highest penetration of oil, an artificial gas cap is created by injecting gas which drives the updip oil down-structure to producing wells. The injected gas is usually recovered by updip wells after the oil is depleted. In one-well reservoirs, alternate cycles of gas injection and oil production in the single well are utilized to recover updip oil.

Water injection is utilized primarily to maintain reservoir pressure. Pressure maintenance conserves reservoir-drive energy and prevents undesirable expansion of primary and artificially created gas caps. Carefully combining both gas and water injection has successfully controlled the levels of fluid contacts in many reservoirs and resulted in the most efficient oil recovery and utilization of well bores.

Summary

The Grand Isle Block 16 field had an original accumulation of recoverable oil estimated as 277 million bbl. About 200 million bbl had been produced to the end of 1973. The shallow piercement salt dome, under 40 to 60 ft (12–18 m) of water, was discovered in 1946 during an early period of offshore exploration. Reflection seismograph data were responsible for the discovery. The initial discovery of a small oil reservoir in November 1948 was followed by 7 years of additional offshore exploration on the structure with poor economic results. Major reserves were finally established in 1955 on the south half of the dome, and development of the giant field was completed by 1971.

Block 16 is one of several large oil-productive piercements located on the regional Bay Marchand–Timbalier Bay–Caillou Island salt-ridge trend. On the north is a related, parallel salt-withdrawal syncline called the "Terrebonne trough." The salt ridge and withdrawal syncline are reflected at Block 16 field by down-to-north faulting, steep north dip, and thickening of strata from the south to north. Deposition of the producing section contemporaneous with domal growth is indicated by thinning toward the salt.

The oil is found in 26 upper Miocene sandstone

members located in several peripheral reservoir segments which are separated by radially oriented faults. The most prolific segment—fault block "E"—is on the southeast flank; oil is trapped in 19 sandstones between −6,500 and −11,000 ft (−1,980 and −3,350 m). Updip reservoir seals are located at the contact with the salt or with abnormally pressured shale, or at places where sandstone grades into shale.

The upper Miocene reservoir sandstones are typically friable and have 29 percent porosity, 500–800 md permeability, and flow rates in the range of 500–800 BOPD. Periodic sand consolidation by use of plastics is required in many wells to maintain production of oil and to control unwanted entry of sand into well bores. A few exceptional wells have produced up to 4,000 BOPD where open-hole gravel-pack techniques were used for sand control in single-zone, highly permeable reservoirs with strong water drive.

Development of the multiple reservoirs of Block 16 field from 23 production platforms required careful planning and efficient use of the limited number of available well bores. Therefore, many wells have dual and triple completions. Secondary-recovery techniques such as water and gas injection have been applied extensively to increase well-bore effectiveness and ultimate oil production from the field.

Selected References

Frey, M. G., and W. H. Grimes, 1968, Bay Marchand–Timbalier Bay–Caillou Island salt complex, Louisiana, *in* M. T. Halbouty, ed., Geology of giant petroleum fields: AAPG Mem. 14, p. 277-291.

Steiner, R. J., 1973, Grand Isle Block 16 field, *in* Offshore Louisiana oil and gas fields: New Orleans and Lafayette Geol. Socs., p. 65-70.

Tinsley Oil Field, Yazoo County, Mississippi[1]

M. F. SHELTON, JR.[2]

Abstract The first commercial oil field in Mississippi was discovered near the small town of Tinsley in Yazoo County, in 1939, and nearly 500 wells have been drilled on and around the Tinsley structure since that time

Cumulative oil production from the Upper Cretaceous Woodruff, Perry, Stevens, Lammons, McGraw, and Brumfield sandstones was 195,848,466 bbl through December 31, 1974. During October 1974, daily production from 167 wells was 6,757 bbl of oil, and 57 wells injected 50,493 bbl of water per day into five different waterflood projects. The first full-scale waterflood was begun at Tinsley field in 1968.

History

Surface indications of a structure and faulting in the Tinsley area were recognized in 1938 and reported in a Mississippi State Geological Survey Memorandum released on April 12, 1939, by William C. Morse, State Geologist. This release was based on a report made by Frederic F. Mellen, Field Geologist, who supervised the field investigations.

Seismic surveys confirmed the presence of this structure, and Tinsley field was discovered on September 1, 1939, when the Union Producing Company[3] No. 1 Woodruff was completed as an oil well in the Woodruff sandstone of Late Cretaceous age. This was the first commercial oil production in the state of Mississippi, and the discovery was the forerunner of intensive seismic, leasing, and drilling programs that resulted in numerous significant discoveries in Mississippi, such as Baxterville, Heidelberg, Eucutta, and Brookhaven, in the 1940s.

Prior to 1939, only two wells had been drilled in Yazoo County, Mississippi. A total of 474 wells has been drilled to date in the development of this field.

The first full-scale secondary-recovery project was put into operation at Tinsley field in 1968. Five waterflood projects are presently in operation. Total production of oil through 1974 was 195,848,466 bbl. Average production for the last part of 1974 was 6,700 BOPD.

Location and General Geology

The Tinsley field is located in T9–11N, R2–3W, Yazoo County, Mississippi. It is approximately 10 mi (16 km) southwest of Yazoo City and 35 mi (56 km) north of Jackson (Fig. 1).

The field lies in an area of Pleistocene loess bluffs that extend north-south along the eastern edge of the Yazoo River alluvial plain. The loess hills form a rather rugged topography in the field area; ground elevations vary as much as 100 ft (30 m) in short distances. The field structure is a faulted anticline that is elongated north-south; it is 7.5 mi (12 km) long and 3 mi (4.8 km) wide. The anticline lies near the north edge of the Mississippi Interior salt basin. Figure 1 shows part of this salt-basin area and the geologic and geographic setting of the Tinsley field. Figure 2 shows the productive limits of the field and the trace of the two major faults that divide the field into three segments.

During deposition of the Late Cretaceous Eutaw Formation and Selma Chalk,[4] this area was a shallow marine sea. The Sharkey uplift was forming to the west, a positive Paleozoic area lay to the north and northeast, and open sea and the Jackson dome were present to the south. The Sharkey uplift and the Jackson dome were areas of igneous activity during deposition of the Eutaw. Igneous sills have been penetrated in wells on the Sharkey uplift. Rhyolite with salt-filled vugs was seen in a core from the Valley Park field (Issaquena County). In the area directly north of the Jackson dome, a considerable thickness of water-laid volcanic beds has been drilled in the Eutaw interval. Volcanic materials are found in the Eutaw Formation throughout the Jackson dome–Sharkey uplift–Tinsley field area. An example is the very ashy Perry sandstone, which produces at Tinsley field.

[1]Manuscript received, May 23, 1975.

[2]Pennzoil Producing Company, Shreveport, Louisiana 71106.
I thank Pennzoil Producing Company for permission to publish this paper. F. L. Burgess and Tom Hambleton, Pennzoil Producing Company, reviewed the manuscript and made suggestions for its improvement. Pat Hudson and Theresa Willingham typed the manuscript, and Maurice Milam drafted the illustrations.

[3]This company was a predecessor of the Pennzoil Producing Company, a subsidiary of The Pennzoil Company.

[4]Although the Selma has been raised to group status (Belt et al, 1945), the older usage as a formation is retained in this paper. *Editor.*

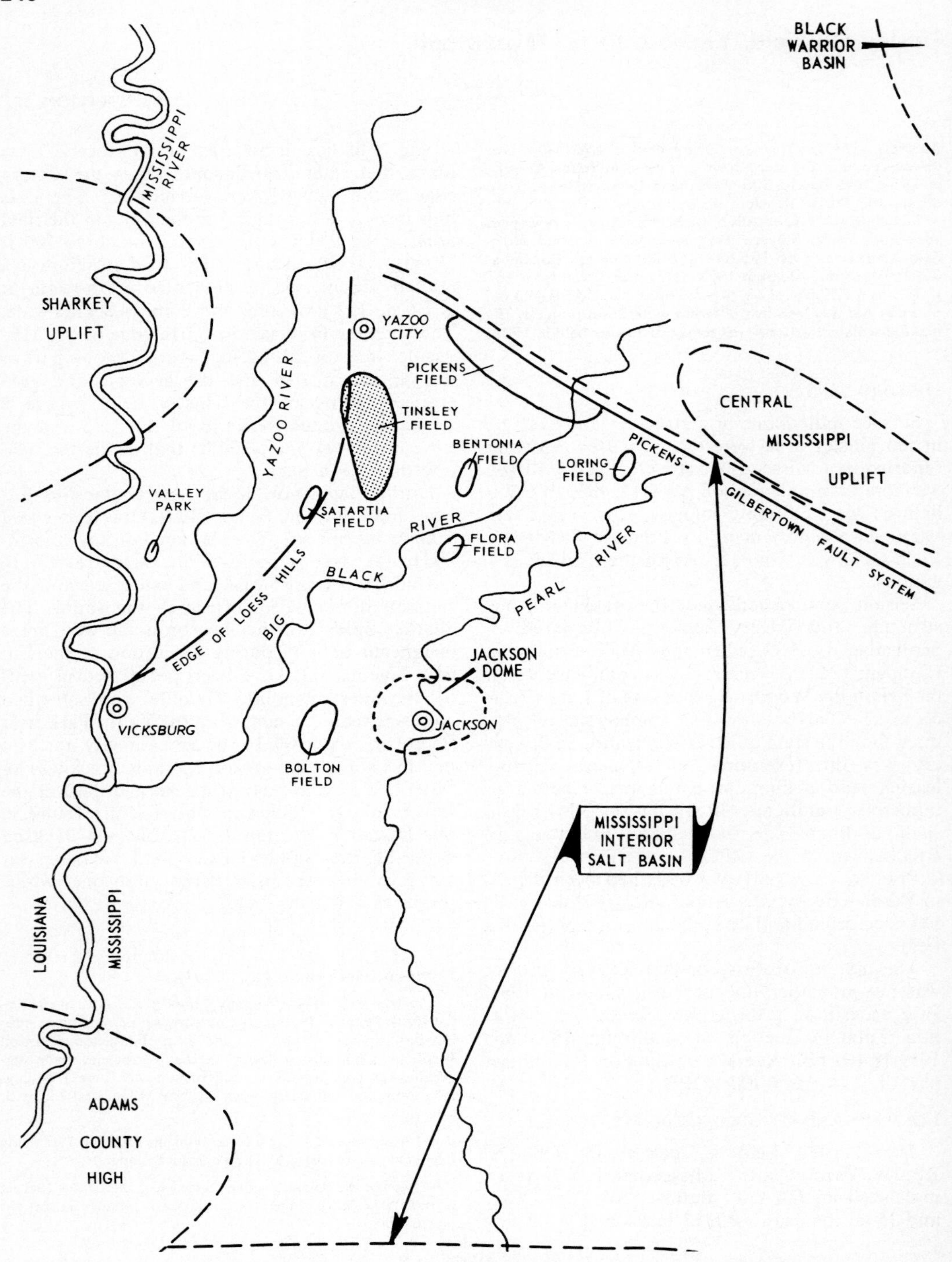

FIG. 1—Southwestern Mississippi, geographic and geologic features in relation to Tinsley field.

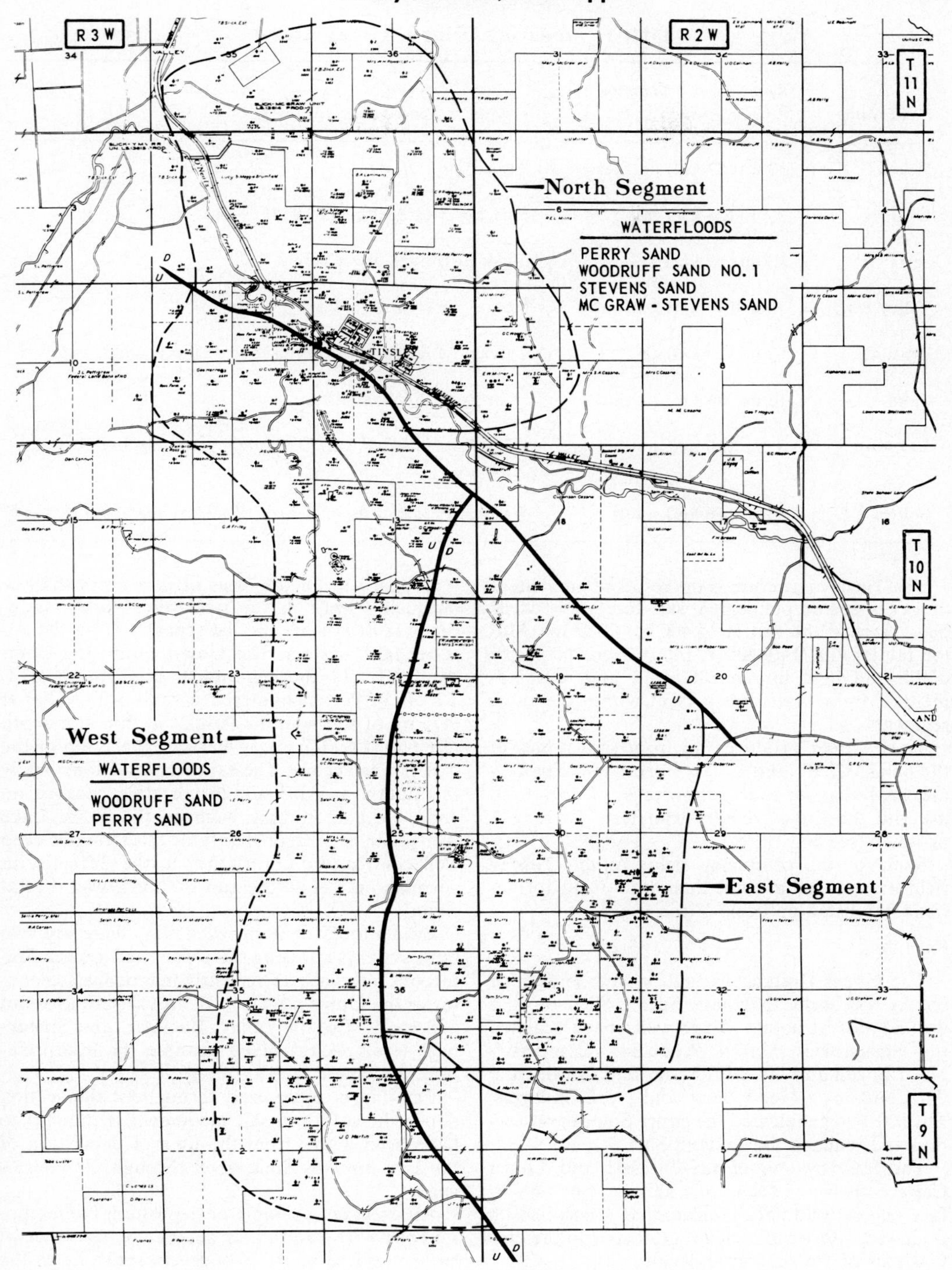

FIG. 2—Productive area and faulting, Tinsley field.

Table 1. Producing Fields in Tinsley Area

Field Name	*Approximate Distance from Tinsley*	*Producing Formations*
Pickens	33 mi (53 km) northeast	Selma and Eutaw
Loring	35 mi (56 km) east	Smackover (sandstone facies; H_2S gas)
Sartartia	12 mi (19 km) west	Sligo, Hosston, Cotton Valley
Valley Park	30 mi (48 km) southwest	Mooringsport and Rodessa
Bentonia	12 mi (19 km) east	Tuscaloosa, Washita, Fredericksburg, Paluxy, and Hosston
Flora	15 mi (24 km) southeast	"Gas Rock" (Selma, oil)
Jackson	35 mi (56 km) southeast	"Gas Rock" (Selma, gas). Depleted; presently used for gas storage
Bolton	35 mi (56 km) south	Washita, Fredericksburg, Paluxy, Rodessa, Sligo, and Hosston

The Tinsley structure is the result of movement of a deep-seated piercement salt dome. Two wells have reached the salt at 14,400 ft (4,389 m). Major faults with 150–400 ft (46–122 m) of throw divide the field into north, west, and east segments. Minor faults are present within the three segments.

Tinsley field produces oil from sandstones in the Selma Chalk and Eutaw Formation. The Selma Chalk has an average thickness of 265 ft (80 m), and the Eutaw is approximately 775 ft (236 m) thick (see Fig. 3).

Some of the producing fields in the Tinsley field area, together with their producing formations, are listed in Table 1 (see Fig. 1).

Faulting

The Upper Cretaceous fault system is defined by the 474 wells that have been drilled on and around this structure. Four wells have reached the Smackover: Union Producing Company's No. 1 Logan and No. 21 Stevens, and Continental Oil Company's No. 1 Berry and No.1 Childress. The last two penetrated the entire Smackover section and reached the Louann Salt.

The fault system of the Jurassic and Lower Cretaceous is not accurately defined, but subsurface and seismic data indicate that a north-south fault with a throw of 1,500 ft (460 m) or more cuts that part of the section at Tinsley.

As mapped on the basis of data from shallow producing wells in the field, the Upper Cretaceous fault system may be separate from the Jurassic fault system. The Union Producing Company No. 44 Stevens (north segment, Sec. 1, T10N, R3W) encountered a fault with 1,500 ft (475 m) of throw in the Paluxy section at a depth of 6,650 ft (2,027 m); it dies out at the top of the Lower Cretaceous. The sandstone sections below the Eutaw are difficult to correlate because no good correlation unit, such as the Ferry Lake Anhydrite, is present. The wide scattering of deep well control also contributes to the difficulty in correlations below the top of the Lower Cretaceous.

Even though it is believed that there are two fault systems affecting the Upper and Lower Cretaceous intervals, as presently mapped, a steepening of the Upper Cretaceous fault system to about 50° at the Cotton Valley, Buckner, and Smackover levels would make possible the interpretation of a single fault system.

Faulting is recognized throughout the section above the salt, and salt movement is thought to have been active from the time of deposition of the Late Jurassic Smackover through the Pleistocene.

An excellent example of deposition contemporaneous with faulting is shown by the interval from the base of the Woodruff sandstone to the top of the Perry sandstone. In the north segment of the field, which is downthrown to the two main faults, the thickness of this interval is 375 ft (114 m). A thickness of 300 ft (91 m) is seen in the east segment, which is upthrown to the north segment. A still thinner interval (265 ft or 80 m) is present in the west segment, which is upthrown on both faults.

Productive Sandstones

Figure 3, a type log, shows the productive sandstones. In Figure 4, these are shown on a generalized east-west cross section through the east and west segments of the field. Figure 5 is a stratigraphic section of the field. The Woodruff sandstone and the Stevens sandstone (Eutaw), which was found productive in only one well, are the only sandstones that produce in the east segment. The other sandstones that are productive in the north and west segments are present but are salt-water bearing in the east segment.

Table 2 shows the yearly production of oil and salt water for the field through 1974.

Woodruff Sandstone

Stratigraphy and lithology—This productive zone is a sandstone within the Selma Chalk which occurs about 90 ft (27 m) below its top. This sandstone grades into a normal chalk section west and north of the pinchout line. The Woodruff sandstone pinchout line runs north-south on the west side of the field and then turns east across the north flank of the structure (Fig. 6). To the east and south, the Woodruff sandstone grades into a sandy chalk. This facies is productive in Flora field, 15 mi (24 km) southeast. Farther south this zone grades into vuggy chalk, which was productive in the Jackson gas field. The chalk in the Jackson gas field is in the form of a reef which is draped over the igneous neck of the extinct Jackson Dome volcano.

The Woodruff sandstone is a fine- to medium-grained, light gray, calcareous sandstone with interbedded thin, gray, nonporous limestone beds. The east field segment has a 5–10-ft (1.5–3 m) calcareous shale section near the center of the sandstone interval. The sandstone is found in each well south and east of the strandline, except where it has been removed by faulting.

The thin limestone beds are less numerous near the top of the Woodruff interval. The sandstone nearest the pinchout line contains 20 percent limestone layers, whereas the thick sandstones of the east segment (up to 100 ft or 30 m) contain up

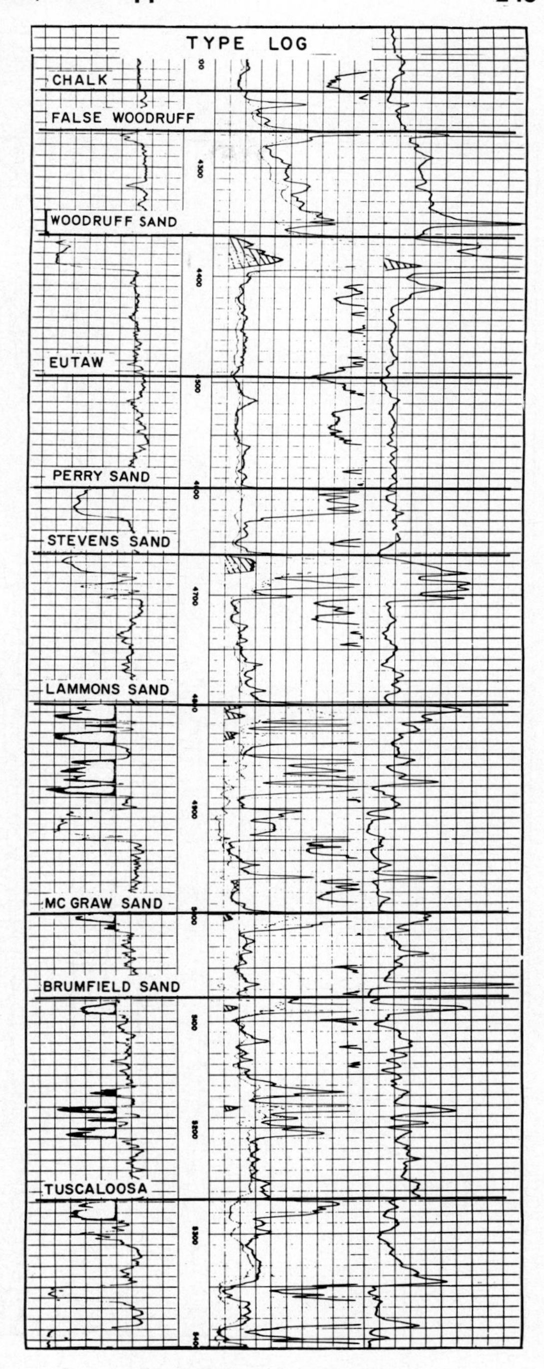

Fig. 3—Type log, Tinsley field.

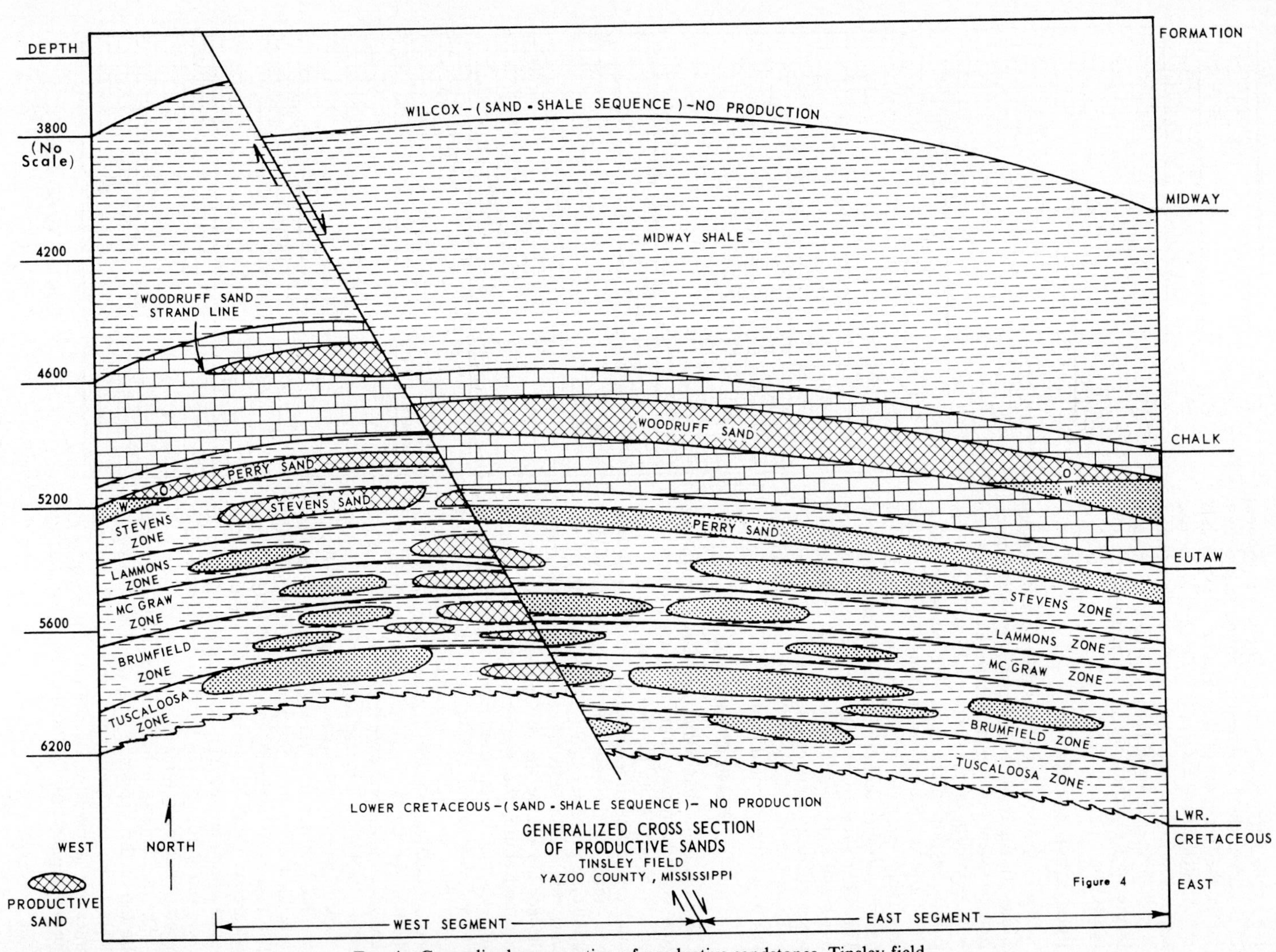

FIG. 4—Generalized cross section of productive sandstones, Tinsley field.

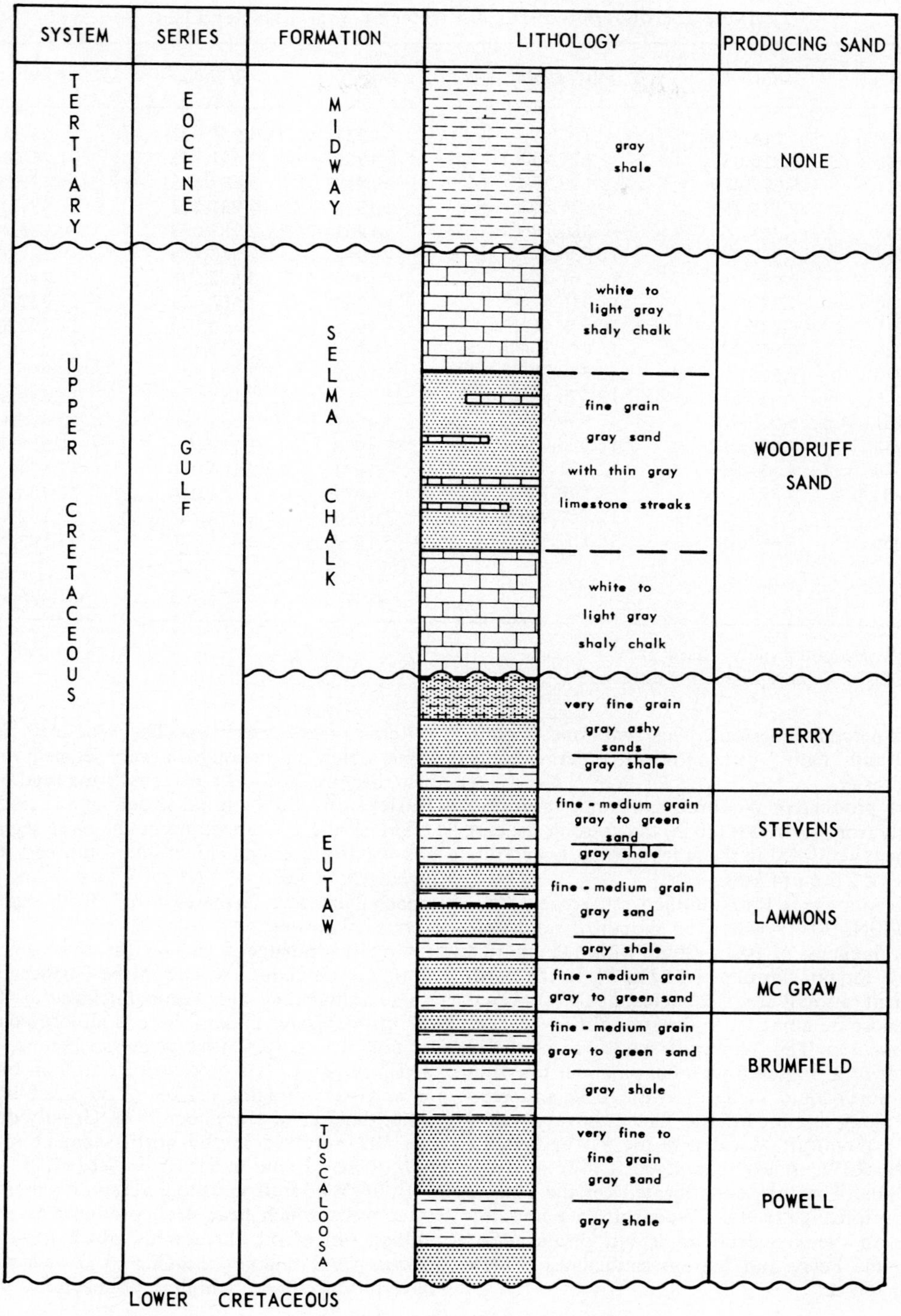

SYSTEM	SERIES	FORMATION	LITHOLOGY	PRODUCING SAND
TERTIARY	EOCENE	MIDWAY	gray shale	NONE
UPPER CRETACEOUS	GULF	SELMA CHALK	white to light gray shaly chalk	
			fine grain gray sand with thin gray limestone streaks	WOODRUFF SAND
			white to light gray shaly chalk	
		EUTAW	very fine grain gray ashy sands / gray shale	PERRY
			fine - medium grain gray to green sand / gray shale	STEVENS
			fine - medium grain gray sand / gray shale	LAMMONS
			fine - medium grain gray to green sand	MC GRAW
			fine - medium grain gray to green sand / gray shale	BRUMFIELD
		TUSCALOOSA	very fine to fine grain gray sand / gray shale	POWELL

LOWER CRETACEOUS

FIG. 5—Stratigraphic section, Tinsley field.

Table 2. Production of Oil and Water by Years, Tinsley Field

Year	Oil (bbl)	Water (bbl)	Year	Oil (bbl)	Water (bbl)
1939	114,972		1957	3,956,027	11,647,282
1940	4,210,022	179,400	1958	3,491,955	11,260,080
1941	15,277,270	1,126,382	1959	3,403,613	11,723,364
1942	28,179,758	5,191,265	1960	3,231,232	11,591,025
1943	17,264,156	10,384,190	1961	2,983,679	10,773,201
1944	11,802,335	12,936,577	1962	2,843,728	10,191,661
1945	9,400,014	12,361,364	1963	2,852,598	10,790,897
1946	7,995,302	11,915,675	1964	2,668,423	11,032,305
1947	6,738,397	10,527,461	1965	2,450,129	11,193,930
1948	6,047,852	11,007,104	1966	2,323,415	11,239,815
1949	5,568,681	12,333,305	1967	2,274,647	12,250,994
1950	5,187,517	12,520,894	1968	2,167,973	12,719,549*
1951	5,029,505	11,949,729	1969	2,156,519	13,515,111*
1952	4,925,918	12,000,664	1970	2,282,911	17,928,908*
1953	4,541,955	12,121,507	1971	2,566,096	15,634,077*
1954	4,325,628	12,540,204	1972	3,114,070	18,736,409
1955	4,519,277	12,655,297	1973	3,053,706	22,162,404
1956	4,465,656	12,194,871	1974**	2,433,530	23,188,281
			Total	195,848,466	421,525,182

*Waterflood project begun.
**Last 2 months production estimated.

to 55 percent limestone. This limestone is gray, hard, and "tight," with no effective permeability or porosity.

The productive Woodruff sandstone thins uniformly from the 100-ft (30 m) thickness in the east segment westward to the pinchout line, over a distance of 2.5 mi (4 km).

Exceptions are found in the west segment (Sec. 11, T10N, R3W), where the Woodruff sandstone has a thickness of 100 ft (30 m) within 0.5 mi (0.8 km) of the pinchout line (see Fig. 9). In the north segment, a small area (120 acres) has 40 ft (12 m) of sandstone adjacent to the strandline.

Structure—The Woodruff sandstone structure is shown on Figure 6 with the pinchout line running north-south along the west flank and east-west along the north flank. The discovery well—No. 1 Woodruff—located in the SW ¼, Sec. 13, T10N, R3W, is less than 1,400 ft (427 m) from this line. The two wells located on the 80-acre tract offsetting the No. 1 Woodruff have no sandstone in the Woodruff zone, but produced oil from the Perry and Stevens sandstones (Eutaw Formation).

Reservoir characteristics—The Woodruff sandstone, which is productive in every segment of the field, has produced 82.6 percent of the total oil at Tinsley—161.8 million bbl to January 1, 1975.

The Woodruff sandstone in the west segment had water encroachment on the south end moving across a 40-acre tract in 3 to 4 years. Encroachment rates in the east and north segment were even slower.

Gravity drainage occurs in the west segment along the pinchout line and in the northern part where a small area with a gross thickness of 100 ft (30 m) of Woodruff sandstone is almost isolated by faulting and pinchout of the sandstone. One well in this area has produced 2.6 million bbl of oil and was averaging 152 bbl of oil plus 2 bbl of water daily as of December 1974. Gravity drainage has occurred in the north segment in the NW ¼, Sec. 2, and in the SE ¼, Sec. 12.

In the Woodruff sandstone reservoirs, there are three wells which have each produced over 2.5 million bbl of oil, three wells which have produced over 2 million bbl each, and 27 wells which have produced over 1 million bbl each.

Table 3. Average Reservoir Characteristics of Productive Sandstones, Tinsley Field

Productive Sandstones	*Woodruff*	*Perry*	*Stevens*	*Lammons*	*McGraw*	*Brumfield*	*Tuscaloosa*
Proven acreage	8,930	5,425	1,500	550	500	450	160
Average depth (ft)	4,800	5,200	5,300	5,400	5,780	5,850	5,950
Average thickness (ft)	26	21	20	20	11	16	12
Average porosity (%)	26	29	26	25	27	29	28
Average permeability (md)	289	47	450	510	310	850	400
Water saturation (%)	20	40	25	25	25	25	35
Factor, original formation volume	1.075	1.18	1.19	1.25	1.25	NA	NA
Gravity (°API)	33	33	36	46	44	45	45
Viscosity (Cp at reservoir temperature)	3.8	1.5	NA	NA	NA	NA	NA
Original bottomhole pressure (psi)	2,025	2,050	2,100	2,400	2,540	2,575	2,620
Original gas/oil ratio	100	300	110	300	200	300	300
Reservoir temp. (°F)	164	170	174	180	184	185	186
Orig. oil in place (bbl/acre-ft)	1,500	1,145	1,275	1,165	1,260	1,205	1,090
Oil produced thru 1974 (million bbl)	161.8	15.4	9.8	4	2	2.6	0.25
Average recovery, (bbl/acre-ft)	695	110	325	360	365	360	130
Average recovery (% oil in place)	46	10	26	31	29	30	12

NA = Not available.

Low gas and water saturations, good porosity and permeability, and water encroachment and gravity drainage have resulted in excellent recoveries. As an example, the Woodruff sandstone in the east segment will ultimately yield 750 bbl per acre-foot or 53.5 percent of the oil in place.

A portion of the oil from the Woodruff sandstone might be considered secondarily recovered oil, because several pore volumes of water have been moved through this reservoir in the downdip areas of the east and west segments.

Table 3 lists the area, depth, and reservoir characteristics of the Woodruff sandstone.

Perry Sandstone

Stratigraphy and lithology—The Perry sandstone is the uppermost sandstone in the Eutaw Formation and is present about 100 ft (30 m) below the top of that formation.

The Perry is a blanket-type sandstone, and is present in every well on the Tinsley structure except where it is missing as a result of faulting.

The Perry is a fine- to very fine-grained, light gray sandstone; it is in part shaly and silty, as well as glauconitic. The dominant lithologic characteristic of the sandstone is its high content of bentonitic ash. Because of the ash, this sandstone has high porosity, low average permeability, and a high content of connate water.

In part of the north segment, the Perry sandstone has a 5-ft (1.5 m) interval near the top with permeability that ranges from 60 to 879 md. The average permeability of the entire Perry sandstone reservoir in the north segment is 49 md. In the west segment, the sandstone permeability is more uniform and averages 47 md.

In the early development of the field, the Perry was largely bypassed for primary completions but, with the development of a sand-fracturing

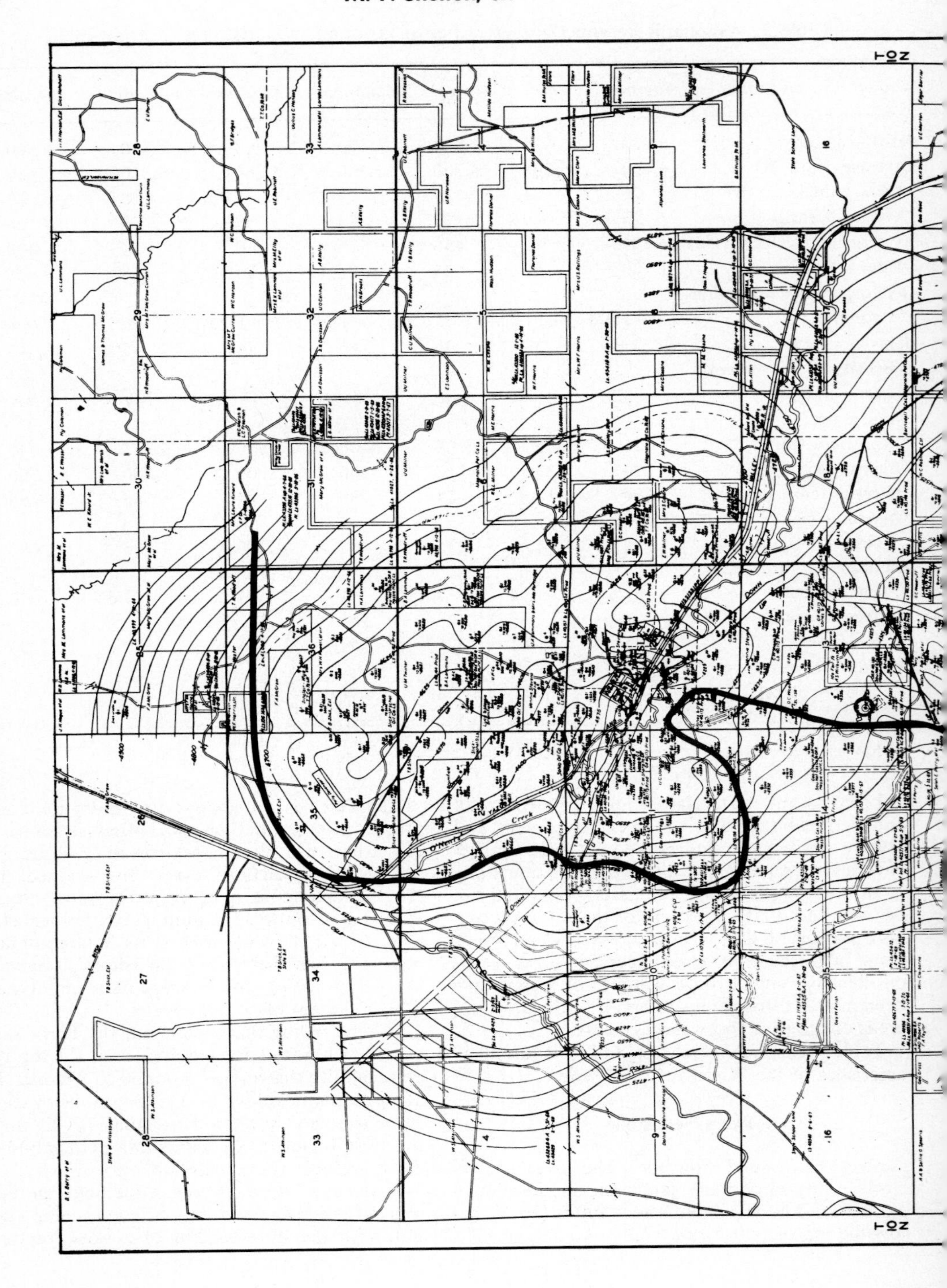

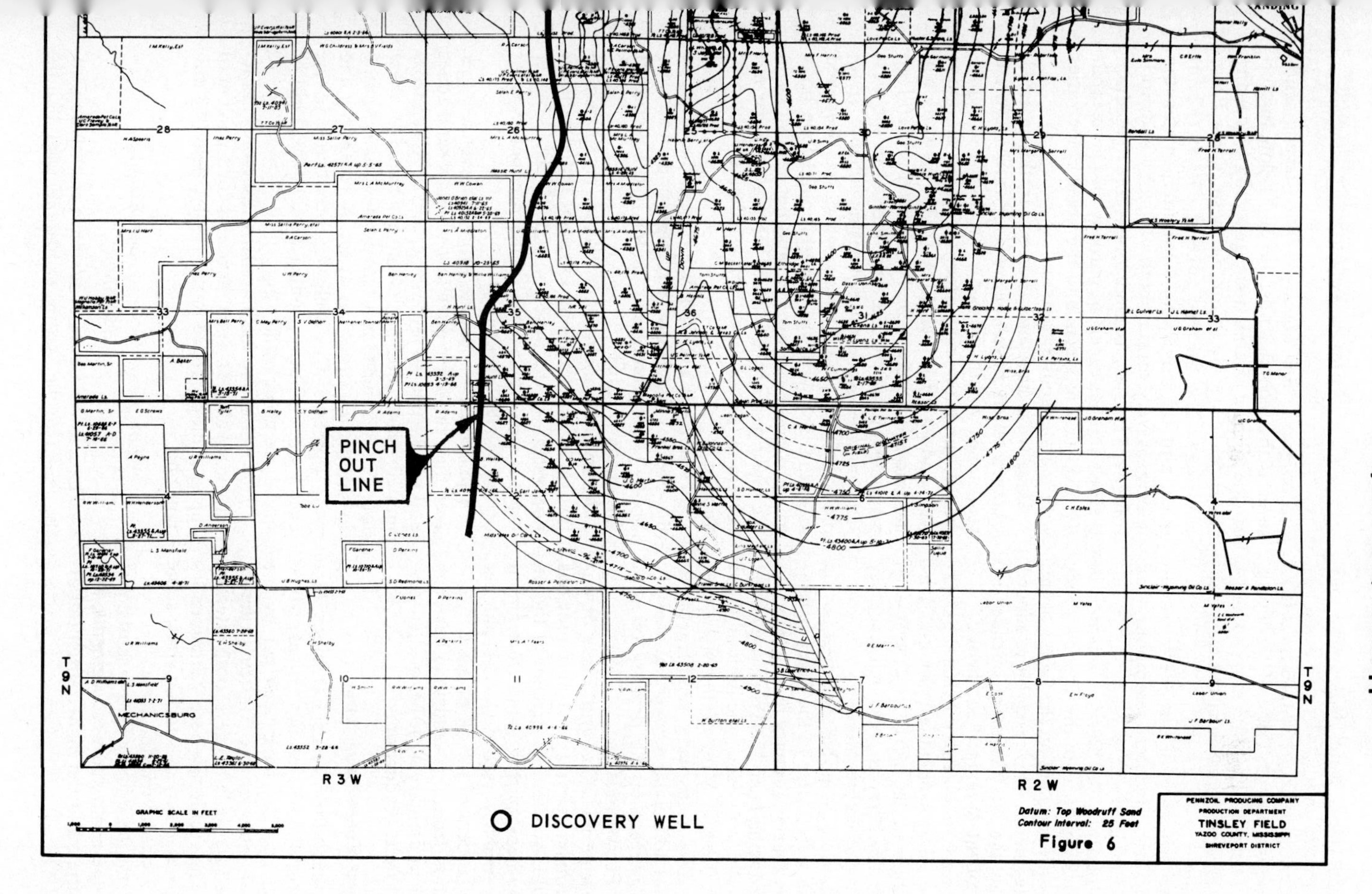

FIG. 6—Structure map of top of Woodruff sandstone. C.I. = 25 ft.

process, additional wells were drilled and old wells were recompleted in it.

Sandstone thickness of about 20 ft (6 m) is fairly uniform except on the west side of the north segment, where it thins to 12 ft (3.7 m). The highly permeable zone is not present in the west half of the north segment.

Structure—The structure of the Perry sandstone has a configuration closely resembling that of the Woodruff sandstone (Fig. 6).

Unlike the Woodruff, the Perry sandstone is present in the downdip areas of the Tinsley structure with an oil-water contact on the flanks of the west and north segments. This reservoir produced oil in 12 wells in the west segment outside the productive area of the Woodruff sandstone.

Reservoir characteristics—Table 3 lists the characteristics of the Perry sandstone reservoir. This sandstone produces in the north and west segments. Although the Perry is present in the east segment, it is not productive there.

Some water encroachment has occurred in this reservoir, but the water drive in the Perry is relatively ineffective. During primary production, the average well producing from the Perry sandstone decreased gradually to its economic limit of about 5 BOPD and 4 BWPD. Most of the wells that produced from this zone were sand-fractured, with a consequent additional recovery of possibly as much as 30 percent of the primary oil produced from the Perry sandstone.

Because of low permeability and a lack of water drive, the primary recovery per acre-foot has been only 110 bbl, or about 10 percent of the oil in place. The southeast part of the north segment has had better primary recovery—159 bbl per acre-foot, or 15 percent of the oil in place.

Total production from this sandstone has been 15.4 million bbl to January 1, 1975, or 7.86 percent of the total field production.

Lower Eutaw Sandstones

These sandstones—the Stevens, Lammons, McGraw, and Brumfield—are not blanket-type sandstones (Fig. 3). They have been cut by the major and minor faults on the Tinsley structure and are separated by shale barriers, both laterally and vertically. The result has been numerous locally productive reservoirs of varying sizes, some of which lack an effective water drive.

Production from these lower Eutaw sandstones (Fig. 3) is from the north and west segments, and is primarily restricted to the higher structural areas.

Stevens Sandstone

Stratigraphy and lithology—The Stevens sandstone is the second sandstone unit below the top of the Eutaw Formation, generally occurring about 50 ft (15 m) below the base of the Perry sandstone. The Stevens zone is 170 ft (52 m) thick, and the sandstone distribution within the interval is very erratic, both vertically and horizontally. The Stevens sandstone is fine to medium grained and has excellent porosity and permeability. It is gray to green, and the green color varies with the glauconite content.

Some wells have no sandstone in this interval, whereas others have one, two, or three different sandstones within the zone. Thickness of a productive sandstone may exceed 50 ft (15 m) in one well, and an offset well 1,000 ft (300 m) away will have no sandstone. A well completed in a 20-ft (6 m) sandstone in the Stevens interval in the west segment was lost because of casing collapse. In an old well 400 ft (122 m) away, which was deepened after the first well was lost, this sandstone body was not present. However, another sandstone in the lower part of the Stevens interval that was not present in the collapsed well was found to be productive.

Structure—A map is not included inasmuch as the Stevens sandstone structure conforms very closely to the structure of the top of the Woodruff sandstone (Fig. 6).

Reservoir characteristics—Wells completed in this sandstone have had excellent initial production rates, on the order of 300 BOPD. A water drive is present at some of the productive areas on the Tinsley structure.

The Stevens sandstone produces over a larger area than the other three lower Eutaw sandstones and has yielded about 53.3 percent of the approximately 18,400,000 bbl of oil attributed to these four Eutaw sandstones. This production is limited to the higher structural areas of the north and west segments. One well was productive in the east segment, but it produced only 14,950 bbl of oil before abandonment.

This sandstone has excellent reservoir characteristics (Table 3), and it accounts for 5 percent of the total oil produced in the Tinsley field (9.8 million bbl).

Lammons, McGraw, and Brumfield Sandstones

These lower Eutaw sandstones are lithologically similar to the Stevens sandstone; also, like the

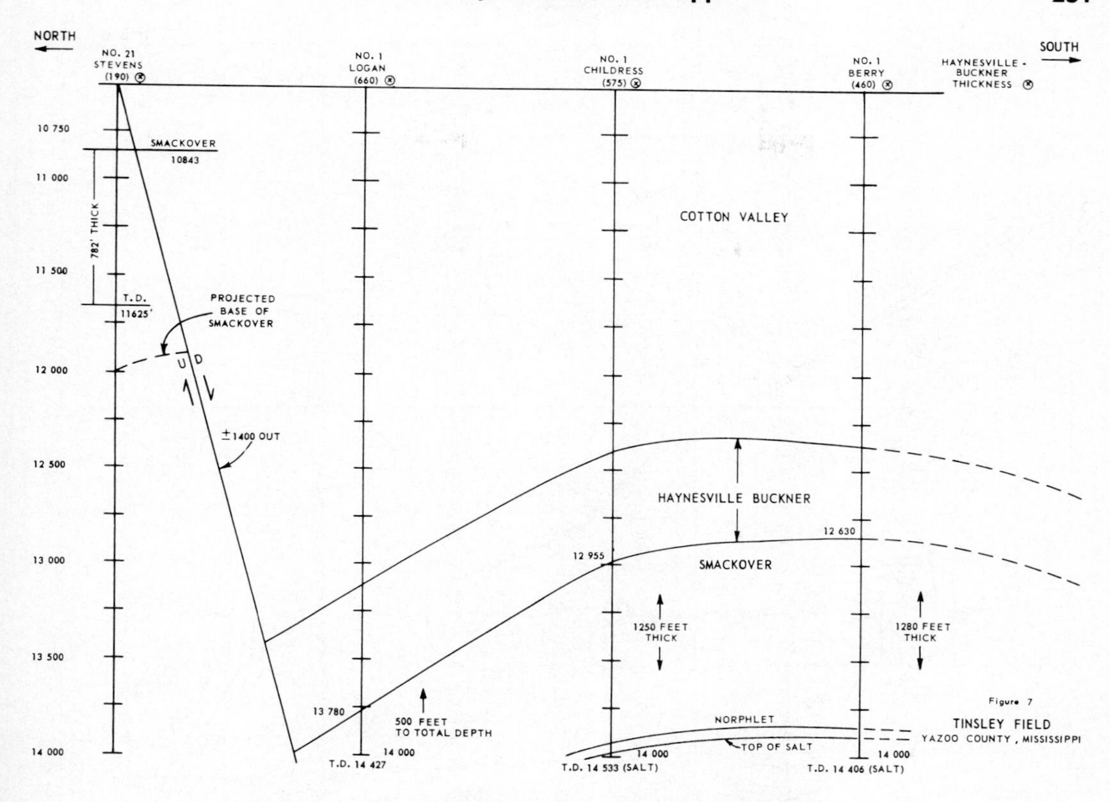

FIG. 7—Generalized cross section of Jurassic section in Tinsley field.

Stevens, their occurrence is erratic. The productive area of these sandstones is limited to the highest structural positions in the north segment and in the north part of the west segment.

Reservoir characteristics of the sandstones are excellent (Table 3). Cumulative production from these three sandstones is 8.6 million bbl of oil, or 4.4 percent of the total field production.

Powell Sandstone

The Powell sandstone in the Tuscaloosa is a fine- to very fine-grained, slightly ashy sandstone with excellent porosity and permeability.

Production was limited to the highest part of the structure in the north part of the west segment near the fault and in a small area in the north segment.

Cumulative production from this zone was 250,000 bbl of oil and 319,222 bbl of water over a rather short productive life. Production came from one well in the north segment and two wells in the west segment.

Lower Cretaceous to Smackover

Four wells on the Tinsley structure have penetrated all of the Lower Cretaceous–to–Smackover section and three have penetrated part of it.

There has been no indication of commercial production and no significant shows have been encountered, except in one well in the north segment. The Union Producing Company No. 44 Stevens (Sec. 1, T10N, R3W) had numerous shows in the Paluxy and Hosston. Completion attempts failed to establish commercial production, although logs, cores, and analyses indicated the presence of sandstones of sufficient quantity and quality to produce. Faulting may have ruptured these reservoirs and permitted the oil in them to migrate vertically and be trapped in the Eutaw sandstones productive in the north segment.

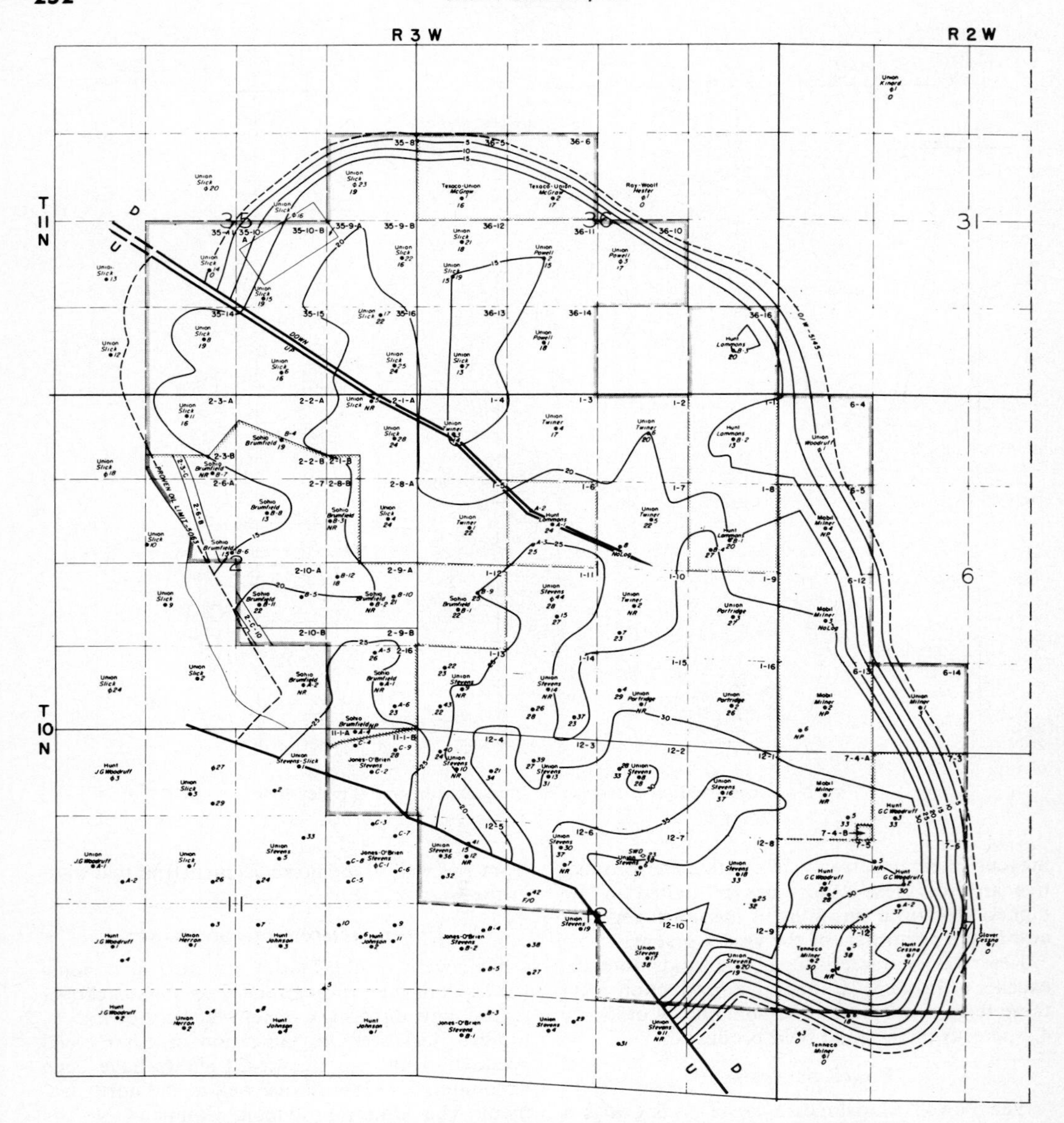

FIG. 8—Isopach of net oil-productive sandstone, Perry sandstone, Tinsley field north segment. C.I. = 5 ft.

Smackover Formation

The Smackover Formation at Tinsley field has a thickness of 1,250 ft (381 m) in the two wells that penetrated the entire formation. Two other wells reached the Smackover but did not penetrate the full section.

In both the Continental Oil Company No. 1 Berry and No. 1 Childress, a 65-ft (20 m) section of the Norphlet Formation was penetrated before salt was encountered.

The Union Producing Company No. 1 Logan (Sec. 23, T10N, R3W) and the Continental Oil Company No. 1 Berry (Sec. 25, T10N, R3W) tested H_2S gas before being abandoned.

A schematic cross section (Fig. 7) shows the relation to the Haynesville-Buckner-Smackover sections, as well as faulting in the four wells.

WATERFLOODS

Perry Sandstone, North Segment

In November 1965, a pilot project was begun in the Perry sandstone of the north segment with the injection of water into three wells. Response to this injection came after 500,000 bbl of salt water had been injected. This pilot area was expanded to include about 70 percent of the north segment, and the first full-scale waterflood was started in December 1968. This flood was expanded further in January 1970 to include all of the Perry sandstone in the north segment. This unit covers 2,472 surface acres (10 km^2; Fig. 8).

Table 4 shows the cumulative oil and water production, injection volumes, and daily rates for October 1974 for all of the secondary projects in the Tinsley field. Waterfloods in operation on January 1, 1975, in the Tinsley field are shown in Figure 2.

McGraw-Stevens, North Segment

Both of these Eutaw sandstones are developed on the north flank of the Tinsley structure. The trap for this hydrocarbon accumulation is formed by gradation into shale in conjunction with a small (75± ft or 23± m) sealing fault that runs east-west.

Flooding of the McGraw sandstone was begun in December 1969; the Stevens sandstone was included in the flood as of July 1970, and it is being flooded simultaneously with the McGraw.

This unit includes 339 acres (McGraw) and 290 acres (Stevens) in Secs. 35 and 36, T11N, R3W.

Stevens 13–38, North Segment

This two-well flood (No. 13 and No. 38 Stevens) is located in Sec. 12, T10N, R3W, in the north segment. These wells had produced more than 700,000 bbl of oil and 20,000 bbl of water from the "A" and "B" sandstones of the Stevens zone prior to the start of the flood in April 1969.

Woodruff Sandstone, North Segment

This Woodruff sandstone reservoir averages about 7 ft (2.1 m) in thickness and is located on the north flank of the Tinsley structure. The area being flooded is bounded on the north and west by the pinchout of the sandstone and on the south by a small sealing fault. A stationary, inef-

Table 4. Tinsley Field Waterflood Production and Injection Through October 1974

Reservoir	Segment	Date Installed	Oil (bbl)	Water (bbl)	Injected Water (bbl)	No. Wells Producing	No. Wells Injecting	Average Daily Oil (bbl)	Average Daily Water (bbl)	Average Daily Injection (bbl)
Perry	North	10-1-68	2,125,022	10,674,525	27,684,619	23	20	844	7,566	10,852
Woodruff	North	11-1-71	187,426	1,012,183	3,494,433	3	4	87	1,230	2,333
Perry	West	11-1-70	868,557	1,495,033	11,696,698	17	19	872	24,677	7,603
Woodruff	West	11-1-70	1,734,388	17,822,876	33,560,128	17	11	1,520	23,899	27,575
Perry-Woodruff	West	11-1-70	1,404,983	10,407,263	0	15	(Commingled wells, production allocated)			
Stevens 13-28	North	4-1-69	0	0	1,595,389	1	1	25	6	596
McGraw-Stevens	North	12-1-69	205,187	466,103	2,234,074	1	2	31	576	1,534

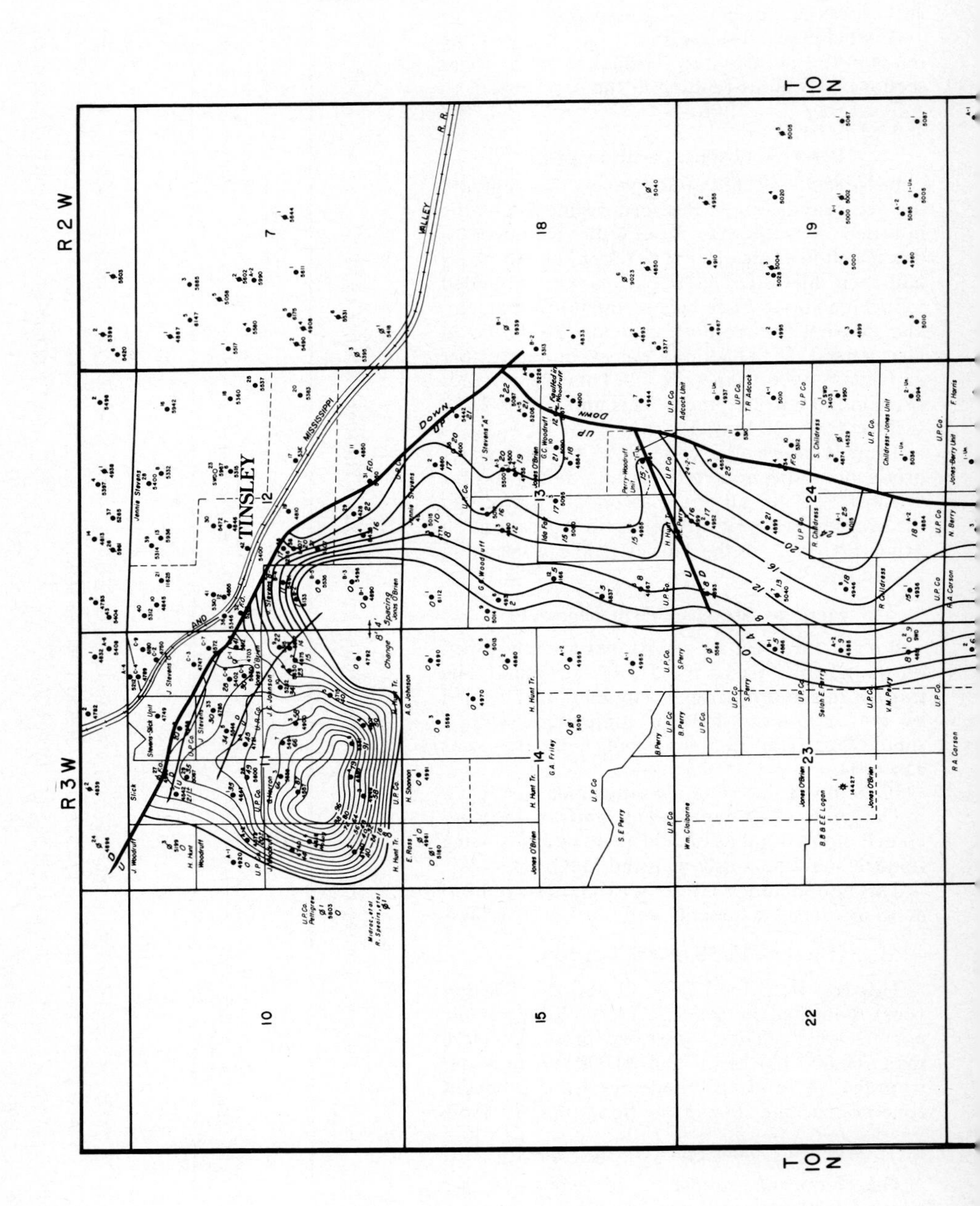
R 2 W
R 3 W
T 10 N
T 10 N
TINSLEY
MISSISSIPPI
VALLEY
AND
R.R.
DOWN
UP
Spacing Change
10
15
22
7
18
19
12
13
24
11
14
23

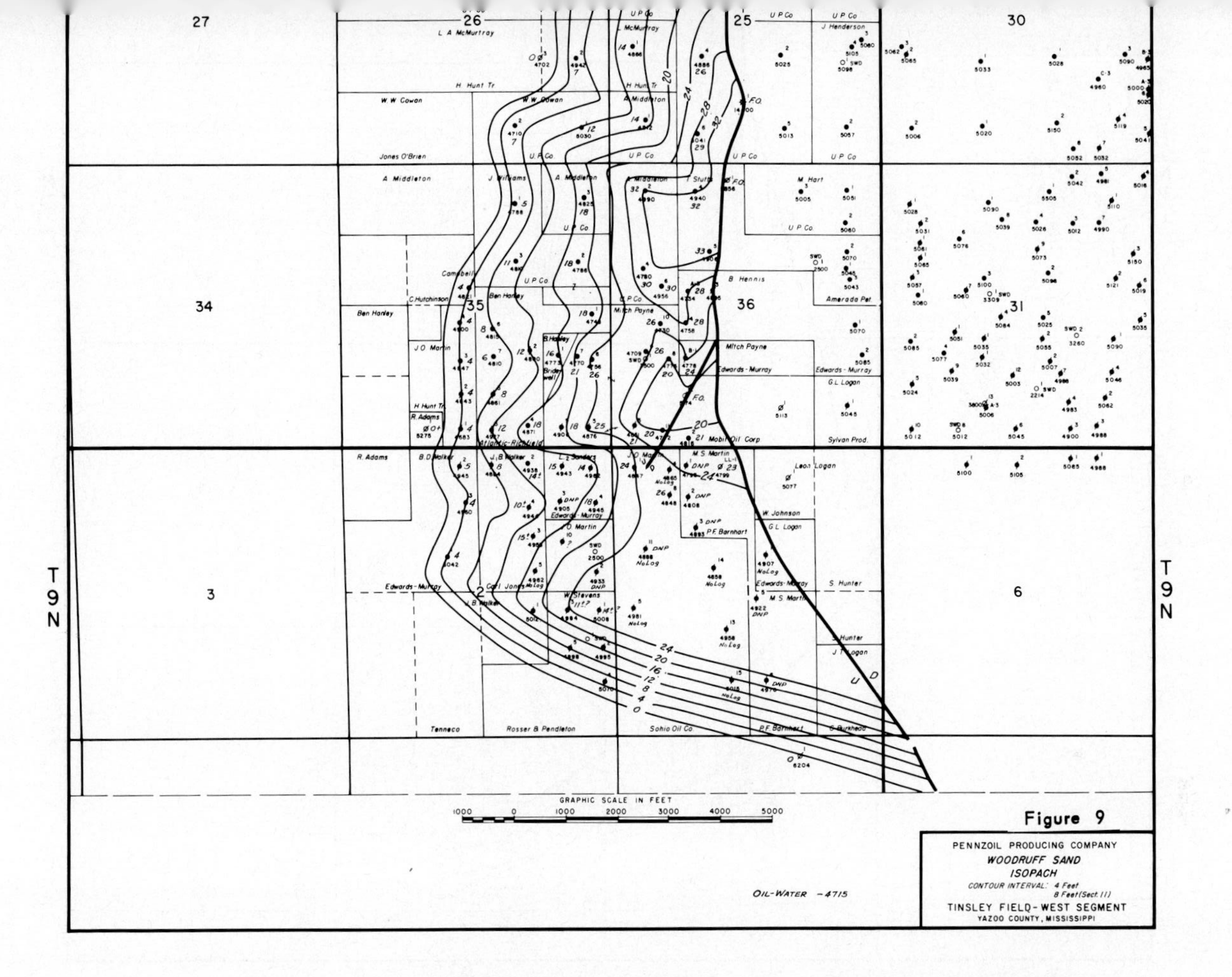

FIG. 9—Isopach of net oil-productive sandstone, Woodruff sandstone, Tinsley field west segment. C.I. = 4 ft; 8 ft in Sec. 11.

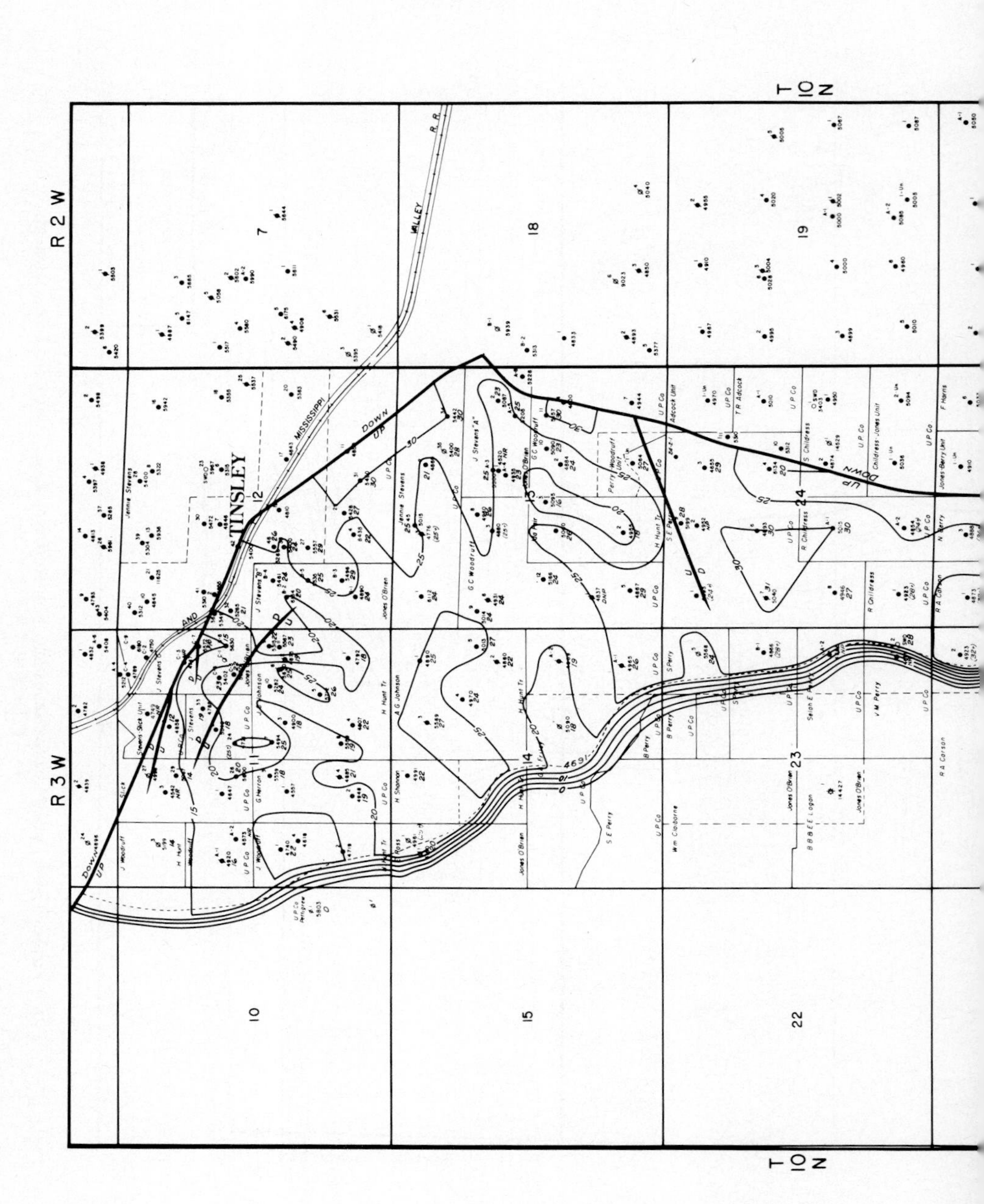
R 2 W
R 3 W
T 10 N
T 10 N
TINSLEY
MISSISSIPPI
VALLEY
R R
AND
DOWN
UP
7
18
19
12
24
10
15
22
23
14
11

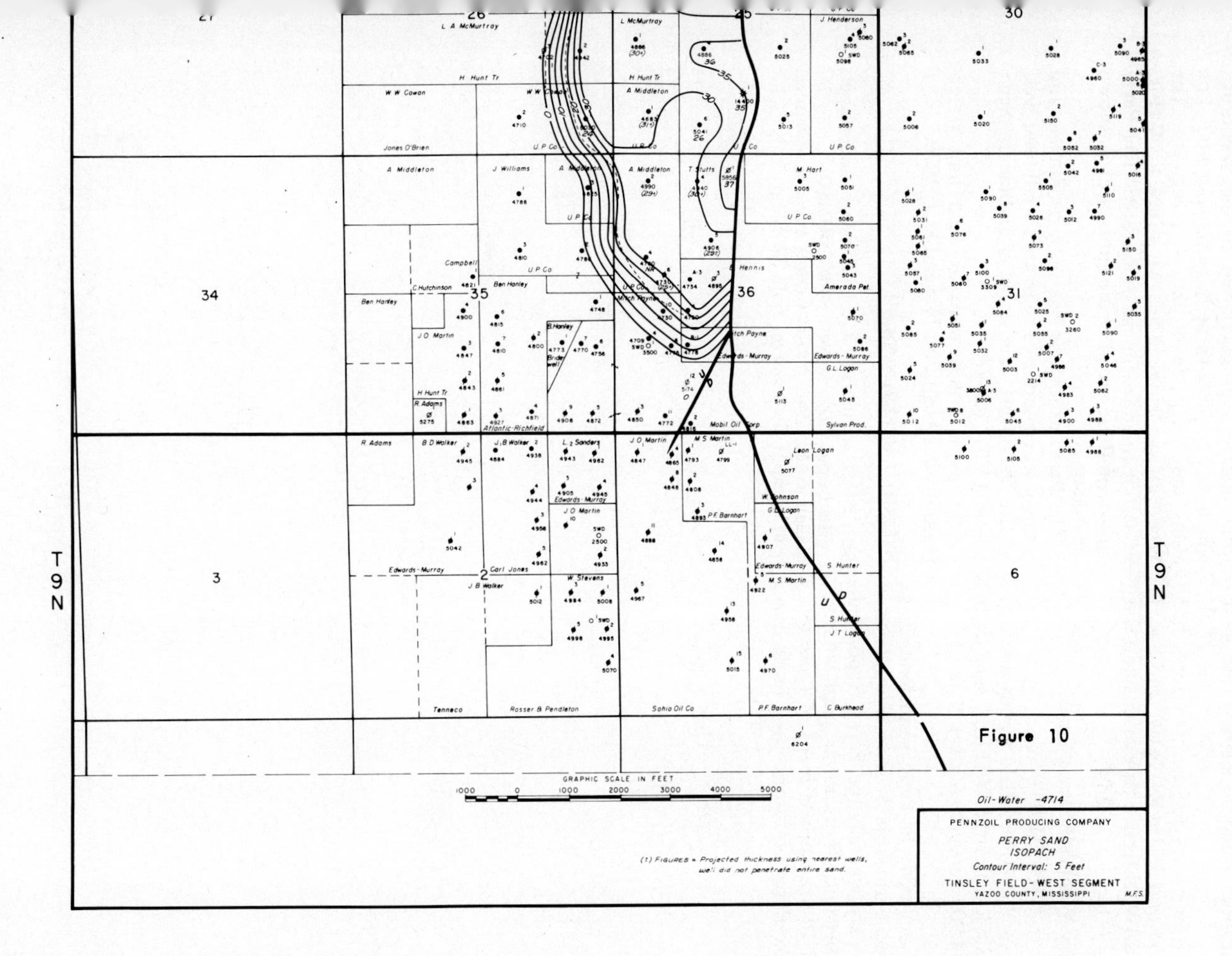

FIG. 10—Isopach of net oil-productive sandstone, Perry sandstone, Tinsley field west segment. C.I. = 5 ft.

fective water front is found on the east side of the reservoir.

The flood area covers 625 surface acres (2.5 km^2) in Secs. 35 and 36, T11N, R3W, and Sec. 1 and 2, T10N, R3W. Effective date of this unit was November 1, 1971.

Perry-Woodruff Sandstones, West Segment

The largest of the Tinsley field floods involves 4,600 acres (18.6 km^2) of the Woodruff sandstone reservoir and 4,900 acres (19.8 km^2) of the Perry sandstone.

Response from both of these sandstones has been good; the permeable Woodruff sandstone responded quickly whereas the Perry responded more slowly because of its lower permeability. Figures 9 and 10 are isopach maps of these two sandstones. The Perry sandstone has a fairly uniform thickness as compared to the Woodruff, which wedges out to zero at the pinchout line along the west side of the west segment. The Woodruff sandstone in Sec. 11, T10N, R3W (Fig. 9), is not included in the flood area. The Perry sandstone flood includes the entire area as shown on Figure 10.

The Perry sandstone has been productive and is being waterflooded in the area south and west of the Woodruff sandstone strandline in Secs. 11–14, T10N, R3W.

These sandstones were unitized and waterflooding was begun on November 1, 1970.

Selected References

Belt, W. E., et al, 1945, Geologic map of Mississippi: Mississippi Geol. Survey, scale 1:500,000.

Mellen, F. F., 1940, Yazoo County mineral resources: Mississippi Geol. Survey Bull. 39, 132 p.

Citronelle Oil Field, Mobile County, Alabama[1]

EVERETT EAVES[2]

Abstract The Citronelle field was discovered in 1955 by the Zack Brooks Drilling Company No. 1 Donovan, SW 1/4, NW 1/4, Sec. 25, T2N, R3W, Mobile County, Alabama. The well produced from the lower Glen Rose Formation at a depth of 10,879 ft (3,315.9 m). During the next 10 years, 434 productive wells were drilled. The productive limits completely enveloped the town of Citronelle, 32 mi (51.5 km) north of Mobile, Alabama. Forty-acre spacing, low gas-oil ratio, and rapid bottomhole-pressure drop, necessitating pumping of all wells, resulted in slow and spasmodic development. Unitization of 139 wells for waterflood was initiated in 1961, and a saltwater-injection program proved successful. Later, fresh water from the Wilcox Formation was used for injection fluids. By May 1966 all wells were unitized, and on December 31, 1973, the field had produced over 107 million bbl of oil. The structure, which lies in the southwest Alabama salt basin, is residual in origin. The field, as well as the townsite, is topographically high. The presence of this high led to the drilling of two shallow tests. Although both wells were dry, the fact that they appeared to be structurally high led to the drilling of the deeper discovery well. The structure at the producing-depth datum is that of a simple, moderately flat-topped, ovate dome, but the field presents a complex meander-belt pattern with 52 productive sandstone zones in 330 separate reservoirs. The entire field was eventually unitized under three agreements as though there were three separate reservoirs. All the productive wells produce through electrically driven downhole pumps. Production for the field reached its peak in the spring of 1972 when all three injection programs were utilizing all wells and all economically floodable sandstone reservoirs. Thirty-three percent of the oil in place had been produced by the end of 1973. Profitable production is expected to continue for another 8 years. At present, the Citronelle Operators Unit is studying the economics of a tertiary flood using carbon dioxide as a miscible agent.

INTRODUCTION

The town of Citronelle, Alabama, is located on a topographic high. This high is the surface expression of a deep-seated salt mass which, in relation to the surrounding area, is still growing. The presence of the surface structure was known to the oil industry for many years. The subsurface anomaly became apparent following the drilling of two shallow tests on the east flank of the structure (Smith County Oil Co. No. 1 Odom, Sec. 24, T2N, R3W, and Sullivan No. 1 Thompson, Sec. 31, T2N, R2W) and a lower Glen Rose test on the southwest flank (Humble No. 1 School Board, Sec. 16, T1N, R3W). All were dry but appeared to be high with respect to the available subsurface control.

During 1953 and 1954, Gulf Oil Corporation began acquiring a block of oil and gas leases in T1–2N, R2–3W, in and east of the townsite. Exploration enthusiasm was lacking, however, inasmuch as many dry holes had been drilled following the Pollard field development, 30 mi (48.3 km) to the east. Also, the Alabama oil and gas laws stated, "All oil wells must be located on a drilling unit consisting of at least forty surface contiguous acres." Wider spacing required landowner approval. Scarcity of pipe was another negative factor.

Interest in drilling a deep test increased following the discovery of production in the Rodessa (lower Glen Rose) in the Soso field, which had geologic characteristics similar to those of the Citronelle structure. The Soso field, in southern Mississippi, is located 60 mi (95.6 km) northwest of the Citronelle structure and on strike with it.

Rather than drill a deep well itself, Gulf Oil encouraged the drilling with a generous contribution. Additional financial help was received from Central Oil, Magnolia Petroleum Company (Mobil), Sun Oil, Superior Oil, Seaboard, the G.M. & O. Railroad, and T. K. Jackson of Mobile, Alabama. With such encouragement, Zack Brooks Drilling Co. of Eldorado, Arkansas, drilled the No. 1 Donovan (SW 1/4, NW 1/4, Sec. 25, T2N, R3W) through the Rodessa Formation, proving six or more productive sandstones in the 830 ft (253 m) of section below the base of the Ferry Lake Anhydrite. Discovery of oil production was confirmed by means of a drill-stem test made on August 16, 1955, following the recovery of saturated sandstones in conventional cores.

A thick, saltwater-bearing, blanket-type sandstone separated the two intervals of productive

[1]Manuscript received, May 24, 1974; revised, October 14, 1974. Published with permission of the Citronelle Operators Committee.

[2]Consulting geologist, Shreveport, Louisiana 71101.

The writer expresses his appreciation to the operators in the Citronelle field for their cooperation in releasing the information used in this paper, and to Philip E. LaMoreaux, State Geologist of Alabama, for supplying continuous production and injection totals covering more than 18 years. The writer also thanks Beryl Barre for drafting the illustrations.

section; in order to get a maximum allowable, the operators chose to complete the well dually with two separate strings of tubing run through 7-in. (17.8 cm) casing. Dual completion through one string of tubing was unlawful at that time.

After discovery, new locations were staked in alternate 40-acre tracts, as operators hoped to obtain 80-acre spacing. An earnest attempt was made to induce the Alabama legislature to revise the spacing laws, but a wider spacing bill failed when more than 100 landowners appeared at the state capitol to lobby against the bill. Thus 40-acre spacing controlled the development and economics of the field.

Surface Geology

The Citronelle oil field is within the eastern limits of the Gulf Coast salt basin where it offlaps the southern margin of the Appalachian Mountains in central Alabama and southern Georgia. It lies 32 mi (51.5 km) north of Mobile, Alabama, a seaport located at the north end of Mobile Bay (Fig. 1).

Covering the surface from the center of the field southward to the coast is the Citronelle Formation (known as the "Willis" in Texas.) It is a fluvial deposit which has been considered to be Pliocene (Otvos, 1973), Pleistocene (Aronow, 1974), or transitional between the two. According to Stringfield and LaMoreaux (1957), the formation contains certain flora of Pliocene age; they have observed that the oldest marine terrace of Pleistocene age in Florida is underlain by the Citronelle Formation. However, "recent discovery of abundant vertebrate fossils indicates that deposition of this formation began in the mid-Pliocene and continued into the pre-Nebraskan Pleistocene" (Isphording and Lamb, 1971).

Stringfield and LaMoreaux's (1957) description of the formation included the following: "Clay, silty, grayish brown to dusky brown, weathering pale yellowish brown, lenticular, carbonaceous, in parts fine sandy; contains scattered mica, abundant plant fragments. . . . Sand, light gray, weathered moderate red to moderate pink along strong cross-bedding, lenticular, medium- to coarse-grained. . . . Sandy clay. . . . Sand, white, weathering to moderate reddish orange, massive, lenticular, medium- to coarse-grained and in places granular, slightly micaceous; contains stringers of dark heavy minerals."

The Citronelle Formation in this area is draped unconformably on the Hattiesburg Clay, which consists chiefly of white, pink, and gray sandstone and locally indurated clay.

About 14 mi (22.5 km) east of the Citronelle field lies a delta 4–7 mi (6.4–11.3 km) wide; it stretches from Jackson, Alabama, on the north, southward to Mobile Bay, a distance of 52 mi (83.7 km). This delta, which accepts and spreads the waters of the Black Warrior, Tombigbee, and Alabama Rivers,[3] has a recent alluvial cover over its entire length, and averages about 5.5 mi (8.9 km) in width. The delta overlies a major fault system. The north end of this fault system is exposed and easily recognized on the surface at Jackson, Alabama. From there it disappears southward under the delta deposits toward Mobile Bay and the Gulf of Mexico.

The high bluffs extending along the eastern bank of the delta and of Mobile Bay appear to be surface escarpments of the fault system. Mobile Bay is on the downthrown side or, perhaps, represents a downdropped segment.

Geologic Structure

The Citronelle structure is, at every mappable horizon, a broad, slightly flat-topped, ovate anticline. It is residual in that it is the result of downwarping or subsidence of a salt basin, extensively on the east and north and moderately on the west. Normal dip is toward the Gulf of Mexico on the south. The structure has a north-northwest to south-southeast axis. Steepest dip is on the east, and most gentle dip is on the south. Closure is present on all the formations that have been mapped (Figs. 4–6). The structure has no appreciable gravity expression.

There is no faulting associated with the structure, no major truncations, nor any noticeable facies changes down to and through the producing formation. Although 434 wells have been drilled, no oil shows have been logged in any of the shallow formations which produce throughout the Gulf Coast (i.e., Wilcox, Annona, Eutaw, Tuscaloosa, Paluxy, and Mooringsport.

Surface elevations in the field range from 111 to 346 ft (33.8–105.5 m) above sea level.

A geologic section (Fig. 2) and a type electric log (Fig. 3) show average thicknesses of and depths to the formations penetrated in the Citronelle field. The structural pattern and the formations are surprisingly consistent (see Figs. 4–6). Only the recognition of the Glen Rose–Hosston and Hosston–Cotton Valley contacts are the subjects of disagreement among lithologists. The en-

[3]These three rivers join and then spread across the delta, where they are known as the "Mobile" and the "Tensaw" as they enter Mobile Bay.

tire producing "Donovan" section is probably equivalent to the lower Glen Rose (Rodessa, Pine Island, Sligo) of Louisiana and Mississippi. The Rodessa and Sligo are prolific oil- and gas-bearing formations in Arkansas, Texas, Louisiana, and Mississippi.

It is my opinion that the hydrocarbon accumulation in the Rodessa section was concentrated to the east of the present structure during deposition of the lower Glen Rose, prior to the growth of the Citronelle structure and formation of the Jackson-Mobile fault system (also called "Delta fault system"). The faulting, probably caused by the deposition of flood overburden, initiated the downwarping of the deep salt basin between Citronelle and the "Delta fault system" and directed a westward migration of the hydrocarbons, but not before most of the gas had been dissipated through the fault planes. Thus, the reservoirs in the present field are undersaturated, an abnormality compared with other, high-gas-ratio Rodessa reservoirs. Because of their blanket deposition, the sandstone lenses in the saltwater zone were probably flushed of any hydrocarbon accumulation before faulting or salt movement.

The Wiggins anticline, an irregularly shaped anticlinal ridge, extends westward from the Citronelle structure into central southern Mississippi. This feature has an east-west length of approximately 58 mi (93.3 km). It appears to have developed during the Early Cretaceous, prior to any Late Cretaceous deposition.

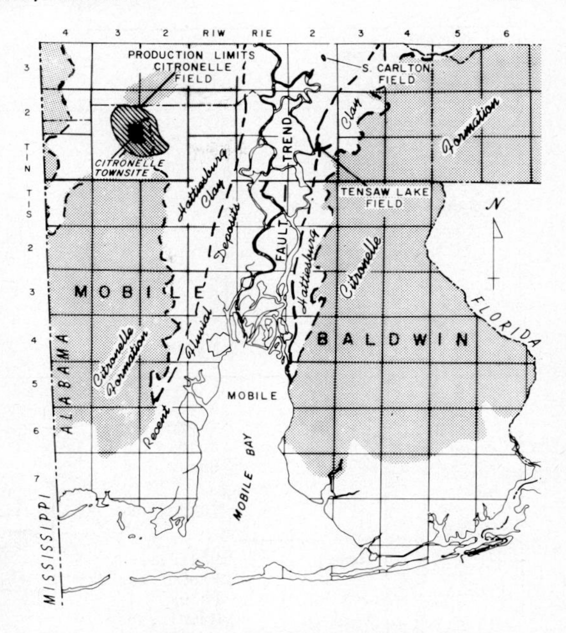

FIG. 1—Southwest corner of Alabama, showing surface geology and southward-flowing rivers, which meander over a long, wide delta. This delta overlies a complex fault system. Citronelle field envelops Citronelle townsite. Dashed line denotes depositional limits of Hattiesburg Clay.

RESERVOIR

Though the Citronelle structure is a simple "textbook" anticline, the reservoir is complicated for primary oil production, and extremely so for secondary production.

Production is found in the channel sandstones that were deposited and preserved in narrow, irregularly shaped meander belts which cut through and across a slightly positive area in a brackish-water embayment. A gently oscillating sea level caused the infilling sediment to be entrapped rather than flushed out, and the slow but continual growth of this structure influenced the shape and direction of each reservoir sandstone during each depositional cycle.

The same environmental conditions evidently continued throughout "Donovan" deposition, for there is constant conformity even of the lowermost productive sandstone, or its equivalent, with the overlying Ferry Lake Anhydrite marker.

The reservoir rock in the "Donovan" section consists of poorly sorted siliciclastic sediments: a small amount of gray to brown broken shale irregularly interbedded with siltstone, asphaltene, and an abundance of mica; and sandstone lenses that are highly undersaturated, "tight," micaceous, and silty. The sand grains are subangular and vary in composition, grain size, and shape. The individual sandstones generally grade from coarse grained at the base to fine grained at the top.

The siltstone and medium- to fine-grained sandstone range from low-porosity, nonpermeable, slaty siltstone, to clean, porous, permeable, subangular sandstone. There is a high degree of heterogeneity throughout the section. The high mica content indicates that the source of the sediments was probably the Paleozoic and Precambrian complex of the Southern Appalachian system. The siltstones and claystones show vertical bedding disturbances caused by plant roots. The presence of oyster shells suggests that deposition took place in a shallow, brackish-water embayment. There is very little calcareous material present; therefore, only a few wells were acidized in an attempt to increase production.

SYSTEM	SERIES	GROUP OR FORMATION		LITHOLOGY	THICKNESS (FT)
Quaternary	Pleistocene	Citronelle			1,600
Tertiary	Pliocene				
	Miocene	Undifferentiated		Sandstone and shale	
	Oligocene	Undifferentiated			
	Eocene	Jackson		Sandstone and shale	1,500
		Claiborne			
		Wilcox			
	Paleocene	Midway			1,300
Cretaceous	Upper	Navarro		Chalk	1,300
		Taylor			
		Selma			
		Eutaw		Sandstone	50
		Tuscaloosa	Upper	Sandstone and shale	900
			Marine		
			Lower		
	Lower	Dantzler		Massive sandstone section	3,800
		Fredericksburg			
		Paluxy			
		Mooringsport		Shale	
		Ferry Lake		Anhydrite	100
		Rodessa	"Donovan"	Sandstone and shale	850
		Pine Island			
		Sligo			
		Hosston		Sandstone and shale	2,100
Jurassic		Cotton Valley		Sandstone and shale	2,200
		Haynesville		Sandstone, shale, and ls.	2,100
		Buckner		Shale and anhydrite	300
		Smackover		Limestone	200
		Norphlet		Sandstone and shale	1,000

FIG. 2—Type geologic section showing formation thickness in Citronelle field.

The "Donovan" sandstone zones were numbered from 1 to 43 (some were divided into an upper member, A, and a lower member, B), starting with the shallowest, and the zones were correlated as to their equivalent stratigraphic position in the overall producing column. Any one zone may have had more than one time-equivalent meander bed and, consequently, may form more than one reservoir.

The producing column was divided into four intervals. Group IV includes sand zones 1 through 11; Group III includes sand zones 12 through 18; Group II includes sand zones 19 through 32; and Group I includes sand zones 34 through 43. Each group has average porosity, average permeability, and calculated oil in place as shown in Table 1.

Considerable credit has been given to M. D. Wilson and J. R. Warne, consulting stratigraphers retained by the Citronelle Operators Unit, and to Webb Holland, Unit Reservoir Engineer, for their reservoir work and projections. Fifty-two separate productive sandstone zones were studied, none of which blankets the field; 330 separate reservoirs were mapped and evaluated. One hundred reservoirs have been flooded and, overall, the projections have been very satisfactory. Four maps, chosen at random from over 45, show the thickness, distribution, and interconnection of an individual reservoir sandstone (Fig. 7). Certain areas are indicated where one stream has eroded downward into an underlying reservoir unit or laterally into an adjacent, previously deposited sandstone, thereby connecting the reservoirs.

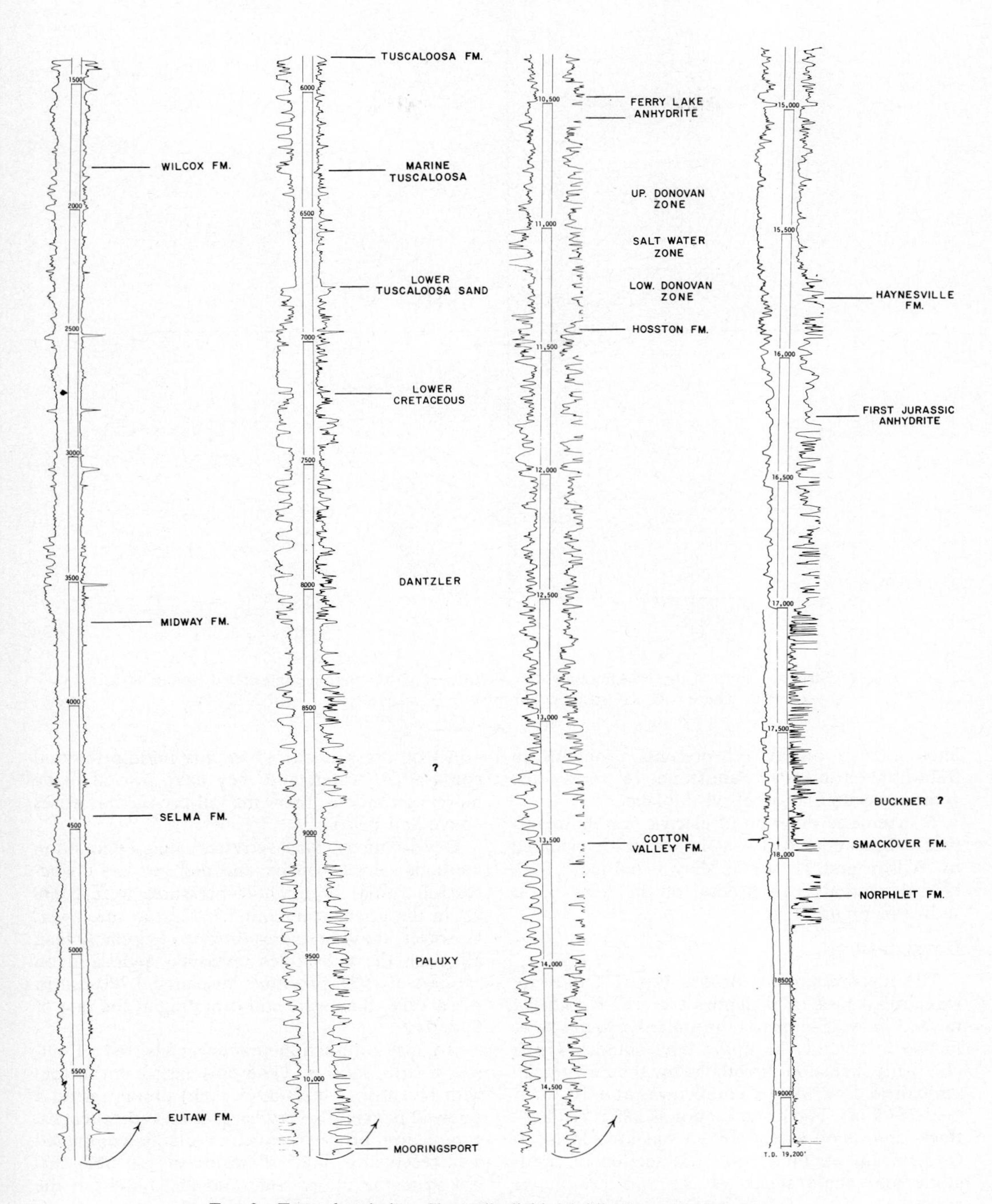

FIG. 3—Type electric log, Citronelle field, Mobile County, Alabama.

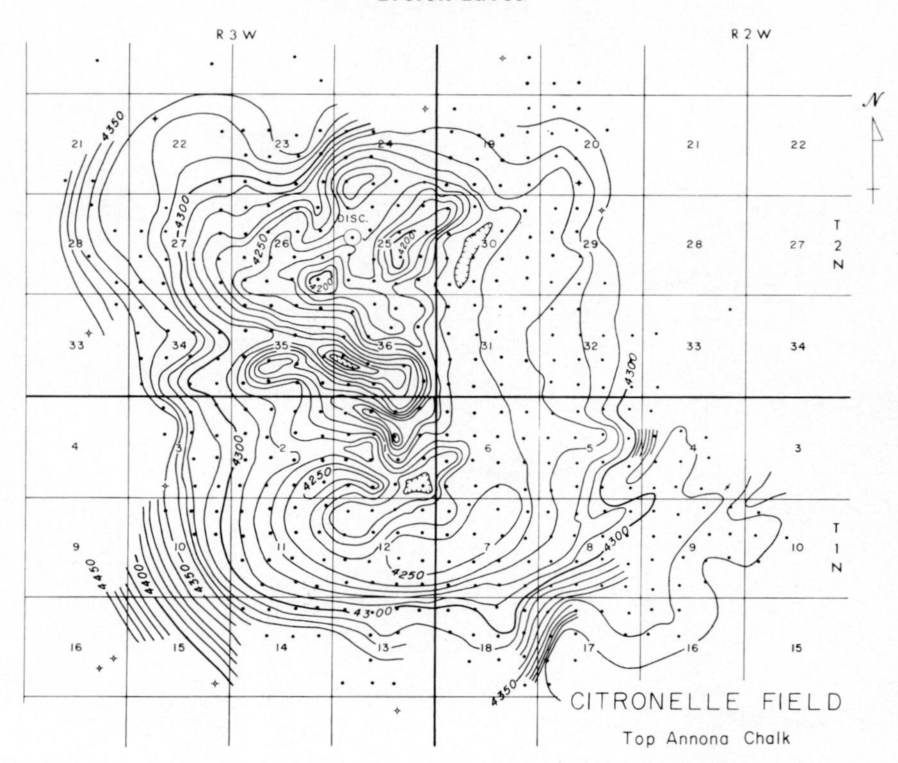

FIG. 4—Structure map of top of Annona Chalk. Weathering and channeling created erratic surface of crest. There was no subsidence in history of structure. C.I. = 10 ft (3.3 m).

Such a situation always presents problems in fluid-flow efficiency. Sandstone 16 shows the greatest accumulation of oil in place.

A schematic diagram of electric-log characteristics of meander-belt sandstones, as determined by Wilson and Warne, is shown in Figure 8. A typical vertical cross section of the reservoir is shown in Figure 9.

DEVELOPMENT

The discovery well, Brooks No. 1 Donovan, was drilled to a total depth of 11,517 ft (3,510.4 m) and proved to have oil-productive sandstones in two intervals. The upper one includes 450 ft (137.2 m) of section directly below the Ferry Lake Anhydrite. Its oil-water contact was at −10,787 ft (−3,287.9 m). The lower section is 240 ft (73.2 m) thick, and its oil-water contact was at −11,145 ft (−3,397 m). A 140-ft (42.7 m) section of sandstone and shale separates the two productive zones.

All of the sandstones in this middle interval contain salt water, and they have proved to be more continuous than the oil-producing lenses above and below.

Development was erratic; dual-completion methods were expensive and more or less unsuccessful. Initial bottomhole pressures were 5,030 psi in the upper zone and 5,351 psi in the lower. However, these pressures dropped extremely fast, although the allowables remained high, and on January 1, 1957, pressures measured 2,700 psi. In April 1958, they were still dropping at the rate of 5 lb/day.

Original production practice consisted of setting a string of 7-in. (17.8 cm) casing on bottom with two strings of 2-in. (5.1 cm) tubing set on a sidewall packer—a very hazardous and expensive completion program. Such a dually completed well received a daily allowable of 550 bbl—300 bbl from the upper zone and 250 bbl from the lower. These allowables were reduced to 200 bbl and 150 bbl, respectively, on September 1, 1956. Only a few wells were capable of sustaining these allowables.

The two oil reservoirs and intervening saltwater zone appeared to be interconnected downdip, because all three registered bottomhole-pressure drop coincident with primary withdrawal. Water drive in both oil zones was minimal. Oil/water contacts varied in different sandstones as much as 30 ft (9.1 m).

Fire hazards, paved streets, pipelines, tank-battery locations, rights-of-way, pollution, and well testing within the city presented many physical, economic, and political problems. Fourteen wells were drilled the first year, but only one rig was running on August 16, 1956, a year after discovery of the field.

Many wells were produced from only one zone until completion regulations were changed by the Oil and Gas Board in 1960, permitting wells to be dually completed through one string of tubing.

After the regulation changes, completion costs dropped from $300,000 to $180,000 per well, and drilling and completion time decreased from 45 to 30 days. Gulf Oil completed a 38-mi (61.2 km), 8-in. (20.3 cm) crude line to a terminal at the port of Mobile, Alabama, on September 1, 1956, which eliminated a trucking charge that ranged between $0.25 and $0.30/bbl. There were 133 wells completed by the end of 1958, and 314 by April 1962. All wells had to be pumped. It was soon realized that beam pumps were insufficient; electrically driven downhole pumps were a necessity. This factor increased lifting costs, but blended into later production procedure that was necessitated for field-wide unitization and improved economics.

The necessity of unitization to carry out secondary recovery became apparent, and the first unitization meeting was called in September 1957. Core Laboratories presented a program on October 15, 1958 (Core Laboratories, 1958). There

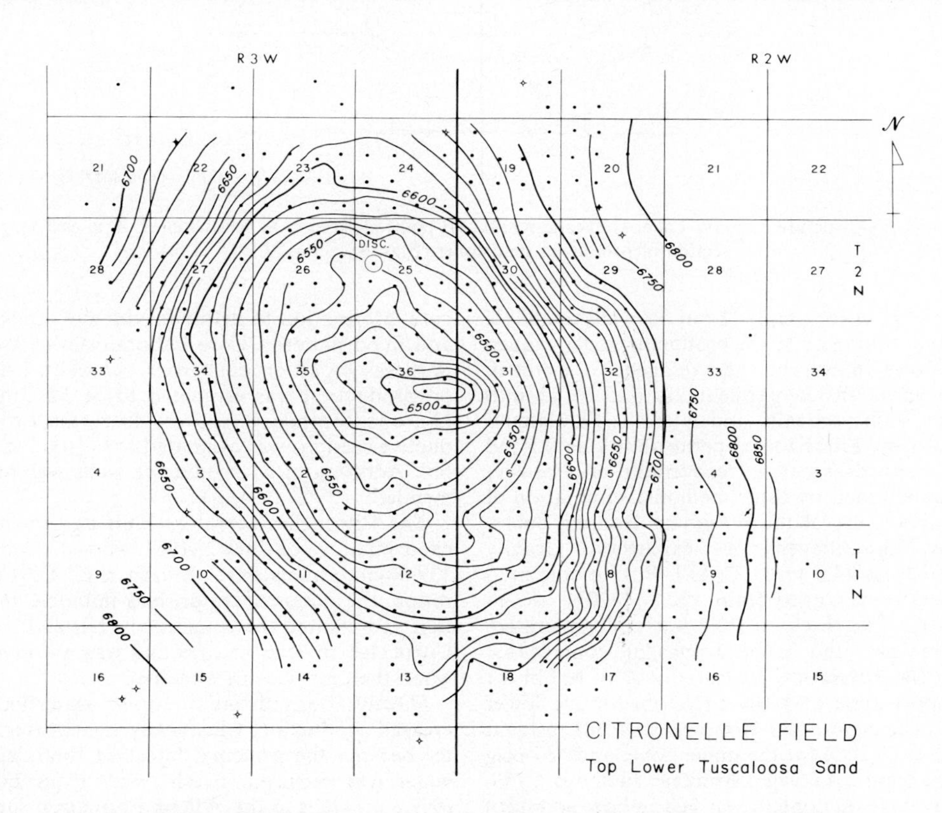

FIG. 5—Structure map of top of lower Tuscaloosa sandstone. C.I. = 10 ft (3.3 m).

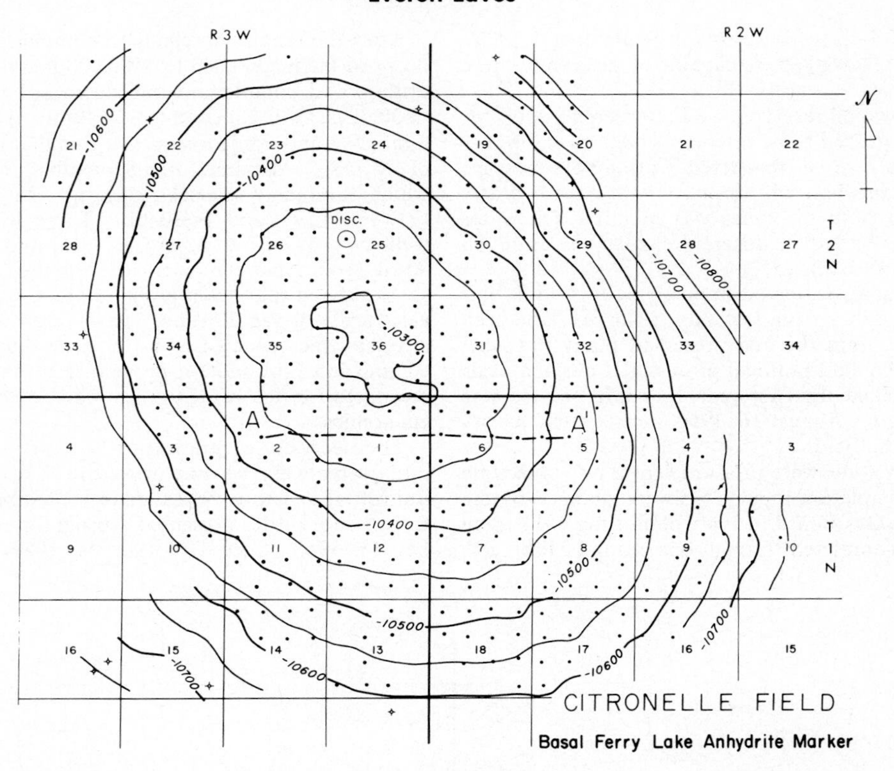

FIG. 6—Structure of Ferry Lake Anhydrite marker, at point close to base of section. *A–A′* is east-west cross section of producing sandstones (Figs. 9). C.I. = 50 ft (15.2 m).

were 3,192 core samples from 55 wells available for study. Porosity in the producing zones ranged from 10 to 16 percent and averaged 13.3 percent. Wide variations in permeability (0.5 to 75 md) existed both vertically and laterally. Sandstones with as low a horizontal permeability as 0.5 md were mapped as having recoverable oil. Porosity was determined by three methods: summation of total fluids, use of the Washburn-Bunting porosimeter, and Stevens gas expansion. Gravity ranged from 44.5 to 45.7° API. Bottomhole mud temperatures ranged from 190 to 220°F (87.8–104.4°C). The reservoir fluid was undersaturated. No free gas, and hence no gas/oil contact, existed. Gas saturation measured 244 cu ft/bbl for the upper zone and 104 cu ft/bbl for the lower zone. The saturation pressure was 581 psig at 242°F (116.7°C) for the upper zone, and 487 psig for the lower. Average shrinkage factor is 1.119. The average oil content was 303 bbl per acre-foot of sand. Gamma-ray logs were used along with all core information to define productive sandstones and flood patterns. Core Laboratories' (1958) report was based on data from 112 wells. Primary oil production was estimated to be 52.1 million bbl, and secondary recovery from the two productive sections was estimated to be 106.9 million bbl. Addition of 21 productive wells was recommended.

The Citronelle Operators Unit agreement became effective on June 1, 1961, when 138 wells on 139 40-acre tracts were unitized. Gulf Oil Corporation, the major operator, had initiated unitization and chaired the committees, but had sold its Citronelle interests in 1959 and was not involved when the unit became effective.

Shortly after unitization, a pilot waterflood increased production. Chemically treated fresh water became the principal injection fluid; all salt water was recycled. Fresh water from 2,000-ft (609.6 m) wells in the Wilcox Formation supplies all the injection water needed.

Although many of the wells had penetrated from 8 to 12 potentially productive sandstones, no more than two or three were used for production or water injection. The well-conversion and workover program was immense, as can be easily understood by studying the cross section (Fig. 9) and the four sandstone maps (Fig. 7). Each economically floodable foot of sandstone was included in the program.

On October 5, 1972, the field had produced 100 million bbl of oil and, at the end of 1973, total production was 107,203,492 bbl. December 1973 production was 436,290 bbl from 326 producing wells; 63 injection wells, 4 saltwater-disposal wells, and two shallow freshwater-supply wells were in operation (Fig. 10).

A liquid-extraction plant was completed and operated by an outside firm in 1964, and later was sold to Cities Service, although they had no production in the field.

In July 1961 a deep test was drilled on the east flank of the Citronelle structure to total depth of 19,200 ft (5,852.2 m). The test proved to be dry in the Hosston, Cotton Valley, Smackover, and the upper 900 ft (274.3 m) of the Norphlet (which at present is the deepest producing formation in the Gulf Coast area). The well was plugged back and completed in the upper "Donovan" sandstones.

Table 1. Producing Intervals, Citronelle Field, Alabama

	Average Porosity (%)	*Average Permeability (%)*	*Calculated Oil in Place (bbl)*
Group IV	12.37	4.97	47,002,363
Group III	14.22	22.92	151,411,061
Group II	12.76	7.31	42,452,291
Group I	13.60	11.91	79,025,533
Av. and Total	13.24	11.78	319,891,248

UNITIZATION

The original unitization agreement became effective June 1, 1961, with participation of each quarter-quarter section dependent on (1) a completed producing well with at least 6-months production history; (2) a Microlog acre-foot factor, which was the total aggregate thickness of the various producing columns in the Rodessa section as interpreted from the Microlog; and (3) an "oil-in-place equivalent acre-foot factor," which was based on the porosity and permeability characteristics of various zones or columns in which each productive sandstone was correlated. The oil and gas laws required 75 percent approval by both the operators and royalty owners.

Adjustment of assets among the operators was based on various factors: whether 7-in. (17.8 cm) or 5 ½-in. (13.9 cm) casing was used, 2 ½-in. (6.4 cm) or 2-in. (5.1 cm) tubing, number of tanks in battery, etc.

Operations were carried on by a manager and a staff of 66, who answered to an Operators Committee. Engineering, geologic, auditing, legal, executive, and insurance subcommittees were appointed. Subcommittee members were furnished by operators with interest in the unit, and were not part of the unit staff.

Although 190 or more tracts qualified on the effective date, only 138 wells were committed to the agreement. Many of the operators, if not all, questioned the success of a secondary program, and 42 qualifying wells were held out to await results.

Either saltwater or freshwater fluid injection was the only choice; there was no gas available. The erratic nature of the reservoir presented many problems. However, when the flood proved successful in both production and economics, most operators wanted to be included in the secondary-recovery program.

Two operators on the east flank chose to create their own units. On April 1, 1964, Sun Oil Company became the operator of the Southeast Citronelle Unit, which included 38 wells, and on March 1, 1965, Ancora Corporation became the operator of the East Citronelle Unit, which included 31 tracts with 27 wells.

On October 1, 1962, the Citronelle Operators Unit added 94 wells, and on May 26, 1966, this unit took in all the remaining productive wells in the field. This last enlargement was contested in the courts by the Ancora Corporation, and was delayed many months before being approved.

Thus, the field was to have three separate waterflood developments: the Citronelle Operators Unit, 341 wells; Sun's Southeast Citronelle Unit, 38 wells; and Ancora's East Citronelle Unit, 27 wells (Fig. 11).

There were some critical problems concerning boundary-line injection and withdrawal agreements, but none were ever signed, and each operation was conducted as if there were three separate reservoirs. This was the only blemish in an

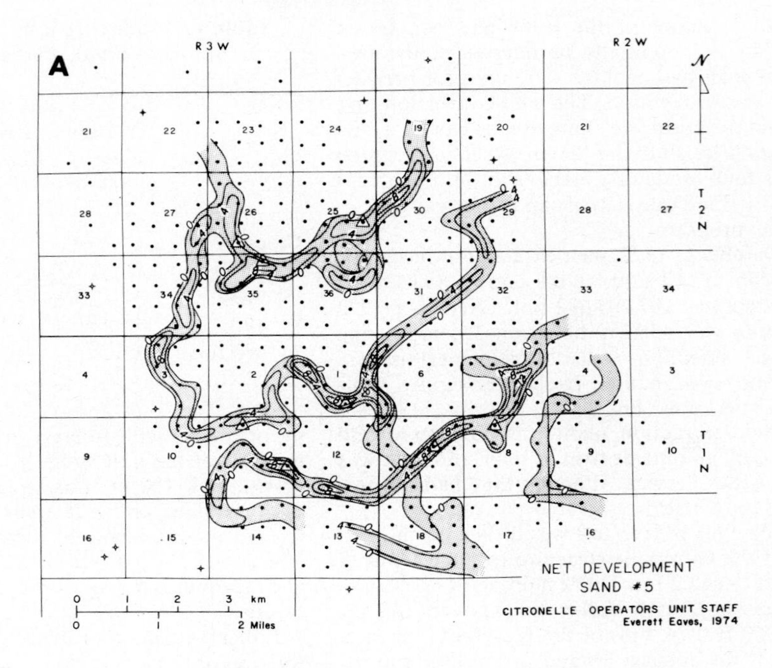

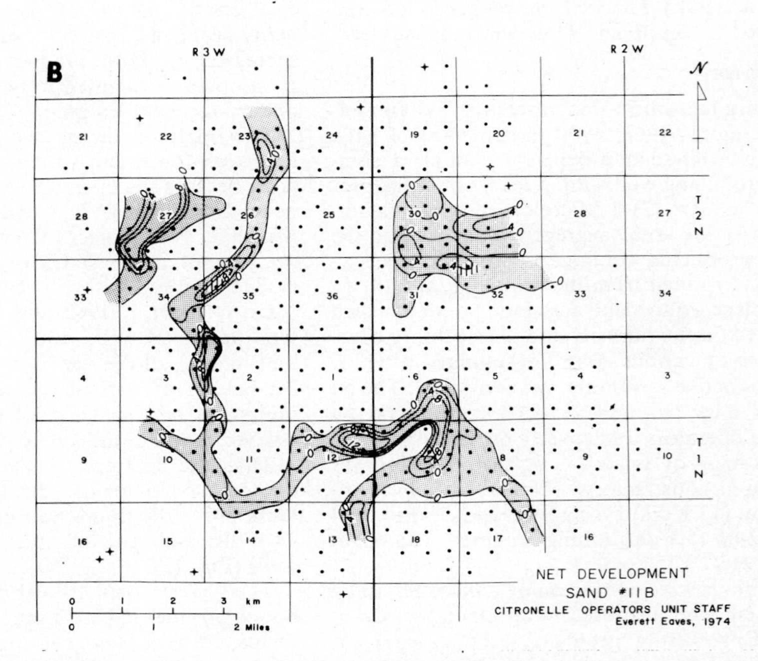

FIG. 7—Isoporosity maps of "Donovan" sandstones showing certain areas of lateral (horizontal hachures) c
A, sandstone 5; **B**, sandst

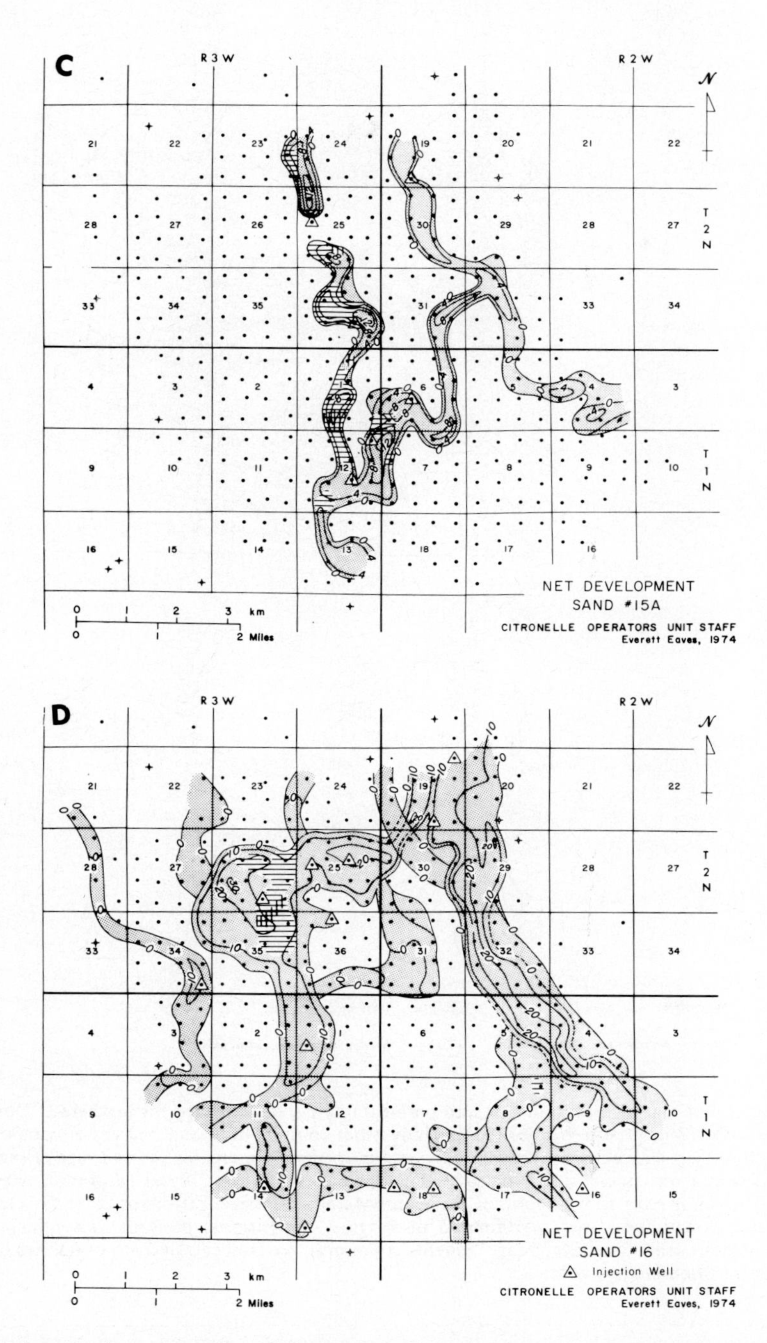

achures) erosion creating a connection of fluid-flow paths with either overlying or underlying sandstone. one 15-A; **D**, sandstone 16.

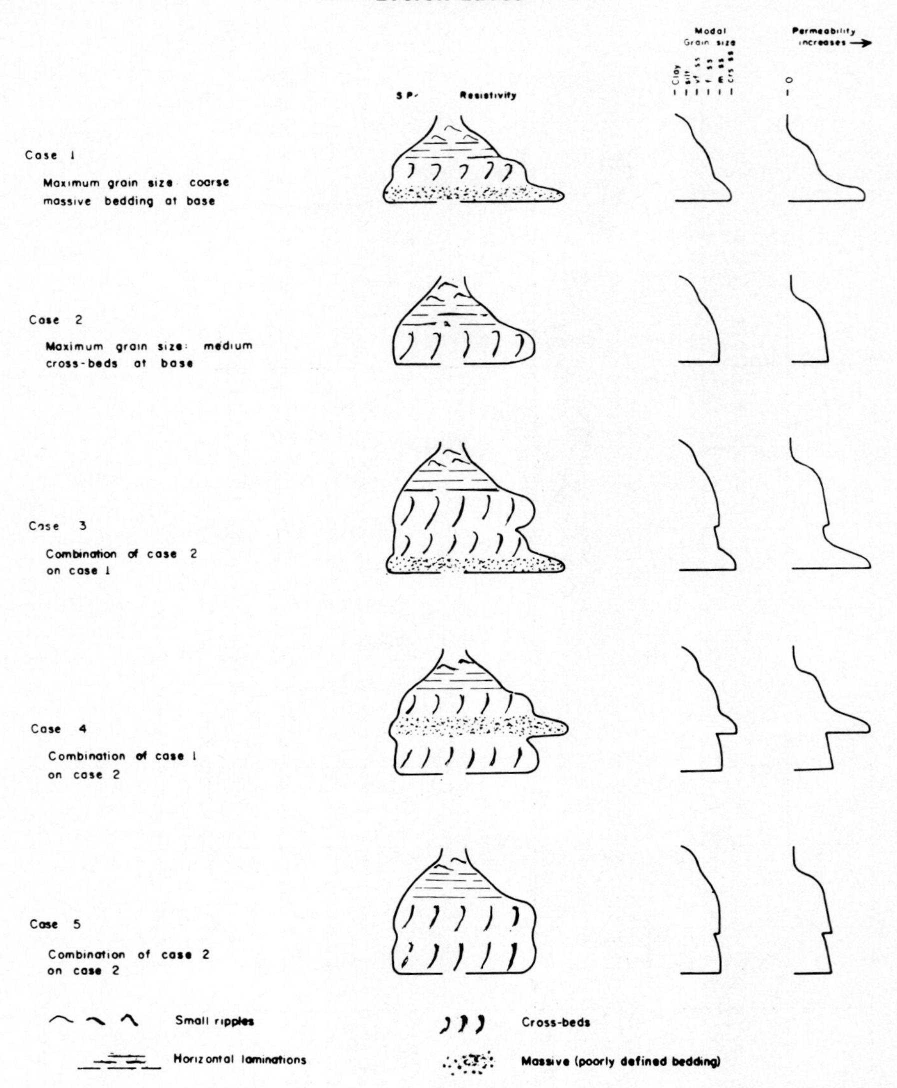

FIG. 8—Schematic diagram of log shapes, grain size, and permeability curves expected for unaltered "Donovan" meander-belt sandstones. After Wilson and Warne (1964). Many other combinations and degrees of compounding are possible. SP deflection reflects clay content. Primary clay content generally increases as grain size of sandstone decreases. Resistivity decreases with increasing clay content and water saturation. Water saturation is partly a function of permeability, which in turn is a function of clay content and grain size. Effects of secondary clay and chemical cements are not shown; they are superimposed on original depositional patterns. Secondary clay is commonly scattered throughout sandstone. Secondary cements commonly are concentrated at boundaries of sandstone beds, causing resistant impermeable zones.

→

FIG. 9—Typical cross section of upper and lower "Donovan" sandstone reservoirs in Citronelle field. Location is shown as *A–A′* in Figure 6. Datum is base of Ferry Lake Anhydrite.

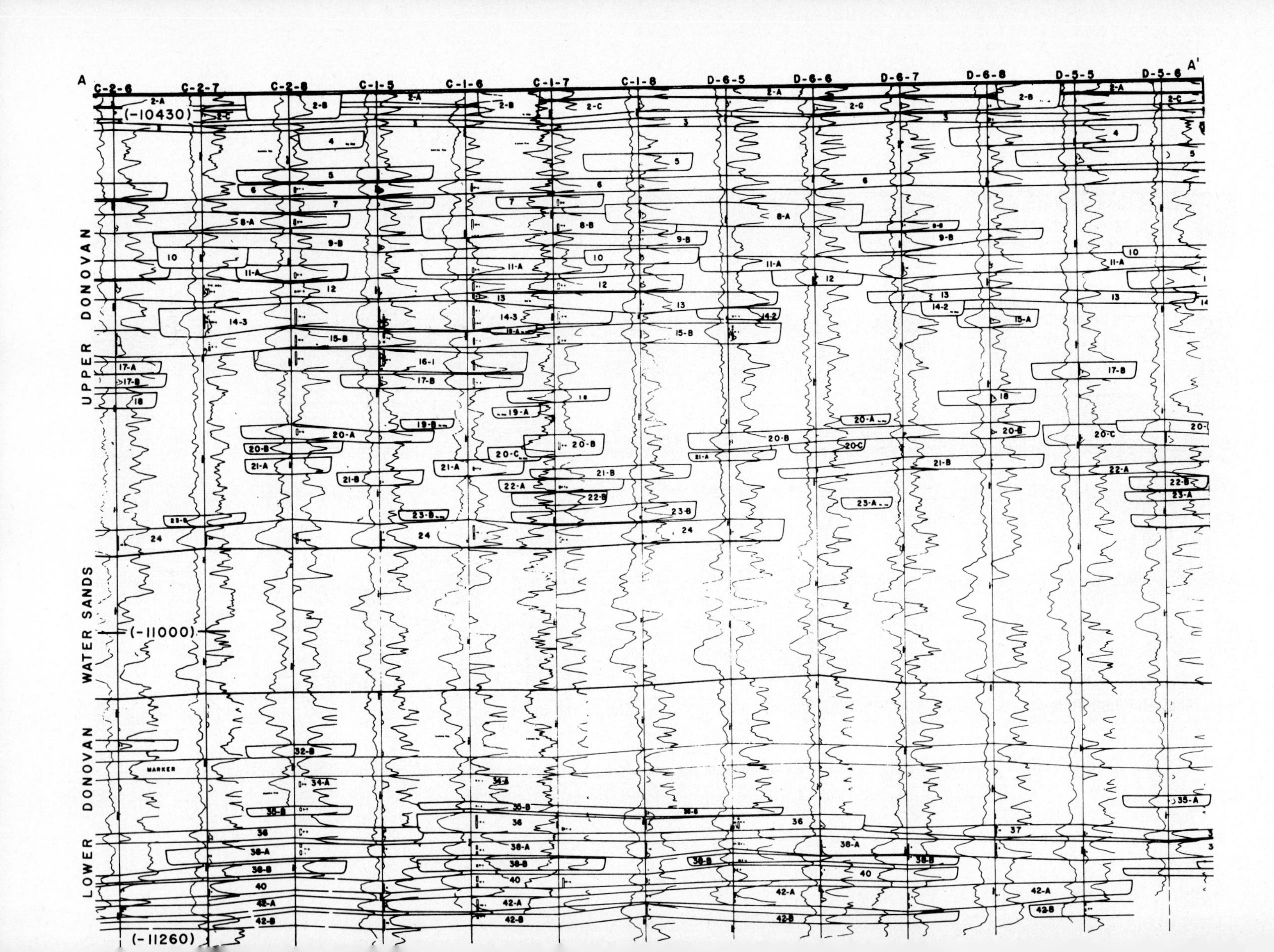

A
A'
C-2-6
C-2-7
C-2-8
C-1-5
C-1-6
C-1-7
C-1-8
D-6-5
D-6-6
D-6-7
D-6-8
D-5-5
D-5-6
UPPER DONOVAN
WATER SANDS
LOWER DONOVAN
(-10430)
(-11000)
(-11260)
MARKER

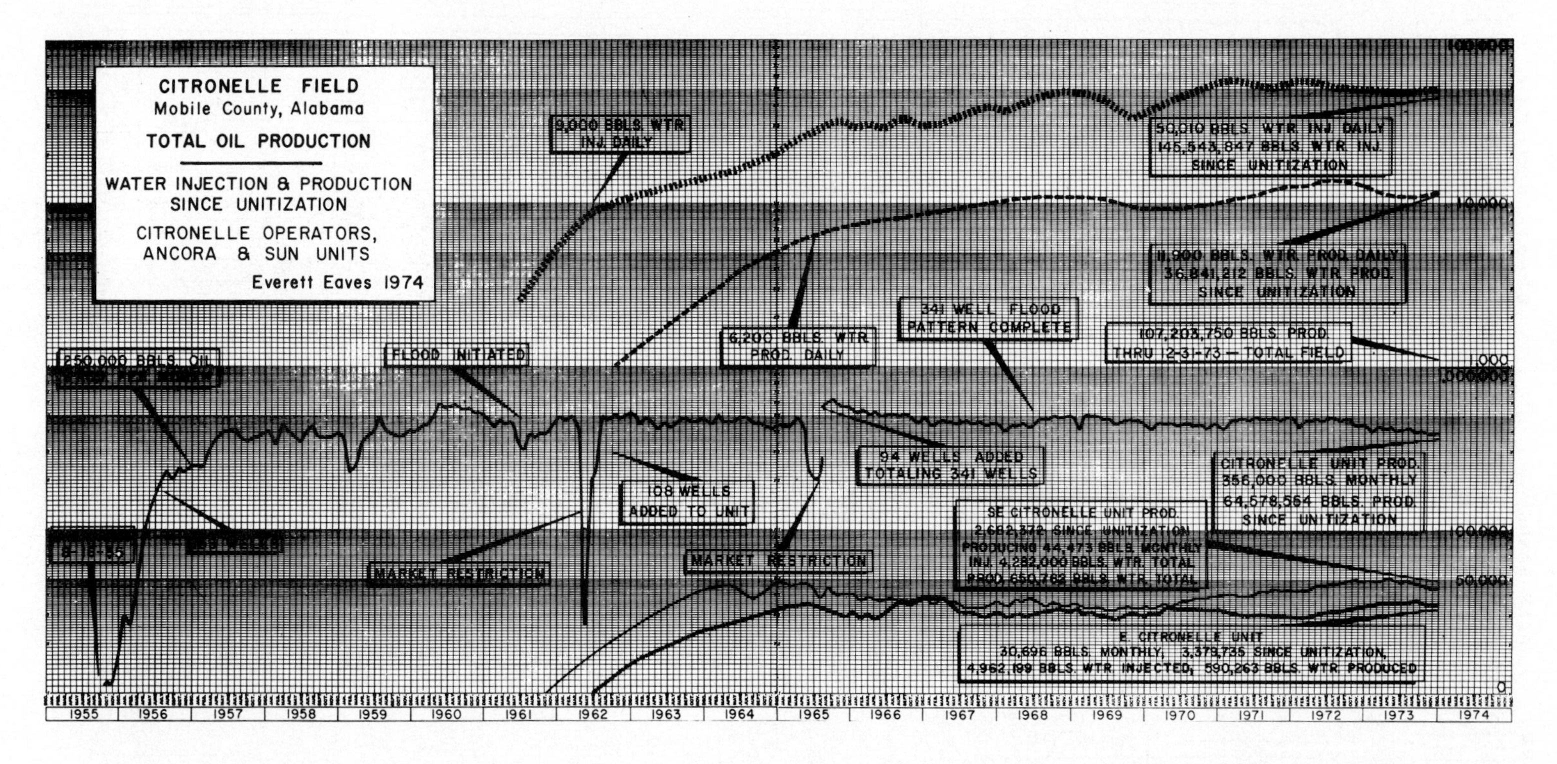

FIG. 10—Set of curves showing monthly rate and accumulation of oil production, water production, and injection through December 31, 1973.

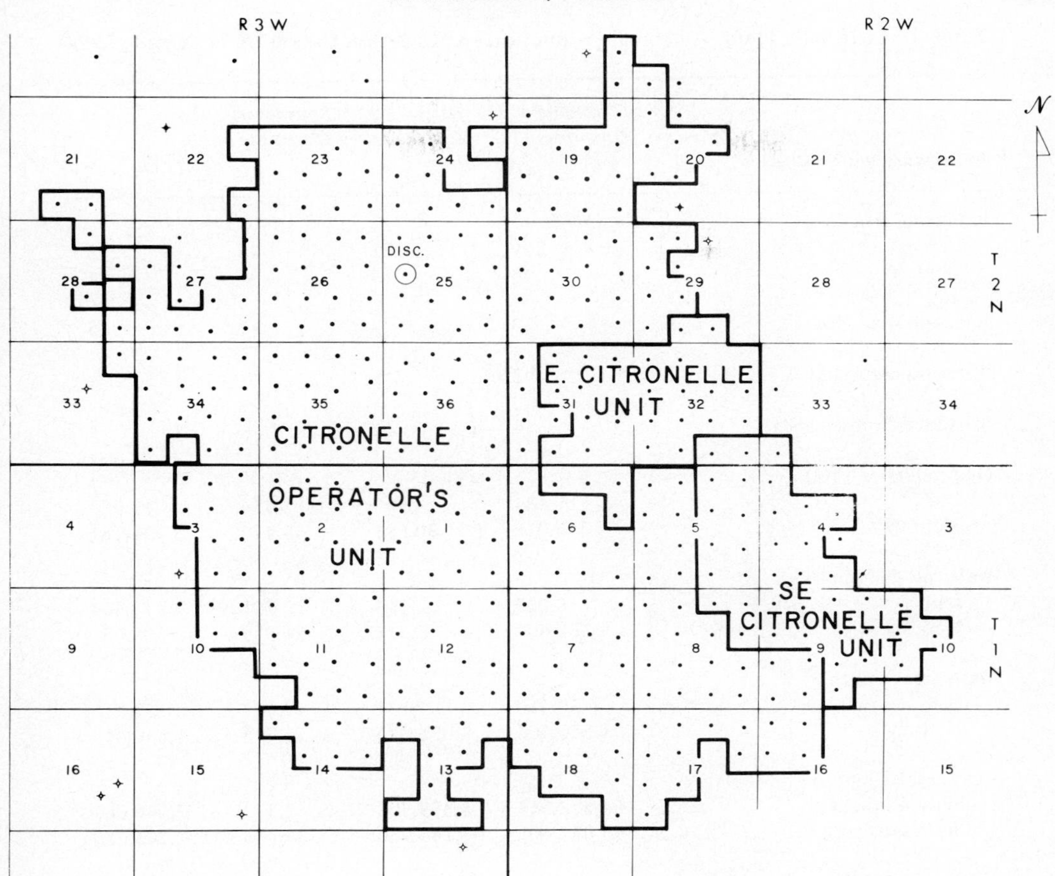

FIG. 11—Citronelle oil field and limits of three unitized areas: Citronelle Operators Unit, 341 wells; Southeast Citronelle Unit, 38 wells; and East Citronelle Unit, 27 wells, 31 tracts.

otherwise very successful secondary-recovery program.

The peak number of producing wells for the field was 412. In all, 434 productive wells have been drilled in the structure, but sporadic development and the abandonment of some wells after only a few months' production resulted in a varying number of producing wells at different times during the history of the field. The Citronelle Operators Unit included over 500 operators and 2,000 royalty owners. One tract in the middle of the field, SW ¼, SE ¼, Sec. 36, T2N, R3W, was never drilled because of landowner-operator problems, but adjustment was made with the unit and no one suffered loss of revenue because of offset withdrawal.

Estimated recovery per acre-foot and for the field has varied within the separate units. Lack of boundary-line accord and cooperation resulted in different injection volumes and pressures and, consequently, different calculations. The Citronelle Operators Unit, consisting of 341 tracts on 13,640 acres out of a total of 410 tracts on 16,400 acres, had been calculated as having 319,000,000 bbl of oil in place. Since discovery, 95,037,000 bbl (29.8 percent) had been produced from these tracts through December 31, 1973. Total recoverable oil is placed at 124,628,000 bbl. This leaves 29,591,000 bbl (9.28 percent) to be produced between January 1, 1974, and the present projected date of abandonment on December 31, 1984. Increase in the price of crude or a tertiary flood could increase these estimates.

Operation of the Citronelle Operators Unit was switched from an employed staff to Mobil Oil Corporation on August 1, 1967, and then back to

Table 2. Citronelle Field, Alabama, Production and Injection Data as of January 1, 1974

Well Status and Production	*Citronelle Operators Unit*	*East Citronelle Unit*	*Southeast Citronelle Unit*	*Total*
Producing wells				
Active	229	21	28	278
Shut in	49	3	6	58
Injection wells, water	58	3	4	65
Water-supply wells	2	1	1	4
Saltwater-disposal wells	4	-	-	4
Oil production (bbl)				
Daily	11,484	990	1,434	13,908
Monthly	356,012	30,696	44,473	431,181
Water production (bbl)				
Daily	11,900	113	791	12,804
Monthly	368,607	3,515	24,539	396,661
Water injection (bbl)				
Daily	50,010	1,141	1,765	52,916
Monthly	1,550,306	35,375	54,738	1,640,419
Cumulative oil produced (bbl)				
Since unitization	64,578,554	3,379,735	4,682,372	72,640,661
Since discovery	95,037,401	5,146,230	6,534,130	107,203,750
Cumulative water produced (bbl)				
Since unitization	36,841,212	590,263	650,762	38,082,237
Cumulative water injected (bbl)				
Since unitization	145,543,847	4,962,199	2,281,909	152,787,955

a staff operation on July 1, 1973. The original unitized flood program, which consisted of a complete production-injection pattern for all floodable reservoirs, using all the wells in the unit, was completed during Mobil's operating period, and the return to staff operation was agreeable to all.

The production-injection curves since discovery are shown in Figure 10. Statistics as of January 1, 1974, are shown in Table 2.

Pollution Control

Development of the field involved many operators and many more operating interests. The last wells drilled were chiefly farmouts completed on a very tight budget.

At the time of complete unitization, over 80 tank batteries existed, flow lines ran everywhere, and some locations were covered with weeds. Electrical hookups were hazardous and fires were numerous.

The problems for consolidation and clean-up programs were staggering. However, they were not ignored; by 1971, over $500,000 had been spent on a general clean-up program, and $1,400,000 additional on tank-battery and flow-line consolidation. Since the program started, fire insurance premiums have been reduced three times. The operators believe they have met all of the industry requirements for safety and appearance.

Summary

The Citronelle field, because of its 40-acre spacing and its undersaturated reservoirs, was a disappointment to the operators, especially during primary production, and cannot be considered a great success as to return on dollar invested. Unitization of the field, however, and excellent stratigraphic interpretation of complicated multi-sandstone reservoirs resulted in a successful secondary-recovery program. The number of wells drilled, although expensive, presented greater efficiency in production and helped change earlier disappointments into a satisfactory investment.

Recent seismograph investigation of the area by some of the operators may lead to deeper drilling. The Hosston, Cotton Valley, Smackover, and Norphlet produce oil and gas in other areas and cannot be eliminated here as possible productive formations. These formations are not unitized, however, and any exploration will be by individuals, not by the units.

The Citronelle Operators Unit is studying the merit of a tertiary flood in the "Donovan" sandstones using carbon dioxide as a miscible agent. Such a program has proved economically successful in certain other areas. The cost of generating the carbon dioxide at Citronelle no doubt will be the deciding factor.

References Cited

Aronow, Saul, 1974, Review of geology of the Mississippi-Alabama coastal area and nearshore zone: AAPG Bull., v. 58, no. 1, p. 157-158.

Core Laboratories, Inc., 1958, Reservoir engineering study of the Donovan sand reservoirs, Citronelle field, Mobile Co., Alabama: Unpub. Rept.

Isphording, W. C., and G. M. Lamb, 1971, Age and origin of the Citronelle Formation in Alabama: Geol. Soc. America Bull., v. 82, no. 3, p. 775-779.

Otvos, E. G., Jr., 1973, Geology of the Mississippi-Alabama coastal area and nearshore zone: New Orleans Geol. Soc. Field Trip Guidebook, 67 p.

Stringfield, V. T., and P. E. LaMoreaux, 1957, Age of Citronelle Formation in Gulf coastal plain: AAPG Bull., v. 41, no. 4, p. 742-757.

Wilson, M. D., and J. R. Warne, 1964, Sand continuity study, Citronelle field, Mobile Co., Alabama: Unpub. Rept. Prepared for Unit Manager, Citronelle Unit, Nov. 25, 1964.

Jay Field, Florida—A Jurassic Stratigraphic Trap[1]

R. D. OTTMANN,[2] P. L. KEYES,[3] and M. A. ZIEGLER[4]

Abstract The first Jurassic oil discovery in Florida was made in June 1970, in Santa Rosa County near Jay, 35 mi (56.3 km) north of Pensacola. Current estimates indicate recoverable reserves in the Smackover Formation of 346 million STB of oil and 350 Bcf of gas. The accumulation occurs on the south plunge of a large subsurface anticline, and the updip trap is formed by a facies change from porous dolomite to dense micritic limestone.

The Smackover consists of a lower transgressive interval of laminated algal-mat and mud-flat deposits and an upper regressive section of hard-pellet grainstones. Early dolomitization and freshwater leaching have provided a complex, extensive, high-quality reservoir. Irregular distribution of facies presents difficult problems in development drilling, unitization, and pressure-maintenance programs.

Hydrogen sulfide content of the hydrocarbons requires expensive processing facilities and well investment. A typical completed well costs $650,000, and an additional $200,000 is required for flow-line and inlet separation facilities. Add to this $550,000 for plant facilities to sweeten the oil for market, and each well investment approaches $1,400,000.

Daily production from Jay field is 93,500 bbl from 89 wells. The rapid development of this field resulted from a drilling program coordinated with modular plant design.

INTRODUCTION

Jay field is the most significant discovery in the United States since the Prudhoe Bay field discovery of 1968. Current estimates indicate recoverable reserves of 346 million STB of oil and 350 Bcf of gas. Since discovery of Jay field in Santa Rosa County, Florida, 35 mi (56.3 km) north of Pensacola, industry has drilled more than 200 wildcats in southwestern Alabama and the Florida Panhandle. Most Jurassic traps in this area are very subtle, and the key to exploration lies in the ability to combine the prediction of occurrence of favorable facies with the delineation of low-relief structures. Without question, many significant oil and gas accumulations remain to be found in this area and in other areas with similar subtle relationships. This discovery has indeed proved that large undiscovered accumulations still exist in the United States.

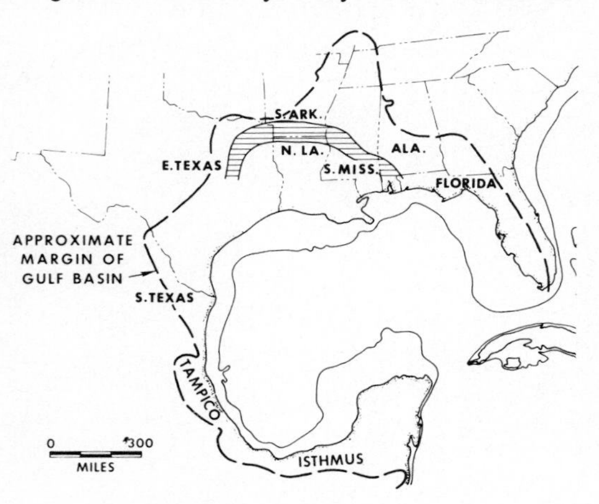

FIG. 1—Smackover producing trend, northern Gulf basin.

HISTORY OF JURASSIC SMACKOVER EXPLORATION

Exploration of the Jurassic Smackover Formation can be divided into four major periods. The first includes the years 1937–1950, in which 34 fields, containing about 250 million bbl of oil in reservoirs between 6,000 and 12,000 ft (1,830 and 3,658 m), were discovered in southern Arkansas and northern Louisiana (Fig. 1). These fields occur in the updip part of the Jurassic embayment and produce from a widespread carbonate grainstone facies. Shallow depths and the effectiveness of conventional seismic methods provided the keys to this early period of successful exploration.

During the period from 1950 through 1960, attention shifted to the East Texas basin, where exploration met with only limited success because of restricted distribution of reservoir facies and the inability to define by seismic means deep structures in the central part of the basin.

Interest in Smackover exploration was rejuvenated from 1960 to 1968 as a result of the development of common-depth-point (CDP) seismic techniques, which extended structural definition to as deep as 25,000 ft (7,620 m), and the development of capability to drill to these depths. Dur-

[1]Manuscript received, October 17, 1974. Reprinted in revised form with permission from Gulf Coast Association of Geological Societies Trans., 1973, v. 23, p. 146-157.

[2]Exxon Co., U.S.A., Midland, Texas 79701

[3]Exxon Co., U.S.A., Houston, Texas 77001.

[4]American Arabian Oil Co., Dhahran, Saudi Arabia.

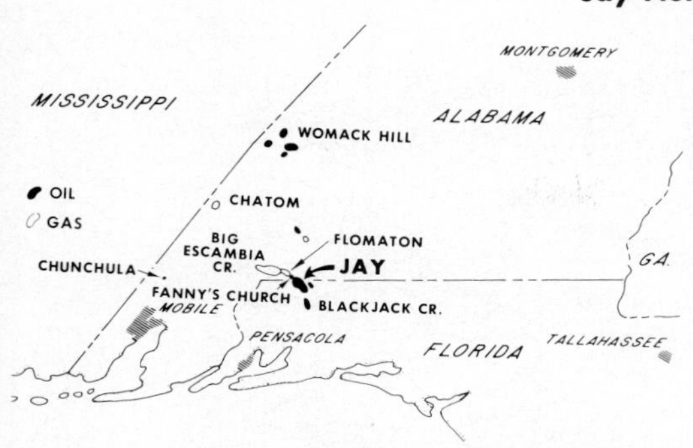

FIG. 2—Index map of Jay field area.

ing this period, exploration shifted to southern Mississippi, where 38 new fields were discovered with reservoirs at depths ranging from 12,000 to 18,000 ft (3,658 to 5,486 m).

Exploration for Jurassic Smackover accumulations in Alabama and Florida was a logical extension along trend of the discoveries of the first three periods. The fourth period began in 1968 with a Norphlet discovery on a well-defined structure at Flomaton in Escambia County, Alabama (Fig. 2), 90 mi (145 km) east of the closest Jurassic production. Discovery of the Jay field followed in 1970 when dolomitized Smackover was found on a structural nose approximately 7 mi (11.3 km) south of Flomaton. Other significant discoveries during this period were Womack Hill, Chatom, Big Escambia Creek, Fanny's Church, and Chunchula fields in Alabama and the Blackjack Creek field in Florida.

REGIONAL STRATIGRAPHY

Figure 3 shows the Jurassic section penetrated in southern Alabama and the Florida Panhandle. The Louann Salt, of probable Middle Jurassic age, is overlain by Norphlet clastic rocks. Above the Norphlet is the Smackover, a carbonate section which is overlain in turn by the evaporites of the Buckner. On top of the Buckner is the upper Haynesville Formation, which consists of red, fine- to coarse-grained clastic units interbedded with evaporites. West of Jay the upper Haynesville Formation also contains carbonate rocks. This sequence is capped by the coarse-grained, in places gravelly, clastic beds of the Cotton Valley Group.

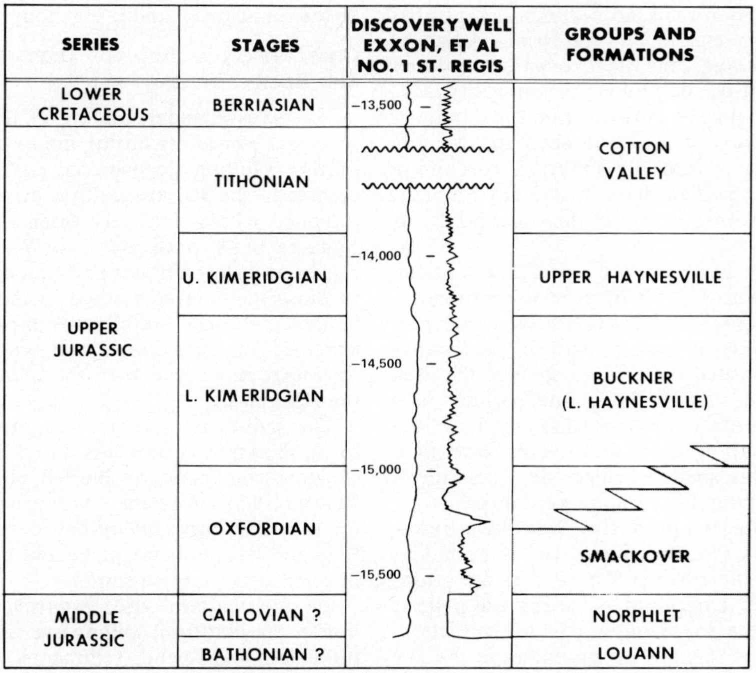

SERIES	STAGES	DISCOVERY WELL EXXON, ET AL NO. 1 ST. REGIS	GROUPS AND FORMATIONS
LOWER CRETACEOUS	BERRIASIAN	−13,500	COTTON VALLEY
UPPER JURASSIC	TITHONIAN		COTTON VALLEY
	U. KIMERIDGIAN	−14,000	UPPER HAYNESVILLE
	L. KIMERIDGIAN	−14,500	BUCKNER (L. HAYNESVILLE)
	OXFORDIAN	−15,000 −15,500	SMACKOVER
MIDDLE JURASSIC	CALLOVIAN ?		NORPHLET
	BATHONIAN ?		LOUANN

FIG. 3—Stratigraphic nomenclature, Jay field, Alabama and Florida.

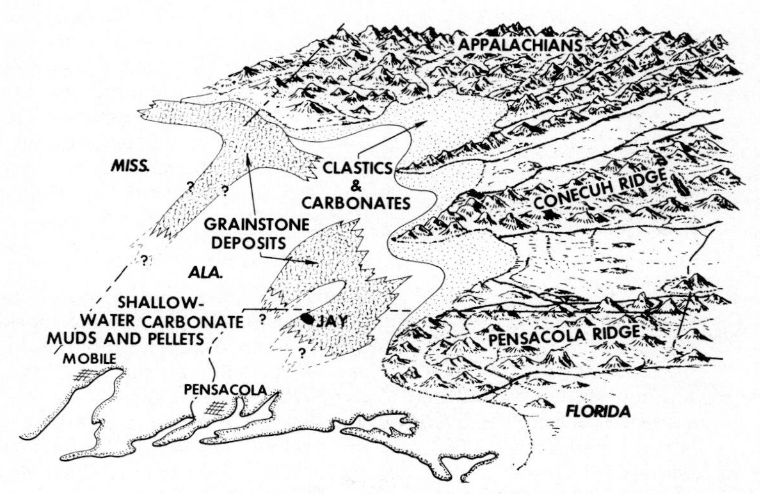

FIG. 4—Jay field, regional setting at end of Smackover deposition.

The Appalachian structural trend (Fig. 4) and other pre-Jurassic tectonic elements, such as the Conecuh and Pensacola ridges, had significant influence on the distribution of Jurassic sediments and were the source areas for basal Jurassic clastic material and for clastic sediments deposited during the "Smackover" transgression. Adjacent to the Appalachians, the Smackover is composed of a mixture of clastic and carbonate rocks. Major production in the Jurassic trend is obtained basinward from the area characterized by this mixed lithology, primarily in areas containing grainstone facies. The depositional environment of these grainstones is one of the focal points of this study.

As shown in Figure 5, the updip limits of the Louann Salt and the Smackover Formation reflect the topography of the pre-Jurassic surface. The shallow-water carbonate muds of the Smackover were deposited on a broad shelf in the area of the Florida Panhandle—a shelf which narrowed considerably toward Mississippi. Widespread distribution of shallow-water carbonate material on this shelf provided ideal conditions for the development of stratigraphic traps.

In Mississippi, the producing Smackover reservoir is generally an oolitic facies. In southern Alabama and in the Florida Panhandle, the grainstone facies is composed of hardened pellets; oolites constitute a very minor part of the section. Development of this low-energy facies in the Jay area can be attributed to the dissipation of wave energy over a wide shelf and to the absence of topographic features which would have created a focal point for the generation of oolites.

Figure 5 also shows the distribution of the Cotton Valley clastic beds, which overlap the pre-Jurassic surface for a considerable distance east of the Smackover updip pinchout.

COMPARISON OF JURASSIC DEPOSITION AND RECENT ENVIRONMENT

To increase understanding of the sedimentary processes prevailing during the Late Jurassic, and of the resulting depositional patterns, we have compared the Jay area with a modern carbonate province which it closely resembles—the Great Bahama Bank, particularly the Andros platform. Facies and depositional patterns here are similar to those in the area studied in the Florida Panhandle, although the climate during the closing phase of the Smackover sedimentary cycle was obviously more arid than the present climate of the Bahamas.

The sedimentary pattern of the Andros platform, shown in Figure 6, is based on descriptions of Holocene facies by Newell et al (1959) and Purdy (1963). The bank edges are shown to drop off abruptly into surrounding deeper water, and the bank margin is characterized by the presence of coralgal carbonate sand.

On the western side, extensive oolite shoals border the platform, spilling over more landward, protected, hard-pellet sediments. This sediment type represents a facies of carbonate sand composed predominantly of fecal pellets. The pellets

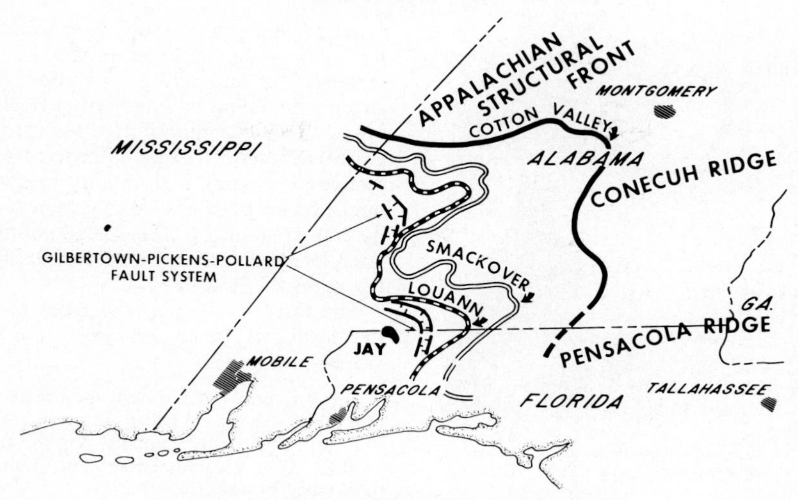

FIG. 5—Jay field, regional depositional limits.

are grains which became indurated either in a very shallow subtidal environment or during intermittent exposure at times of low tide on intensely burrowed stable flats. Such stable flats are distributed extensively along the protected interior side of all the bank islands.

The areas of grapestone with aggregate grains are located at the northern and southern ends of the platform. The area containing pelletoid carbonate muds is on the western or leeward side of Andros Island.

A good correlation exists between the facies distribution and varied sediment types of the Smackover at Jay and the stable-flat area west of Joulter Cays. The detailed facies distribution in this area, mapped by Purdy (1963) and others, is shown in Figure 7; the reef-covered bank margin is adjacent to the Tongue of the Ocean on the right and the northern tip of Andros Island at the lower left.

The reefs and the associated coralgal-sand-covered lagoon characterize the deeper marine environment of deposition on the east. In the Joulter Cays area, sediments accumulate in less than 3 ft (1 m) of water in the shallow shoals. The seaward side of these shoals is bordered by oolite bars and tidal deltas. These bars and tidal deltas protect the adjacent shallow stable-flat area, where grapestone bars, small emerging islands, and tidal flats are developed. Tidal channels are superimposed on these deposits.

The sediments of the shallow stable-flat areas are intensely burrowed by shrimplike decapods, which produce countless burrow mounds of digested and extruded sand-size pellets and skeletal fragments. The grass blades on such stabilized flats are inhabited by numerous species of Foraminifera, Bryozoa, and other organisms; algal balls or oncolites are also present. Identical lithologies with comparable fauna and flora have been observed in the lower part of the Smackover Formation.

Smackover Deposition

The postulated sedimentary conditions during deposition of the Smackover in the Jay area are shown in the block diagram in Figure 8. A coastal complex is shown with two very shallow embayments flanked by shallow subtidal slopes which outline the adjacent, burrowed stable flats. Above this sequence and closer to shore lie the algal mats of the closing phase of carbonate deposition. The arid supratidal flats represent the landward areas where evaporites were being formed while carbonate deposition was occurring farther seaward.

The basal Smackover carbonate rocks are characterized by a sequence of alternating laminites and pelletal-oncoidal limestone, grading upward to a more micritic pelletal unit. The laminites reflect a stromatolitically bedded sequence deposited in an intertidal to low supratidal environment. The burrowed pelletal-oncoidal deposits seem to correlate best with sediments deposited in modern grass-covered, stable-flat environments. Near the middle of the Smackover interval, a micritic

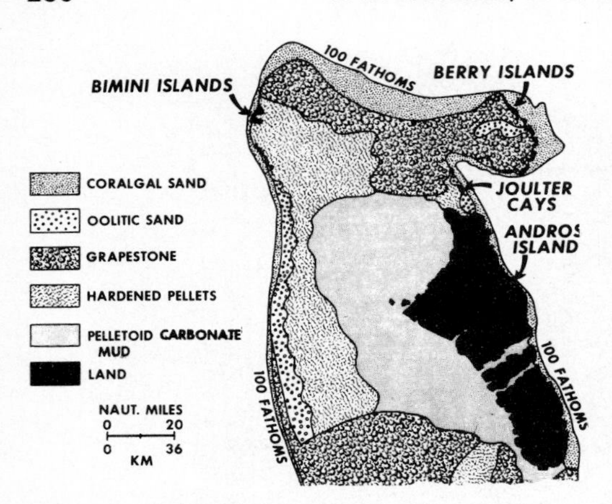

FIG. 6—Carbonate depositional pattern, Andros platform, Great Bahama Banks.

skeletal limestone marks the maximum extent of the marine transgression. The leached pelletal dolomite interval in the upper half of the Smackover represents deposition during the regressive cycle in an environment intermediate between shallow stable flats and more marine conditions. The uppermost Smackover laminites reflect renewed deposition in supratidal and intertidal environments.

The cuts of the block diagram depict the vertical and lateral distribution to the various sediment types. The patterns reflect a slow marine transgression during deposition of the lower Smackover. The regressive phase, which began after the deposition of the micritic skeletal limestone, is characterized by an infill of topographic lows and by shallow marine embayments.

Distinguishing features of the four major rock types making up the Smackover are illustrated in the photomicrographs of Figure 8:

1. Algal stromatolite, with laminated bedding;
2. Pelletal-oncoidal micrite, with the symmetry of an algal oncolite;
3. Fine skeletal micrite, with fossil hash;
4. Pelletal grainstone, shown with ovoid pellets of varied sizes.

Smackover Porosity

In Figure 9, the present porosity distribution is shown in relation to the depositional component parts of the Smackover. The same block diagram is used as that in Figure 8. Porosity formation by dolomitization is associated with the presence above the Smackover of the evaporite deposits of the Buckner Formation. Hypersaline waters, resulting from isolation and evaporation of seawater along the shallow coastline, percolated through the underlying formation by flood recharge. Leaching by continental fresh waters followed as the coastline shifted seaward. In areas of grain carbonate rocks, the process was most effective in creating high-quality reservoir rock. In micritic and pelletal-oncoidal facies, some porosity was created, but areas of dolomitization and later leaching are more localized, mainly along the paths of fluid flow.

The three major porosity types and their relative locations are illustrated by the photomicrographs in Figure 9.

1. Grain moldic dolomite is the leached end product of a dolomitized hard-pellet grainstone.
2. Intercrystalline dolomite represents the leached product of a dolomitized micritic pelletal limestone (packstone to wackestone).
3. Leached matrix is the result of dolomitization of an algal stromatolite.

An example of the porosity distribution and its relation to the lithology of the Smackover Formation in the Jay field area can be seen in Figure 10, which is a reproduction of part of a density log from the Exxon-LL&E No. 1 Jones McDavid, the confirmation well for the field. The Smackover

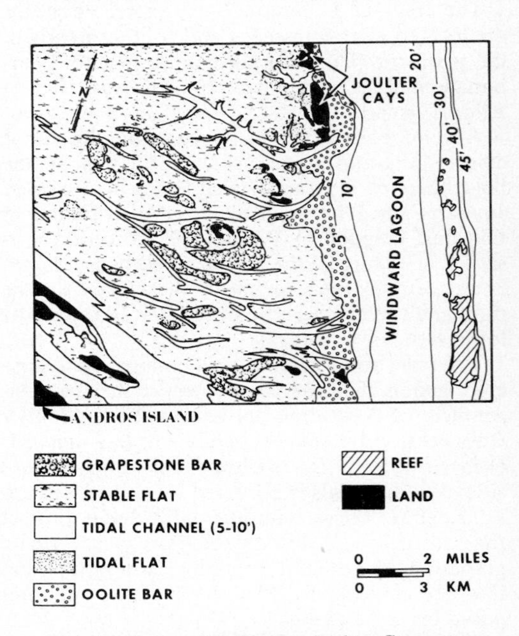

FIG. 7—Facies distribution, Joulter Cays area, Andros platform, Great Bahama Banks.

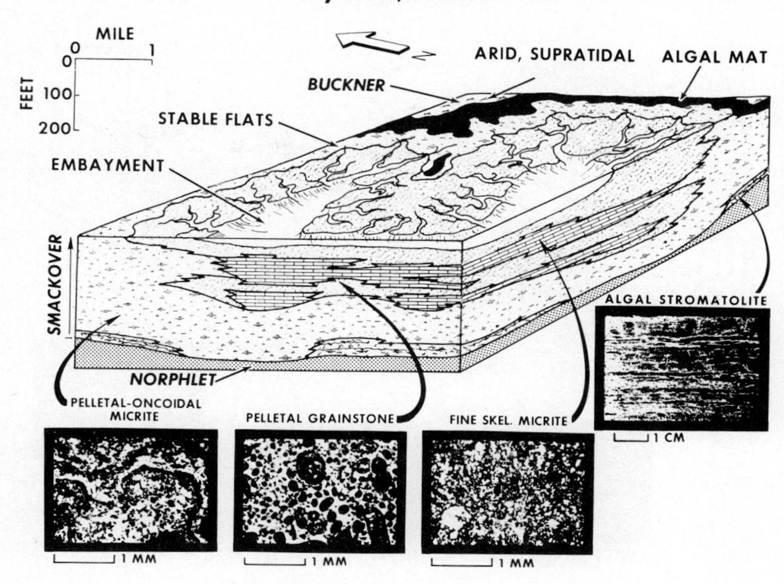

FIG. 8—Smackover facies, Jay field area.

has a thickness of slightly over 370 ft (113 m). The bottom contact of the Smackover Formation with the fine-grained Norphlet sandstone is typically sharp, whereas the transition at the top into the Buckner evaporites is gradational.

The facies distribution shown is similar to that presented in the block diagrams; the laminites and pelletal-oncoidal deposits are at the base, and the micritic skeletal limestone is in the middle. The main porosity is associated with the hard-pellet grainstone facies in the upper part of the Smackover. This zone of secondary moldic porosity was created by freshwater leaching in a superficially dolomitized pelletal limestone. Development of patchy, leached matrix porosity in the lower transgressive part of the Smackover is associated with possible intermittent surface exposure. The homogeneous character of the main Smackover reservoir rock is demonstrated by the smooth density curve. The lower portion of the reservoir, however, is broken by "tight" intervals, which reflect the occurrence of the laminites.

STRUCTURE AND TRAP

The accumulation at Jay field occurs on a south plunge of a large subsurface anticline, and the updip trap is formed by a facies change. The map of the structure at the top of the Smackover (Fig. 11) shows a combination structural-stratigraphic trap with 420 ft (128 m) of oil column extending over an area 7 mi (11.3 km) long and 3 mi (4.8 km) wide. There is only 100 ft of structural relief on the Jay nose above the saddle separating it from the large Flomaton structure adjacent on the north. The Flomaton anticline, which produces gas from the Norphlet, rises another 350 ft (107 m) to its crest. Other structurally significant features at Jay include the steep northeast flank, the Gilbertown-Pickens-Pollard fault on the east, and the small closures along the crest of the nose. The latter trap oil in the underlying Norphlet, but reserves are only about 1 percent of those of the Smackover.

Updip termination of favorable Smackover reservoir rocks occurs almost perpendicular to strike and reflects a boundary between depositional environments that remained constant for an extended period. North of the saddle, the rock is a dense micritic limestone; on the south, dolomitized grain carbonate rock is present. The productive area covers roughly 14,400 acres (58.4 km^2).

Cross sections *A–A′* and *B–B′* (Fig. 12) show the irregular distribution of the porosity and its relation to present structure. The porosity buildups, mainly in the hard-pellet facies, may be explained as resulting from the normal depositional process of filling the shallow embayments. Another explanation for the irregular thicknesses of

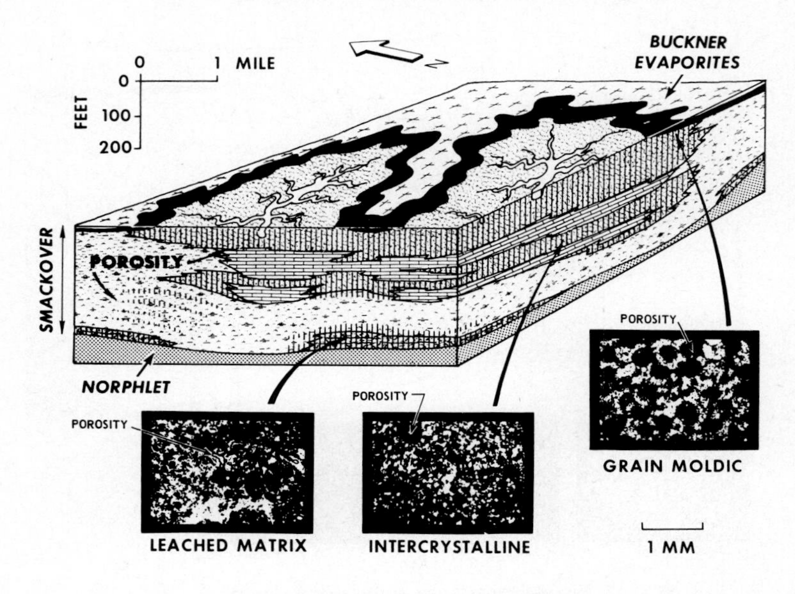

FIG. 9—Smackover porosity, Jay field area. Present porosity distribution is indicated by vertical hachures.

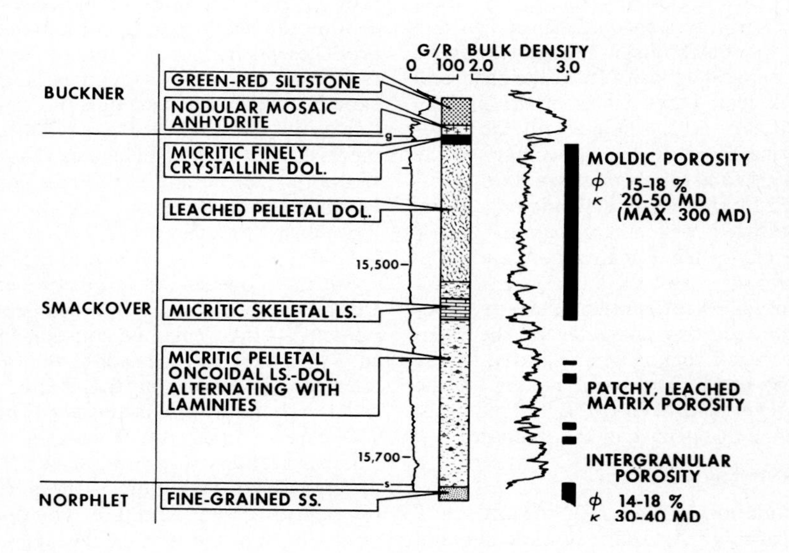

FIG. 10—Log of sedimentary sequence in Exxon-LL&E No. 1 Jones McDavid well, Jay field.

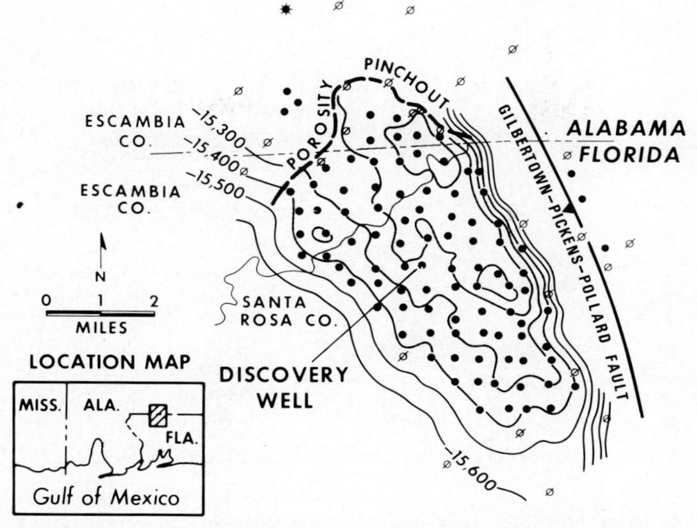

FIG. 11—Jay field, structure map of top of Smackover. C.I. = 100 ft.

porous rock would be the natural selection of environments by the organisms contributing to this facies and buildups resulting from extended stable conditions.

RESERVOIR PROPERTIES

An isopach map of the net feet of "pay" with greater than 8 percent porosity is shown in Figure 13. On the northwest end of the field, the productive limit corresponds to the porosity pinchout. Elsewhere, it follows an oil-water contact at −15,475 ft (−4,717 m). The contact varies in individual wells, which creates difficulties in totaling net "pay." As an example, the oil-water contact in two adjacent wells on the west flank—the LL&E No. 2–2 and No. 35–2 McDavid Lands—differed in elevation by 100 ft (30 m). It is possible that this irregular oil-water contact may be due to the limited volume of water associated with this reservoir. In an irregular development of porosity lenses, the presence of a limited quantity of water certainly would provide such a condition. Current reservoir data indicate a water volume of only one fourth of the oil volume.

On the north, the abruptness of the porosity pinchout was documented when the LL&E No. 33–3 McDavid Lands encountered only 9 ft (2.7 m) of "pay" in the original hole and 109 ft (32.2 m) in a horizontal sidetrack of 700 ft (213.4 m). A similar situation occurred on the east flank, where the Amerada Hess No. 40–16 Findley encountered no "pay" only 2,500 ft (761 m) from a well with 243 ft (74 m) of "pay."

The thickest net-"pay" interval is on the eastern, or shoreward, side of the field (Fig. 13); it thins abruptly toward the northeast and more gradually toward the southwest. It is apparent that the net-"pay" thickness corresponds to the distribution of the hardened pellets in the section. Subsequent dolomitization provided greater porosity and permeability on the east side, where hardened pellets are concentrated, than in any other area of the field. One well, the Exxon No. 5–2 St. Regis, drilled rock with permeability greater than 8,000 md. Average permeability for the reservoir is 35 md. Porosity ranges up to 31 percent and the average is 13 percent. Average "pay" thickness is 100 ft (30.5 m).

HYDROGEN SULFIDE

Crude oil in the Jay field contains 9 percent hydrogen sulfide, 3 percent carbon dioxide and nitrogen, and 88 percent hydrocarbons. Free sulfur, which has been found in impermeable parts of the Smackover, is possibly an alteration product of anhydrite. It is logical, therefore, to postulate that free sulfur was present in some of the sediments which subsequently were altered by dolomitization, and it was this free sulfur that furnished the source of the hydrogen sulfide in the crude. The free sulfur apparently was randomly distributed. Both south and east of Jay, sweet

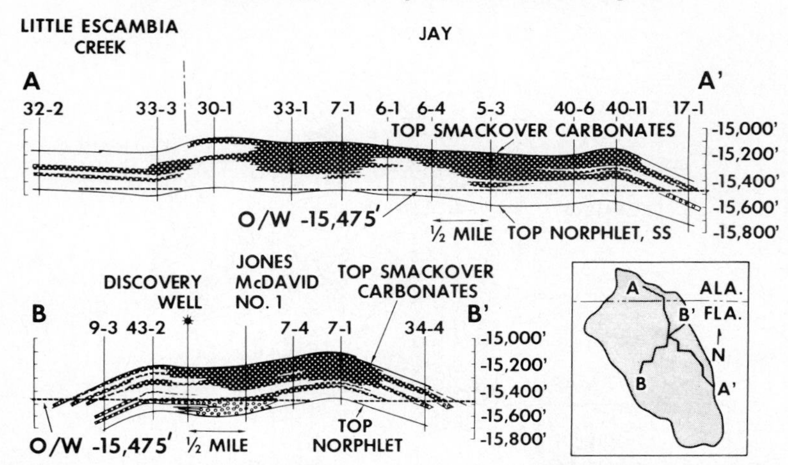

FIG. 12—Porosity-distribution cross sections of Little Escambia Creek and Jay fields.

crude is produced from Norphlet sandstones, which underlie the Smackover. In other characteristics, the crude is similar to that from the Smackover.

PRODUCTION

Production results from rock and fluid expansion. In general, the productivity of wells in this field will be restricted more by the 3 ½-in. (8.89 cm) tubing through which they produce than by reservoir quality. Original bottomhole pressure was 7,850 lb, or about 700 lb above saltwater gradient; saturation pressure was approximately 5,000 lb lower—2,830 lb. At reservoir conditions, the Jay crude is a mobile, low-viscosity liquid that will flow 2 ½ times more freely than formation water.

A comparison of the Jay Smackover reservoir with some well-known fields around the world is shown in Figure 14. Middle East field "A," productive from a Cretaceous dolomite, has reserves of approximately 3 billion bbl of oil. Middle East field "B" produces from lower Eocene carbonate rock and has reserves of more than 2.5 billion bbl of oil. Wasson field, in West Texas, produces from a Permian dolomite and has reserves of 1.5 billion bbl. The East Texas field, with original reserves of more than 5 billion bbl of recoverable oil, is included as a point of reference for a sandstone reservoir.

Of the fields compared, Jay shows the most rapid decrease in water saturation as height above free-water level is attained. In addition, a consistent relation of decreasing water saturation with increasing permeability exists at Jay. Even at 2 md permeability, the reservoir contains an irreducible water saturation of less than 15 percent.

This relation reflects less porosity volume occupied by connate water in the "pay" zone. It can be assumed that these low saturations are evidence of a clean reservoir (without typical water-retaining constituents) and will result in lower residual oil saturations at depletion.

These unique properties possibly result from a uniform rhombohedral porosity throughout the matrix which will contribute to a higher yield of oil per acre-foot. Field-wide unitization became effective March 1, 1974. Pressure-maintenance operations in the form of water injection will recover twice as much oil as will be produced from primary recovery.

DRILLING

Initially, drilling was difficult because of the Buckner salt and evaporite section overlying the Smackover, and because of the hydrogen sulfide in the hydrocarbons. Strict safeguards had to be employed at all times. As a result, a directional survey was run above the Smackover in each hole to provide a target in the event that the well blew out. To penetrate the Buckner required increasing the salt content of the mud to a level which would minimize solution but still provide a medium in which logs could be run.

Operating costs in the Jay area are high. Well costs range from $650,000 for a producing well to $375,000 for a dry hole. Flow-line costs vary with

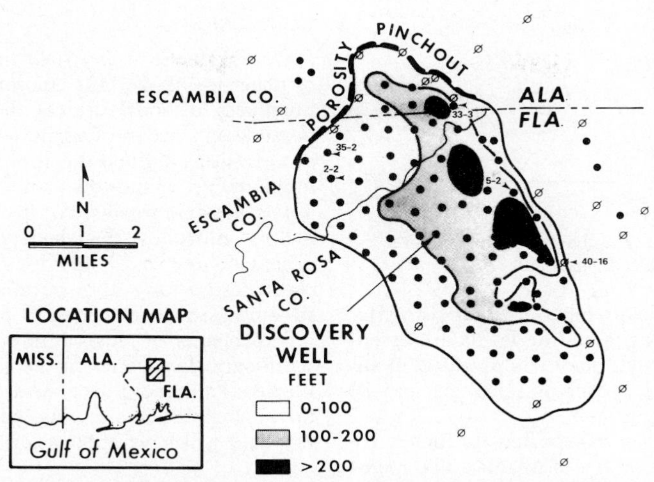

FIG. 13—Jay field, isopach map of Smackover net "pay" with porosity greater than 8 percent.

length, but an average cost with inlet separation facilities is $200,000. When each well's share of a 12,000 bbl/day processing facility costing $7 million is added, the investment per well reaches $1.4 million. Based on field rules established in Alabama and Florida providing for 160-acre spacing and allowables prior to unitization of 1,000 bbl/day per well, and considering current prices of $8.08/bbl for crude, 44.5¢/Mcf for 1,350-BTU gas, and $13/LT for sulfur, revenue equaled investment for a producing well in slightly less than 5 months.

To date, 89 producing wells and 18 dry holes have been drilled in the Jay field. Major operators include Exxon, Louisiana Land and Exploration, Amerada Hess, Sun, Chevron, and Moncrief and Young.

Total field investment to date, including the dry holes and processing facilities, but not the exploration and leasing costs, amounts to approximately $132 million.

PROCESSING

The flow through a sweetening facility similar to those at Jay is shown schematically in Figure 15. Each full well stream enters two separators in series, from which the oil is fed to an oil-stabilization unit where heat is added and the hydrogen sulfide is vaporized from the crude. The sweetened crude oil then goes to the stock tank. The gas from the two separators and the oil-stabilization unit separately enters three gas-treating towers where diethanolamine, a liquid with a natural affinity for hydrogen sulfide, absorbs the hydrogen sulfide from the gas. Sweet gas leaving the gas-treating towers goes to the dehydrator and then to the gas market. Sour gas stripped from the diethanolamine is carried to a Claus sulfur-recovery unit, where 96 percent of the sulfur is recovered by heating and cooling the acid gas

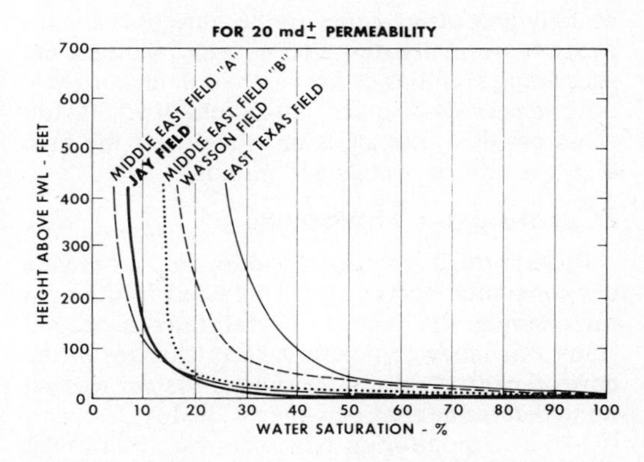

FIG. 14—Reservoir comparison of Jay field and several other fields. Capillary pressure, plotted as height above free-water level, is compared with permeabilities of about 20 md and porosities of 15-20 percent.

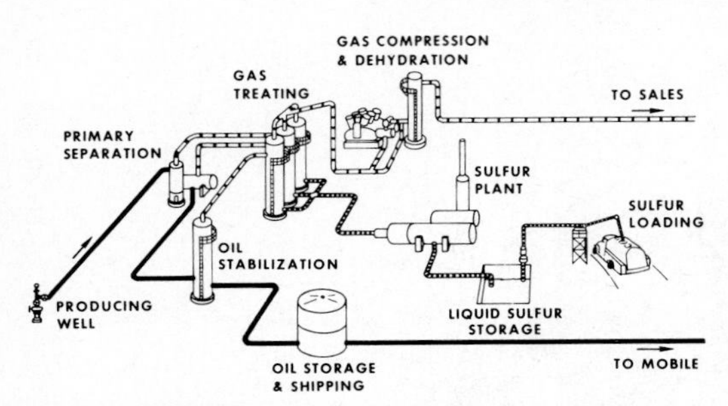

FIG. 15—Schematic flow diagram of production facilities at Jay field.

through several critical temperature ranges. After treatment, the hydrogen sulfide levels are essentially zero. A 12,000-bbl plant will produce, in addition to the oil, 12 MMcf of sales gas and 80 long tons of sulfur per day.

The treated crude is transported through a 16-in. (40.6 cm) line from Jay to Mobile. The major part of the gas is sold to Florida Gas Transmission Company. Sulfur is sold to Freeport Sulphur Company and is trucked in liquid state to Mobile. Cumulative production from the Jay field as of June 1, 1974, was 58 million bbl of oil and 70 Bcf of gas.

One of the key decisions in developing a field that requires processing facilities is determining the proper size of plant to be built. The specific sizes constructed result from the location and size of the operator's leases, the economy of the operation, and the pressure of competitive production. Six 12,000-bbl and three 6,500-bbl plants have been built at Jay, providing a total of 91,500 bbl of daily processing capacity. Because of the extra capacity normally designed in such facilities, the processing facilities at Jay can handle the production, as permitted under unitization, of 93,500 bbl of oil per day. The gross investment for the nine plants is approximately $50 million.

Environmental Monitoring

Prior to the beginning of production, a baseline environmental survey was conducted in the area surrounding the plants. A year later a second study was made to determine any changes in the environment. At the same time, a system was set up to determine routinely the air quality. Currently, 10 air-monitoring stations are established throughout the area to report the sulfur dioxide and hydrogen sulfide concentrations to the state on a weekly basis.

Conclusion

Development of Jay field has required a level of expertise above that commonly required. The timeliness of most critical decisions was made possible as a result of early field definition. Step-out locations defined the approximate productive limits within 18 months of discovery. As a result, it was possible to start construction of the plants before many of the development wells were drilled. These early decisions and the modular plant design have allowed this field to come on stream in an extremely short period of time.

The results of this venture have led to other significant discoveries in the southern Alabama–Florida Panhandle area, and it is probable that the geologic understanding gained from study of Jay field will have application in other areas.

Selected References

Bathurst, R. G. C., 1969, Bimini Lagoon, *in* H. G. Multer, Field guide to some carbonate rock environments, Florida Keys and western Bahamas: Miami Geol. Soc., Rosenstiel School of Marine and Atmos. Sci., Univ. Miami, Miami, Florida, p. 62-69.

Keyes, P. L., 1971a, Jurassic geology of Flomaton field area of southern Alabama (abs.): AAPG Bull., v. 55, no. 2, p. 347.

———1971b, Geology of the Jurassic, Flomaton-Jay area, Alabama and Florida (abs.): Gulf Coast Assoc. Geol. Socs. Trans., v. 21, p. 30.

Kinsman, D. J. J., 1969, Modes of formation, sedimentary association, and diagnostic features of shallow water and supratidal evaporites: AAPG Bull., v. 53, no. 4, p. 830-840.

Logan, B. W., et al, 1970, Carbonate sedimentation and environments, Shark Bay, Western Australia: AAPG Mem. 13, 223 p.

Newell, N. D., et al, 1959, Organism communities and bottom facies, Great Bahama Bank: Am. Mus. Nat. History Bull., v. 117, art. 4, p. 179-228.

Ottmann, R. D., P. L. Keyes, and M. A. Ziegler, 1973, Jay field—a Jurassic stratigraphic trap (abs.): AAPG Bull., v. 57, no. 4, p. 748.

Purdy, E. G., 1963, Recent calcium carbonate facies of the Great Bahama Bank—1: Jour. Geology, v. 71, p. 334-335; pt. 2: Jour. Geology, v. 71, p. 472-497.

Traverse, A., and R. N. Ginsburg, 1966, Palynology of the surface sediments of Great Bahama Bank, as related to water movement and sedimentation: Marine Geology, v. 4, no. 6, p. 417-459.

Sunoco-Felda Field, Hendry and Collier Counties, Florida[1]

A. N. TYLER and W. L. ERWIN[2]

Abstract Sunoco-Felda field is located on the South Florida shelf, on the northeastern flank of the South Florida embayment. Production is principally from a stratigraphically trapped oil accumulation in a reefoidal, algal-plate, gastropod-bearing limestone mound in the Sunniland Limestone of Early Cretaceous age. The discovery well was drilled in July 1964 by Sun Oil Company on the basis of a combination of regional subsurface geology and geophysical work. The oil reservoir is about 11,475 ft (3,500 m) below the surface and is characterized by excellent vuggy porosity ranging upward to 28 percent; maximum permeability reaches 665 md. The field has a 34-ft (10 m) oil column and encompasses a surface area of approximately 4,500 acres (18 km^2). Inplace oil reserves are estimated to be 44 million bbl. The South Florida shelf area is sparsely drilled and offers great potential for the discovery of additional fields the size of Sunoco-Felda field. The subtle expression of this type of low-relief feature in the subsurface requires the complete coordination and application of sophisticated geological and geophysical techniques in order to provide a successful and economically attractive exploration program.

INTRODUCTION

Sunoco-Felda field is located in the southern part of the Florida Peninsula in Hendry and Collier Counties, about 25 mi (40 km) east of Fort Myers (Fig. 1). The oil accumulation is principally stratigraphic; it is in a reefoidal, algal-plate, gastropod-bearing limestone mound in the Sunniland Limestone of Early Cretaceous age. The top of the reservoir is at approximately 11,475 ft (3,500 m). Sun Oil Company drilled the discovery well in July 1964 on the basis of a combination of subsurface geology and seismic data. At the time of the discovery, only four wells had been drilled in Hendry County to the Sunniland Limestone in the search for oil. Indeed, the whole state of Florida had only one producing oil field within its borders—Sunniland field, about 15 mi (24 km) southeast of Sunoco-Felda, which produces from the Sunniland and was discovered in December 1943 by Humble Oil & Refining Company (now Exxon, U.S.A.). Forty Mile Bend field had been found in Dade County in 1954 but was abandoned in 1955. It, too, produced from the Sunniland Limestone.

The Felda prospect had been an area of interest to oil explorationists since the drilling in June 1954 of the Commonwealth No. 3 Red Cattle, in Sec. 25, T45S, R28E, Hendry County. This wildcat well, drilled with financial help from Sun Oil Company and Humble Oil & Refining Company, was located on a seismic anomaly. It recovered 1,090 ft (332 m) of oil and 9,060 ft (2,761 m) of salt water on a drill-stem test in the Sunniland Limestone. Since all information indicated this well to be marginal at best, a completion was not attempted and the hole was plugged and abandoned. After two additional attempts to discover a commercial oil field in the area, interest subsided and activity ceased. No further drilling took place in the area until 10 years later, when Sun Oil Company reassembled and enlarged its block of acreage and drilled the discovery well. Later, during the development of the field, a commercial oil well was completed by Sun approximately 1,000 ft (300 m) southwest of the abandoned Commonwealth No. 3 Red Cattle well.

PHYSIOGRAPHY

The surface is a featureless sandy plain with a maximum relief of only 10 ft (3 m) in the field proper. The higher elevations are generally pine- and palmetto-covered terraces of Pleistocene-Holocene sandstones and sands; the lower areas are marshy ponds and sloughs. Average surface elevation is about 32 ft (10 m) above sea level. There is no evident surface expression of the underlying oil-bearing feature. Surface drainage is northwest to the Caloosahatchee River. The area is directly north of the Big Cypress Swamp and just west of the Everglades.

Agriculture is the predominant economic activity of the area; beef cattle and citrus fruit are the principal year-round products. During the fall and winter months the vegetable-growing industry predominates. Much of the cultivated land is eventually turned into improved pasture providing lush and nutritious forage for a rapidly growing cattle industry.

[1]Manuscript received, January 29, 1975.

[2]Sun Oil Company, Dallas, Texas 75230.

Acknowledgment is given to Sun Oil Company for permission to publish this paper. The writers gratefully express appreciation to the many Sun Oil Company people who advised and assisted in its preparation, and especially to John A. Means for his technical assistance and helpful suggestions.

General Geology

The field is situated on the South Florida shelf, on the northeastern flank of the South Florida embayment, southwestward from the Peninsular arch (Fig. 2). Regional dip on the shelf proper at the Sunniland level is to the southwest at the rate of about 20 ft/mi (3.7 m/km). In specific areas where low-relief depositional structures or patch reefs are present, the local dip may reach 60–80 ft/mi (11–15 m/km). The entire shelf area, however, is characterized by uniform gentle dip with little or no evidence of structural deformation.

Depositional conditions across the shelf area appear to have undergone little change since Early Cretaceous time. This environment was one of shallow, clear, subtropical seas covering, in general, a slowly subsiding sea bottom. Lithologic uniformity of the resultant thick section of carbonate rock is evidence that the rate of deposition was approximately equal to the rate of subsidence. Local transgressions and regressions of the sea did occur, however, as evidenced by the presence of cyclic depositional sequences within the section and the existing types of lithology, which range from marine limestones to a dolomite-evaporite facies.

The Sunniland oil reservoir at Sunoco-Felda field appears to be a localized reef buildup of a part of a regional carbonate bank. Production is from an algal-plate, foraminiferal, pelletal limestone—a stacked biostromal unit. A quiet-water environment existed at the time of deposition. Tidal channels or passes probably cut the north-

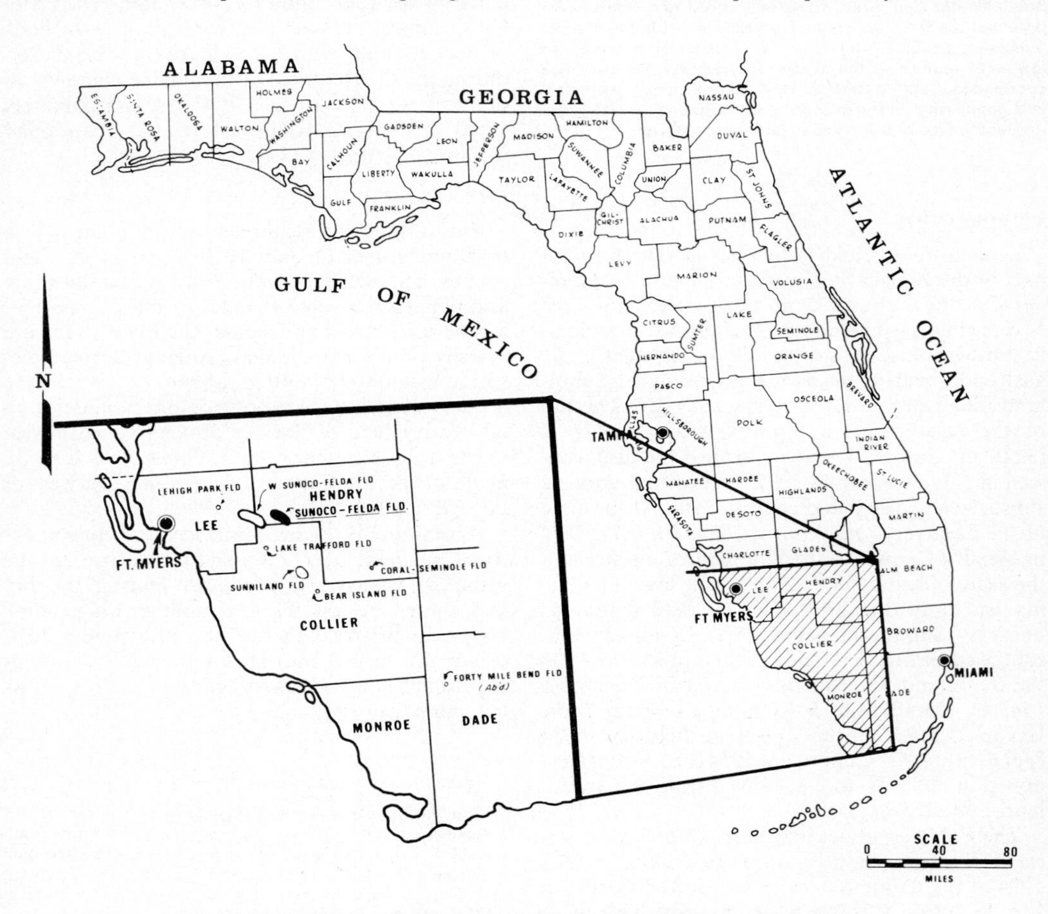

Fig. 1—Index map of Florida showing location of Sunoco-Felda field.

west-southeast-trending carbonate bank, separating the local reef pods. These channels eventually filled with carbonate muds, resulting in a loss of permeability and effectively separating one porous algal mound from another (Fig. 3).

Sunoco-Felda field is composed of two of these individual reef pods, both of which have a generally north-south trend. The most easterly pod contains by far the most porous and best preserved reservoir rock and is the significant oil-producing portion of the field. The Commonwealth No. 3 Red Cattle well, which tested the first oil in the area, is located on the western reef pod.

STRATIGRAPHY

The stratigraphic section below the surface Pleistocene-Holocene sandstones and sands consists predominantly of carbonate rocks and evaporites ranging in age from Early Cretaceous to Holocene. Figure 4 is a composite log of the sedimentary section.

Basement rock, although not penetrated in Sunoco-Felda field, was present at a depth of 15,670 ft (4,776 m) in the Humble No. 1 Lehigh Acres Development Corporation well, in Sec. 14, T45S, R27E, Lee County. This well is essentially along strike with, and approximately 8 mi (13 km) west-northwest of, Sunoco-Felda field. The basement complex found in this well—a gabbro-type mafic igneous rock—and the overlying 3,000 ft (915 m) of Lower Cretaceous carbonate rocks and evaporites are assumed to be similar to the section underlying Sunoco-Felda field. The deepest well in the field, Sunoco-Felda Unit No. 30-1, in the NE 1/4, Sec. 30, T45S, R29E, Hendry County, was

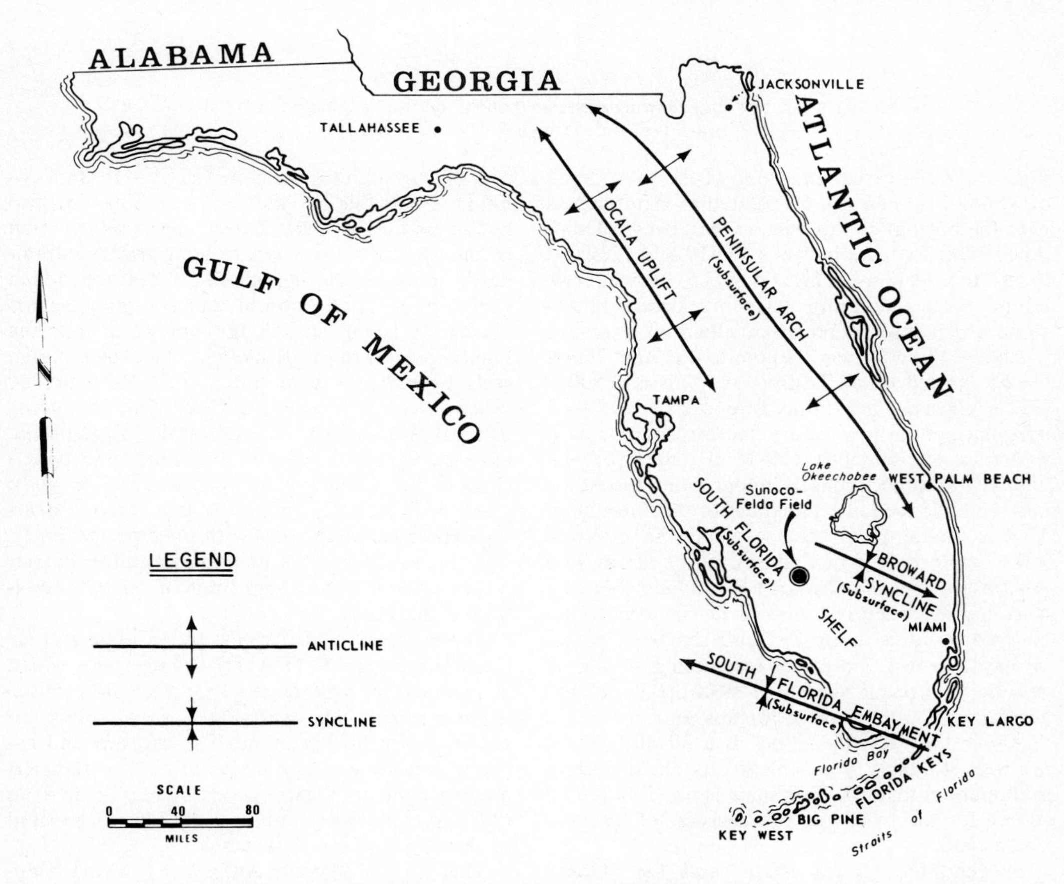

FIG. 2—Map of Florida Peninsula showing major structural features (after Applin and Applin, 1965).

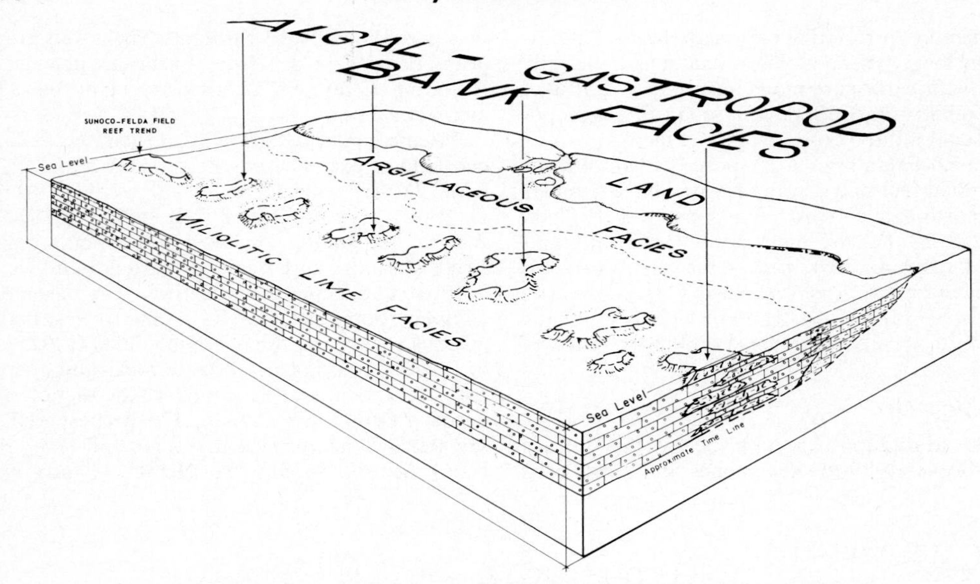

FIG. 3—Schematic drawing showing postulated shelf conditions during time of deposition of Sunniland, Sunoco-Felda reef trend.

in Lower Cretaceous limestone at the total depth of 12,686 ft (3,866 m). Of particular significance, near the bottom of this well, is the "Brown Dolomite," which occurs between 12,410 and 12,620 ft (3,783 and 3,847 m). This is a buff to brown, crystalline porous dolomite with interbedded limestone which could make an excellent oil reservoir.

Above the "Brown Dolomite" is the Punta Gorda Anhydrite of Trinity age. This is a 500-ft (152 m) massive gray anhydrite interbedded with irregular laminations of argillaceous limestone. It underlies the 50–60-ft (15–18 m) dark brown, highly fractured, dense, lithographic limestone unit which is the lowest member of the Sunniland. This is the oil-productive zone at the one-well Lake Trafford field in Collier County, about 9 mi (14 km) southwest of Sunoco-Felda field. Because it is highly fractured, this zone is usually recovered in cores as broken and shattered pieces of hard, brown limestone resembling a pile of rubble. As a result, the name "Rubble Zone" has come into common usage for this unit.

Above the "Rubble Zone" is a 30–40-ft (9–12 m) bed of black, very calcareous shale and/or carbonate mudstone. This shale is possibly the oil source for the overlying oil reservoir of Sunoco-Felda field.

Between the "Black Shale" and the "Upper Massive Anhydrite" lies a 150±-ft (45± m) section of Sunniland Limestone (Fig. 5). In Sunoco-Felda field this is the oil-producing interval known as the "Roberts Zone." The lower portion of the "Roberts" is a tan to light gray, medium-hard, miliolid-bearing, chalky limestone which generally is slightly porous and has poor permeability. It is on this chalky limestone that the highly porous and permeable, bioclastic patch reefs developed; these patch reefs form the oil reservoir at Sunoco-Felda field. The producing interval is composed of gastropods, algal plates, assorted bioclastic debris, and limestone pellets (Figs. 6–9). These porous reef patches, or pods, grade both laterally and updip into miliolid-bearing carbonate mudstones with poor permeability. This facies change forms a permeability barrier which provides the updip limit of oil accumulation in the field.

The oil reservoir is capped by an impermeable brown limestone, 9–11 ft (2.7–3.3 m) thick, which in the local area effectively separates the producing zone from an overlying porous, saltwater-bearing, rudistid limestone. The caprock and rudistid limestone merge into one porous and permeable zone in West Sunoco–Felda field, 6 mi (9.6 km) to the west, and form an integral portion of the oil reservoir of that field.

The "Upper Massive Anhydrite," which is approximately 100 ft (30 m) thick, overlies the

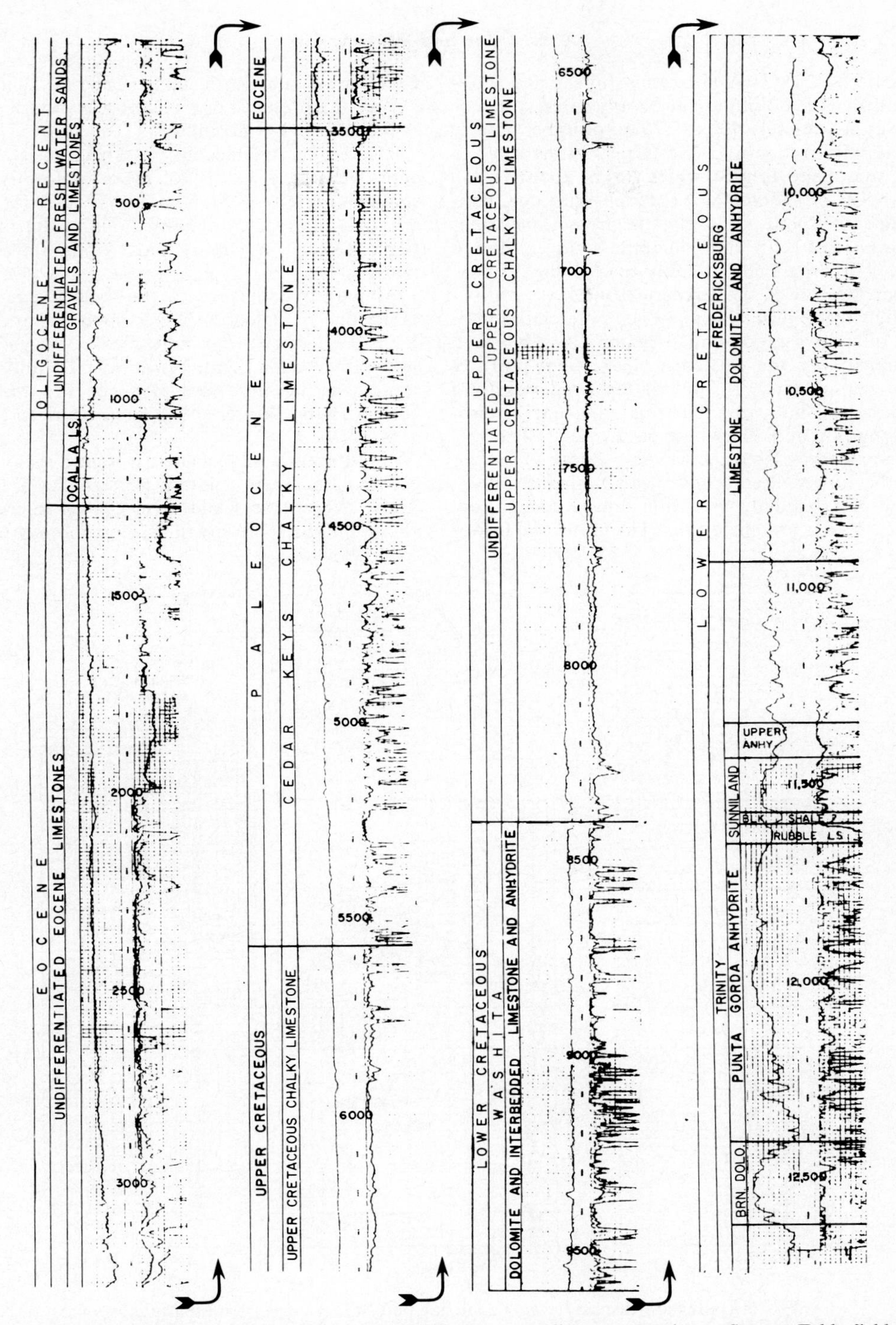

FIG. 4—Composite log showing age and general lithology of sedimentary section at Sunoco-Felda field.

"Roberts Zone." Argillaceous limestones interbedded with dolomites and anhydrites compose the approximately 400 ft (120 m) of upper Trinity section from the top of the "Upper Anhydrite" to the top of the Trinity, which lies at about 10,930 ft (3,330 m). Strata of Fredericksburg age, consisting of about 1,330 ft (405 m) of limestones interbedded with thin dolomites, shales, and a few anhydrite beds, directly overlie the Trinity. Near the top of the Fredericksburg, at approximately 9,600 ft (2,925 m), noncommercial asphaltic oil shows are found occasionally. This fact, coupled with the existence of excellent porosity and permeability in some of the Fredericksburg limestones, indicates that this zone has oil-producing potential elsewhere on the shelf. The section extending from the Fredericksburg to the top of the Lower Cretaceous is predominantly dolomite interbedded with thin limestones, anhydrites, and a few shale beds. The top of the Lower Cretaceous, at about 8,400 ft (2,560 m), is identified by the presence of a bed of waxy, dark green shale, 3–6 ft (1–2 m) thick.

The Upper Cretaceous section, consisting of approximately 3,100 ft (945 m) of uniformly soft, chalky limestone, is found at about 5,300 ft (1,615 m). This interval commonly drills at the rate of 100 ft (30 m) per hour, under optimum conditions.

Above the Cretaceous is the thick Cedar Keys Limestone of Paleocene age, which lies at about 3,500 ft (1,067 m). This soft, chalky, vuggy limestone, interbedded with hard dolomites and limestones, is the time-equivalent of the Midway Group of the Western Gulf area (Puri and Vernon, 1964).

Undifferentiated Eocene carbonate rocks, topped by the cream-colored, porous Ocala Limestone, overlie the Cedar Keys. This interval is about 2,400 ft (730 m) thick, and the top of the

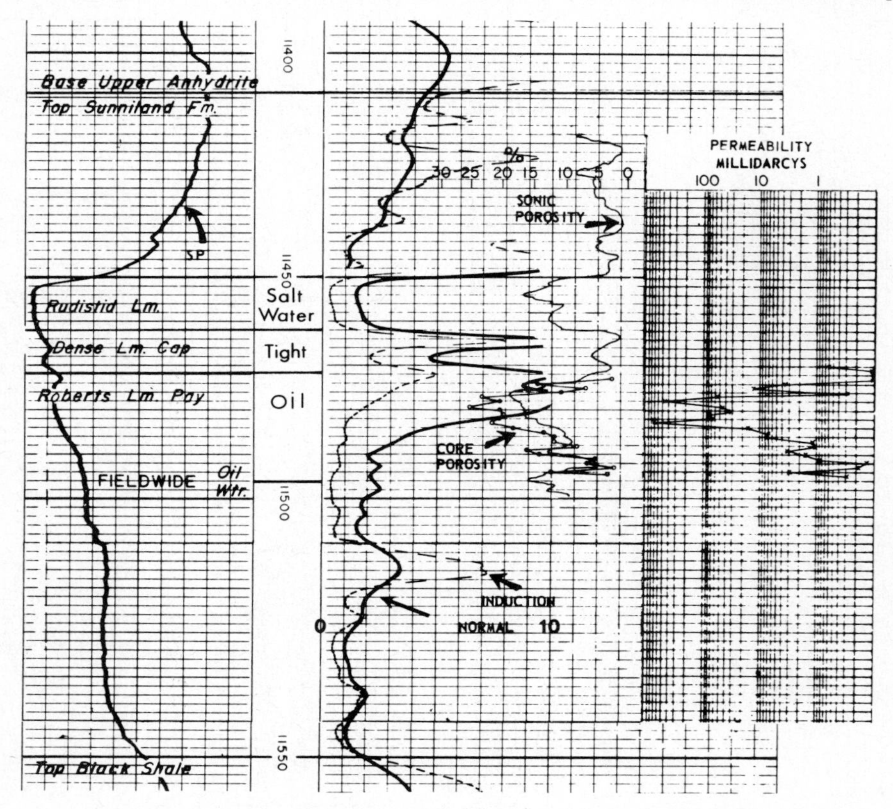

FIG. 5—Induction-electric log with sonic-log porosity and core permeability plots of "Roberts" limestone producing zone in typical Sunoco-Felda field well.

Ocala Limestone is about 1,100 ft (335 m) below the surface. These units are principally hard limestones and dolomites interbedded with soft, chalky, argillaceous limestones. In the depth interval of 2,000–3,000 ft (610–915 m), soft, chalky limestones are commonly leached out, leaving a section of caverns and ledges (Fig. 10). This 1,000-ft (300 m) interval, called the "Boulder Zone" by the oil industry, is the most hazardous interval of the entire section to drill. The drilling difficulty is caused primarily by the numerous voids and ledges encountered. There is a great variation in the size of the voids; the largest encountered in the Sunoco-Felda No. 32-2 well was 90 ft (27 m) from roof to floor. During drilling operations, the ledges form traps for the accumulation of broken pieces of formation, which may wedge between the drill pipe and a protruding ledge, causing the drill pipe to stick and twist off. Attempts to recover the parted drill pipe commonly meet with failure and the hole must be redrilled. The bottom of the "cavernous" zone is normally at about 3,000 ft (915 m), and protection casing is set at about 3,500 ft (1,065 m) to seal off this section.

FIG. 6—Algal-plate–gastropod carbonate bank typical of "Roberts Zone" oil reservoir (×1).

FIG. 7—Typical bioclastic grainstone facies of "Roberts Zone" (×3).

Unconsolidated sands, gravels, and carbonate muds of Oligocene to Holocene age extend from the top of the Ocala Limestone to the surface. These beds contain brackish to fresh water which is protected from contamination by the surface pipe.

FIELD GEOLOGY AND RESERVOIR CHARACTERISTICS

The "Roberts" producing zone of the Sunniland Limestone in Sunoco-Felda field has a 34-ft (10 m) oil column between the oil-water contact at −11,444 ft (−3,488 m) and the highest structural point of the producing zone. Porosities and permeabilities in the reservoir range up to 28 percent and 665 md, respectively. Development drilling was essentially complete by July 1966, when 25 productive oil wells had been drilled on a spacing pattern of 160 acres. Currently there are 17 active oil wells. All but one of the abandoned oil completions have been converted to saltwater-injection wells. Approximately 4,500 productive acres (7 km^2) lie within the confines of the field.

The structural configuration of the field is shown in Figures 11 and 12. The downdip limit of oil accumulation is governed by the oil-water contact, and the updip limit is formed by a facies change from the porous and permeable limestone

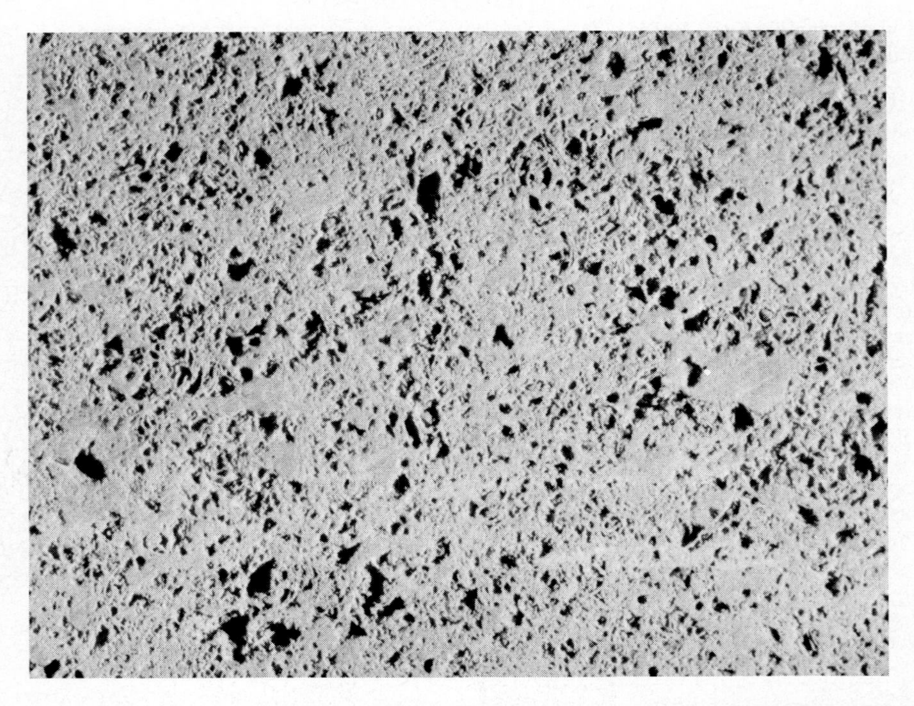

FIG. 8—Core slab of "Roberts" limestone producing section composed of gastropods, algal plates, rudistid fragments, and bioclastic debris in a recrystallized limestone matrix (×3).

reef to a chalky, miliolid-bearing, impermeable limestone. The best porosity is found basinward on the reef front, indicating that perhaps wave action and a higher level of energy played a part in forming the excellent porosity present in the reservoir rock.

Reservoir conditions are considerably poorer in the westernmost of the two Sunoco-Felda reef pods. Limited permeabilities and lower porosities in this area indicate a more complicated biostromal fabric.

Reservoir Data

Reservoir data for the "Roberts" limestone oil reservoir and other field information are summarized in Table 1.

The crude oil produced at Sunoco-Felda field is a black paraffinic-naphthenic oil containing 3.084 lb of salt per barrel and 3.42 percent sulfur. Although the oil is highly contaminated with salt, this does not appear to affect either production or equipment adversely. The oil contains an unusually small amount of gas in solution, and the average well has an initial potential gas/oil ratio of approximately 100 cu ft/bbl. Initial production rates from individual wells range up to 427 BOPD.

Drilling and Completion

A typical Sunoco-Felda well is spudded with a 26-in. (66 cm) hole and drilled to a depth of 100 ft (30 m) using a light-weight, fluffy mud with lost-circulation material to hold back the surface sands. Twenty-inch (50.8 cm) casing is set at this point and cemented to the surface to protect freshwater zones and prevent erosion and caving around the surface location. The same type of mud is used for drilling a 17 ½-in. (44.5 cm) hole into the top of the Ocala Limestone at approximately 1,100 ft (335 m), and 13 ⅜-in. (33.9 cm) casing is set. This casing string is also cemented to the surface to protect all freshwater zones and prevent the shallow sand and gravel beds from sloughing into the hole when circulation is lost while drilling the "Boulder Zone."

A 12 ¼-in. (31.1 cm) hole is then drilled approximately 3,500 ft (1,065 m), penetrating completely the cavern-and-ledge interval of the Eocene limestones. Use of a special sealed-bearing tungsten-carbide bit for drilling this hazardous in-

FIG. 9—Core slab of "Roberts Zone," algal-plate–gastropod-bearing limestone showing excellent vuggy porosity (×3).

terval generally makes it possible to penetrate the entire zone without changing bits. Another casing string (9 5/8-in. or 24.4 cm) is run and cemented to 3,500 ft. An 8 3/4-in. (22.2 cm) hole is drilled to the core point with constant addition of fresh water into the drilling-fluid system to compensate for the inability of the open hole to hold a full column of water. Because of lost-circulation problems in drilling the section below the 9 5/8-in. casing, well-sample recovery of the section from this point to the "Roberts Zone" is poor to nonexistent.

Most of the wells in the Sunoco-Felda field were completed in open hole after a minimum penetration of the oil "pay" with the core barrel. In these wells, a string of 5 1/2-in. (13.9 cm) casing was cemented at the top of the porous zone and completion was made from the open hole after a light acid treatment. Later completions were made in a more conventional manner. After coring and drilling through the producing interval, 5 1/2-in. casing was cemented on bottom and perforated. Although the early wells would flow initially, production rates were inadequate and rod pumping facilities were installed immediately upon completion. At present, the installation of downhole centrifugal pumps is under way for those wells capable of producing large volumes of fluid.

Unitization of the royalty interests was concluded on October 1, 1968 (Sun is the only operator in the field), and the field is currently being produced under a water-injection pressure-maintenance program in an effort to combat declining bottomhole pressures. Both produced and extraneous water are injected into the oil reservoir at an average rate of 10,000 bbl/day. Response to the injection program has created a rise in working fluid levels in some of the producing wells, resulting in shallower pump depths and a more efficient operation. It is too early in the life of the waterflood to determine its effectiveness, but the response to date is encouraging.

CONCLUSION

The Sunoco-Felda field reef trend of the Sunniland Limestone along the ancient shoals of the South Florida shelf offers great potential for the discovery of future oil reserves. Sunoco-Felda field is an example of the type, size, and quality of

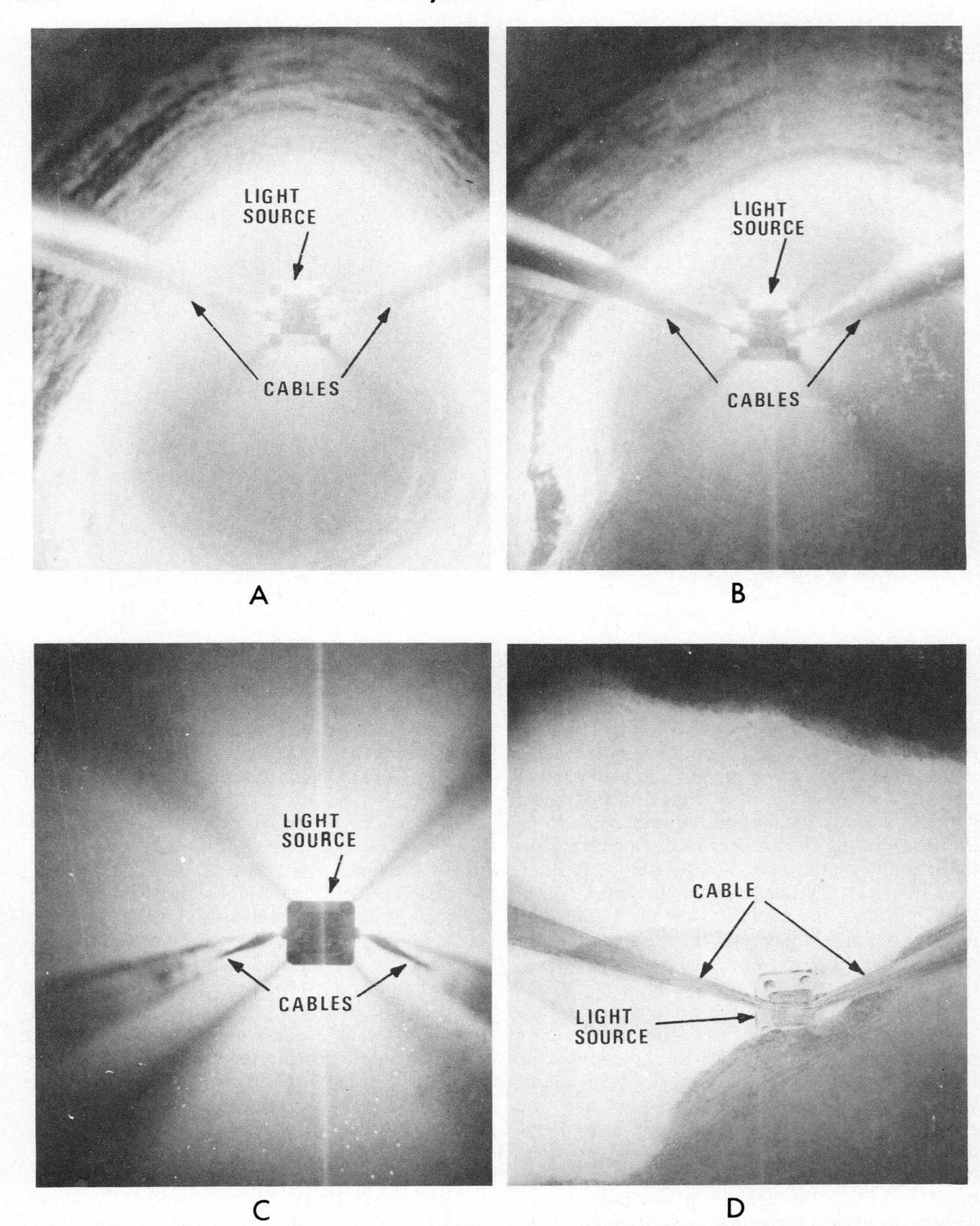

FIG. 10—Downhole photographs of "Boulder Zone" taken in clear water looking downward between depths of 2,400 and 2,500 ft (732–762 m). **A.** All sides of egg-shaped hole visible. **B.** Only one-half side of enlarged hole visible within light range. **C.** No hole sides visible within light range; camera and light source opposite cavern. **D.** Irregular hole with protruding limestone ledge.

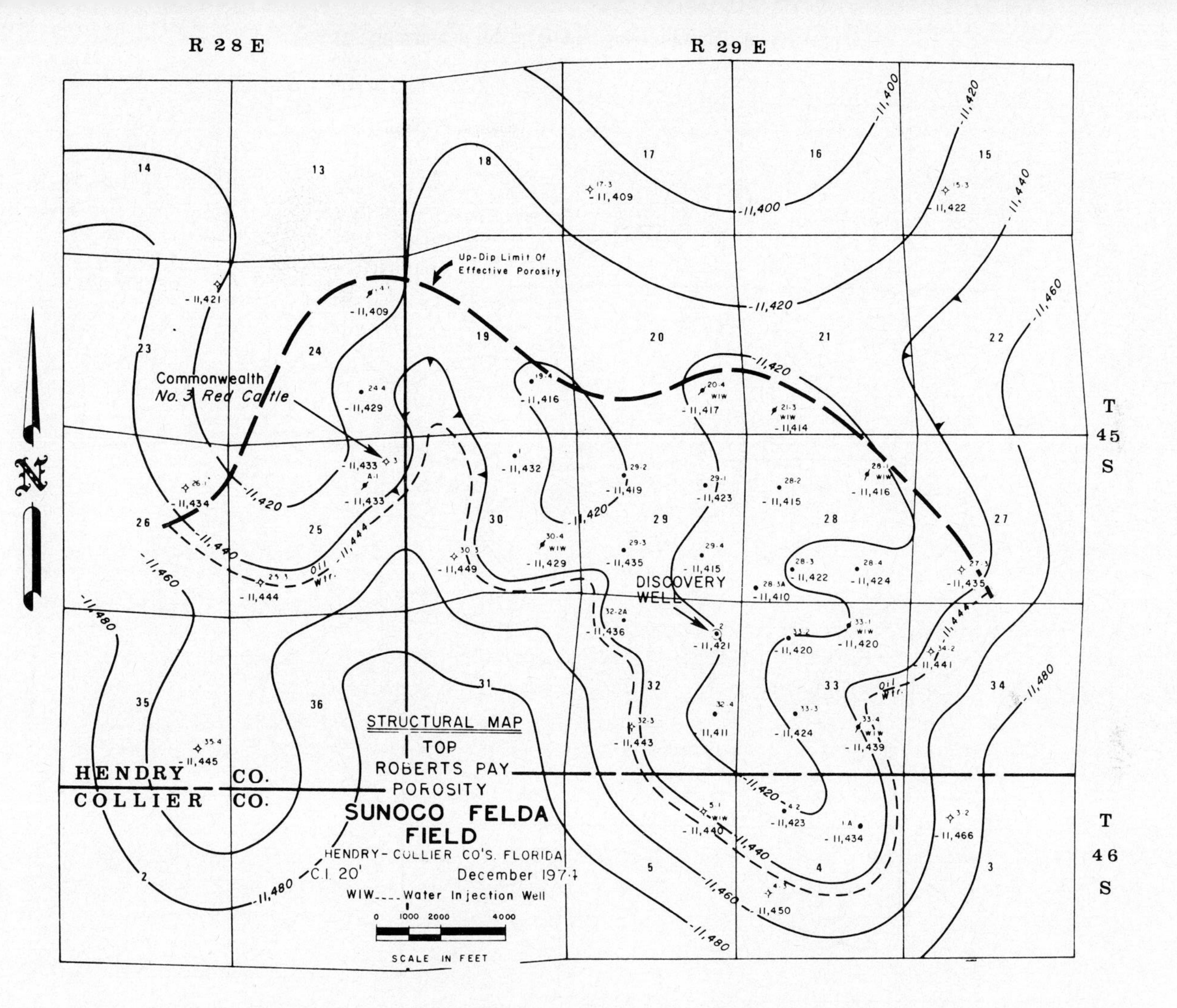

FIG. 11—Structure of top of “Roberts” limestone porous zone, Sunoco-Felda field. C.I. = 20 ft.

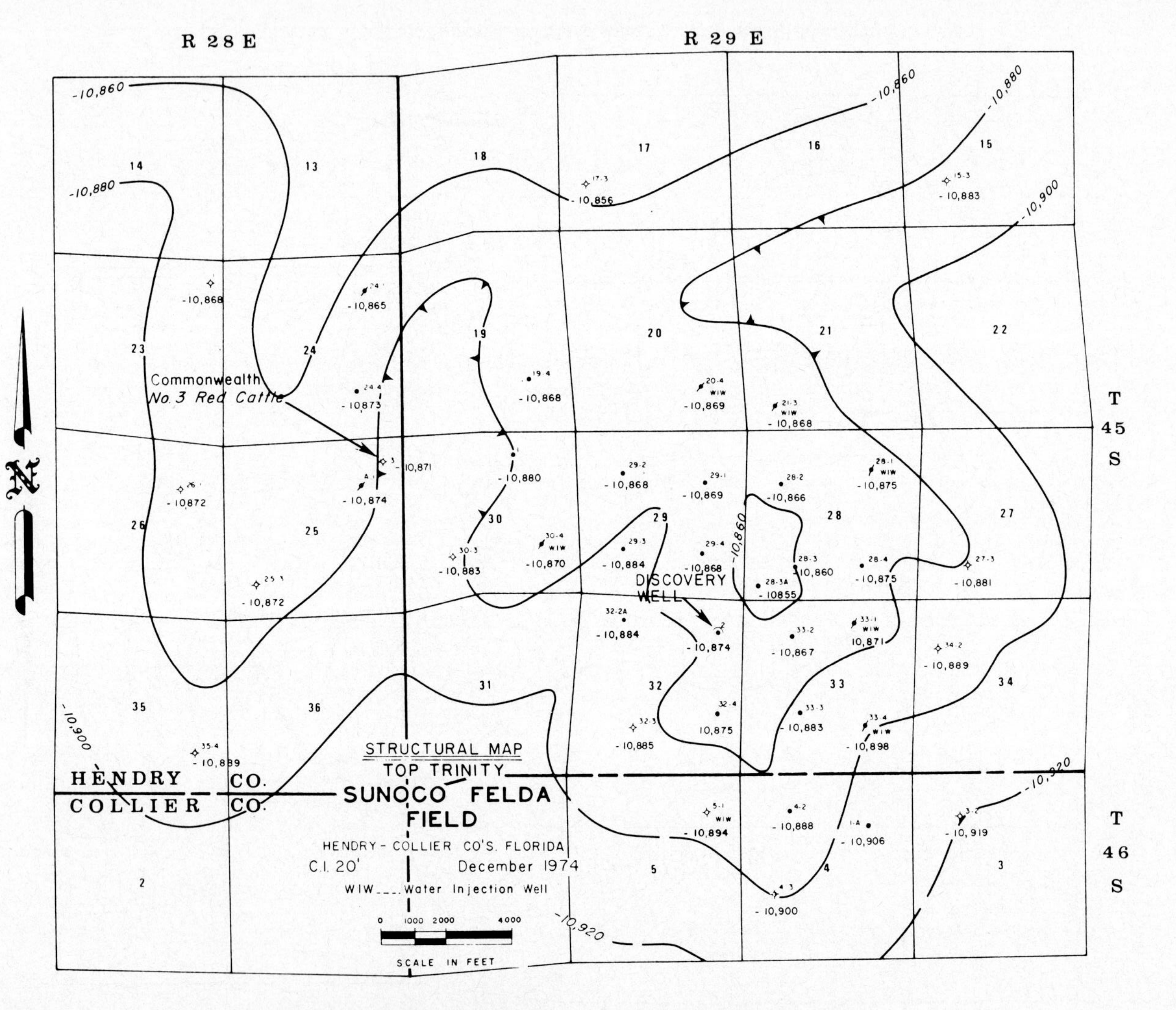

FIG. 12—Structure of top of Trinity, Sunoco-Felda field. C.I. = 20 ft.

Table 1. Reservoir and Field Data, Sunoco-Felda Field

Proved acreage (ac)	4,500
Well spacing (ac)	160
No. producing wells (orig.)	25
Av. depth to producing zone (ft)	11,475
Av. producing-zone thickness (ft)	13
Av. porosity (%)	18
Av. permeability (md)	60
Av. water saturation (%)	38
Gravity of oil (°API)	25
Viscosity of oil (cp)	3.0
Formation volume factor	1.1/1
Gas-solution ratio (cu ft/bbl)	88
Type of drive	Limited bottom water
Orig. bottomhole pressure (psia)	5,166 @ -11,421 ft
Reservoir temperature (°F)	195
Orig. oil in place (bbl)	44,000,000
Cum. oil production to 1-1-75 (bbl)	7,400,000
Av. daily oil production (bbl)	1,700

stratigraphic-trap oil reservoirs that should be prevalent all along the shelf proper. However, the subtle expression of these patch reefs in the subsurface and the limited amount of geologic well control along the trend make locating them a high-risk venture. The entire shelf area is very sparsely drilled, but the use of complete and detailed stratigraphic and environmental subsurface studies, coupled with carefully selected and sophisticated geophysical programs, should make exploration for additional reserves of this nature economically attractive. Exploratory drilling along the trend is fairly active and, as subsurface information becomes more readily available, the South Florida shelf and embayment area should emerge as a significant oil-producing province.

References Cited

Applin, P. L., 1960, Significance of changes in thickness and lithofacies of the Sunniland Limestone, Collier County, Florida, *in* Short papers in the geological sciences: U.S. Geol. Survey Prof. Paper 400-B, p. 209-211.

——— and E. R. Applin, 1944, Regional subsurface stratigraphy and structure of Florida and southern Georgia: AAPG Bull., v. 28, no. 12, p. 1673-1753.

——— and ——— 1965, The Comanche Series and associated rocks in the subsurface in central and south Florida: U.S. Geol. Survey Prof. Paper 447, p. 1-84.

Banks, J. E., 1960, Petroleum in Comanche (Cretaceous) section, Bend area, Florida: AAPG Bull., v. 44, no. 11, p. 1737-1748.

Puri, H. S., and R. O. Vernon, 1964, Summary of the geology of Florida and a guidebook to the classic exposures: Florida Geol. Survey Spec. Pub. 5, 291 p.

Sears, S. O., 1974, Facies interpretations and diagenetic modifications of the Sunniland Limestone, south Florida: Southeastern Geology, v. 15, no 4, p. 177-191.

Bibliography of North American Oil and Gas Fields from AAPG Publications[1]

E. M. TIDWELL[2]

In the 57 years from 1918 through 1974, data on many North American oil and gas fields have been published by AAPG. The amount of data, quality of information, and usefulness of material contained in these field papers have varied throughout the years. In trying to compile a bibliography of field papers, it was difficult to determine how much or how little to include. The final breakdown was made rather arbitrarily, and scanning the actual article may be the only way to determine its particular usefulness.

Any field that is mentioned by name in the title of a paper—whether that paper is a one-paragraph note or a 25-page article—is included in the alphabetical list and indicated by an asterisk. Other field listings, some of which may give as much information as the preceding type, are from articles that have the field name in a heading or on an illustration. A few fields were not included because the data were so scant that usefulness seemed minimal. Unfortunately, some material may have been overlooked because the field name did not appear in a title or heading; use of AAPG comprehensive indexes[3] is recommended to find this material.

Some very important sources of data that are *not* included are the annual "Development Papers." Although papers of the development type appeared in earlier issues of the *Bulletin*, it was not until 1938 that they were presented on an annual basis. For that reason, fields mentioned in development-type papers in the earlier volumes of the *Bulletin are* included in this bibliography.

Excluded from the field listings, mainly because they are so numerous, are field data from some excellent tables that have appeared in various publications. For example, tables on giant oil fields and gas fields, respectively, compiled by Halbouty et al, appear between pages 504 and 505 in *Memoir* 14. These tables actually present more information about some fields than is given in some of the articles listed in the bibliography. Again, AAPG comprehensive indexes should be used in the search for field data that may have appeared only in tabular form.

The bibliography is divided into two parts. The first is an alphabetical listing, by name, of all fields. The information given includes: (a) the state(s) or province(s) in which the field is located; (b) the county (counties) or parish(es) where the field is located; (c) the year of publication; (d) the publication in which the article appears (abbreviated as indicated below) or the volume number of the *Bulletin*; (e) the page number(s) of the article; and (f) whether the field is mentioned in the title (an asterisk indicates that it is).

The abbreviations used for publications are listed by date of publication, as follows:

1926	Salt	Geology of Salt Dome Oil Fields
1929	Struc 1	Structure of Typical American Oil Fields, Vol. I
1929	Struc 2	Structure of Typical American Oil Fields, Vol. II
1931	Alta	Stratigraphy of Plains of Southern Alberta
1934	Prob	Problems of Petroleum Geology
1935	Gas	Geology of Natural Gas
1936	GC	Gulf Coast Oil Fields
1936	Mex	Geology of the Tampico Region, Mexico
1941	Strat	Stratigraphic Type Oil Fields
1948	Struc 3	Structure of Typical American Oil Fields, Vol. III
1954	WCan	Western Canada Sedimentary Basin
1956	S. Ok 1	Petroleum Geology of Southern Oklahoma, Vol. I

[1]This compilation was suggested by the editor of this volume, Jules Braunstein, who gave invaluable advice regarding the selection of types of data to be included.

[2]AAPG, Tulsa, Oklahoma 74101.

Nancy Greeson, AAPG Editorial Staff Secretary, and Peggy Rice, Book Editor, assisted in preparation of this bibliography.

[3]AAPG Comprehensive Indexes for 1917–1945, 1946–1955, 1956–1965, and 1966–1970.

1958	Hab	Habitat of Oil
1959	S. Ok 2	Petroleum Geology of Southern Oklahoma, Vol. II
1965	Mem 4	Fluids in Subsurface Environments
1968	Mem 9	Natural Gases of North America
1970	Mem 14	Geology of Giant Petroleum Fields
1972	Mem 16	Stratigraphic Oil and Gas Fields—Classification, Exploration Methods, and Case Histories
1972	Mem 18	Underground Waste Management and Environmental Implications

In the alphabetical list, fields that are prefixed by a direction are alphabetized under the field name (e.g., Baxter Basin Middle, Baxter Basin North, Baxter Basin South). Exceptions to this are where the direction is actually part of the geographic name, such as South Bend, South Pass, and East Texas. Two fields named only by location (T25N, R8E; "2-4") are listed at the end of the bibliography. Part 1 starts on page 302.

The second part of the bibliography is a listing of fields by areas. (Reference must be made to the alphabetical list for the bibliographic sources.) This listing is divided into three sections: Canada, Mexico, and United States. Within each country, the areas are divided into states or provinces and, in the United States, there is a further subdivision by counties. Fields are listed by name under the location; if a field area covers more than one county or state, it is listed under each. For example, Hugoton-Panhandle field is listed under Kansas, Oklahoma, and Texas; furthermore, under each of those states it is listed under each county in which it is located. Lima-Indiana field covers many counties in northwestern Ohio and east-central Indiana; in the county listings it is shown without mention of the counties as the first item under each of those states. Offshore Louisiana fields appear at the end of the Louisiana listings.

Part 2 of the bibliography starts on page 329.

Part 1. Alphabetical List of Fields

Field	*State*	*County*	*Year*	*Pub.*	*Pages*	
Ackman	Nebraska	Red Willow	1962	V. 46	2079–2089	*
Adena	Colorado	Morgan	1957	V. 41	839– 847	
			1958	Hab	328– 343	
			1968	Mem 9	899– 927	
Aetna	Arkansas	Franklin, Logan	1968	Mem 9	1668–1681	*
Ajax	Louisiana	De Soto, Natchitoches	1942	V. 26	1255–1276	
Alamo	Veracruz		1936	Mex	220– 222	
Alazan	Veracruz		1936	Mex	217– 218	
Albert	Kansas	Barton, Rush	(see Otis-Albert)			
Aldrich	Kansas	Ness	1945	V. 29	564– 567	*
Alexander	Alberta		1968	Mem 9	705– 712	*
Alida	Saskatchewan		1958	Hab	149– 177	
Aliso	California	Los Angeles	1948	Struc 3	24– 37	*
Alma North	Oklahoma	Stephens	1956	S. Ok 1	282– 293	*
Alma-Smelter	Arkansas	Crawford	1935	Gas	533– 574	
Altamira	Tamaulipas		1936	Mex	195– 196	
Altus	Oklahoma	Jackson	1959	S. Ok 2	165– 179	*
Amatlan Northern	Veracruz		1936	Mex	213	
Amatlan Southern	Veracruz		1936	Mex	214	
Amelia	Texas	Jefferson	1939	V. 23	1635–1665	*
Amory	Mississippi	Monroe	1935	Gas	853– 879	
Aneth	Utah	San Juan	1960	V. 44	1541–1569	
			1968	Mem 9	1327–1356	
Anse La Butte	Louisiana	St. Martin	1943	V. 27	1123–1156	*
Antelope	North Dakota	McKenzie	1968	Mem 9	1304–1326	
Antioch Southwest	Oklahoma	Garvin	1950	V. 34	386– 422	
			1951	V. 35	582– 606	
Apache	Oklahoma	Caddo	1941	V. 25	2194	*
			1945	V. 29	100– 105	*
			1951	V. 35	582– 606	
Apco	Texas	Pecos	1948	Struc 3	399– 418	*
Appling	Texas	Calhoun, Jackson	1968	Mem 9	233– 263	
Arbuckle	California	Colusa	1968	Mem 9	646– 652	*
Arcadia-Coon Creek	Oklahoma	Logan, Oklahoma	1948	Struc 3	319– 340	*
Arch	Wyoming	Sweetwater	1968	Mem 9	817– 827	
Ardmore Southwest	Oklahoma	Carter	1959	S. Ok 2	262– 273	*
Armena	Alberta		1954	WCan	452– 463	*
Arroya Grande	California	San Luis Obispo	1934	Prob	177– 234	
Artemis-Himyar	Kentucky	Knox	1935	Gas	915– 947	
Artesia	New Mexico	Eddy	1929	Struc 1	112– 123	*
			1935	Gas	417– 457	
Ashland	Oklahoma	Coal, Pittsburg	1935	Gas	511– 532	
Ashland	Kentucky	Boyd	1935	Gas	915– 947	
Ashley Creek	Utah	Uintah	1951	V. 35	1000–1037	
Aspermont	Texas	Stonewall	1940	V. 24	1839–1840	*
Athens	Louisiana	Claiborne	1942	V. 26	1255–1276	
Atoka Penn	New Mexico	Eddy	1968	Mem 9	1394–1432	
Augusta	Kansas	Butler	1921	V. 5	421– 424	
			1935	Gas	483– 509	
			1948	Struc 3	213– 224	*
Austin	Michigan	Mecosta	1938	V. 22	129– 174	
			1941	Strat	237– 266	
Avery Island	Louisiana	Iberia	1959	V. 43	944– 957	
Aviator's	Texas	Webb	1923	V. 7	532– 545	
			(also see Schott-Aviator)			
Aylesworth	Oklahoma	Bryan, Marshall	1956	S. Ok 1	373– 391	
Badger Creek	Colorado	Adams	1955	V. 39	630– 648	

Field	State	County	Year	Pub.	Pages	
Baker-Glendive	Montana	Fallon	(see Cedar Creek)			
Baldwin Hills	California	Los Angeles	(see Inglewood)			
Ball	Texas	Freestone	1935	Gas	651– 681	
Bannatyne	Montana	Chouteau	1929	V. 13	779– 797	
Barataria	Louisiana	Jefferson	1941	V. 25	322– 323	*
Barbers Hill	Texas	Chambers	1925	V. 9	958– 973	*
			1926	Salt	530– 545	*
			1930	V. 14	719– 741	*
Barker Creek	Colorado	La Plata	1960	V. 44	1541–1569	
	New Mexico	San Juan	1968	Mem 9	1327–1356	
Barnhart	Texas	Reagan	1942	V. 26	387– 388	*
Barron	Texas	Limestone	1935	Gas	651– 681	
Bastian Bay	Louisiana	Plaquemines	1968	Mem 9	376– 581	
Batson	Texas	Hardin	1925	V. 9	1277–1282	*
			1926	Salt	524– 529	*
Battleship	Wyoming	Jackson	1968	Mem 9	840– 855	
Baxter Basin Middle	Wyoming	Sweetwater	1968	Mem 9	803– 816	
Baxter Basin North	Wyoming	Sweetwater	1935	Gas	323– 339	*
			1968	Mem 9	803– 816	
Baxter Basin South	Wyoming	Sweetwater	1935	Gas	323– 339	*
			1968	Mem 9	803– 816	
Baxterville	Mississippi	Lamar	1968	Mem 9	1176–1228	
Bay Marchand	Louisiana	Lafourche	1970	Mem 14	277– 291	*
Bay Sainte Elaine	Louisiana	Terrebonne	1959	V. 43	2470–2480	*
Bayou Blue	Louisiana	Iberville	1957	V. 41	1915–1951	*
Bayou Segnette	Louisiana	Jefferson	1968	Mem 9	376– 581	
Bazette	Texas	Navarro	1935	Gas	651– 681	
Beaver	Kansas	Barton	1953	V. 37	300– 313	
Beaver Creek	Michigan	Crawford, Kalkaska	1968	Mem 9	1761–1797	
Beaver Lodge	North Dakota	Williams	1953	V. 37	2294–2302	
			1957	V. 41	2493–2507	*
			1958	Hab	149– 177	
			1968	Mem 9	1304–1326	
Becher East	Ontario		1949	V. 33	153– 188	
Becher West	Ontario		1949	V. 33	153– 188	
Beehive Bend	California	Glenn	(see Willows-Beehive Bend)			
Bell Creek	Montana	Carter, Powder River	1968	V. 52	1869–1887	*
			1968	V. 52	1888–1898	*
			1970	Mem 14	128– 146	*
			1972	Mem 16	367– 375	*
Bell Lake	New Mexico	Lea	1968	Mem 9	1394–1432	
Bellevue	Louisiana	Bossier, Webster	1922	V. 6	179– 192	
			1922	V. 6	247– 251	*
			1923	V. 7	645– 652	*
			1929	Struc 2	229– 253	*
			1938	V. 2	1658–1681	*
Bellshill Lake	Alberta		1959	V. 43	880– 889	*
Belridge	California	Kern	1934	Prob	177– 234	
			1934	Prob	735– 760	
			1934	Prob	785– 805	
Belridge North	California	Kern	1934	Prob	177– 234	
			1934	Prob	735– 760	
			1934	Prob	785– 805	
Ben Bolt	Texas	Jim Wells	1939	V. 23	1237–1238	*
Bend Unit	Utah	Uintah	1968	Mem 9	174– 198	
Benton	Illinois	Franklin	1948	V. 32	745– 766	*
Bernard East	Texas	Wharton	1968	Mem 9	340– 358	
Bernard Prairie	Texas	Wharton	1968	Mem 9	340– 358	
Bernard West	Texas	Wharton	1968	Mem 9	340– 358	
Bergton	Virginia	Rockingham	1955	V. 39	317– 328	*

Field	*State*	*County*	*Year*	*Pub.*	*Pages*	
Bethany	Louisiana	Caddo	1922	V. 26	179– 192	
	Texas	Harrison, Panola				
Bethany-Longstreet	Louisiana	Caddo, De Soto	1968	Mem 9	1142–1146	*
Bethel	Texas	Anderson	1968	Mem 9	1008–1012	*
Big Hill	Texas	Jefferson	1926	Salt	497– 500	*
Big Lake	Texas	Reagan	1926	V. 10	365– 381	*
			1929	Struc 2	500– 541	*
			1930	V. 14	798– 806	*
			1934	Prob	347– 363	
			1934	Prob	399– 427	
			1935	Gas	417– 457	
			1942	V. 26	1398–1409	
Big Mineral	Texas	Grayson	1959	S. Ok 2	53– 100	
Big Piney-La Barge	Wyoming	Lincoln, Sublette	1968	Mem 9	780– 797	
			(also see La Barge)			
Big Sand Drew	Wyoming	Fremont	1928	V. 12	1137–1146	*
			1934	V. 18	1454–1492	
Big Sandy	Kentucky	Floyd, Johnson, Knott, Magoffin, Martin, Pike	1953	V. 37	282– 299	*
Big Sinking	Kentucky	Lee	1941	Strat	166– 207	*
Big Slough	Louisiana	Catahoula	(see Larto Lake)			
Big Springs	Nebraska	Deuel	1968	Mem 9	899– 927	
Big Springs West	Nebraska	Deuel	1968	Mem 9	899– 927	
Big Wall	Montana	Mussel Shell	1966	V. 50	2245–2259	
Billings	Oklahoma	Noble	1940	V. 24	2006–2018	*
Billy Creek	Wyoming	Johnson	1934	V. 18	1454–1492	
			1935	Gas	297– 303	*
Birch Creek Unit	Wyoming	Sublette	1966	V. 50	2176–2184	*
Bisti	New Mexico	San Juan	1957	V. 41	906– 922	*
			1961	V. 45	315– 329	*
			1963	V. 47	193– 228	*
			1972	Mem 16	610– 622	*
Black Jack	Texas	Aransas	1968	Mem 9	233– 263	
Black Lake	Louisiana	Natchitoches	1968	Mem 9	1152–1156	*
			1972	Mem 16	481– 488	*
Black Hollow	Colorado	Weld	1958	Hab	328– 343	
Black Mountain	Wyoming	Hot Springs	1947	V. 31	797– 823	
Blackfoot	Texas	Anderson	1950	V. 33	1750–1755	
Blackwell	Oklahoma	Kay	1929	Struc 1	158– 175	
Blackwell South	Oklahoma	Kay	1929	Struc 1	158– 175	
Blake	Texas	Brown	1941	Strat	548– 563	
Blankenship	Kansas	Butler, Greenwood	1921	V. 5	421– 424	
Blue Creek	West Virginia	Jackson	1949	V. 33	336– 345	
Blue Ridge	Texas	Fort Bend	1925	V. 9	304– 316	
			1926	Salt	600– 612	
Blue Rock-Salt Creek	Ohio	Muskingum	1950	V. 34	1874–1886	
Blue Springs	Missouri	Jackson	1941	V. 25	1405–1409	
Blue Springs East	Missouri	Jackson	1941	V. 25	1405–1409	
Blue Springs Northeast	Missouri	Jackson	1941	V. 25	1405–1409	
Boggy Creek	Texas	Anderson, Cherokee	1932	V. 16	584– 600	
			1935	Gas	651– 681	
Boling	Texas	Wharton	1935	Gas	683– 740	
Bonanza	Wyoming	Big Horn	1966	V. 50	2197–2220	
Bone Camp	Tennessee	Morgan	1929	Struc 1	243– 255	
Bonita	Texas	Montague	1940	V. 24	1838–1839	
Bonnsville	Texas	Jack, Wise	1968	Mem 9	1446–1454	
Border-Red Coulee	Alberta		1931	V. 15	1161–1170	
	Montana	Toole	1931	Alta	33– 42	
			1934	Prob	695– 718	
			1935	Gas	245– 276	
			1941	Strat	267– 326	

Field	State	County	Year	Pub.	Pages	
Bornholdt	Kansas	McPherson, Rice	1948	Struc 3	225– 248	
Bossier	Louisiana	Bossier	(see Elm Grove)			
Bow Island	Alberta		1935	Gas	1– 58	
Bowdoin	Montana	Phillips, Valley	1935	Gas	245– 276	
Bowerbank	California	Kern	1968	Mem 9	113– 134	
Bowers	Texas	Montague	1943	V. 27	20– 37	*
Bowes	Montana	Blaine	1935	Gas	245– 276	
			1963	V. 47	1943–1951	*
Bowlegs	Oklahoma	Seminole	1929	Struc 2	315– 361	
Boxelder	Montana	Blaine, Hill	1935	Gas	245– 276	
Boxer	California	Morgan	1972	Mem 16	383– 388	*
Boyd-Peters	Michigan	St. Clair	1972	Mem 16	460– 471	*
Bradford	New York	Cattaraugus	1929	Struc 2	407– 442	*
	Pennsylvania	McKean				
Brea-Olinda	California	Los Angeles, Orange	1934	Prob	177– 234	
			1935	Gas	113– 220	
			(also see Olinda)			
Breckenridge	Texas	Stephens	1934	Prob	347– 363	
Breedlove	Texas	Martin	1971	V. 55	403– 411	*
Brentwood	California	Contra Costa	1968	Mem 9	104– 112	*
Breton Sound Block 36	Louisiana	(Offshore)	1968	Mem 9	376– 581	
Brock West	Oklahoma	Carter	1959	S. Ok 2	249– 261	*
Brookhaven	Mississippi	Lincoln	1950	V. 34	1517–1529	*
			1968	Mem 9	1176–1226	
Brooks	Alberta		1935	Gas	1– 58	
Broomfield	Michigan	Isabella	1935	Gas	787– 812	
			1938	V. 22	129– 174	
Brown-Bassett	Texas	Terrell	1965	Mem 4	257– 279	
			1968	Mem 9	1394–1432	
Brown-Cude	Texas	Caldwell	(see Buchanan)			
Brownsville	Ontario		1949	V. 33	153– 188	
Bruner	Oklahoma	Tulsa	1927	V. 11	933– 944	
			1929	Struc 1	211– 219	
Bryan	Wyoming	Big Horn, Park	(see Garland)			
Bryson	Texas	Jack	1932	V. 16	179– 188	*
			1941	Strat	539– 547	*
Buchanan	Texas	Caldwell	1932	V. 16	741– 768	
Buckeye	Texas	Matagorda	1935	V. 19	378– 400	*
			1936	GC	734– 756	*
			1940	Gladwin	1950–1982	*
Buena Vista Hills	California	Kern	(see Midway-Sunset)			
Buffalo	Texas	Leon	1935	Gas	651– 681	
Buffalo North	Oklahoma	Harper	1974	V. 58	447– 463	
Burbank	Oklahoma	Kay, Osage	1921	V. 5	502	*
			1924	V. 8	584– 592	*
			1927	V. 11	1045–1054	*
			1929	Struc 1	220– 229	*
			1932	V. 16	881– 890	
			1934	Prob	581– 627	
			1941	V. 25	1175–1179	*
Burbank South	Oklahoma	Osage	1937	V. 21	560– 579	*
			1941	V. 25	1175–1179	*
Burkett-Seeley	Kansas	Greenwood	1923	V. 7	482– 487	*
Burrton	Kansas	Harvey, Reno	1935	V. 24	1779–1797	
			1940	Gas	459– 482	
Bush City	Kansas	Anderson	1926	V. 10	568– 580	
			1941	Strat	43– 56	*
Buttonwillow	California	Kern	1935	Gas	113– 220	
Byron	Wyoming	Big Horn	1935	Gas	291	
Cabin Creek	West Virginia	Boone, Kanawha	1927	V. 11	705– 720	*
			1929	Struc 1	462– 475	*

Field	State	County	Year	Pub.	Pages
Cacalilao	Veracruz		1936	Mex	183
Cacalilao Eastern	Veracruz		1936	Mex	183
Cacalilao North-Central	Veracruz		1936	Mex	191
Cacalilao West-Central	Veracruz		1936	Mex	191
Cache	Colorado	Montezuma	1967	V. 51	1959–1978
Caddo	Louisiana	Caddo	1918	V. 2	61– 69
			1922	V. 6	179– 192
			1929	Struc 2	183– 195
Caillou Island	Louisiana	Terrebonne	1970	Mem 14	277– 291
Cairo	Arkansas	Union	1950	V. 34	1954–1980
Calavera	Veracruz		1936	Mex	191
Calder	California	Kern	1951	V. 35	1070–1073
Calgary	Alberta		1968	Mem 9	1238–1284
Calhoun	Arkansas	Columbia	1945	V. 29	459
Calhoun	Louisiana	Jackson, Lincoln, Ouachita	1973	V. 57	301– 320
Calloway-Henry	Texas	Stephens	1935	Gas	609– 649
Camargo	Tamaulipas		1949	V. 33	1351–1384
Camden Gore	Ontario		1949	V. 33	153– 188
Cameron	Oklahoma	Leflore	1935	Gas	511– 532
Cameron Meadows	Louisiana	Cameron	1932	V. 16	255– 256
Camp	Oklahoma	Carter	1956	S. Ok 1	174– 185
Campbell-Davis Creek	West Virginia	Kanawha	1938	V. 22	175– 188
Campbell West	Oklahoma	Major	1972	Mem 16	568– 578
Campbells Creek	West Virginia	Boone	1949	V. 33	336– 345
Campton	Kentucky	Wolfe	1927	V. 11	477– 492
			1929	Struc 1	73– 90
Camrick	Oklahoma	Beaver, Texas	1968	Mem 9	1525–1538
	Texas	Ochiltree			
Camrose	Alberta		1954	WCan	452– 463
Canadian	Wyoming	Jackson	1968	Mem 9	840– 855
Caracol	Veracruz		1936	Mex	185
Carbon	Alberta		1968	Mem 9	726– 730
Carbonero	Colorado	Garfield	1935	Gas	363– 384
Carmi	Kansas	Pratt	1944	V. 28	125– 126
Carolina-Texas	Texas	Webb	1923	V. 7	532– 545
			1929	Struc 1	389– 408
Carter-Knox	Oklahoma	Grady, Stephens	1959	S. Ok 2	198– 219
			1968	Mem 9	1476–1491
Carterville-Sarepta	Louisiana	Bossier, Webster	1938	V. 22	1473–1503
Carthage	Texas	Panola	1968	Mem 9	1020–1059
Casmalia	California	Santa Barbara	1934	Prob	177– 234
Castle River-Waterton	Alberta		1959	V. 43	992–1025
Cat Canyon	California	Santa Barbara	1934	Prob	177– 234
Cat Canyon West	California	Santa Barbara	1949	V. 33	32– 51
Cat Creek	Montana	Garfield, Petroleum	1921	V. 5	252– 275
			1934	Prob	695– 718
			1947	V. 31	797– 823
Cato	Texas	Montague	1956	S. Ok 1	355– 372
Cayuga	Texas	Anderson	1935	Gas	651– 681
Cedar Creek	Montana	Fallon, Wibaux	1935	Gas	245– 276
	North Dakota	Bowman	1947	V. 31	797– 823
			1968	Mem 9	1304–1326
Cedar Creek	Texas	Bastrop	(see Yoast)		
Cedar Point	Texas	Chambers	1938	V. 22	1601–1602
Cement	Oklahoma	Caddo, Grady	1960	V. 44	210– 226
			1961	V. 45	1971–1993
Centerville	Kansas	Linn	1923	V. 7	103– 113
Centralia-Sandoval	Illinois	Clinton, Marion	1929	Struc 2	115– 141
Cerritos	Veracruz		1936	Mex	181– 182
Cerro Azul	Veracruz		1936	Mex	216

Field	*State*	*County*	*Year*	*Pub.*	*Pages*	
Cerro Viejo	Veracruz		1936	Mex	219	
Chanute	Kansas	Neosho	1941	Strat	57– 77	*
Chapel Hill	Texas	Smith	1938	V. 22	1107	*
Chapapote Nunez	Veracruz		1936	Mex	219	
Chapman	Texas	Williamson	1932	V. 16	741– 768	
Chatham	Ontario		1949	V. 33	153– 188	
Chetopa	Kansas	Labette	1941	V. 25	1934–1939	*
Chickasha	Oklahoma	Caddo, Grady	1961	V. 45	1971–1993	
Chiconcillo	Veracruz		1936	Mex	210– 212	
Chijol	Veracruz		1936	Mex	193	
Chiles Ranch	Oklahoma	Coal	(see Coalgate)			
Chinampa North	Veracruz		1936	Mex	210– 212	
Chinampa South	Veracruz		1936	Mex	213	
Chowchilla	California	Madera	1968	Mem 9	113– 134	
Church Buttes	Wyoming	Sweetwater, Uinta	1968	Mem 9	798– 802	*
Church-McElroy	Texas	Crane	1935	Gas	417– 457	
Church Run	Pennsylvania	Crawford	1941	Strat	507– 538	
Circle Ridge	Wyoming	Fremont	1951	V. 35	1000–1037	
Clara Couch	Texas	Crockett	1968	Mem 9	1394–1432	
Clare	Michigan	Clare	1938	V. 22	129– 174	
Clarke Lake	British Columbia		1963	V. 47	467– 483	*
Clarks	Louisiana	Caldwell	1942	V. 26	1255–1276	
Clarksville	Arkansas	Johnson	1935	Gas	533– 574	
Clay	Kentucky	Clay	1927	V. 11	477– 492	
			1929	Struc 1	73– 90	
Clay Creek	Texas	Washington	1931	V. 15	43– 60	*
			1936	V. 20	68– 90	*
			1936	GC	757– 779	*
Clear Creek	Utah	Carbon, Emery	1955	V. 39	385– 421	
			1968	Mem 9	928– 945	
Clearwater	Kansas	Sumner	1935	Gas	459– 482	
Cleveland	Texas	Liberty	1938	V. 22	1274–1277	*
Clodine	Texas	Fort Bend	1941	V. 25	2057–2058	*
Coalgate	Oklahoma	Coal	1935	Gas	511– 532	
Coalinga East Extension	California	Fresno	1958	Hab	99– 112	
			1968	Mem 9	113– 134	
Coalinga Eastside	California	Fresno	1931	V. 15	829– 836	*
			1934	Prob	177– 234	
			1934	Prob	735– 760	
			1934	Prob	785– 805	
			1934	Prob	953– 985	
			1940	V. 24	1940–1949	
Coalinga Westside	California	Fresno	1934	Prob	177– 234	
			1934	Prob	735– 760	
			1934	Prob	785– 805	
			1934	Prob	953– 985	
Coffee Bay	Louisiana	La Fourche	1968	Mem 9	376– 581	
Coffeyville	Kansas	Montgomery	1929	Struc 1	49– 51	*
Cogdell	Texas	Kent, Scurry	1970	Mem 14	185– 203	
Cole	Texas	Duval, Webb	1929	Struc 1	389– 408	
Coleman	Texas	Archer	1941	V. 25	428– 429	*
Coles Levee	California	Kern	1968	Mem 9	113– 134	
Collegeport	Texas	Matagorda	1968	Mem 9	359– 367	*
Colony	Kansas	Anderson	1923	V. 7	103– 113	
			1935	Gas	483– 509	
Columbia West	Texas	Brazoria	1921	V. 5	212– 251	*
			1929	Struc 2	451– 469	*
			1942	V. 26	1441–1466	*
Conroe	Texas	Montgomery	1935	Gas	683– 740	
			1936	V. 20	736– 779	*
			1936	GC	789– 832	*

Field	*State*	*County*	*Year*	*Pub.*	*Pages*	
Coon Creek	Oklahoma	Logan	(see Arcadia-Coon Creek)			
Cooper Creek	West Virginia	Kanawha	1938	V. 22	175– 188	
Copley	West Virginia	Gilmer, Lewis	1927	V. 11	581– 600	*
			1929	Struc 1	440– 461	*
Corcovado	Veracruz		1936	Mex	185	
Corning	California	Tehama	1968	Mem 9	635– 638	*
Corning South	California	Tehama	1968	Mem 9	635– 638	*
Corral Canyon	New Mexico	Eddy	1965	Mem 4	294– 307	
Cote Blanche Island	Louisiana	St. Mary	1959	V. 43	2592–2622	
Cotton Lake South	Texas	Chambers	1941	V. 25	1899–1920	*
Cotton Valley	Louisiana	Webster	1924	V. 8	244– 246	*
			1925	V. 9	875– 885	*
			1930	V. 14	983– 996	*
			1938	V. 22	1603	*
Cottonwood Creek	Wyoming	Washakie	1957	V. 41	823– 838	*
			1967	V. 51	2122–2132	*
Covert-Sellers	Kansas	Marion	1929	Struc 1	60– 72	
Cowden North	Texas	Ector	1941	V. 25	593– 629	*
Cowley	Wyoming	Big Horn	1966	V. 50	2197–2220	
Coyanosa	Texas	Pecos	1965	Mem 4	257– 279	
			1968	Mem 9	1394–1432	
Coyote Creek	Wyoming	Crook, Weston	1968	V. 52	2116–2122	
Coyote East	California	Orange	1935	Gas	113– 220	
Coyote West	California	Orange	1935	Gas	113– 220	
Craigend	Alberta		1968	Mem 9	1238–1284	
Cranfield	Mississippi	Adams, Franklin	1968	Mem 9	1176–1226	
Cree-Flowers	Texas	Roberts	1968	Mem 9	1525–1538	
Creole	Louisiana	(Offshore)	1948	Struc 3	281– 298	*
Crinerville	Oklahoma	Carter	1927	V. 11	1067–1085	*
			1929	Struc 1	192– 210	*
Cromwell	Oklahoma	Okfuskee, Seminole	1929	Struc 2	300– 314	*
			1934	Prob	581– 627	
Cross Cut	Texas	Brown	1941	Strat	548– 563	*
Crystal	Michigan	Montcalm	1938	V. 22	129– 174	
Cumberland	Oklahoma	Bryan, Marshall	1948	Struc 3	341– 358	*
Cunningham	Kansas	Kingman, Pratt	1935	Gas	459– 482	
			1937	V. 21	500– 524	*
			1940	V. 24	1779–1797	
Currie	Texas	Navarro	1923	V. 7	25– 36	*
			1926	V. 10	61– 71	*
			1929	Struc 1	304– 388	
Curry	Texas	Stephens	1934	Prob	581– 627	
Cushing	Oklahoma	Creek	1929	Struc 2	396– 406	*
			1934	Prob	581– 627	
Cut Bank	Montana	Glacier	1934	Prob	695– 718	
			1935	Gas	245– 276	
			1941	Strat	327– 381	*
			1947	V. 31	797– 823	
			1951	V. 35	1000–1037	
Cuyama South	California	Santa Barbara	1958	Hab	79– 98	
Cymric	California	Kern	1948	Struc 3	38– 57	*
Cypress Bayou	Louisiana	La Salle	1942	V. 26	1255–1276	
Dale	Texas	Caldwell	1932	V. 16	741– 768	
Daly-East Cromer	Manitoba		1958	Hab	149– 177	
Damon Mound	Texas	Brazoria	1918	V. 2	16– 37	
			1925	V. 9	505– 535	*
			1926	Salt	613– 643	*
Darrow Dome	Louisiana	Ascension	1938	V. 22	1412–1422	*
Darst	Texas	Guadalupe	1933	V. 17	16– 37	*
Davenport	Oklahoma	Lincoln	1941	Strat	386– 407	*

Field	State	County	Year	Pub.	Pages	
Davis-Gardner	Texas	Coleman	1972	Mem 16	406– 414	*
Dawn	Ontario		1949	V. 33	153– 188	
D'Clute	Ontario		1949	V. 33	153– 188	
De Soto	Louisiana	De Soto	1918	V. 2	61– 69	
De Soto-Red River	Louisiana	De Soto, Red River	1922	V. 6	179– 192	
Deep River	Michigan	Arenac	1948	Struc 3	299– 304	*
Deerfield	Michigan	Monroe	1948	Struc 3	305– 318	*
Del Monte	Texas	Zavala	1947	V. 31	772– 773	*
Del Valle	California	Los Angeles	1941	V. 25	1159–1166	
			1942	V. 26	188– 196	*
Delaware Extension	Oklahoma	Nowata	1929	Struc 2	362– 364	*
Delhi	Louisiana	Franklin, Madison, Richland	1972	Mem 16	548– 558	*
Delta Farms	Louisiana	LaFourche	1953	V. 37	2649–2676	*
Depew	Oklahoma	Creek	1929	Struc 2	365– 377	*
Desert Springs	Wyoming	Sweetwater	1968	Mem 9	817– 827	
Desert Springs West	Wyoming	Sweetwater	1968	Mem 9	817– 827	
Diamond M	Texas	Borden, Scurry	1970	Mem 14	185– 203	
Dickson	Tennessee	Dickson	1935	Gas	853– 879	
Dineh-bi-Keyah	Arizona	Apache	1968	V. 52	2045–2057	*
Dixie	Louisiana	Caddo	1930	V. 14	743– 763	*
Dominguez	California	Los Angeles	1928	V. 12	625– 650	
			1934	Prob	177– 234	
			1935	Gas	113– 220	
Dora	Oklahoma	Seminole	1939	V. 23	692– 698	*
			1941	Strat	408– 435	*
Dorcheat	Arkansas	Columbia	1940	V. 24	738– 740	*
Dos Bocas	Veracruz		1936	Mex	208	
Douglas (= Douglass)	Kansas	Butler	1921	V. 5	421– 424	
Driscoll	Texas	Duval	1933	V. 17	816– 826	*
			1936	GC	620– 630	*
Dry Creek	Montana	Carbon	1934	Prob	695– 718	
			1935	Gas	245– 276	
Duncan West	Oklahoma	Stephens	1956	S. Ok 1	319– 326	*
Durant East	Oklahoma	Bryan	1968	Mem 9	1467–1475	*
Dwyer	Montana	Sheridan	1966	V. 50	2260–2268	
Eagle Springs	Nevada	Nye	1967	V. 51	2133–2145	*
Earlsboro	Oklahoma	Pottawatomie, Seminole	1929	Struc 2	315– 361	
East Texas	Texas	Cherokee, Gregg, Rusk,	1931	V. 15	843– 847	*
		Smith, Upshur	1932	V. 16	907– 923	*
			1933	V. 17	757– 792	*
			1941	Strat	600– 640	*
			1972	Mem 18	331– 340	*
Ebano	San Luis Potosi		1936	Mex	193– 194	
Edgerly	Louisiana	Calcasieu	1925	V. 9	497– 504	*
			1926	Salt	470– 477	
Edison	California	Kern	1941	Strat	1– 8	*
			1948	Struc 3	58– 85	*
Edison West	California	Kern	1953	V. 37	797– 820	*
Edmond West	Oklahoma	Canadian, Kingfisher, Logan,	1946	V. 30	1797–1829	*
		Oklahoma	1948	Struc 3	359– 398	*
Edna	California	San Luis Obispo	(see Arroya Grande)			
Edna	Texas	Jackson	1941	V. 25	104– 119	*
Edwards	Kansas	Ellsworth, Rice	1948	Struc 3	225– 248	
Egan	Louisiana	Acadia	1968	Mem 9	376– 581	
Egypt	Texas	Wharton	1968	Mem 9	340– 358	
El Barco	Veracruz		1936	Mex	183	
El Capitan	California	Santa Barbara	1934	Prob	177– 234	
El Dorado (or Eldorado)	Arkansas	Union	1922	V. 6	179– 192	
			1922	V. 6	193– 198	*

Field	*State*	*County*	*Year*	*Pub.*	*Pages*
El Dorado (cont'd)			1922	V. 6	358– 367
			1923	V. 7	350– 361
			1942	V. 26	1255–1276
El Segundo	California	Los Angeles	1936	V. 20	939– 950
Elbing	Kansas	Butler	1921	V. 5	421– 424
			1929	Struc 1	60– 72
Eldorado	Kansas	Butler	1921	V. 5	421– 424
			1929	Struc 2	160– 167
Elk Basin	Montana	Carbon	1929	Struc 2	577– 588
	Wyoming	Park	1935	Gas	291
			1947	V. 31	797– 823
			1948	V. 32	52– 67
			1966	V. 50	2197–2220
			1969	V. 53	2094–2113
Elk City	Kansas	Montgomery	1935	Gas	483– 509
Elk Hills	California	Kern	1929	Struc 2	44– 61
			1934	Prob	177– 234
			1934	Prob	399– 427
			1934	Prob	735– 760
			1934	Prob	785– 805
			1934	Prob	953– 985
			1935	Gas	113– 220
Elk-Poca	West Virginia	Kanawha	1938	V. 22	175– 188
			(also see Sissonville)		
Elk Springs-Winter Valley	Colorado	Moffat	1968	Mem 9	856– 865
Elliott	Kentucky	Elliott	1927	V. 11	477– 492
			1929	Struc 1	73– 90
Ellisburg	Pennsylvania	Potter	1938	V. 22	241– 266
			1949	V. 33	305– 335
Elm Grove	Louisiana	Bossier	1918	V. 2	61– 69
			1922	V. 6	179– 192
Elsmore	Kansas	Allen	1923	V. 7	103– 113
Elton North	Louisiana	Allen	1968	Mem 9	376– 581
Elwood	California	Santa Barbara	1934	Prob	177– 234
			1935	Gas	113– 220
Embar	Texas	Andrews	1965	Mem 4	225– 242
Emma	Texas	Andrews	1965	Mem 4	280– 293
Empire Abo	New Mexico	Eddy	1972	Mem 16	472– 480
Encill	Kansas	Elk	1935	Gas	483– 509
Enos Creek	Wyoming	Hot Springs	1935	Gas	291
Eola	Louisiana	Avoyelles	1940	V. 24	701– 715
			1941	V. 25	1363–1395
Eola	Oklahoma	Garvin	1950	V. 34	2176–2199
			1964	V. 48	1555–1567
Esperson	Texas	Liberty	1934	V. 18	1632–1654
			1935	Gas	683– 740
			1936	GC	857– 879
Eylau	Texas	Bowie	1968	Mem 9	1074–1084
Fairbanks	Texas	Harris	1939	V. 23	686– 688
Fairport	Kansas	Russell	1929	Struc 1	35– 48
Farmington	Pennsylvania	Tioga	1931	V. 15	671– 688
			1935	Gas	949– 987
Farnham	Utah	Carbon	1935	Gas	363– 384
Fashing	Texas	Atascosa, Karnes	1968	Mem 9	976– 981
Fayette	Alabama	Fayette	1935	Gas	853– 879
Ferris	Wyoming	Carbon	1929	Struc 2	636– 666
Ferris Middle	Wyoming	Carbon	1935	Gas	305– 322
Ferris West	Wyoming	Carbon	1935	Gas	305– 322
Ferronales	Veracruz		1936	Mex	192
Fiddler Creek	Wyoming	Weston	1951	V. 35	1000–1037

Field	*State*	*County*	*Year*	*Pub.*	*Pages*	
Figure Lake	Alberta		1968	Mem 9	1238–1284	
Fisk	Texas	Coleman	1929	V. 13	1214	*
Fitts	Oklahoma	Pontotoc	1936	V. 20	951– 974	*
Flat Canyon	Utah	Emery	1955	V. 39	385– 421	
Flat Gap-Win-Ivyton	Kentucky	Johnson, Magoffin	1935	Gas	915– 947	
Florence	Colorado	Fremont	1929	Struc 2	75– 92	*
			1951	V. 35	1000–1037	
Florence	Kansas	Marion	1929	Struc 1	60– 72	
Florence-Urschel	Kansas	Marion	(see Urschel)			
Flower Acres	Texas	Bowie	1968	Mem 9	1074–1084	
Floyd	Kentucky	Floyd	1935	Gas	915– 947	
Foremost	Alberta		1935	Gas	1– 58	
Fort Collins	Colorado	Larimer	1958	Hab	328– 343	
Fort Norman	N.W.T.		1921	V. 5	85– 86	*
Fort St. John	British Columbia		1968	Mem 9	683– 697	
Fort Saskatchewan	Alberta		1968	Mem 9	713– 720	*
Foster-Reno-Oil City	Pennsylvania	Crawford, Venango	1941	Strat	507– 538	
Fouke	Arkansas	Miller	1948	Struc 3	5– 23	*
Fox	Oklahoma	Carter	1922	V. 6	367– 369	*
Fox	Texas	Refugio	1934	V. 18	519– 530	*
			1936	GC	664– 675	*
Fox-Bush	Kansas	Butler	1921	V. 5	421– 424	
Francisco	Indiana	Gibson	1929	Struc 2	115– 141	*
Frannie	Wyoming	Park	1966	V. 50	2197–2220	
Frederick West	Oklahoma	Tillman	1959	S. Ok 2	180– 186	*
Friendswood	Texas	Harris	1938	V. 22	1602–1603	*
Frobisher	Saskatchewan		1958	Hab	149– 177	
Fruitvale	California	Kern	1934	Prob	177– 234	
			1934	Prob	953– 985	
Fryburg	North Dakota	Billings	1958	Hab	149– 177	
Fullerton	Texas	Andrews	1944	V. 28	1541–1542	*
Funny Louis Bayou	Louisiana	La Salle	1942	V. 26	1255–1276	
Garber	Oklahoma	Garfield	1929	Struc 1	176– 191	*
			1934	Prob	347– 363	
Garden City	Louisiana	St. Mary	1968	Mem 9	376– 581	
Garland	Wyoming	Big Horn, Park	1934	Prob	347– 363	
			1935	Gas	291	
			1947	V. 31	797– 823	
Garnett	Kansas	Anderson	1923	V. 7	103– 113	
			1935	Gas	483– 509	
Garrucho	Veracruz		1936	Mex	185	
Gay	West Virginia	Jackson	1941	Strat	806– 829	*
Geneseo	Kansas	Rice	1948	Struc 3	225– 248	
Gilbert Creek	West Virginia	Mingo	1964	V. 48	465– 486	
Gilbertown	Alabama	Choctaw	1948	Struc 3	1– 4	*
			1953	V. 37	245– 249	*
Glenmora	Louisiana	Rapides	1952	V. 36	146– 159	*
Glenn	Oklahoma	Creek	1927	V. 11	1055–1065	*
			1929	Struc 1	230– 242	*
			1932	V. 16	881– 890	
Glenmary	Tennessee	Scott	1929	Struc 1	243– 255	
Glenrock South	Wyoming	Converse	1954	V. 38	2119–2156	*
			1972	Mem 16	415– 427	*
Glick	Kansas	Comanche, Kiowa	1968	Mem 9	1576–1582	*
Golden Eagle	Wyoming	Hot Springs	1935	Gas	291	
Goleta	California	Santa Barbara	1934	Prob	177– 234	
Goldsmith	Texas	Ector	1939	V. 23	1525–1552	*
			1965	Mem 4	280– 293	
Gomez	Texas	Pecos	1968	Mem 9	1394–1432	

Field	State	County	Year	Pub.	Pages
Goose Creek	Texas	Harris	1918	V. 2	16– 37
			1925	V. 9	286– 297
			1926	Salt	546– 557
			1927	V. 11	729– 745
Goose River	Alberta		1968	V. 52	21– 56
Gordon Creek	Utah	Carbon	1955	V. 39	385– 421
Gotebo North	Oklahoma	Kiowa, Washita	1968	Mem 9	1492–1508
Government Wells	Texas	Duval	1935	V. 19	1131–1147
			1936	GC	631– 647
Grabs	Kansas	Harper	(see Spivey-Grabs)		
Graham	Oklahoma	Carter	1924	V. 8	593– 620
Gramp's	Colorado	Archuleta	1946	V. 30	561– 580
			1948	Struc 3	86– 109
Grand Cane	Louisiana	De Soto	1942	V. 26	1255–1276
Grand Valley-Triumph	Pennsylvania	Warren	1941	Strat	507– 538
Granny's Creek	West Virginia	Clay	1929	Struc 2	571– 576
Grass Creek	Wyoming	Hot Springs	1929	Struc 2	623– 635
			1934	Prob	347– 363
Gratiot	Michigan	Gratiot	1935	Gas	787– 812
Greasewood	Colorado	Weld	1932	V. 16	256– 257
			1941	Strat	19– 42
Greensburg Consolidated	Kentucky	Green, Taylor	1972	Mem 16	579– 584
Greenwich	Kansas	Sedgwick	1939	V. 23	643– 662
Greenwood	Colorado	Baca	1968	Mem 9	1557–1566
	Kansas	Morton			
	Oklahoma	Texas			
Grenora	North Dakota	Williams	1966	V. 50	2260–2268
Greta	Texas	Refugio	1934	V. 18	519– 530
			1935	V. 19	544– 559
			1936	GC	648– 663
			1936	GC	664– 675
Griffithsville	West Virginia	Lincoln	1929	Struc 2	571– 576
Grimes	California	Colusa, Sutter	1972	Mem 16	428– 439
Grizzly Bluff	California	Humboldt	1968	Mem 9	68– 75
Groesbeck North	Texas	Limestone	1929	Struc 1	304– 388
Groesbeck South	Texas	Limestone	1929	Struc 1	304– 388
Grove	Texas	Eastland	1935	Gas	609– 649
Gwinville	Mississippi	Jefferson Davis	1968	Mem 9	1176–1226
Hackberry East	Louisiana	Cameron	1931	V. 15	247– 256
Half Moon Bay	California	San Mateo	1934	Prob	177– 234
Halverson	Wyoming	Campbell	1967	V. 51	705– 709
Ham Gossett	Texas	Kaufman	1954	V. 38	306– 318
Hamilton	Colorado	Moffat	(see Moffat)		
Handy	Texas	Grayson	1959	S. Ok 2	53– 100
Hansford	Texas	Hansford	1968	Mem 9	1525–1538
Hardin	Texas	Liberty	1936	V. 20	1122–1123
			1941	Strat	564– 599
Hardtner	Kansas	Barber	1968	Mem 9	1566–1569
Harmattan-Elkton	Alberta		1968	Mem 9	1238–1284
Harper Ranch	Kansas	Clark	1974	V. 58	447– 463
Harrisburg	Nebraska	Banner	1957	V. 41	839– 847
Harrison	Pennsylvania	Potter, Tioga	1949	V. 33	305– 335
Hartburg Northwest	Texas	Newton	1960	V. 44	458– 470
Hastings	Texas	Brazoria	1972	Mem 18	331– 340
Hatton	Saskatchewan		(see Medicine Hat)		
Hawkins	Texas	Wood	1941	V. 25	898– 899
			1946	V. 30	1830–1856
Haynesville	Louisiana	Claiborne	1921	V. 5	629– 633
			1922	V. 6	142
			1922	V. 6	179– 192

Field	State	County	Year	Pub.	Pages	
Haynesville (cont'd)			1942	V. 26	1255–1276	
			1944	V. 28	333– 340	
			1971	V. 55	566– 580	*
Haynesville North	Louisiana	Claiborne	1968	V. 52	92– 128	*
Hazlit Creek	Mississippi	Wilkinson	1972	Mem 16	318– 328	*
Healdton	Oklahoma	Carter	1921	V. 5	469– 474	*
			1968	V. 52	3– 20	*
			1970	Mem 14	255– 276	*
Hebron	Pennsylvania	Potter	1936	V. 20	1019–1027	*
			1938	V. 22	241– 266	
			1949	V. 33	305– 335	
Heinz	Kansas	Rice	1953	V. 37	300– 313	
Hemphill	Louisiana	La Salle	1942	V. 26	1255–1276	
Henderson	Texas	Clay	1940	V. 24	1495	
Hendrick	Texas	Winkler	1930	V. 14	923– 944	*
			1935	Gas	417– 457	
Henne-Winch-Fariss	Texas	Jim Hogg	1929	Struc 1	389– 408	
Henrietta	Texas	Clay	(see Petrolia)			
Hershey	Texas	Pecos	1965	Mem 4	257– 279	
Hewitt	Oklahoma	Carter	1929	Struc 2	290– 299	*
			1934	Prob	581– 627	
			1956	S. Ok 1	154– 161	*
Hewitt West	Oklahoma	Carter	1956	S. Ok 1	162– 173	*
Hiawatha	Colorado	Moffat	1929	Struc 2	93– 114	
	Wyoming	Sweetwater	1930	V. 14	1013–1040	
			1935	Gas	341– 361	*
Hiawatha West	Colorado	Moffat	1930	V. 14	1013–1040	
Hidden Dome	Wyoming	Washakie	1935	Gas	291	
High Island	Texas	Galveston	1936	V. 20	560– 611	*
			1936	GC	909– 960	*
Hilbig	Texas	Bastrop	1935	V. 19	206– 220	*
			1935	V. 19	1023–1037	
Hitchcock	Texas	Galveston	1941	Strat	641– 660	*
Hitesville Consolidated	Kentucky	Union	1948	V. 32	2063–2082	*
Hobbs	New Mexico	Lea	1932	V. 16	51– 90	*
			1934	Prob	347– 363	
			1935	Gas	417– 457	
Hoffman	Texas	Duval	1940	V. 24	2126–2142	*
Hogback	New Mexico	San Juan	1929	V. 13	117– 151	
			1960	V. 44	1541–1569	
			1968	Mem 9	1327–1356	
Hogshooter	Oklahoma	Washington	1919	V. 3	212– 216	*
Holly	Louisiana	De Soto	1945	V. 29	96– 100	*
Hollywood	Louisiana	Terrebonne	1968	Mem 9	376– 581	
Holt	Texas	Montague	1956	S. Ok 1	355– 372	
Homeglen-Rimbey	Alberta		1968	Mem 9	2034–2036	*
Homer	Louisiana	Claiborne	1922	V. 6	179– 192	
			1929	Struc	196– 228	*
			1934	Prob	399– 427	
Homer	Ohio	Knox, Licking	1929	Struc 1	124– 137	
Hoover Southeast	Oklahoma	Garvin, Murray	1964	V. 48	1555–1567	*
Horseshoe Canyon	New Mexico	San Juan	1972	Mem 16	620– 622	*
Houma	Louisiana	Terrebonne	1918	V. 2	16– 37	
			1935	Gas	683– 740	
			1968	Mem 9	376– 581	
Houston South	Texas	Harris	1945	V. 29	210– 214	*
Howell	Michigan	Livingston	1968	Mem 9	1761–1797	
Hugoton-Panhandle	Kansas	Finney, Grant, Hamilton,	1935	Gas	385– 415	*
		Haskell, Kearny, Morton,	1939	V. 23	1054–1067	*
		Seward, Stanton, Stevens	1940	V. 24	1779–1797	

Field	*State*	*County*	*Year*	*Pub.*	*Pages*	
Hugoton-Panhandle	Oklahoma	Texas	1940	V. 24	1798–1804	*
(cont'd)	Texas	Carson, Collingsworth, Gray,	1941	Strat	78– 104	*
		Hansford, Hartley,	1968	Mem 9	1539–1547	*
		Hutchinson, Moore, Potter,	1970	Mem 14	204– 222	*
		Sherman, Wheeler	(also see Panhandle field)			
Hull	Texas	Liberty	1935	Gas	683– 740	
Hull-Silk	Texas	Archer	1941	Strat	661– 679	*
Humble	Texas	Harris	1917	V. 1	60– 84	*
Hungerford	Texas	Wharton	1968	Mem 9	340– 358	
Hungerford North	Texas	Wharton	1968	Mem 9	340– 358	
Huntington Beach	California	Orange	1922	V. 6	303– 316	
			1924	V. 8	41– 46	*
			1928	V. 12	625– 650	
			1934	V. 18	327– 342	*
			1934	Prob	177– 234	
			1934	Prob	953– 985	
			1935	Gas	113– 220	
Huntsville	Alabama	Madison	1935	Gas	853– 879	
Ibex	Texas	Stephens	1934	Prob	581– 627	
Iles	Colorado	Moffat	1929	Struc 2	93– 114	
Indian Basin	New Mexico	Eddy	1968	Mem 9	1394–1432	
Indian Creek	Kentucky	Knox	1935	Gas	915– 947	
Inglewood	California	Los Angeles	1928	V. 12	625– 650	
			1934	Prob	177– 234	
			1934	Prob	953– 985	
			1935	Gas	113– 220	
Inscho	Oklahoma	Tulsa	1927	V. 11	933– 944	
			1929	Struc 1	211– 219	
Iowa	Louisiana	Calcasieu, Jefferson Davis	1932	V. 16	255– 256	*
			1959	V. 43	2592–2622	
Irma	Arkansas	Nevada	1929	Struc 1	1– 17	*
Isleta	Veracruz		1936	Mex	178	
Isonville	Kentucky	Elliott	1935	Gas	915– 947	
Jackson	Mississippi	Hinds, Rankin	1935	Gas	881– 896	*
			1968	Mem 9	1176–1228	
Janet	Kentucky	Powell	1935	Gas	915– 947	
Jardin	Veracruz		1936	Mex	220– 222	
Jefferson Island	Louisiana	Iberia, Vermilion	1953	V. 37	433– 443	*
Jennings	Louisiana	Acadia	1926	V. 10	72– 92	*
			1926	Salt	398– 418	*
			1935	V. 19	1308–1329	*
			1936	GC	961– 982	*
			1943	V. 27	1102–1122	*
Jennings	Texas	Zapata	1929	Struc 1	389– 408	
Jerusalem	Arkansas	Conway	1968	Mem 9	1682–1692	*
Jesse	Oklahoma	Coal, Pontotoc	1938	V. 22	1560–1578	*
Joe's Lake	Texas	Tyler	1940	V. 24	701– 715	
Joffre	Alberta		1959	V. 43	311– 328	
Johe Ranch	California	Kern	1968	Mem 9	113– 134	
Johnsons Bayou	Louisiana	Cameron	1968	Mem 9	376– 581	
Joseph Lake	Alberta		1950	V. 34	1802–1806	*
			1954	WCan	452– 463	*
Joyce Creek	Wyoming	Sweetwater	1968	Mem 9	803– 816	
Judkins	Texas	Ector	1935	Gas	417– 457	
Jumping Pound	Alberta		1959	V. 43	992–1025	
Katy	Texas	Fort Bend, Harris, Waller	1946	V. 30	157– 180	*
Kelly-Snyder	Texas	Scurry	1970	Mem 14	185– 203	
Kemnitz	New Mexico	Lea	1970	V. 54	2317–2335	*
Kenai	Alaska		1968	Mem 9	49– 68	
Kent Bayou	Louisiana	Terrebonne	1964	V. 48	1705–1725	*

Field	State	County	Year	Pub.	Pages	
Keokuk	Oklahoma	Pottawatomie, Seminole	1939	V. 23	220– 245	*
Kern Front	California	Kern	1941	Strat	9– 18	*
Kern River	California	Kern	1934	Prob	177– 234	
			1934	Prob	785– 805	
Kettleman Hills	California	Fresno, Kings	1929	V. 13	1479–1483	*
(Middle and North)			1933	V. 17	1161–1193	*
			1934	V. 18	435– 475	*
			1934	V. 18	1454–1492	
			1934	Prob	177– 234	
			1934	Prob	399– 427	
			1934	Prob	735– 760	
			1934	Prob	785– 805	
			1934	Prob	953– 985	
			1935	Gas	113– 220	
			1940	V. 24	1940–1949	
Kevin-Sunburst	Montana	Toole	1923	V. 7	263– 276	*
			1929	V. 13	779– 797	
			1929	Struc 2	254– 268	*
			1934	Prob	347– 363	
			1934	Prob	695– 718	
			1935	Gas	245– 276	
Kibler	Arkansas	Crawford	1935	Gas	533– 574	
Kimball	Ontario		1949	V. 33	153– 188	
Kingsville	Ontario		1949	V. 33	153– 188	
Kinta	Oklahoma	Haskell	1968	Mem 9	1636–1643	*
KMA	Texas	Wichita	1940	V. 24	1494–1495	*
Knox City North	Texas	Knox	1972	Mem 16	453– 459	*
Knox Creek	Kentucky	Pike	1964	V. 48	465– 486	
	Virginia	Buchanan				
Kraft-Prusa	Kansas	Barton, Ellsworth, Russell	1948	Struc 3	249– 280	*
			1953	V. 37	300– 313	
Kutz Canyon	New Mexico	San Juan	1935	Gas	363– 384	
La Barge	Wyoming	Lincoln, Sublette	1947	V. 31	797– 823	
La Cima	Veracruz		1936	Mex	192	
La Gloria	Texas	Brooks, Jim Wells	1968	Mem 9	233– 263	
La Rosa	Texas	Refugio	1941	V. 25	300– 317	*
Lacey	Oklahoma	Blaine, Kingfisher	(see Star-Lacey)			
Lake Arthur	Louisiana	Jefferson Davis	1968	Mem 9	376– 581	
Lake Arthur Southwest	Louisiana	Cameron	1972	Mem 16	389– 398	*
Lake Basin	Montana	Stillwater	1935	Gas	245– 276	
Lake Enfermer	Louisiana	Lafourche	1968	Mem 9	376– 581	
Lake Raccourci	Louisiana	Lafourche, Terrebonne	1968	Mem 9	376– 581	
Lake Washington	Louisiana	Plaquemines	1959	V. 43	2592–2622	
Lance Creek	Wyoming	Niobrara	1929	Struc 2	604– 613	*
Landa Northeast	North Dakota	Bottineau	1958	Hab	149– 177	
Larremore	Texas	Caldwell	1930	V. 14	917– 922	*
Larto Lake	Louisiana	Catahoula	1942	V. 26	1255–1276	
Lathrop	California	San Joaquin	1968	Mem 9	653– 663	*
Lavaca	Arkansas	Sebastian	1935	Gas	533– 574	
Laverne	Oklahoma	Beaver, Harper	1968	Mem 9	1509–1524	*
			(also see Mocane-Laverne)			
Lea County East	New Mexico	Lea	1935	Gas	417– 457	
Leading Creek	West Virginia	Lewis	1970	V. 54	758– 782	
Leaton	Michigan	Isabella	1938	V. 22	129– 174	
Leduc	Alberta		1949	V. 33	572– 602	*
			1950	V. 34	295– 312	*
			1954	WCan	415– 431	*
			1958	V. 42	1– 93	
Lee-Estill-Powell	Kentucky	Estill, Lee, Powell	1927	V. 11	477– 492	
			1929	Struc 1	73– 90	

Field	State	County	Year	Pub.	Pages	
Leroy North	Louisiana	Vermilion	1968	Mem 9	376– 581	
Lewis	Kansas	Edwards	1935	Gas	459– 482	
Liberty South	Texas	Liberty	1951	V. 35	1939–1977	*
			1968	Mem 9	330– 339	*
Lick Creek	Louisiana	Claiborne	1971	V. 55	51– 63	*
Lima-Indiana	Indiana Ohio	(Various counties in east central Indiana and northeast Ohio)	1934	Prob	521– 529	*
Limon	Veracruz		1936	Mex	193– 194	
Lindsay	Oklahoma	Garvin, McClain	1950	V. 34	386– 422	
Lips	Texas	Roberts	1968	Mem 9	1525–1538	
Lirrette	Louisiana	Terrebonne	(see Houma)			
Lisbon	Louisiana	Claiborne, Lincoln	1939	V. 23	281– 324	*
Lisbon	Utah	San Juan	1968	Mem 9	1371–1388	*
Lisbon North	Louisiana	Claiborne	1942	V. 26	1255–1276	
			1944	V. 28	333– 340	
Lissie South	Texas	Wharton	1968	Mem 9	340– 358	
Little Beaver	Colorado	Washington	1955	V. 39	155– 188	*
			1955	V. 39	630– 648	*
			1957	V. 41	839– 847	
Little Buffalo Basin	Wyoming	Hot Springs Park	1934	V. 18	1454–1492	
			1935	Gas	292	
			1947	V. 31	797– 823	
Little Creek	Louisiana	La Salle	1942	V. 26	1255–1276	
			1944	V. 28	333– 340	
Little Grass Creek	Wyoming	Hot Springs	1935	Gas	292	
Little Lost Soldier	Wyoming	Sweetwater	1923	V. 7	131– 146	
			1929	Struc 2	636– 666	
Little Missouri	North Dakota	Bowman	1968	Mem 9	1304–1326	
Little Polecat	Wyoming	Park	1935	Gas	292	
Little River	Oklahoma	Seminole	1929	Struc 2	315– 361	
Little Worm Creek	Wyoming	Sweetwater	1968	Mem 9	803– 816	
Llano de Silva	Veracruz		1936	Mex	193	
Lloydminster	Saskatchewan		1959	V. 43	311– 328	
Lockport	Louisiana	Calcasieu	1935	Gas	683– 740	
Lompoc	California	Santa Barbara	1934	Prob	177– 234	
			1935	Gas	113– 220	
			1949	V. 33	32– 51	
Lone Grove Southwest	Oklahoma	Carter	1951	V. 35	582– 606	
			1956	S. Ok 1	144– 153	*
Long Beach	California	Los Angeles	1922	V. 6	303– 316	
			1924	V. 8	403– 423	*
			1928	V. 12	625– 650	
			1929	Struc 2	62– 74	*
			1930	V. 14	997–1011	
			1934	Prob	177– 234	
			1934	Prob	953– 985	
			1935	Gas	113– 220	
Long Lake	Texas	Anderson	1935	Gas	651– 681	
Longton	Kansas	Chautauqua, Elk	1935	Gas	483– 509	
Lopez	Texas	Duval, Webb	1941	Strat	680– 697	*
Los Angeles City	California	Los Angeles	1934	Prob	177– 234	
Lost Hills	California	Kern	1934	Prob	177– 234	
			1934	Prob	735– 760	
			1934	Prob	785– 805	
Lost Soldier	Wyoming	Carbon, Sweetwater	1947	V. 31	797– 823	
			1949	V. 33	1998–2010	*
Lovedale	Oklahoma	Harper, Woods	1974	V. 58	447– 463	
Luling	Texas	Caldwell	1924	V. 8	775– 788	*
			1925	V. 9	632– 654	*
			1929	Struc 1	256– 281	*

Field	*State*	*County*	*Year*	*Pub.*	*Pages*	
Luling (cont'd)			1932	V. 16	206– 209	*
			1934	Prob	399– 427	
Lusk Strawn	New Mexico	Eddy, Lea	1968	V. 52	66– 81	*
Lyons	Kansas	Rice	1940	V. 24	1779–1797	
			1948	Struc 3	225– 248	
Lytton Springs	Texas	Caldwell	1926	V. 10	935– 975	
			1932	V. 16	741– 768	
Maddux Ranch	California	Kern	1968	Mem 9	113– 134	
Madill North	Oklahoma	Marshall	1959	S. Ok 2	274– 286	*
Madison	Kansas	Greenwood	1929	Struc 2	150– 159	*
Magnolia City	Texas	Jim Wells	1939	V. 23	1238	*
Mahoney	Wyoming	Carbon	1935	Gas	305– 322	
			1951	V. 35	1000–1037	
Main Pass Block 35	Louisiana	(Offshore)	1970	V. 54	783– 788	*
			1972	V. 56	554– 558	*
Maine Prairie	California	Solano	1968	Mem 9	79– 84	*
Malahide	Ontario		1949	V. 33	153– 188	
Malden	Ontario		1949	V. 33	153– 188	
Maljamar	New Mexico	Lea	1935	Gas	417– 457	
Mansfield	Arkansas	Scott, Sebastian	1935	Gas	533– 574	
Manvel	Texas	Brazoria	1935	Gas	683– 740	
Marapos	Colorado	Rio Blanco	(see Thornburg)			
Maricopa Flat	California	Kern	1931	V. 15	689– 696	*
Marine	Illinois	Madison	1948	Struc 3	153– 188	*
Markham	Texas	Matagorda	1935	Gas	683– 740	
Martin	Kentucky	Martin	1935	Gas	915– 947	
Martinsville	Illinois	Clark	1929	Struc 2	115– 141	*
Marysville	California	Sutter	(see Sutter Buttes)			
Massard Prairie	Arkansas	Sebastian	1935	Gas	533– 574	
Mata de Chapapote	Veracruz		1936	Mex	184	
Maxie	Mississippi	Forrest	1968	Mem 9	1176–1226	
Mayes South	Texas	Chambers	1947	V. 31	495– 499	*
Mayfield	Ohio	Cuyahoga	1949	V. 33	1731–1746	*
Maysville	Oklahoma	Garvin	1950	V. 34	386– 422	
McCallum	Colorado	Jackson	1934	V. 18	1454–1492	
			1935	Gas	363– 384	
McCallum North	Wyoming	Jackson	1968	Mem 9	840– 855	
McCallum South	Wyoming	Jackson	1968	Mem 9	840– 855	
McCamey	Texas	Upton	1934	Prob	869– 889	
McElroy	Texas	Crane, Upton	(see Church-McElroy)			
McFaddin-O'Connor	Texas	Refugio, Victoria	1934	V. 18	519– 530	*
			1936	GC	664– 675	*
McKittrick	California	Kern	1929	Struc 1	18– 22	*
			1933	V. 17	1– 15	*
			1934	Prob	177– 234	
			1934	Prob	953– 985	
McLouth	Kansas	Jefferson, Leavenworth	1942	V. 26	133– 135	*
McMullin Ranch	California	San Joaquin	1968	V. 52	1152–1161	*
McPherson	Kansas	McPherson	1935	Gas	459– 482	
Means	Texas	Andrews	1965	Mem 4	280– 293	
Medicine Bow	Wyoming	Carbon	1947	V. 31	797– 823	
Medicine Hat-Hatton	Alberta		1935	Gas	1– 58	
	Saskatchewan		1968	Mem 9	731– 735	*
Medicine Lodge	Kansas	Barber	1935	Gas	459– 482	
			1940	V. 24	1779–1797	
Melstone	Montana	Mussel shell	1966	V. 50	2245–2259	
Menudillo	Veracruz		1936	Mex	185	
Mercedes	Texas	Hidalgo	1935	V. 19	1226–1231	*
Merigale	Texas	Wood	1945	V. 29	1779–1780	*
Mersea	Ontario		1949	V. 33	153– 188	

Field	*State*	*County*	*Year*	*Pub.*	*Pages*	
Mervine	Oklahoma	Kay	1929	Struc 1	158– 175	*
Mexia	Texas	Limestone	1921	V. 5	419– 421	*
			1923	V. 7	226– 236	*
			1929	Struc 1	304– 388	
			1934	Prob	581– 627	
Mexia-Groesbeck	Texas	Limestone	1935	Gas	651– 681	
Mexican Hat	Utah	San Juan	1951	V. 35	1000–1037	
Midale	Saskatchewan		1958	Hab	149– 177	
Middle Dome	Wyoming	Washakie	1966	V. 50	2197–2220	
Middlemist	Colorado	Adams	1955	V. 39	630– 648	*
Midland	Kentucky	Muhlenberg	1972	Mem 16	585– 598	*
Midland	Louisiana	Acadia	1968	Mem 9	376– 581	
Midway	Arkansas	Lafayette	1942	V. 26	1289–1291	*
Midway-Sunset	California	Kern, San Luis Obispo	1934	Prob	177– 234	
			1934	Prob	735– 760	
			1934	Prob	785– 805	
			1935	Gas	113– 220	
Milbur Wilcox	Texas	Burleson, Milam	1972	Mem 16	399– 405	*
Miller Creek	Wyoming	Crook	1968	V. 52	2116–2122	
Milroy	Oklahoma	Carter, Stephens	1959	S. Ok 2	220– 226	*
Minerva	Texas	Milam	1924	V. 8	632– 640	*
Minnehik-Buck Lake	Alberta		1968	Mem 9	1238–1284	
Mision	Tamaulipas		1949	V. 33	1351–1384	
Mirando	Texas	Zapata	1921	V. 5	625– 626	*
Mirando City	Texas	Webb	(see Schott-Mirando City)			
Mirando Valley	Texas	Zapata	1923	V. 7	532– 545	
Mocane-Laverne	Oklahoma	Beaver, Harper	1968	Mem 9	1525–1538	
			(also see Laverne)			
Moffat	Colorado	Moffat	1929	Struc 2	93– 114	
Molino	Veracruz		1936	Mex	223	
Monroe	Louisiana	Morehouse, Ouachita, Union	1918	V. 2	61– 69	
			1922	V. 6	179– 192	
			1923	V. 7	565– 574	*
			1935	Gas	741– 772	*
			1968	Mem 9	1161–1168	*
Montebello	California	Los Angeles	1934	Prob	177– 234	
			1935	Gas	113– 220	
Monte Christo	Texas	Hidalgo	1968	Mem 9	233– 263	
Moody Gulch	California	Santa Clara	1934	Prob	177– 234	
Morgan	Kentucky	Morgan	1935	Gas	915– 947	
Morrison	Oklahoma	Pawnee	1927	V. 11	1087–1096	*
			1929	Struc 1	148– 157	*
Mount Holly	Arkansas	Union	1942	V. 26	1255–1276	
			1944	V. 28	326– 332	
Mount Pleasant	Michigan	Midland	1935	Gas	787– 812	
Mount Poso	California	Kern	1934	Prob	177– 234	
			1958	Hab	99– 112	
Mud Lake East	Louisiana	Cameron	1968	Mem 9	376– 581	
Muldon	Mississippi	Monroe	1968	Mem 9	1693–1701	
Muralla	Texas	Duval	1939	V. 23	1237	*
Murdock	Texas	Kenedy	1968	Mem 9	233– 263	
Murphy Dome	Wyoming	Hot Springs	1958	Hab	293– 306	
			1966	V. 50	2197–2220	
Mush Creek-Skull Creek	Wyoming	Weston	1950	V. 34	1850–1865	*
Music Mountain	Pennsylvania	McKean	1941	Strat	492– 506	*
Muskegon	Michigan	Muskegon	1932	V. 16	153– 168	*
			1935	Gas	787– 812	
Mustang Island	Texas	Nueces	1968	Mem 9	264– 270	*
Mykawa	Texas	Harris	1935	Gas	683– 740	
Naborton	Louisiana	De Soto	(see De Soto)			

Field	*State*	*County*	*Year*	*Pub.*	*Pages*	
Nacata	Veracruz		1936	Mex	192	
Navarro Crossing	Texas	Houston	1938	V. 22	1600–1601	*
Neale	Louisiana	Beauregard	1940	V. 24	2036–2037	*
			1968	Mem 9	376– 581	
Nebo	Louisiana	La Salle	1942	V. 26	1255–1276	
			1944	V. 28	333– 340	
Nevis	Alberta		1968	Mem 9	1238–1284	
New Harmony	Illinois	White	1942	V. 26	1594–1607	*
	Indiana	Posey				
New Haven	Michigan	Gratiot	1938	V. 22	129– 174	
New Hardin	Texas	Liberty	1935	V. 19	1389	*
New Hope	Texas	Franklin	1945	V. 29	836– 839	*
			1968	Mem 9	1069–1073	*
New Iberia	Louisiana	Iberia	1918	V. 2	16– 37	
New Milton	West Virginia	Doddridge	1970	V. 54	758– 782	
Newburg	North Dakota	Bottineau	1972	Mem 16	633– 642	*
Nichols	Kansas	Kiowa	1968	Mem 9	1582–1587	*
Nigger Creek	Texas	Limestone	1926	V. 10	997– 998	*
			1929	Struc 1	304– 388	
			1929	Sturc 1	409– 420	*
			1934	Prob	399– 427	
Nikkel	Kansas	Harvey, McPherson	1941	Strat	105– 117	*
Nitchie Gulch	Wyoming	Sweetwater	1968	Mem 9	803– 816	
Nocona North	Texas	Montague	1956	S. Ok 1	355– 372	
Noodle Creek	Texas	Jones	1941	Strat	698– 721	*
Norman Wells	N.W.T.		1948	Struc 3	86– 109	*
Norris-Red Oak	Oklahoma	Latimer, Le Flore	(see Red Oak)			
North Fork	Wyoming	Johnson	1967	V. 51	705– 709	*
Northville	Michigan	Oakland, Washtenaw, Wayne	1968	Mem 9	1761–1797	
Nowood	Wyoming	Washakie	1966	V. 50	2197–2220	
O'Connor	Texas	Refugio	(see McFaddin-O'Connor)			
O'Hern	Texas	Duval, Webb	1941	Strat	722– 749	*
Oil City	California	Fresno	1934	Prob	177– 234	
Oklahoma City	Oklahoma	Oklahoma	1929	V. 13	1387–1394	*
			1930	V. 14	1515–1533	*
			1932	V. 16	957–1020	*
			1932	V. 16	1021–1028	*
			1934	Prob	347– 363	
			1934	Prob	399– 427	
			1934	Prob	581– 627	
			1970	Mem 14	223– 254	*
Old Ocean	Texas	Brazoria, Matagorda	1968	Mem 9	295– 305	*
Olds	Alberta		1968	Mem 9	1238–1284	
Olinda	California	Orange	(see Brea-Olinda)			
Okotoks	Alberta		1968	Mem 9	1238–1284	
Olla	California	La Salle	1941	V. 25	747– 750	*
			1942	V. 26	1255–1276	
Olla South	Louisiana	La Salle	1942	V. 26	1255–1276	
Olympic	Oklahoma	Hughes, Okfuskee	1938	V. 22	1579–1587	*
			1941	Strat	456– 472	*
Omaha	Illinois	Gallatin	1948	Struc 3	189– 212	*
Opelika	Texas	Henderson	1968	Mem 9	1005–1007	*
Opelousas	Louisiana	St. Landry	1968	Mem 9	376– 581	
Oneida-Burning Springs	Kentucky	Clay	1935	Gas	915– 947	
Orange	Texas	Orange	1936	V. 20	531– 559	*
			1936	GC	880– 908	*
			1939	V. 23	602– 603	*
Oregon Basin	Wyoming	Park	1935	Gas	292	
			1947	V. 31	797– 823	
			1947	V. 31	1431–1453	*
			1966	V. 50	2197–2220	

Field	*State*	*County*	*Year*	*Pub.*	*Pages*	
Orth	Kansas	Rice	1953	V. 37	300– 313	
Osage	Wyoming	Weston	1941	Strat	847– 857	*
			1947	V. 31	797– 823	
Oscar	Oklahoma	Jefferson	1956	S. Ok 1	355– 372	
Otis	Kansas	Barton, Rush	1935	Gas	459– 482	
			1940	V. 24	1779–1797	
Otis-Albert	Kansas	Barton, Rush	1968	Mem 9	1588–1615	*
Overisel	Michigan	Allegan	1968	Mem 9	1761–1797	
Owsley	Kentucky	Owsley	1927	V. 11	477– 492	
			1929	Struc 1	73– 90	
Ozona	Texas	Crockett	1968	Mem 9	1394–1432	
Paciencia y Aguacate	Veracruz		1936	Mex	179	
Page	Texas	Schleicher	1941	V. 25	630– 636	*
Palacine South	Oklahoma	Stephens	1959	S. Ok 2	187– 197	*
Paloma	California	Kern	1940	V. 24	742– 744	*
Panhandle	Texas	Carson, Collingsworth, Gray,	1935	V. 19	1089–1109	*
		Hansford, Hartley,	1935	Gas	385– 415	
		Hutchinson, Moore, Potter,	1939	V. 23	983–1053	*
		Sherman, Wheeler	1970	Mem 14	204– 222	*
			(also see Hugoton-Panhandle field)			
Panuco	Veracruz		1928	V. 12	395– 441	*
			1936	Mex	173– 177	
Parkman	Saskatchewan		1972	Mem 16	502– 510	*
Paso Real	Veracruz		1936	Mex	220– 222	
Patrick	Texas	San Patricio	1968	Mem 9	233– 263	
Patrick Draw	Wyoming	Sweetwater	1968	Mem 9	817– 827	
Patterson	Kansas	Kearny	1942	V. 26	400– 401	*
Patton	Arkansas	Lafayette	1942	V. 26	1255–1276	
			1944	V. 28	326– 332	
Pauls Valley	Oklahoma	Garvin	1956	S. Ok 1	337– 354	*
Payton	Texas	Pecos, Ward	1942	V. 26	1632–1646	*
Peabody	Kansas	Marion	1929	Struc 1	60– 72	
Pearson Switch	Oklahoma	Pottawatomie	1929	Struc 2	315– 361	
Pecos Valley	Texas	Pecos	1935	Gas	417– 457	
Pembina	Alberta		1957	V. 41	937– 949	*
			1959	V. 43	311– 328	
			1968	Mem 9	698– 704	*
Penescal	Texas	Kenedy	1968	Mem 9	233– 263	
Penwell	Texas	Ector	1965	Mem 4	280– 293	
Perkins Lake	California	Butte	1968	Mem 9	76– 78	*
Perry	Kentucky	Perry	1935	Gas	915– 947	
Pershing	Oklahoma	Osage	1934	Prob	581– 627	
Peters	Michigan	St. Clair	1966	V. 50	327– 350	*
			(also see Boyd-Peters)			
Petersburg	Texas	Hale	1948	V. 32	780– 789	*
Peters Point	Utah	Carbon	1968	Mem 9	174– 198	
Petrolia	Texas	Clay	1929	Struc 2	542– 555	*
Piceance Creek	Colorado	Rio Blanco	1935	Gas	363– 384	
Pickton	Texas	Hopkins	1945	V. 29	1777–1779	*
Pierce Junction	Texas	Harris	1935	Gas	683– 740	
Pierson	Manitoba		1958	Hab	149– 177	
Pincher Creek	Alberta		1959	V. 43	992–1025	
Pine Creek	Alberta		1968	Mem 9	1238–1284	
Pine Island	Louisiana	Caddo	1925	V. 9	171– 172	*
			1929	Struc 2	168– 182	*
			1934	Prob	581– 627	
Piqua	Kansas	Allen, Woodson	1940	V. 24	1779–1797	
Pitchfork	Wyoming	Park	1934	V. 18	1454–1492	
			1967	V. 51	2115–2121	*
Pistol Ridge	Mississippi	Forrest, Pearl River	1968	Mem 9	1176–1226	

Field	State	County	Year	Pub.	Pages	
Pittsburg	Texas	Camp	1940	V. 24	2032–2033	*
Placedo	Texas	Victoria	1935	V. 19	1693–1694	*
Playo del Rey	California	Los Angeles	1934	Prob	177– 234	
			1934	Prob	399– 427	
			1934	Prob	953– 985	
			1935	V. 19	172– 205	*
			1935	Gas	113– 220	
			1968	Mem 9	169– 173	*
Pleasant Valley	California	Fresno	1951	V. 35	619– 623	*
Pledger	Texas	Brazoria	1935	Gas	683– 740	
Polecat	Wyoming	Park	1935	Gas	292	
Ponca	Oklahoma	Kay	1929	Struc 1	158– 175	
Pondera	Montana	Pondera, Teton	1929	V. 13	779– 797	
			1934	Prob	695– 718	
			1951	V. 35	1000–1037	
Poplar East	Montana	Roosevelt	1953	V. 37	2294–2302	
Port Acres	Texas	Jefferson	1968	Mem 9	368– 375	*
			1972	Mem 16	329– 341	*
Port Arthur	Texas	Jefferson	1968	Mem 9	368– 375	*
			1972	Mem 16	329– 341	*
Porter	Michigan	Midland	1944	V. 28	173– 196	*
Potrero	California	Los Angeles	1935	Prob	177– 234	
			1934	Prob	953– 985	
Potrero del Llano	Veracruz		1936	Mex	217– 218	
Potrero Hills	California	Solano	1939	V. 23	1230–1231	*
Powell	Texas	Navarro	1929	Struc 1	304– 388	
Powell's Lake	Kentucky	Union	1948	V. 32	34– 51	*
Poza Rica	Veracruz		1949	V. 33	1385–1409	*
Powder Wash	Colorado	Moffat	1938	V. 22	1020–1047	*
			1947	V. 31	797– 823	
Propp	Kansas	Marion	1935	Gas	459– 482	
Provost	Alberta		1968	Mem 9	721– 725	*
Prudhoe Bay	Alaska		1972	Mem 16	489– 501	*
Pulaski	New York	Oswego	1938	V. 22	79– 99	
Puente	California	Los Angeles	1934	Prob	177– 234	
Puckett	Texas	Pecos	1965	Mem 4	257– 279	
			1968	Mem 9	1394–1432	
Quebracha	Veracruz		1936	Mex	197– 200	
Quinton	Oklahoma	Haskell, Pittsburg	1935	Gas	511– 532	
Quitman	Texas	Wood	1948	Struc 3	419– 431	*
Raccoon Bend	Texas	Austin	1933	V. 17	1459–1491	*
			1935	Gas	683– 740	
			1936	GC	676– 708	*
			1946	V. 30	1306–1307	*
Rainbow	Alberta		1968	V. 52	1925–1955	*
			1970	V. 54	2260–2281	*
Rainbow Bend	Kansas	Cowley	1925	V. 9	974– 982	*
			1929	Struc 1	52– 59	*
Rainbow City	Arkansas	Union	1928	V. 12	763– 764	*
Ramsey	Oklahoma	Payne	1940	V. 24	1995–2005	*
			1955	V. 40	122– 139	*
Randado	Texas	Jim Hogg	1929	Struc 1	389– 408	
Randlett Southwest	Oklahoma	Cotton	1956	S. Ok 1	311– 318	*
Rangely	Colorado	Rio Blanco	1929	Struc 2	93– 114	
			1948	Struc 3	132– 152	*
Rattlesnake	New Mexico	San Juan	1929	V. 13	117– 151	
			1947	V. 31	731– 771	*
Rattlesnake Hills	Wyoming	Benton	1934	V. 18	847– 859	*
Raven Creek	Wyoming	Campbell	1972	Mem 16	511– 519	*
Rayne	Louisiana	Acadia	1968	Mem 9	376– 581	

Field	*State*	*County*	*Year*	*Pub.*	*Pages*	
Recluse	Wyoming	Campbell	1972	Mem 16	376– 382	*
Red Bird	Kentucky	Bell	1935	Gas	915– 947	
Red Coulee	Alberta		(see Border-Red Coulee)			
Red Fish Bay	Texas	Nueces	1968	Mem 9	264– 270	*
Red Fork	Oklahoma	Creek, Pawnee, Tulsa	1941	Strat	473– 491	*
Red Oak	Oklahoma	Latimer, Le Flore	1935	Gas	511– 532	
Red Oak-Norris	Oklahoma	Latimer, Le Flore	1968	Mem 9	1644–1657	
Red River-Crichton	Louisiana	Caddo	1918	V. 2	61– 69	
Red Springs	Wyoming	Hot Springs	1947	V. 31	797– 823	
Red Wash	Utah	Uintah	1958	Hab	344– 365	
			1968	Mem 9	174– 198	
			1972	Mem 16	342– 353	*
Red Wash-Walker Hollow	Utah	Uintah	1957	V. 41	923– 936	*
			1962	V. 46	690– 694	*
Redondo	California	Los Angeles	(see Torrance)			
Reed City	Michigan	Lake, Osceola	1968	Mem 9	1761–1797	
Refugio	Texas	Refugio	1931	V. 15	954– 964	*
			1934	V. 18	519– 530	*
			1936	GC	664– 675	*
			1938	V. 22	1184–1216	*
Redwater	Alberta		1958	V. 42	1– 93	
			1959	V. 43	311– 328	
Reynosa	Tamaulipas		1949	V. 33	1351–1384	
Richardson	West Virginia	Calhoun	1941	Strat	806– 829	*
Richburg	New York	Allegany	1929	Struc 2	269– 289	
Richey	Montana	Dawson, McCone	1953	V. 37	2294–2302	
Richfield	California	Los Angeles	1922	V. 6	303– 316	
			1934	Prob	177– 234	
			1934	Prob	953– 985	
			1935	Gas	113– 220	
Richland	Louisiana	Richland	1928	V. 12	985– 993	*
			1931	V. 15	939– 952	*
			1935	Gas	773– 786	*
			1968	Mem 9	1156–1161	*
Richland	Texas	Navarro	1929	Struc 1	304– 388	
Ringwald	Kansas	Rice	1953	V. 37	300– 313	
Rio Bravo	California	Kern	1940	V. 24	1330–1333	*
Rio Tamesi	Tamaulipas		1936	Mex	197	
Rio Vista	California	Contra Costa, Sacramento, Solano	1968	Mem 9	93– 101	*
Robberson	Oklahoma	Garvin	1923	V. 7	625– 644	*
Rochester	Alberta		1968	Mem 9	1238–1284	
Rock River	Wyoming	Carbon	1929	Struc 2	614– 622	*
Rogers	Texas	Montague	1940	V. 24	1836–1838	*
Rojo Caballos	Texas	Pecos	1965	Mem 4	257– 279	
Rose Hill	Virginia	Lee	1948	Struc 3	452– 479	*
Rosecrans	California	Los Angeles	1928	V. 12	625– 650	
			1934	Prob	177– 234	
			1934	Prob	953– 985	
			1935	Gas	113– 220	
Rothwell	Kentucky	Menifee	1935	Gas	915– 947	
Round Mountain	California	Kern	1934	Prob	177– 234	
Russell Ranch	California	Santa Barbara	1958	Hab	79– 98	
Ruston	Louisiana	Lincoln	1945	V. 29	226– 227	*
			1968	Mem 9	1138–1142	*
Saber Bar	Colorado	Logan, Weld	1966	V. 50	2112–2118	*
Sabinsville	Pennsylvania	Tioga	1938	V. 22	241– 266	
Sabre	Texas	Reeves	1965	Mem 4	294– 307	
Sage Creek	Wyoming	Big Horn, Park	1960	V. 50	2197–2220	
Saginaw	Michigan	Saginaw	1927	V. 11	959– 966	*
			1929	Struc 1	105– 111	*

Field	State	County	Year	Pub.	Pages	
St. Charles Ranch	Texas	Aransas	1968	Mem 9	233– 263	
St. Louis	Oklahoma	Pottawatomie	1929	Struc 2	315– 361	*
Salem	Illinois	Marion	1939	V. 23	1352–1373	*
Salinas	Veracruz		1936	Mex	185	
Sallyards	Kansas	Butler, Greenwood	1921	V. 5	276– 281	*
Salt Creek	California	Kern	1947	V. 31	1674–1677	*
Salt Creek	Wyoming	Natrona	1924	V. 8	492– 504	*
			1929	Struc 2	589– 603	*
			1934	Prob	399– 427	
			1947	V. 31	797– 823	
			1966	V. 50	2185–2196	*
			1970	Mem 14	147– 157	*
Salt Flat	Texas	Caldwell	1930	V. 14	1177–1185	*
			1930	V. 14	1401–1423	*
			1934	Prob	347– 363	
			1934	Prob	399– 427	
Salt Lake	California	Los Angeles	1934	Prob	177– 234	
Salt Wells	Wyoming	Sweetwater	1968	Mem 9	803– 816	
San Emidio Nose	California	Kern	1972	Mem 16	297– 312	*
San Geronimo	Veracruz		1936	Mex	209	
San Miguel	Veracruz		1936	Mex	210	
San Miguelito	California	Ventura	1951	V. 35	2542–2560	*
San Salvador	Texas	Hidalgo	1968	Mem 9	233– 263	
Sand Flat	Texas	Smith	1944	V. 28	1647–1648	*
Sand River	Colorado	Morgan	1957	V. 41	839– 847	
Sandoval	Illinois	Clinton, Marion	(see Centralia-Sandoval)			
Sandusky Oil Creek	Texas	Grayson	1959	S. Ok 2	53– 100	
Sanish	North Dakota	McKenzie	1968	V. 52	57– 65	*
Santa Fe	Oklahoma	Stephens	1956	S. Ok 1	234– 243	*
Santa Fe Springs	California	Los Angeles	1922	V. 6	303– 316	
			1924	V. 8	178– 194	*
			1930	V. 14	997–1011	
			1934	Prob	177– 234	
			1934	Prob	399– 427	
			1934	Prob	953– 985	
			1935	Gas	113– 220	
Santa Maria	California	Santa Barbara	1934	Prob	177– 234	
			1935	Gas	113– 220	
Santa Maria Valley	California	Santa Barbara	1939	V. 23	45– 81	*
			1949	V. 33	32– 51	
Santiago	California	Kern	1947	V. 31	2063–2067	*
Saratoga	Texas	Hardin	1925	V. 9	263– 285	*
			1926	Salt	501– 523	*
Sarepta	Louisiana	Bossier, Webster	(see Carterville-Sarepta and Spring Hill-Sarepta)			
Sargent	California	Santa Clara	1934	Prob	177– 234	
Satsuma	Texas	Harris	1939	V. 23	686– 688	*
Savanna Creek	Alberta		1959	V. 43	992–1025	
Saxet	Texas	Nueces	1930	V. 14	1351	*
			1934	V. 18	519– 530	*
			1936	GC	664– 675	*
			1940	V. 24	1805–1835	*
Sayre	Oklahoma	Beckham	1924	V. 8	347– 349	*
			1935	Gas	385– 415	
Scarars	California	Ventura	1934	Prob	177– 234	
Scenery Hill	Pennsylvania	Washington	1929	Struc 2	443– 450	*
Schimmel-Batts	Texas	Bastrop	1932	V. 16	741– 768	
Schott-Aviator	Texas	Webb	1929	Struc 1	389– 408	
			(also see Aviator's)			
Schott-Mirando City	Texas	Webb	1923	V. 7	532– 545	

Field	*State*	*County*	*Year*	*Pub.*	*Pages*
Screwbean Northeast	Texas	Reeves	1965	Mem 4	294– 307
Schuler	Arkansas	Union	1942	V. 26	1255–1276
			1942	V. 26	1467–1516
			1944	V. 28	326– 332
Schuler East	Arkansas	Union	1942	V. 26	1255–1276
			1944	V. 28	326– 332
Scurry	Texas	Scurry	1970	Mem 14	185– 203
Seal Beach	California	Los Angeles	1928	V. 12	625– 650
			1935	Gas	113– 220
Searight	Oklahoma	Seminole	1929	Struc 2	315– 361
Seay	Oklahoma	Jefferson	1956	S. Ok 1	355– 372
Seeley	Kansas	Greenwood	(see Burkett-Seeley)		
Seeligson	Texas	Jim Wells, Kleberg	1954	V. 38	96– 117
Segno	Texas	Polk	1938	V. 22	1274–1277
Seminole	Oklahoma	Seminole	1929	Struc 2	315– 361
(formerly Seminole City)			1934	Prob	347– 363
Semitropic	California	Kern	1936	V. 20	939– 950
Sewell-Eddleman	Texas	Young	1942	V. 26	204– 216
Seymour	Texas	Baylor	1941	Strat	760– 775
			1956	V. 40	879– 889
Seymour East	Texas	Baylor	1956	V. 40	879– 889
Shawnee Creek	Wyoming	Converse	1951	V. 35	1000–1037
Sheridan	Texas	Colorado	1968	Mem 9	306– 329
Sherman	Texas	Grayson	1959	S. Ok 2	53– 100
Shields	Texas	Coleman	(see Fisk)		
Shiels Canyon	California	Ventura	1934	Prob	177– 234
Shinnston	West Virginia	Harrison	1941	Strat	830– 846
Shipley	Texas	Ward	1941	V. 25	425– 427
Shira Streak	Pennsylvania	Butler	1941	Strat	507– 538
Sholem Alechem	Oklahoma	Carter, Stephens	1951	V. 35	582– 606
			1956	S. Ok 1	294– 310
			1959	V. 43	2575–2591
Shongaloo	Louisiana	Webster	1938	V. 22	1473–1503
Shreveport	Louisiana	Bossier, Caddo	1918	V. 2	61– 69
			1938	V. 22	1277–1278
Sidney Southwest	Nebraska	Cheyenne	1968	Mem 9	899– 927
Silver Tip	Wyoming	Park	1951	V. 35	1000–1037
Simi	California	Ventura	1934	Prob	177– 234
Sissonville	West Virginia	Jackson, Kanawha, Putnam	1949	V. 33	336– 345
			1972	Mem 16	313– 317
			(also see Elk-Poca)		
Six Lakes	Michigan	Mecosta, Montcalm	1938	V. 22	129– 174
			1941	Strat	237– 266
Skull Creek	Wyoming	Weston	(see Mush Creek-Skull Creek)		
Slick Wilcox	Texas	DeWitt, Goliad	1948	V. 32	228– 251
Sligo	Louisiana	Bossier	1968	Mem 9	1146–1152
Slocum	Texas	Anderson	1959	V. 43	958– 973
Smith-Ellis	Texas	Brown	1929	Struc 2	556– 570
			1934	Prob	581– 627
Smithland	Tennessee	Lincoln	1935	Gas	853– 879
Smyres	Kansas	Rice	1948	Struc 3	225– 248
Sni-A-Bar	Missouri	Jackson	1941	V. 25	1405–1409
Sonora	Texas	Sutton	1968	Mem 9	1394–1432
South Bend	Texas	Young	1921	V. 5	503– 504
South Mountain	California	Ventura	1924	V. 8	789– 829
			1934	Prob	177– 234
South Pass Block 27	Louisiana	(Offshore)	1961	V. 45	51– 71
Southern Ute Dome	New Mexico	San Juan	1935	Gas	363– 384
Southman Canyon	Utah	Uintah	1968	Mem 9	174– 198
Spanish Camp	Texas	Wharton	1968	Mem 9	340– 358

Field	State	County	Year	Pub.	Pages	
Spencer	West Virginia	Roane	1941	Strat	806– 829	*
Spindletop	Texas	Jefferson	1925	V. 9	594– 612	*
			1926	Salt	478– 496	*
			1935	V. 19	618– 643	*
			1936	GC	309– 334	*
			1937	V. 21	475– 490	*
Spivey-Grabs	Kansas	Harper, Kingman	1968	Mem 9	1569–1576	*
Spraberry	Texas	Dawson	1951	V. 35	899– 915	
Spring	Oklahoma	Jefferson	1956	S. Ok 1	355– 372	
Spring Creek	Tennessee	Overton	1929	Struc 1	243– 255	
Spring Hill-Sarepta	Louisiana	Bossier, Webster	1923	V. 7	546– 554	*
			1923	V. 7	555– 557	*
Spruce Creek	West Virginia	Ritchie	1970	V. 54	758– 782	
Spurrier-Riverton	Tennessee	Pickett	1929	Struc 1	243– 255	
State Line	New York	Allegany	1938	V. 22	241– 266	
	Pennsylvania	Potter	1949	V. 33	305– 335	
Staples	Ontario		1949	V. 33	153– 188	
Star-Lacey	Oklahoma	Blaine, Kingfisher	1972	Mem 16	520– 531	*
Stauffer	Nebraska	Banner	1957	V. 41	839– 847	
Steamboat Butte	Wyoming	Fremont	1948	Struc 3	480– 514	*
Stephens	Arkansas	Columbia, Ouachita	1929	Struc 2	1– 17	*
			1942	V. 26	1255–1276	
			1944	V. 28	326– 332	
Stettler	Alberta		1951	V. 35	865– 884	*
			1954	WCan	432– 451	*
			1958	V. 42	1– 93	
Stony Creek	New Brunswick		1968	Mem 9	1819–1832	*
Strand	California	Kern	1940	V. 24	1333–1388	*
Sugar Creek	Louisiana	Claiborne	1938	V. 22	1504–1518	*
Sugarland	Texas	Fort Bend	1933	V. 17	1362–1386	*
			1935	Gas	683– 740	
			1936	GC	709– 733	*
Sulphur	Louisiana	Calcasieu	1930	V. 14	1079–1086	*
Sulphur Bluff	Texas	Hopkins	1937	V. 21	111– 112	*
Summerland	California	Santa Barbara	1934	Prob	177– 234	
Sumner	Tennessee	Sumner	1929	Struc 1	243– 255	
Sutter Buttes	California	Sutter	1942	V. 26	852– 856	*
Swan Hills	Alberta		1968	Mem 9	1238–1284	
Swanson River	Alaska		1968	Mem 9	49– 64	
Sycamore	West Virginia	Calhoun	1970	V. 54	758– 782	
Table Bluff	California	Humboldt	1968	Mem 9	68– 75	
Table Mesa	New Mexico	San Juan	1929	V. 13	117– 151	
Table Rock	Wyoming	Sweetwater	1968	Mem 9	817– 827	
Talco	Texas	Franklin, Titus	1936	V. 20	978– 979	*
			1948	Struc 3	432– 451	*
Tancoco	Veracruz		1936	Mex	186– 190	
Tanner Creek	West Virginia	Gilmer	1929	Struc 2	571– 576	
Tate Island	Arkansas	Pope	1935	Gas	533– 574	
Tatums	Oklahoma	Carter	1935	V. 19	401– 411	*
			1956	S. Ok 1	186– 206	*
Taylor-Link	Texas	Pecos	1935	Gas	417– 457	
Tepetate	Louisiana	Acadia	1938	V. 22	285– 305	*
Tepetate	Veracruz		1936	Mex	210– 212	
Tepetate West	Louisiana	Jefferson Davis	1948	V. 32	1712–1727	*
Texarkana	Arkansas	Miller	1968	Mem 9	1074–1084	
Thomas	Oklahoma	Kay	1926	V. 10	643– 655	*
			1934	Prob	581– 627	
Thompsons	Texas	Fort Bend	1935	Gas	683– 740	
Thompsonville Northeast	Texas	Jim Hogg, Webb	1966	V. 50	505– 517	*
Thornburg	Colorado	Rio Blanco	1929	Struc 2	93– 114	
			1935	Gas	363– 384	

Field	*State*	*County*	*Year*	*Pub.*	*Pages*
Thornton	California	Sacramento, San Joaquin	1968	Mem 9	85– 92 *
Thornwell	Louisiana	Cameron, Jefferson Davis	1968	Mem 9	582– 596 *
Thornwell South	Louisiana	Cameron, Jefferson Davis	1962	V. 46	2121–2132 *
Thrall	Texas	Williamson	1921	V. 5	657– 660 *
			1932	V. 16	741– 768
Tierra Blanca	Veracruz		1936	Mex	219
Tilbury	Ontario		1935	Gas	59– 88
			1949	V. 33	153– 188
Timbalier Bay	Louisiana	Terrebonne	1970	Mem 14	277– 291 *
Timber Canyon	California	Ventura	1934	Prob	177– 234
Tinsleys Bottom	Tennessee	Clay, Jackson	1929	Struc 1	243– 255
Tioga	North Dakota	Williams	1953	V. 37	2294–2302
Tioga Madison	North Dakota	Burke, Mountrail, Williams	1958	Hab	149– 177
Tioga	Pennsylvania	Tioga	1931	V. 15	925– 937 *
			1938	V. 22	241– 266
Tippett	Texas	Crockett, Pecos	1961	V. 45	1859–1869 *
Todd Deep	Texas	Crockett	1950	V. 34	239– 262 *
Todd Ranch	Texas	Crockett	1940	V. 24	1126–1127 *
Tom O'Connor	Texas	Refugio	1970	Mem 14	292– 300
Tomball	Texas	Harris	1935	Gas	683– 740
Tompkins Hill	California	Humboldt	1968	Mem 9	68– 75
Tonkawa	Oklahoma	Kay, Noble	1924	V. 8	269– 283
			1924	V. 8	284– 300
			1926	V. 10	885– 891
Topila	Veracruz		1936	Mex	180
Torrance	California	Los Angeles	1930	V. 14	997–1011
			1934	Prob	177– 234
			1934	Prob	953– 985
			1935	Gas	113– 220
Toteco	Veracruz		1936	Mex	216
Toto	Texas	Parker	1968	Mem 9	1446–1454
Tow Creek	Colorado	Routt	1929	Struc 2	93– 114
Toyah	Texas	Reeves	1965	Mem 4	257– 279
			1968	Mem 9	1394–1432
Trapp	Kansas	Barton	1953	V. 37	300– 313
Trawick	Texas	Nacogdoches	1968	Mem 9	1013–1019
Tri-County	Indiana	Gibson, Pike, Warrick	1927	V. 11	601– 610
			1929	Struc 1	23– 34
Trico	California	Kern, Kings, Tulare	1968	Mem 9	113– 134
Trout Creek	Louisiana	La Salle	1942	V. 26	1255–1276
Turkey Knob	Kentucky	Owsley	1935	Gas	915– 947
Turner Valley	Alberta		1934	V. 18	1417–1453
			1934	Prob	347– 363
			1935	Gas	1– 58
			1940	V. 24	1620–1640
			1945	V. 29	1156–1168
			1951	V. 35	797– 821
			1954	WCan	397– 414
			1959	V. 43	992–1025
Turtle Bayou	Louisiana	Terrebonne	1964	V. 48	1705–1725
Tuscarora	New York	Steuben	(see Woodhull-Tuscarora)		
Tuscola West	Texas	Taylor	1971	V. 55	1194–1205
Tuskegee East	Oklahoma	Creek	1941	Strat	436– 455
TXL	Texas	Ector	1946	V. 30	118– 119
Tyler South	Texas	Smith	1944	V. 28	1646–1647
Tyrone	New York	Schuyler	1931	V. 15	671– 688
University	Louisiana	East Baton Rouge	1941	Strat	208– 236
Urania	Louisiana	Grant, La Salle, Winn	1929	Struc 1	91– 104
Urschel	Kansas	Marion	1922	V. 6	426– 443
Valadeces	Tamaulipas		1949	V. 33	1351–1384

Field	*State*	*County*	*Year*	*Pub.*	*Pages*	
Van	Texas	Van Zandt	1929	V. 13	1557–1558	*
			1934	Prob	399– 427	
Velma	Oklahoma	Stephens	1921	V. 5	627– 629	*
			1948	V. 32	1948–1979	*
			1951	V. 35	582– 606	
			1956	S. Ok 1	260– 281	*
			1959	V. 43	2575–2591	
Velma Southwest	Oklahoma	Stephens	1956	S. Ok 1	244– 259	*
Velma West	Oklahoma	Stephens	1956	S. Ok 1	221– 233	*
Venice	California	Los Angeles	(see Playa del Rey)			
Ventura Avenue	California	Ventura	1924	V. 8	789– 829	*
			1928	V. 12	721– 742	*
			1929	Struc 2	23– 43	*
			1934	Prob	177– 234	
			1934	Prob	399– 427	
			1934	Prob	953– 985	
			1935	Gas	113– 220	
Vernalis	California	San Joaquin	1968	Mem 9	663– 667	*
Vernon	Michigan	Isabella	1935	Gas	787– 812	
			1938	V. 22	129– 174	
			1941	Strat	237– 266	*
Vernon South	Texas	Wilbarger	1929	Struc 1	293– 303	
Viking	Alberta		1935	Gas	1– 58	
Ville Platte	Louisiana	Evangeline	1940	V. 24	701– 715	
Vinton	Louisiana	Calcasieu	1921	V. 5	339– 340	*
			1928	V. 12	385– 394	*
Virden-Roselea	Manitoba		1958	Hab	149– 177	
Virgil	Kansas	Greenwood	1929	Struc 2	142– 149	*
Voshell	Kansas	McPherson	1933	V. 17	169– 191	*
Walnut Bend	Texas	Cooke	1941	Strat	776– 805	*
Walnut Grove	California	Sacramento	1968	Mem 9	85– 92	*
Ward	Texas	Ward	1935	Gas	417– 457	
Warner	Oklahoma	McIntosh	1935	Gas	511– 532	
Warren	Tennessee	Warren	1935	Gas	853– 879	
Wasco	California	Kern	1939	V. 23	1564–1567	*
Washburn	Texas	La Salle	1942	V. 26	276– 279	*
Washington	Louisiana	St. Landry	1968	Mem 9	376– 581	
Wasson	Texas	Gaines, Yoakum	1943	V. 27	479– 523	*
Wasson East	Texas	Yoakum	1941	V. 25	1880–1897	*
Water Creek	Wyoming	Washakie	1966	V. 50	2197–2220	
Wayne-Dundee	New York	Schuyler	1938	V. 22	241– 266	
Weeks Island	Louisiana	Iberia	1959	V. 43	2592–2622	
Welch	Kansas	Rice	1948	Struc 3	225– 248	
Welland County	Ontario		1935	Gas	59– 88	
Wellington	Colorado	Larimer	1924	V. 8	79– 87	*
			1935	Gas	363– 384	
Wellman	Texas	Terry	1953	V. 37	509– 521	*
Welsh	Louisiana	Jefferson Davis	1925	V. 9	464– 478	*
			1926	Salt	437– 451	*
Wertz	Wyoming	Carbon, Sweetwater	1929	Struc 2	636– 666	
			1935	Gas	305– 322	
			1947	V. 31	797– 823	
West Ranch	Texas	Jackson	1944	V. 28	197– 216	*
Westbrook	Texas	Mitchell	1927	V. 11	467– 476	*
			1929	Struc 1	282– 292	*
Westerose South	Alberta		1968	Mem 9	1238–1284	
Westhope North	North Dakota	Bottineau	1958	Hab	149– 177	
Westhope South	North Dakota	Bottineau	1972	Mem 16	633– 642	*
Wheeler Ridge	California	Kern	1926	V. 10	495– 501	*
			1934	Prob	177– 234	

Field	State	County	Year	Pub.	Pages
Wherry	Kansas	Rice	1941	Strat	118– 138
			1948	Struc 3	225– 248
White Mesa	Utah	San Juan	1959	V. 43	2456–2469
White Point	Texas	San Patricio	1931	V. 15	205– 210
			1934	V. 18	519– 530
			1936	GC	664– 675
White Point East	Texas	Nueces, San Patricio	1941	V. 25	1967–2009
White River	Colorado	Rio Blanco	1929	Struc 2	93– 114
Whitemud	Alberta		1950	V. 34	1795–1801
Whitewater	Manitoba		1958	Hab	149– 177
Whitlash	Montana	Liberty	1934	Prob	695– 718
			1935	Gas	245– 276
Whitson	Texas	Montague	(see Nocona North)		
Whittier	California	Los Angeles	1934	Prob	177– 234
Wild Goose	California	Butte	1968	Mem 9	642– 645
Wildcat Jim North	Oklahoma	Carter	1956	S. Ok 1	207– 220
Williams	Arkansas	Crawford	1935	Gas	533– 574
Williamsburg	Kentucky	Whitley	1935	Gas	915– 947
Willow Lake	Louisiana	Catahoula	1942	V. 26	1255–1276
Willow Springs	Texas	Gregg	1968	Mem 9	1060–1068
Willows-Beehive Bend	California	Glenn	1968	Mem 9	639– 642
Wilmington	California	Los Angeles	1939	V. 22	1048–1079
			1968	Mem 9	164– 168
			1970	Mem 14	158– 184
Wilmington East	California	Los Angeles	1971	V. 55	621– 628
Wilshire Ellenburger	Texas	Upton	1955	V. 39	2484–2504
Windfall	Alberta		1968	Mem 9	1238–1284
Winkleman	Wyoming	Fremont	1947	V. 31	797– 823
Winkler	Texas	Winkler	(see Hendrick)		
Winter Valley	Colorado	Moffat	(see Elk Springs-Winter Valley)		
Woodhull-Tuscarora	New York	Steuben	1949	V. 33	305– 335
Woodnorth	Manitoba		1958	Hab	149– 177
Woodrow	Oklahoma	Jefferson	1956	S. Ok 1	355– 372
Woodside	Utah	Emery	1935	Gas	363– 384
Worsham	Texas	Reeves	1968	Mem 9	1389–1393
Wortham	Texas	Freestone	1929	Struc 1	304– 388
X-Ray	Texas	Erath	1935	Gas	609– 649
Yates	Texas	Pecos	1927	V. 11	635
			1929	V. 13	1509–1556
			1929	Struc 2	480– 499
			1930	V. 14	705– 717
			1934	Prob	347– 363
			1934	Prob	581– 627
			1935	Gas	417– 457
Yoast	Texas	Bastrop	1930	V. 14	1191–1197
			1932	V. 16	741– 768
Zama	Alberta		1972	Mem 16	440– 452
Zenith	Kansas	Stafford	1941	Strat	139– 165
Zacamixtle	Veracruz		1936	Mex	215
T25N, R8E	Oklahoma	Osage	1929	Struc 2	378– 395
"2-4"	Oklahoma	Stephens	1921	V. 5	626– 627

Part 2. Listing by Country, State or Province, and County or Parish

CANADA

Alberta

Alexander
Armena
Bellshill Lake
Bow Island
Brooks
Calgary
Camrose
Carbon
Castle River-Waterton
Craigend
Elkton
Figure Lake
Foremost
Fort Saskatchewan
Goose River
Harmatton-Elkton
Homeglen-Rimbey
Joffre
Joseph Lake
Jumping Pound
Leduc
Medicine Hat-Hatton
Minnehik-Buck Lake
Nevis
Pembina
Okotoks
Olds
Pincher Creek
Pine Creek
Provost
Rainbow
Savanna Creek
Red Coulee
Redwater
Rochester
Stettler
Swan Hills
Turner Valley
Viking
Waterton
Westerose South
Whitemud
Windfall
Zama

British Columbia

Clarke Lake
Fort St. John

Manitoba

Daly-East Cromer
Pierson
Virden-Roselea
Whitewater
Woodnorth

New Brunswick

Stony Creek

Northwest Territories

Fort Norman
Norman Wells

Ontario

Becher East
Becher West
Brownsville
Camden Gore
Chatham
Dawn
D'Clute
Kimball
Kingsville
Malahide
Malden
Mersea
Staples
Tilbury
Welland County

Saskatchewan

Alida
Frobisher
Hatton
Lloydminster
Midale
Parkman

MEXICO

San Luis Potosi
- Ebano

Tamaulipas
- Altamira
- Camargo
- Mision
- Reynosa
- Rio Tamesi
- Valadeces

Veracruz
- Alamo
- Alazan
- Amatlan Northern
- Amatlan Southern
- Cacalilao
- Cacalilao Eastern
- Cacalilao North-Central
- Cacalilao West-Central
- Calavera
- Caracol
- Cerritos
- Cerro Azul
- Cerro Viejo
- Chapapote Nunez
- Chiconcillo
- Chijol
- Chinampa North
- Chinampa South
- Corcovado
- Dos Bocas
- El Barco
- Ferronales
- Garrucho
- Isleta
- Jardin
- La Cima
- Limon
- Llano de Silva
- Mata de Chapaote
- Menudillo
- Molino
- Nacata
- Paciencia y Aguacate
- Panuco
- Pasa Real
- Potrero del Llano
- Poza Rica
- Quebracha
- Salinas
- San Geronimo
- San Miguel
- Tanoco
- Tepetate
- Tierra Blanca
- Toteco
- Topila
- Zacamixtle

UNITED STATES

Alabama

Choctaw County
- Gilbertown

Fayette County
- Fayette

Madison County
- Huntsville

Alaska
- Kenai
- Prudhoe Bay
- Swanson River

Arizona

Apache County
- Dineh-bi-Keyah

Arkansas

Columbia County
- Calhoun
- Dorcheat
- Stephens

Conway County
- Jerusalem

Crawford County
- Alma-Smelter
- Kibler
- Williams

Franklin County
- Aetna

Johnson County
- Clarksville

Arkansas *(cont.)*

Lafayette County
Midway
Patton

Logan County
Aetna

Miller County
Fouke
Texarkana

Nevada County
Irma

Ouachita County
Stephens

Pope County
Tate Island

Scott County
Mansfield

Sebastian County
Lavaca
Mansfield
Massard Prairie

Union County
Cairo
El Dorado
Mount Holly
Rainbow City
Schuler
Schuler East

California

Butte County
Perkins Lake
Wild Goose

Colusa County
Arbuckle
Grimes

Contra Costa County
Brentwood
Rio Vista

Fresno County
Coalinga East Extension
Coalinga Westside
Kettleman Hills Middle
Kettleman Hills North
Pleasant Valley
Oil City

Glenn County
Beehive Bend
Willows

Humboldt County
Grizzly Bluff
Table Bluff
Tompkins Hill

Kern County
Belridge
Belridge North
Bowerbank
Buena Vista Hills
Buttonwillow
Calder
Coles Levee
Cymric
Edison
Edison West
Fruitvale
Johe Ranch
Kern Front
Kern River
Lost Hills
Maddux Ranch
Maricopa Flat
McKittrick
Midway-Sunset
Mount Poso
Paloma
Rio Bravo
Round Mountain
Salt Creek
San Emidio Nose
Santiago
Semitropic
Strand
Trico
Wasco
Wheeler Ridge

Kings County
Kettleman Hills Middle
Kettleman Hills North
Trico

Los Angeles County
Aliso
Baldwin Hills
Del Valle
Dominguez
El Segundo
Inglewood
Long Beach
Los Angeles City
Montebello
Playa del Rey
Potrero
Puente
Redondo
Richfield
Rosecrans
Salt Lake
Santa Fe Springs
Seal Beach
Torrance
Venice

California *(cont.)*

Los Angeles County (cont.)

Whittier
Wilmington
Wilmington East

Madera County

Chowchilla

Morgan County

Boxer

Orange County

Brea
Coyote East
Coyote West
Huntington Beach
Olinda

Sacramento County

Thornton
Walnut Grove

San Joaquin County

Lathrop
McMullin Ranch
Thornton
Vernalis

San Luis Obispo County

Arroya Grande
Edna
Midway-Sunset

San Mateo County

Half Moon Bay

Santa Barbara County

Casmalia
Cat Canyon
Cat Canyon West
Cuyama South
El Capitan
Elk Hills
Elwood
Goleta
Lompoc
Russell Ranch
Santa Maria
Santa Maria Valley
Summerland

Santa Clara County

Moody Gulch
Sargent

Solano County

Maine Prairie
Potrero Hills
Rio Vista

Sutter County

Grimes
Marysville
Sutter Buttes

Tehama County

Corning
Corning South

Tulare County

Trico

Ventura County

San Miguelito
Scarab
Shiels Canyon
Simi
South Mountain
Timbey Canyon
Ventura Avenue

Colorado

Adams County

Badger Creek
Middlemist

Archuleta County

Gramps

Baca County

Greenwood

Fremont County

Florence

Garfield County

Carbonero

Jackson County

McCallum

La Plata County

Barker Creek

Larimer County

Fort Collins
Wellington

Logan County

Saber Bar

Moffat County

Elk Springs
Hamilton
Hiawatha
Hiawatha West
Iles
Moffat
Powder Wash
Winter Valley

Montezuma County

Cache

Morgan County

Adena
Sand River

Rio Blanco County

Marapos
Piceance Creek
Rangely
Thornburg
White River

Colorado *(cont.)*

Routt County

Tow Creek

Washington County

Little Beaver

Weld County

Black Hollow

Greasewood

Saber Bar

Illinois

Clark County

Martinsville

Clinton County

Centralia

Sandoval

Franklin County

Benton

Gallatin County

Omaha

Madison County

Marine

Marion County

Centralia

Salem

Sandoval

White County

New Harmony

Indiana

Lima-Indiana

Gibson County

Francisco

Tri-County

Pike County

Tri-County

Posey County

New Harmony

Warrick County

Tri-County

Kansas

Allen County

Elsmore

Piqua

Anderson County

Bush City

Colony

Garnett

Barber County

Hardtner

Medicine Lodge

Barton County

Albert

Beaver

Kraft-Prusa

Otis

Trapp

Butler County

Augusta

Blankenship

Douglas (= Douglass)

Elbing

Eldorado (or El Dorado)

Fox-Bush

Sallyards

Chautauqua County

Longton

Clark County

Harper Ranch

Comanche County

Glick

Cowley County

Rainbow Bend

Edwards County

Lewis

Elk County

Encill

Longton

Ellsworth County

Edwards

Kraft-Prusa

Finney County

Hugoton

Grant County

Hugoton

Greenwood County

Blankenship

Burkett

Madison

Sallyards

Seeley

Virgil

Hamilton County

Hugoton

Harper County

Grabs

Spivey

Harvey County

Burrton

Nikkel

Haskell County

Hugoton

Jefferson County

McLouth

Kearny County

Hugoton

Patterson

Kingman County

Cunningham

Grabs

Spivey

Kansas *(cont.)*

Kiowa County
- Glick
- Nichols

Labette County
- Chetopa

Leavenworth County
- McLouth

Linn County
- Centerville

Marion County
- Covert-Sellers
- Florence
- Peabody
- Propp
- Urschel

McPherson County
- Bornholdt
- McPherson
- Nikkel
- Voshell

Montgomery County
- Coffeyville
- Elk City

Morton County
- Greenwood
- Hugoton

Neosho County
- Chanute

Ness County
- Aldrich

Pratt County
- Carmi
- Cunningham

Reno County
- Burrton

Rice County
- Bornholdt
- Edwards
- Geneseo
- Heinz
- Lyons
- Orth
- Ringwald
- Smyres
- Welch
- Wherry

Rush County
- Albert
- Otis

Russell County
- Fairport
- Kraft-Prusa

Sedgwick County
- Greenwich

Seward County
- Hugoton

Stafford County
- Zenith

Stanton County
- Hugoton

Stevens County
- Hugoton

Sumner County
- Clearwater

Woodson County
- Piqua

Kentucky

Bell County
- Red Bird

Boyd County
- Ashland

Clay County
- Clay
- Oneida-Burning Springs

Elliott County
- Elliott
- Isonville

Estill County
- Lee-Estill-Powell

Floyd County
- Big Sandy
- Floyd

Green County
- Greensburg Consolidated

Johnson County
- Big Sandy
- Flat Gap-Win-Ivyton

Knott County
- Big Sandy

Knox County
- Artemis-Himyar
- Indian Creek

Lee County
- Big Sinking
- Lee-Estill-Powell

Magoffin County
- Big Sandy
- Flat Gap-Win-Ivyton

Menifee County
- Rothwell

Martin County
- Big Sandy
- Martin

Morgon County
- Morgan

Muhlenberg County
- Midland

Kentucky *(cont.)*

Owsley County
Owsley
Turkey Knob
Perry County
Perry
Pike County
Big Sandy
Knox Creek
Powell County
Janet
Lee-Estill-Powell
Taylor County
Greensburg Consolidated
Union County
Hitesville Consolidated
Powell's Lake
Whitley County
Williamsburg
Wolfe County
Campton

Louisiana

Acadia Parish
Egan
Jennings
Midland
Rayne
Tepetate
Allen Parish
Elton North
Ascension Parish
Darrow Dome
Avoyelles Parish
Eola
Beauregard Parish
Neale
Bossier Parish
Bellevue
Bossier
Carterville
Elm Grove
Sarepta
Shreveport
Sligo
Spring Hill
Caddo Parish
Bethany Longstreet
Caddo
Dixie
Pine Island
Red River-Crichton
Shreveport
Calcasieu Parish
Edgerly
Iowa
Lockport
Sulphur
Vinton
Caldwell Parish
Clarks
Cameron Parish
Cameron Meadows
Hackberry East
Johnsons Bayou
Lake Arthur Southwest
Mud Lake East
Thornwell
Thornwell South
Catahoula Parish
Big Slough
Larto Lake
Willow Lake
Claiborne Parish
Athens
Haynesville
Haynesville North
Homer
Lick Creek
Lisbon
Lisbon North
Sugar Creek
De Soto Parish
Ajax
Bethany-Longstreet
De Soto
De Soto-Red River
Grand Cane
Holly
Naborton
East Baton Rouge Parish
University
Evangeline Parish
Ville Platte
Franklin Parish
Delhi
Grant Parish
Urania
Iberia Parish
Avery Island
Jefferson Island
New Iberia
Weeks Island
Iberville Parish
Bayou Blue
Jackson Parish
Calhoun
Jefferson Parish
Barataria
Bayou Segnette

Louisiana *(cont.)*

Jefferson Davis Parish

Iowa
Lake Arthur
Tepetate West
Thornwell
Thornwell South
Welsh

La Salle Parish

Cypress Bayou
Funny Louis Bayou
Hemphill
Little Creek
Nebo
Olla
Olla South
Trout Creek
Urania

Lafourche Parish

Bay Marchand
Coffee Bay
Delta Farms
Lake Enfermer
Lake Raccouri

Lincoln Parish

Calhoun
Lisbon
Ruston

Madison Parish

Delhi

Moorehouse Parish

Monroe

Natchitoches Parish

Ajax
Black Lake

Ouachita Parish

Calhoun
Monroe

Plaquemines Parish

Bastian Bay
Lake Washington

Rapides Parish

Glenmora

Red River Parish

De Soto-Red River

Richland Parish

Delhi
Richland

St. Landry Parish

Opelousas
Washington

St. Martin Parish

Anse La Butte

St. Mary Parish

Cote Blanche Island
Garden City

Terrebonne Parish

Bay Sainte Elaine
Caillou Island
Hollywood
Houma
Kent Bayou
Lake Raccourci
Lirette
Timbalier Bay
Turtle Bayou

Union Parish

Monroe

Vermilion Parish

Jefferson Island
Leroy North

Webster Parish

Bellevue
Carterville
Cotton Valley
Sarepta
Shongaloo
Spring Hill

Winn Parish

Urania

(Offshore Louisiana)

Breton Sound Block 36
Creole
Main Pass Block 35
South Pass Block 27

Michigan

Allegan County

Overisel

Arenac County

Deep River

Clare County

Clare

Crawford County

Beaver Creek

Gratiot County

Gratiot
New Haven

Isabella County

Broomfield
Leaton
Vernon

Kalkaska County

Beaver Creek

Lake County

Reed City

Livingston County

Howell

Mecosta County

Austin
Six Lakes

Michigan *(cont.)*

Midland County
Mount Pleasant
Porter

Monroe County
Deerfield

Montcalm County
Crystal
Six Lakes

Muskegon County
Muskegon

Oakland County
Northville

Osceola County
Reed City

Saginaw County
Saginaw

St. Clair County
Boyd
Peters

Washtenaw County
Northville

Wayne County
Northville

Mississippi

Adams County
Cranfield

Forrest County
Maxie
Pistol Ridge

Franklin County
Cranfield

Hinds County
Jackson

Jefferson Davis County
Gwinville

Lamar County
Baxterville

Lincoln County
Brookhaven

Monroe County
Amory
Muldon

Pearl River County
Pistol Ridge

Rankin County
Jackson

Wilkinson County
Hazlit Creek

Missouri

Jackson County
Blue Springs
Blue Springs East
Blue Springs Northeast
Sni-A-Bar

Montana

Blaine County
Bowes
Boxelder

Carbon County
Dry Creek
Elk Basin

Carter County
Bell Creek

Chouteau County
Bannatyne

Dawson County
Richey

Fallon County
Cedar Creek

Garfield County
Cat Creek

Glacier County
Cut Bank

Hill County
Boxelder

Liberty County
Whitlash

McCone County
Richey

Musselshell County
Big Wall
Melstone

Petroleum County
Cat Creek

Phillips County
Bowdoin

Pondera County
Pondera

Powder River County
Bell Creek

Roosevelt County
Poplar East

Sheridan County
Dwyer

Stillwater County
Lake Basin

Teton County
Pondera

Toole County
Border
Kevin-Sunburst

Valley County
Bowdoin

Wibaux County
Cedar Creek

Nebraska

Banner County

Harrisburg
Stauffer

Cheyenne County

Sidney Southwest

Deuel County

Big Springs
Big Springs West

Red Willow County

Ackman

Nevada

Nye County

Eagle Springs

New Mexico

Eddy County

Artesia
Atoka Penn
Corrol Canyon
Empire Abo
Indian Basin
Lusk Strawn

Lea County

Bell Lake
Hobbs
Kemnitz
Lea County East
Lusk Strawn
Maljamar

San Juan County

Barker Creek
Bisti
Hogback
Horseshoe Canyon
Kutz Canyon
Rattlesnake
Southern Ute Dome
Table Mesa

New York

Allegany County

Richburg
State Line

Cattaraugus County

Bradford

Oswego County

Pulaski

Schuyler County

Tyrone
Wayne-Dundee

Steuben County

Tuscarora
Woodhull

North Dakota

Billings County

Fryburg

Bottineau County

Landa Northeast
Newburg
Westhope North
Westhope South

Bowman County

Cedar Creek
Little Missouri

Burke County

Tioga Madison

McKenzie County

Antelope
Sanish

Mountrail County

Tioga Madison

Williams County

Beaver Lodge
Grenora
Tioga
Tioga Madison

Ohio

Lima-Indiana

Cuyahoga County

Mayfield

Knox County

Homer

Licking County

Homer

Muskingum County

Blue Rock-Salt Creek

Oklahoma

Beaver County

Camrick
Laverne
Mocane-Laverne

Beckham County

Sayre

Blaine County

Lacey
Star

Bryan County

Aylesworth
Cumberland
Durant East

Caddo County

Apache
Cement
Chickasha

Canadian County

Edmond West

Oklahoma *(cont.)*

Carter County
- Ardmore Southwest
- Brock West
- Camp
- Crinerville
- Fox
- Graham
- Healdton
- Hewitt
- Hewitt West
- Lone Grove Southwest
- Milroy
- Sholem Alechem
- Tatums
- Wildcat Jim North

Coal County
- Ashland
- Chiles Ranch
- Coalgate
- Jesse

Cotton County
- Randlett Southwest

Creek County
- Cushing
- Depew
- Glenn
- Red Fork
- Tuskegee East

Garfield County
- Garber

Garvin County
- Antioch Southwest
- Eola
- Hoover Southeast
- Lindsay
- Maysville
- Pauls Valley
- Robberson

Grady County
- Carter-Knox
- Cement
- Chickasha

Harper County
- Buffalo North
- Laverne
- Lovedale
- Mocane-Laverne

Haskell County
- Kinta
- Quinton

Hughes County
- Olympic

Jackson County
- Altus

Jefferson County
- Oscar
- Seay
- Spring
- Woodrow

Kay County
- Blackwell
- Blackwell South
- Burbank
- Mervine
- Ponca
- Thomas
- Tonkawa

Kingfisher County
- Edmond West
- Lacey
- Star

Kiowa County
- Gotebo North

Latimer County
- Red Oak-Norris

Leflore County
- Cameron
- Red Oak-Norris

Lincoln County
- Davenport

Logan County
- Arcadia
- Coon Creek

Major County
- Campbell West

Marshall County
- Aylesworth
- Cumberland
- Madill North

McClain County
- Lindsay

McIntosh County
- Warner

Murray County
- Hoover Southeast

Noble County
- Billings
- Tonkawa

Nowata County
- Delaware Extension

Ochiltree County
- Camrick

Okfuskee County
- Cromwell
- Olympic

Oklahoma County
- Arcadia
- Coon Creek
- Oklahoma City

Oklahoma *(cont.)*

Osage County

Burbank

Burbank South

Pershing

T25N, R8E

Pawnee County

Morrison

Red Fork

Payne County

Ramsey

Pittsburg County

Ashland

Quinton

Pontotoc County

Fitts

Jesse

Pottawatomie County

Earlsboro

Keokuk

Pearson Switch

St. Louis

Seminole County

Bowlegs

Cromwell

Dora

Earlsboro

Keokuk

Little River

Searight

Seminole

Stephens County

Alma North

Carter-Knox

Duncan West

Milroy

Palacine South

Santa Fe

Sholem Alechem

Velma

Velma Southwest

Velma West

"2-4"

Texas County

Camrick

Greenwood

Hugoton-Panhandle

Tillman County

Frederick West

Tulsa County

Bruner

Inscho

Red Fork

Washington County

Hogshooter

Washita County

Gotebo North

Woods County

Lovedale

Pennsylvania

Butler County

Shira Streak

Crawford County

Church Run

Foster-Reno-Oil City

McKean County

Bradford

Music Mountain

Potter County

Ellisburg

Harrison

Hebron

State Line

Tioga County

Farmington

Harrison

Sabinsville

Tioga

Venango County

Foster-Reno-Oil City

Warren County

Grand Valley-Triumph

Washington County

Scenery Hill

Tennessee

Clay County

Tinsleys Bottom

Dickson County

Dickson

Jackson County

Tinsleys Bottom

Lincoln County

Smithland

Morgan County

Bone Camp

Overton County

Spring Creek

Pickett County

Spurrier-Riverton

Scott County

Glenmary

Sumner County

Sumner

Warren County

Warren

Texas

Anderson County
Bethel
Blackfoot
Boggy Creek
Cayuga
Long Lake
Slocum
Andrews County
Embar
Emma
Fullerton
Means
Aransas County
Black Jack
St. Charles Ranch
Archer County
Coleman
Hull-Silk
Atascosa County
Fashing
Austin County
Raccoon Bend
Bastrop County
Cedar Creek
Hilbig
Schimmel-Batts
Yoast
Baylor County
Seymour
Seymour East
Borden County
Diamond M
Bowie County
Eylaw
Flower Acres
Brazoria County
Columbia West
Damon Mound
Hastings
Manvel
Old Ocean
Pledger
Brooks County
La Gloria
Brown County
Blake
Cross Cut
Smith-Ellis
Burleson County
Milbur Wilcox
Caldwell County
Brown-Cude
Buchanan
Dale
Larremore
Luling
Lytton Springs
Salt Flat
Calhoun County
Appling
Camp County
Pittsburg
Carson County
Panhandle
Chambers County
Barbers Hill
Cedar Point
Cotton Lake South
Mayes South
Cherokee County
Boggy Creek
East Texas
Clay County
Henderson
Henrietta
Petrolia
Coleman County
Davis-Gardner
Fisk
Shields
Collingsworth County
Panhandle
Colorado County
Sheridan
Cooke County
Walnut Bend
Crane County
Church-McElroy
McElroy
Crockett County
Clara Couch
Ozona
Tippett
Todd Deep
Todd Ranch
Dawson County
Spraberry
De Witt County
Slick Wilcox
Duval County
Cole
Driscoll
Government Wells
Hoffman
Lopez
Muralla
O'Hern
Eastland County
Grove

Texas *(cont.)*

Ector County
Cowden North
Goldsmith
Judkins
Penwell
TXL

Erath County
X-Ray

Fort Bend County
Blue Ridge
Clodine
Katy
Sugarland
Thompsons

Franklin County
New Hope
Talco

Freestone County
Ball
Wortham

Gaines County
Wasson

Galveston County
High Island
Hitchcock

Goliad County
Slick Wilcox

Gray County
Panhandle

Grayson County
Big Mineral
Handy
Sandusky Oil Creek
Sherman

Gregg County
East Texas
Willow Springs

Guadalupe County
Darst

Hale County
Petersburg

Hansford County
Hansford
Panhandle

Hardin County
Batson
Saratoga

Harris County
Fairbanks
Friendswood
Goose Creek
Houston South
Humble
Katy
Mykawa
Pierce Junction
Satsuma
Tomball

Harrison County
Bethany

Hartley County
Panhandle

Henderson County
Opelika

Hidalgo County
Mercedes
Monte Christo
San Salvador

Hopkins County
Pickton
Sulphur Bluff

Houston County
Navarro Crossing

Hutchinson County
Panhandle

Jack County
Bonnsville
Bryson

Jackson County
Appling
Edna
West Ranch

Jefferson County
Amelia
Big Hill
Port Acres
Port Arthur
Spindletop

Jim Hogg County
Henne-Winch-Fariss
Randado
Thompsonville Northeast

Jim Wells County
Ben Bolt
La Gloria
Magnolia City
Seeligson

Jones County
Noodle Creek

Karnes County
Fashing

Kaufman County
Ham Gossett

Kenedy County
Murdock Pass
Penescal

Kent County
Cogdell

Texas *(cont.)*

Kleberg County
Seeligson
Knox County
Knox City North
LaSalle County
Washburn
Leon County
Buffalo
Liberty County
Cleveland
Esperson
Hardin
Hull
Liberty South
New Hardin
Limestone County
Barron
Groesbeck North
Groesbeck South
Mexia-Groesbeck
Nigger Creek
Martin County
Breedlove
Matagorda County
Buckeye
Collegeport
Markham
Old Ocean
Milam County
Milbur Wilcox
Minerva
Mitchell County
Westbrook
Montague County
Bonita
Bowers
Cato
Holt
Nocona North
Rogers
Whitson
Montgomery County
Conroe
Moore County
Panhandle
Nacogdoches County
Trawick
Navarro County
Bazette
Currie
Powell
Richland
Newton County
Hartburg Northwest
Nueces County
Mustang
Red Fish Bay
Saxet
White Point East
Ochiltree County
Camrick
Orange County
Orange
Panola County
Bethany
Carthage
Parker County
Toto
Pecos County
Apco
Coyanosa
Gomez
Hershey
Payton
Pecos Valley
Puckett
Rojo Caballos
Taylor-Link
Yates
Polk County
Segno
Potter County
Panhandle
Reagan County
Barnhart
Big Lake
Reeves County
Sabre
Screwbean Northeast
Toyah
Worsham
Refugio County
Fox
Greta
La Rosa
McFaddin
O'Connor
Refugio
Tom O'Connor
Roberts County
Cree-Flowers
Lips
Rusk County
East Texas
San Patricio County
Patrick
White Point
White Point East
Schleicher County
Page

Texas *(cont.)*

Scurry County
- Cogdell
- Diamond M
- Kelly-Snyder
- Scurry

Sherman County
- Panhandle

Smith County
- Chapel Hill
- East Texas
- Sand Flat
- Tyler South

Stephens County
- Breckenridge
- Calloway-Henry
- Curry
- Ibex

Stonewall County
- Aspermont

Sutton County
- Sonora

Taylor County
- Tuscola West

Terrell County
- Brown-Bassett

Terry County
- Wellman

Titus County
- Talco

Tyler County
- Joe's Lake

Upshur County
- East Texas

Upton County
- McCamey
- Wilshire Ellenburger

Van Zandt County
- Van

Victoria County
- McFaddin
- O'Connor
- Placedo

Waller County
- Katy

Ward County
- Payton
- Shipley
- Ward

Washington County
- Clay Creek

Webb County
- Aviator's
- Carolina-Texas
- Cole
- Lopez
- Mirando City
- O'Hern
- Thompsonville Northeast

Wharton County
- Bernard East
- Bernard Prairie
- Bernard West
- Boling
- Egypt
- Hungerford
- Hungerford North
- Lissie South
- Spanish Camp

Wheeler County
- Panhandle

Wichita County
- KMA

Wilbarger County
- Vernon South

Williamson County
- Chapman
- Thrall

Winkler County
- Hendrick
- Winkler

Wise County
- Bonnsville

Wood County
- Hawkins
- Merigale
- Quitman

Yoakum County
- Wasson
- Wasson East

Young County
- Sewell-Eddleman
- South Bend

Zapata County
- Jennings
- Mirando
- Mirando Valley

Zavala County
- Del Monte

Utah

Carbon County
- Clear Creek
- Farnham
- Gordon Creek
- Peters Point

Emery County
- Clear Creek
- Flat Canyon
- Woodside

Utah *(cont.)*

San Juan County

Aneth
Lisbon
Mexican Hat
White Mesa

Uintah County

Ashley Creek
Bend Uintah
Red Wash
Walker Hollow
Southman Canyon

Virginia

Buchanan County

Knox Creek

Lee County

Rose Hill

Rockingham County

Bergton

West Virginia

Boone County

Cabin Creek
Campbells Creek

Calhoun County

Richardson
Sycamore

Clay County

Granny's Creek

Doddridge County

New Milton

Gilmer County

Copley
Tanner Creek

Harrison County

Shinnston

Jackson County

Blue Creek
Copley
Gay
Sissonville

Kanawha County

Cabin Creek
Campbell-Davis Creek
Cooper Creek
Elk-Poca
Sissonville

Lewis County

Leading Creek

Lincoln County

Griffithsville

Mingo County

Gilbert Creek

Putnam County

Elk-Poca
Sissonville

Ritchie County

Spruce Creek

Roane County

Spencer

Wyoming

Benton County

Rattlesnake Hills

Big Horn County

Bonanza
Bryan
Byron
Cowley
Garland
Sage Creek

Campbell County

Halverson
Raven Creek
Recluse

Carbon County

Ferris
Ferris Middle
Ferris West
Lost Soldier
Mahoney
Medicine Bow
Rock River
Wertz

Converse County

Glenrock South
Shawnee Creek

Crook County

Coyote Creek
Miller Creek

Fremont County

Big Sand Draw
Circle Ridge
Steamboat Butte
Winkleman

Hot Springs County

Black Mountain
Enos Creek
Golden Eagle
Grass Creek
Little Buffalo Basin
Little Grass Creek
Murphy Dome
Red Springs

Wyoming *(cont.)*

Jackson County
- Battleship
- Canadian
- McCallum North
- McCallum South

Johnson County
- Billy Creek
- North Fork

Lincoln County
- Big Piney
- La Barge

Natrona County
- Salt Creek

Niobrara County
- Lance Creek

Park County
- Bryan
- Elk Basin
- Frannie
- Garland
- Little Buffalo Basin
- Little Polecat
- Oregon Basin
- Pitchfork
- Polecat
- Sage Creek
- Silver Tip

Sublette County
- Big Piney
- Birch Creek Unit
- La Barge

Sweetwater County
- Arch
- Baxter Basin Middle
- Baxter Basin North
- Baxter Basin South
- Church Buttes
- Desert Springs
- Desert Springs West
- Hiawatha
- Joyce Creek
- Little Lost Soldier
- Little Worm Creek
- Lost Soldier
- Nitchie Gulch
- Patrick Draw
- Salt Wells
- Table Rock
- Wertz

Uinta County
- Church Buttes

Washakie County
- Cottonwood Creek
- Hidden Dome
- Middle Dome
- Nowood
- Water Creek

Weston County
- Coyote Creek
- Fiddler Creek
- Mush Creek
- Osage
- Skull Creek

Index

This index consists of two sections, which appear in the following order:

(1) Author Index and
(2) Keyword Index.

The *author index* is arranged alphabetically according to each author's last name. For papers by more than one author, each author's name appears in the index in alphabetical order. The appearance of an author's name followed by the title of an article does not mean that he is the only author of that article. He may be one of two or more authors of the paper whose title follows his name. The author index does *not* show multiple authors in any single listing.

To locate a reference in the *keyword index*, the reader should begin by thinking of the significant words. Then he should look in the index for the keyword entry for each of those words. The reference codes will direct him to the pages.

The columns on the right-hand side of the keyword index give the page number and a code number (1 or 3) indicating the nature of the source. The code is:

(1) for phrase from title; and
(3) for phrase from abstract, text, table, figure, or figure caption.

The keyword for each entry is located at the left-hand side of the page. The (>) sign indicates the first word in each title or key phrase. The (<) sign indicates the end of the title or key phrase.

Author Index